DEVELOPMENTS IN GEOTECHNICAL ENGINEERING VOL 32

STRATA MECHANICS

ERRATA

P	110	Table 7. Under column 4 for Daw Mill Colliery W 65's substitute 0.26 for Required Setting Load Density and 0.41 for Required Yield Load Density.
P	110	Table 8. Under Daw Mill W 65's substitute 0.52 for Mean RYLD.
P	169	Transpose the captions of Figures 5 and 6.
P	173	Subsection 4 should read: Results so far indicate that bending and mean strains are minimal in the area of the inset/shaft intersection.
PP	210, 211	Transpose the captions of Figures 5 and 6.
P	217, 218	Figures 7, 8, 9 are based on E = 1000 MPa. For E = 2100 MPa the stress multiplying factor is 2.1.

Further titles in this series:

1. G. SANGLERAT/THE PENETROMETER AND SOIL EXPLORATION
2. Q. ZARUBA AND V. MENCL/LANDSLIDES AND THEIR CONTROL
3. E.E. WAHLSTROM/TUNNELING IN ROCK
4. R. SILVESTER/COASTAL ENGINEERING, I and II
5. R.N. YOUNG AND B.P. WARKENTIN/SOIL PROPERTIES AND BEHAVIOUR
6. E.E. WAHLSTROM/DAMS, DAM FOUNDATIONS, AND RESERVOIR SITES
7. W.F. CHEN/LIMIT ANALYSIS AND SOIL PLASTICITY
8. L.N. PERSEN/ROCK DYNAMICS AND GEOPHYSICAL EXPLORATION
   Introduction to Stress Waves in Rocks
9. M.D. GIDIGASU/LATERITE SOIL ENGINEERING
10. Q. ZARUBA AND V. MENCL/ENGINEERING GEOLOGY
11. H.K. GUPTA AND B.K. RASTOGI/DAMS AND EARTHQUAKES
12. F.H. CHEN/FOUNDATIONS ON EXPANSIVE SOILS
13. L. HOBST AND J. ZAJIC/ANCHORING IN ROCK FORMATIONS
14. B. VOIGT (Editor)/ROCKSLIDES AND AVALANCHES, 1 and 2
15. C. LOMNITZ AND E. ROSENBLUETH/SEISMIC RISK AND ENGINEERING DECISIONS
16. C.A. BAAR/APPLIED SALT-ROCK MECHANICS, 1
   The In-Situ Behavior of Salt Rocks
17. A.P.S. SELVADURAI/ELASTIC ANALYSIS OF SOIL-FOUNDATION INTERACTION
18. J. FEDA/STRESS IN SUBSOIL AND METHODS OF FINAL SETTLEMENT CALCULATION
19. Á. KÉZDI/STABILIZED EARTH ROADS
20. E.W. BRAND AND R.P. BRENNER (Editors)/SOFT-CLAY ENGINEERING
21. A. MYSLIVEC AND Z. KYSELA/THE BEARING CAPACITY OF BUILDING FOUNDATIONS
22. R.N. CHOWDHURY/SLOPE ANALYSIS
23. P. BRUUN/STABILITY OF TIDAL INLETS
   Theory and Engineering
24. Z. BAŽANT/METHODS OF FOUNDATION ENGINEERING
25. Á. KÉZDI/SOIL PHYSICS
   Selected Topics
26. H.L. JESSBERGER (Editor)/GROUND FREEZING
27. D. STEPHENSON/ROCKFILL IN HYDRAULIC ENGINEERING
28. P.E. FRIVIK, N. JANBU, R. SAETERSDAL AND L.I. FINBORUD (Editors)/ GROUND FREEZING 1980
29. P. PETER/ CANAL AND RIVER LEVEES
30. J. FEDA/ MECHANICS OF PARTICULATE MATERIALS
   1. The Principles
31. Q. ZARUBA AND V. MENCL/LANDSLIDES AND THEIR CONTROL (2nd edition)

DEVELOPMENTS IN GEOTECHNICAL ENGINEERING VOL 32

STRATA MECHANICS

Proceedings of the Symposium on Strata Mechanics held in
Newcastle upon Tyne, 5–7 April, 1982

Edited by

I.W. FARMER

Department of Mining Engineering, The University of Newcastle upon Tyne,
Newcastle upon Tyne, England

NON SOLUS

ELSEVIER SCIENTIFIC PUBLISHING COMPANY
Amsterdam – Oxford – New York 1982

ELSEVIER SCIENTIFIC PUBLISHING COMPANY
Molenwerf 1
P.O. Box 211, 1000 AE Amsterdam, The Netherlands

Distributors for the United States and Canada:

ELSEVIER SCIENCE PUBLISHING COMPANY INC.
52, Vanderbilt Avenue
New York, N.Y., 10017

ISBN 0-444-42086-X (Vol. 32)
ISBN 0-444-41515-7 (Series)

Printed in The Netherlands

PREFACE

The objective of the Symposium on Strata Mechanics is to improve the design of underground mine structures through a better understanding of the mechanics of rock deformation and fracture induced by excavation. It is intended that the emphasis will be on a mechanical rather than an empirical approach to design.

The justification for the symposium is the evidence that despite extensive research into the interaction between mining methods and strata behaviour, there has not been a radical change in the approach to underground design in stratified rocks. Changes and improvements where they have occurred have usually been a product of the large investment programme in underground machinery and supports. As implied by the widespread use of terms such as 'strata control' and 'ground control', the design of the openings into which this equipment is placed remains largely empirical and non-analytical.

Several examples of this approach to design can be quoted - some described in these proceedings, some described elsewhere. For instance, roof flaking and face spalling in longwall coal mining are associated with vertical fractures induced parallel to and ahead of the longwall face line by lateral expansion of the strata towards the unconfined goaf edge, under the influence of the foward abutment pressure. Nothing can be done to prevent the formation of these fractures but they can develop into cavities very rapidly if the roof is allowed to sag at the face or the effect of the fractures is exacerbated by allowing them to coincide with the major cleat direction. Thus face cavities can be controlled by increasing the support setting pressures - or at least reducing support leakage - and by orientating the face direction correctly. This knowledge is not new, but nevertheless one of the major recent efforts of some strata control engineers has been into investigating methods of filling face roof cavities, rather than eliminating them by correct design.

Another example is floor heave in underground roadways with thick listricated fireclay seatearth. Floor heave occurs because the seatearths at the unsupported floor surface deform in a way which is partly ductile. It is virtually impossible - even with direct invert support - to prevent some floor heave. It is however a process which reduces with time and, provided a degree of deformation is allowed, mobilisation of the residual strength of the fractured floor rocks will eventually create a stable roadway floor. If the floor strata are disturbed before or after this situation is achieved, floor heave in thick seatearths will be reactivated with renewed vigour. The roof rocks will be largely undisturbed, because they are stronger and will tend to deform in a brittle way and supporting arches will rapidly mobilise residual strength. A possible approach to roadway stability is therefore to construct a large roadway profile, suspending all transport systems from the roof, and levelling the floor with fill when deformation has stabilised. Investigations still continue, however, into floor reinforcement and the development of dinting machinery.

There are numerous other examples of difficult mining conditions, often leading to excessive investment and losses in production, which could be quoted - many of which, given an understanding of the mechanics of strata deformation, could be improved with little cost. At the moment many of these problems are attributed to difficult geological conditions.

A review of the literature will show that little is new and that most of the mechanics of strata deformation were described in detail 40 to 50 years ago by engineers such as W.H. Evans and D.W. Phillips. It is even more galling to turn to Lyon Playfair's introductory lecture given after the opening of the School of Mines (later the Royal School of Mines) at the Museum of Practical Geology and Geological Survey in November 1851. In a a discourse entitled 'The study of abstract science essential to the progress of industry' he argues against a perceived English disease which relies on empiricism and common sense in agriculture and manufacturing at the expense of an understanding of the principles on which they depend. He argues that Newton did, 'by the exposition of the laws of gravitation, produce more real practical benefits to industry than all the preceeding ages of empiricism'. The lecture (Playfair 1852) is as relevant to the mining industry today as it was then, and his peroration can usefully be quoted:

'Depend upon it, an empirical knowledge of practice is not the way to succeed in the struggle of individuals or in the struggle of nations. Intellect is on the stretch to get forward and that nation which holds not by it will soon be left behind. For a long time, practice, standing still in the pride of empiricism, and in the ungrateful forgetfulness of what science has done in its development, reared upon its portal the old and vulgar adage, "an ounce of practice is worth a ton of theory". This wretched inscription acted like a Gorgon's head and turned to stone the aspirations of science. Believe it not! For a grain of theory, if that be an expression for science, will, when planted, like the mustard seed of Scripture grow and wax into the greatest of trees...... I have said enough to show you that it is indispensable for this country to have a scientific education in connection with manufactures, if we wish to outstrip the intellectual competition which now, happily for the world prevails in all departments of industry. As surely as darkness follows the setting of the sun, so surely will England recede as a manufacturing nation, unless her industrial population become much more conversant with science than they now are.'

In a year when one of the presidential addresses to a federated institute of the Institution of Mining Engineers has the basic premise that the mining industry has much to learn from Sainsbury the grocers, this is surely as relevant as it was 130 years ago.

Perhaps the symposium will do a little to redress the balance.

I.W. FARMER
NEWCASTLE UPON TYNE, March 1982.

L. Playfair (1852) The study of abstract science essential to the progress of industry. Records of the School of Mines, Museum of Practical Geology and Geological Survey, 1 (1) 23-48.

ACKNOWLEDGEMENTS

The assistance in the preparation of this volume and in the organisation of the Symposium of Ms B. Leggatt, Ms E. Gannie, Mr P.A. Gillanders and especially Ms A. Murphy is gratefully acknowledged. The members of the organising committee were Professor P.B. Attewell, Dr I.W. Farmer, Dr R.J. Fowell, Mr P.A. Gillanders, Dr E.K.S. Passaris, Mr G.M. Pearson, Mr A. Szeki and Professor A.H. Wilson.

CONTENTS

STRATA CONTROL IN HYDRAULIC MINING OF COAL SEAMS AT GREAT DEPTH

Toshiro Isobe

Department of Mineral Resources Development Engineering
Hokkaido University, Sapporo, 060 Japan

Ryuji Kato, Nobuya Fujii

Sunagawa Mine, Mitsui Coal Mine Co.Ltd.
Kamisunagawa, Hokkaido, 073-02 Japan

Kazuhiko Sato

Department of Mineral Resources Engineering
Muroran Institute of Technology, Muroran, 050 Japan

SUMMARY. The microseismic activity has been monitored at Sunagawa Coal Mine to clarify the fracture of rock around underground excavation which is associated with hydraulic mining. The microseismic events were detected at eight seismic stations spread several hundreds meters apart on the surface and underground, and they were analyzed to determine their hypocenters and the seismic energies radiated from their sources. A number of microseismic events concentrated on the coal pillar which had been left in the previous mining panel and bounds on the current panel. The microseismic activity has got gradually moderate as the coal face has gone away from the pillar, although it has been intensive during the period of mining of the blocks adjacent to the pillar. Also the stress relief boring and a reduction in advance rate of coal face have contributed to alleviation of the microseismic activity.

INTRODUCTION

Sunagawa Coal Mine has successfully applied the hydraulic mining technology to steeply inclined coal seams since 1964, and at the present annual coal production using this method has amounted to about 0.7 million tones. The strata control in hydraulic mining has, however, become difficult with increasing mining depth.

During last two decades Sunagawa Coal Mine has mainly concerned three subjects among those related to strata control: the first one is to combat rock burst in mining panel; the second one is to prevent gas outburst occurring in driving cross cut or heading sublevel entry within coal seam; the third one is to maintain access roadways for their service lives. All of them are equally important to safety and productivity in hydraulic mining at great depth, however, the present paper will describe only the strata control for alleviating the hazards of rock burst due to the limitation of available pages.

Before going into detail, the rock burst at the No.1 Pit of the Sunagawa Coal Mine should be mintioned briefly to distinguish it from coal bumps which have been experienced at the other coal mines. The rock burst at the No.1 Pit can be identified as a pillar burst which may be caused by a subben collapse of a coal pillar that appears on the way to the completion of mining a block, or coal remnant that has been left in the previous mining panel. Another source of the rock burst in the No.1 Pit is fracture of strata around tabular excavation that has been formed by hydraulic mining, which often results in ground tremor felt in underground and the surface, and is regarded as majority of rock bursts in this mine. So far as the hydraulic mining is performed by a retreating fully caving method, the formation of coal pillar in an active mining panel is more or less inevitable. At the situation the sublevel entry adjacent to coal pillar may be damaged by coal burst. Therefore the control of coal burst requires to delineate highly stressed zone in front of coal face by monitoring stress state and to reduce the stress level beyond the yielding stress by stress relief techniques. The sublevel entries may not only be damaged by coal burst in an active mining panel, but also may be affected by collapse of coal pillar in mined out area or fracture of strata above and below the excavation previously mined out. Since these fractures can hardly be controlled by current techniques such as stress relief boring, the hazards of rock burst should be minimized by reinforcement of support in sublevel entries and cross cuts.

The present paper will describe the microseismic activity induced by hydraulic mining and discuss the effects of stress relief boring and decrease of advance rate of coal face on the microseismic activity.

MINING OPERATION

Geological Feature and Mining Condition

Fig.1 illustrates the geological cross section of the Bibai coal bearing formation that occurred at the South No.3 860L mining panel of the No.1 Pit. The Bibai formation involving several coal seams strikes almost north to south and dips steeply towards west between 55 and 70 degrees, Ohtsuka (1980). The mining panel spreads about 200 m in the direction of strike and ranges from 940 m to 1040 m below the surface in elevation as shown in Fig.2. The mining panel bounds on the South No.2 panel at the north border and its south border is separated from the South No.4 panel by a pre-existing fault dipping almost vertically. This fault associates several small faults on its north side.

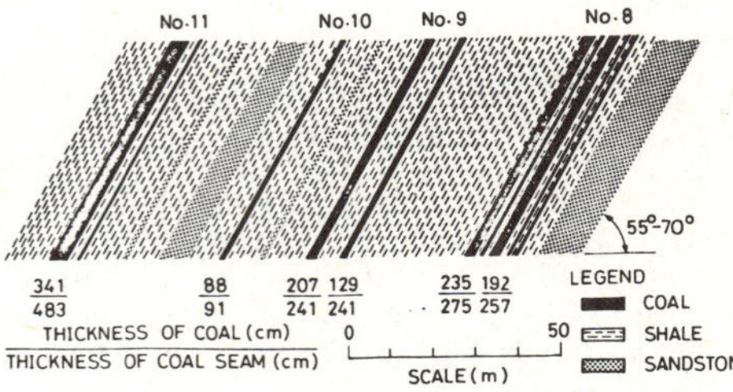

Fig.1 Geological cross section of Bibai coal bearing formation along the 760L Cross Cut of the South No.3 Panel.

The mining panels above the current mining depth were previously mined out by hydraulic mining with fully caving, but a few coal pillars were left to avoid the incidence of pillar burst. The attention has, therefore, been paid to the coal pillar left in the South No.3 760L panel among these remnants because the pillar stability may get worse as the mining area spreads in the 860L panel.

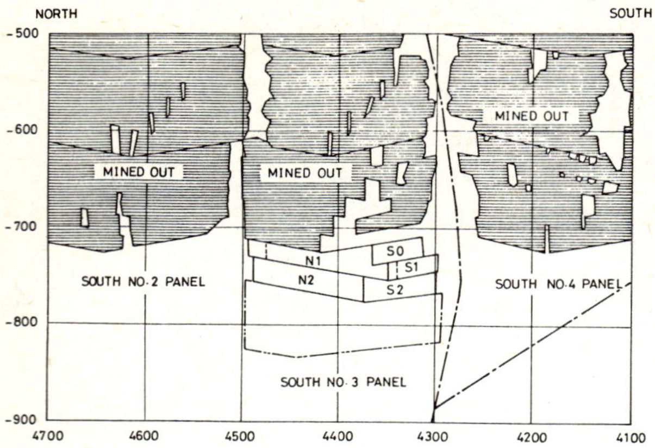

Fig.2 Elevation figure of the mine layout of the South No.3 Panel. The mine geometry of the No.11 seam is projected onto a reference plane parallel to the NS vertical section. The interval of square grid is 100 m.

Working Coal Seams and Srata Control

The mining at the 860L panel was commenced by extracting the No.11 seam which is top seam in the Bibai formation as shown in Fig.1. The extraction of coal by monitor proceeded from the top sublevel towards the lower sublevels following in the order of S0, S1, N1, the remnant of S1, N2 and S2 mining blocks as indicated in Fig.2. After the extraction of the S2 block the mining turned to the lower seams, that is the No.10 seam and the pair seam of No.9. Fig.3 shows the mining activity from June to December 1981 in terms of growth of mining area.

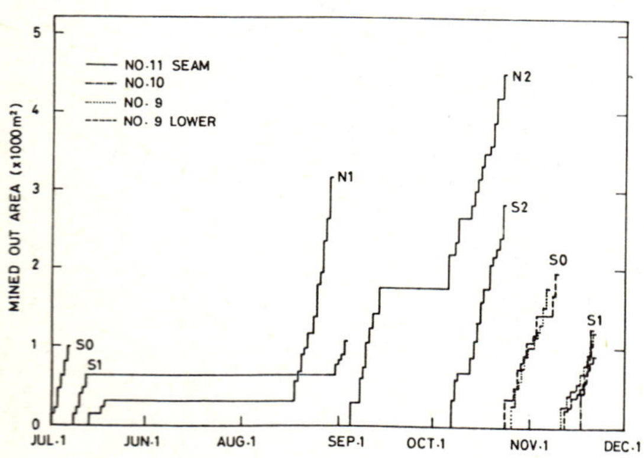

Fig.3 Mining activity at the South No.3 Panel from June to December 1981.

The microseismic event of June 18, 1981, 22:33 gave damage to the sublevel entries in the N1 and N2 blocks without any fatalities and injuries so that the mining of the N1 block was compelled to a pause in order to repair the roadways and to estabilish the peventive measures for damage caused by microseismic event. At first the access roadways have been repaired and reinforced by means of the doubled wood chocks and center wood props instead of the three wood pieces with a center prop support. Secondarily the stress relief boring has been performed to destress the solid coal in advance of the coal face. The spacing between adjacent bore holes has been kept in about 10 m, and bore holes with 100 mm of diameter have been drilled from sublevel entries up to 10 m to 18 m deep within the coal seam. In addition to these preventive measures, the microseismic activity has been monotored by a seismic array and a recording facility.

On August 18, 1981 the mining of the N1 block was resumed, and on September 3 finished without any troubles, although the microseismic activity was enormous as will be described in the later section. The mining of the N2 block was also scarcely interrupted by microseismic activity with an exception of the microseismic event of September 15, 09:32, This event gave a slight damage to the sublevel entry in the N2 block. In addition to the stress relief boring the alternative mining method has been tried to reduce the face advance rate so that the hazards due to the microseismic activity may be alleviated. In this method the N2 and S2 blocks have been simultaneously extracted in every other day by using two monitors so that the face advance rate could be reduced to a half of that in the current mining method. The alternative method has contributed to a sound maintenance of the access roadways for their survice lives.

MICROSEISMIC ACTIVITY

Microseismic Monitoring Facility

Fig.4 illustrates a seismic array installed at the mining area, which consists of eight seismic stations on the surface and underground so that the South No.3 860L mining panel is almost encompassed by them. The seismometer is velocity type of the Katsujima PK-110V which is capable of responding flatly over the frequency range higher than 1.5 Hz.

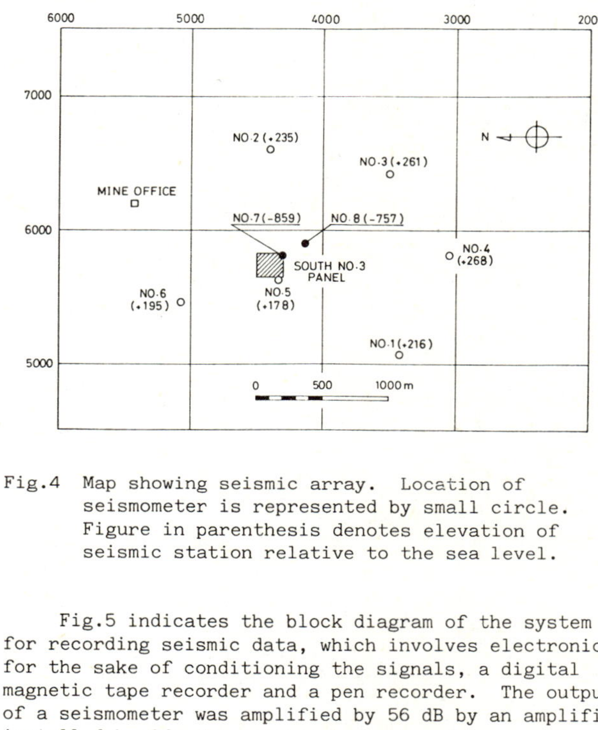

Fig.4 Map showing seismic array. Location of seismometer is represented by small circle. Figure in parenthesis denotes elevation of seismic station relative to the sea level.

Fig.5 indicates the block diagram of the system for recording seismic data, which involves electronics for the sake of conditioning the signals, a digital magnetic tape recorder and a pen recorder. The output of a seismometer was amplified by 56 dB by an amplifier installed beside the seismometer, then it was sent via a shield wire of 2 to 4 km long to the recording system installed at the mine office. Next the signal was digitized by an A/D converter with a rate of 500 samples a second, then it was delayed for 2 seconds by a digital delay device with 16 kbytes memory of Tokyo Sokushin MPL-163. The digital siganl of 12 bits was recorded on the magnetic tape device of Tokyo Sokushin SAMTAC-120G by using trigger method. The microprocessor involved in the device can not only make itself responsible for judging whether the trigger level was exceeded or not, but also can distinguish a seismic signal from artificial noises and last recording exclusively seismic signal for a period until the amplitude of the signal decreases less than 10 per cent of the full scale. The recording device also contains a quartz controlled time encoder by which a seismic datum can be marked its receiving time.

The seismic datum recorded on the magnetic tape was immediately played back by a D/A converter and rerecorded on an oscillograph to analyze its source parameters and seismic energy. After a volume of tape

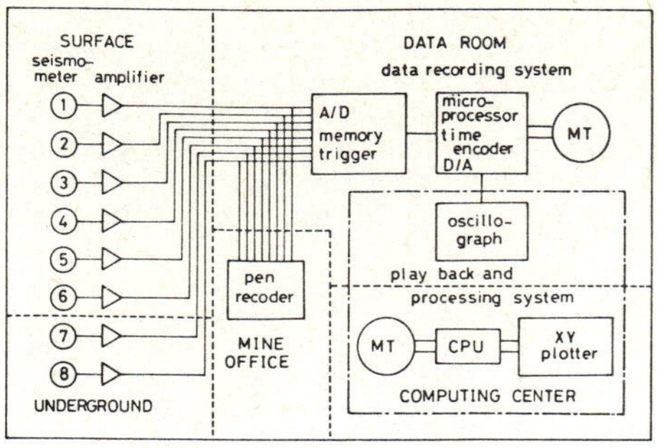

Fig.5 Block diagram of microseismic monitoring facility.

of 2400 feet long was fully recorded, the data were processed directly from the magnetic tape by a digital computer. The source location computer program utilizes the least squares method assuming an anisotropic P-wave velocity model as shown in Table 1. The Richter's scale of magnitude and the seismic energy were evaluated by making use of the Muramatsu's empirical formula and the Gutenberg-Richter's one respectively, Isobe et al (1979).

Table 1. P-wave velocities used in source location

station No.	location NS	EW	elevation	P-wave velocity (m/sec)
1	3,419	5,039	216	3,260
2	4,400	6,603	235	3,460
3	3,496	6,418	261	3,160
4	3,056	5,820	268	3,440
5	4,334	5,630	178	3,560
6	5,074	5,469	195	3,600
7	4,305	5,814	-859	3,925
8	4,141	5,899	-757	3,780

Microseismic Activity

Fig.6 indicates the cumulative seismic energy from August to December 1981. In order to compare seismic energy of a mining block with those in the other blocks each other, the spatial rate of seismic energy release was evaluated in terms of the ratio of the gross seismic energy to the total mining area in the mining block. Table 2 summarizes the spatial rates of seismic enrgy release in all the mining blocks.

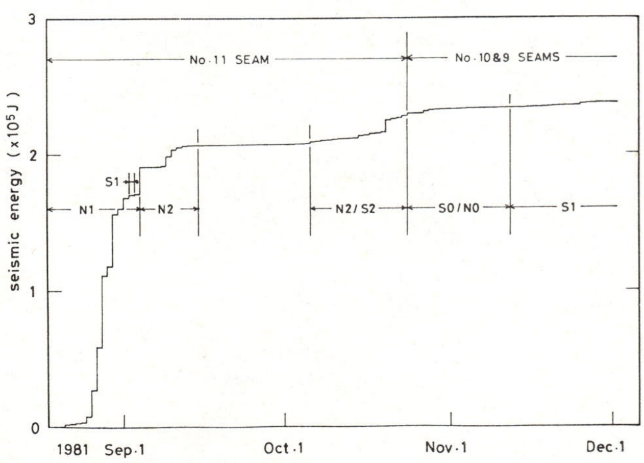

Fig.6 Seismic energy released from the South No.3 Panel.

Table 2. Microseismic activity and related mining activity at the South No.3 Panel.

coal seam No.	mining block	mining period (days)	mined out area (m^2)	number of events	seismic (energy) (kJ)	energy release rate (J/m^2)
	N1 + S1	16	3,305	134	170.6	51.6
11	N2	12	1,755	150	37.2	21.2
	N2 + S2	19	5,586	135	20.2	3.6
9	S0	12	1,768	49	5.7	3.1*
	S0	18	1,962			
10	S1	4	1,250			
9	S1	11	1,207	54	3.2	3.2*
	S1	10	911			

*calculated using the mean value of mined out area.

The magnitude distribution of microseismic event was in a close agreement with the Gutenberg-Richter's formula written by

$$\log N = a - b M$$

where N is cumulative number of events whose magnitudes are not less than M, as shown in Fig.7. The constants of a and b in the formula were determined by the most likelihood method. These results are also summarized in Table 3.

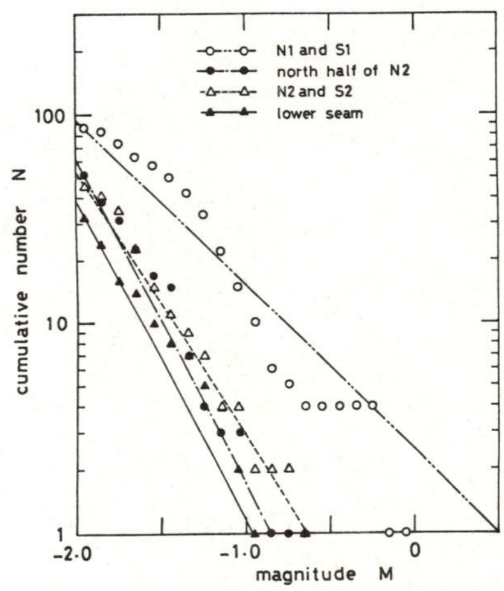

Fig.7 Magnitude distributions of microseismic events occurring the mining of the South No.3 Panel.

Table 3. Magnitude distributions fitted to the Gutenberg-Richter's empirical formula $\log N = a - b\ M$.

coal seam No.	mining block	best estimate b	a	95% limit of confidence b	a
11	N1 + S1	0.78	0.41	0.17	0.32
11	N2	1.56	-1.32	0.42	0.83
11	N2 + S2	1.53	-1.47	0.53	1.03
10	S0	1.28	-0.83	0.37	0.72
9	S1				

The constants b and a were estimated from the data greater than M=-2.0.

The microseismic activity was most intensive at the N1 block among all the mining blocks. However, it is noteworthy that little seismic energy was released for a week after the resumption of mining at the N1 block, and then the microsesmic activity became very active soon after the coal face reached to the portion beneath the coal pillar left in the 760L panel. Fig.8 illustrates foci of seismic events occurring during the mining of the N1 and S1 blocks. During mining of the N1 block seismic events clustered on the coal pillar mentioned above, while seismic events in the S1 block were distributed over the mined out area and rectangular coal pillar which is bounded at the lower edge by the S0 mining block.

Fig.9 shows the distribution of microseismic events occurring during the mining of the north portion in the N2 block. Seismic events concentrated on the north half of the N1 block and the mined out area adjacent to the N1 block. The microsesimic activity lasted for 20 days since September 15, although the mining of the N2 block was in suspension. In the

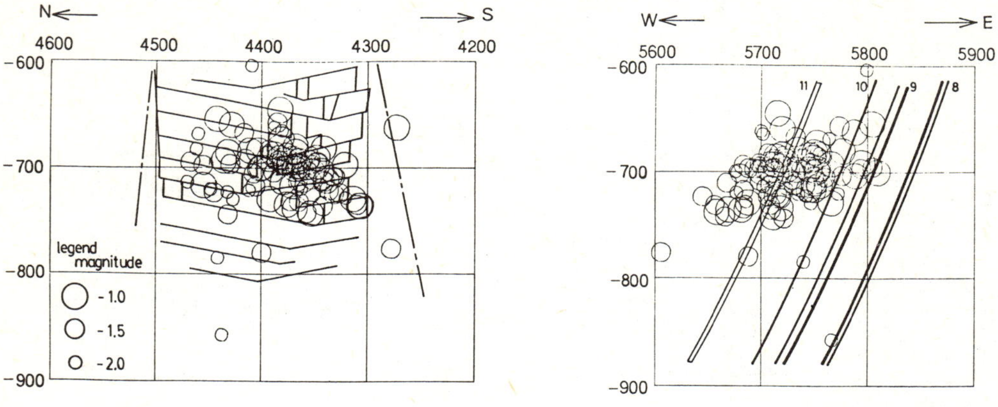

Fig.8 Focal depth distribution of seismic events occurring the mining of the N1 and S1 blocks from Aug.18 to Sept. 3,1981. Seismic event is represented by a circle whose center and radius are the hypocenter and the magnitude of the seismic event respectively. The left is NS vertical cross section onto which the South No.3 Panel is also projected, while the right is EW vertical cross section which includes geological cross section of the Bibai formation.

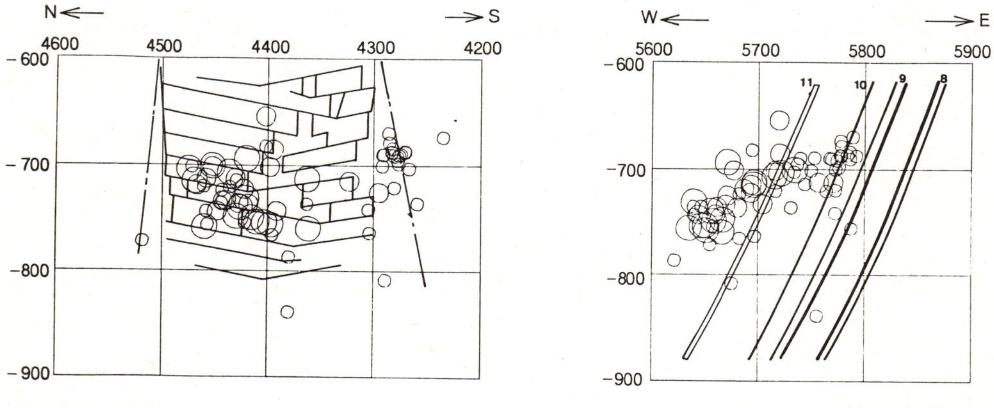

Fig.9 Focal depth distribution of seismic events occurring the mining of the north part of the N2 block from Sept.5 to Sept.15,1981. The foci of the seismic events which were observed during the pause of mining, from Sept.16 to Oct.5, are also shown.

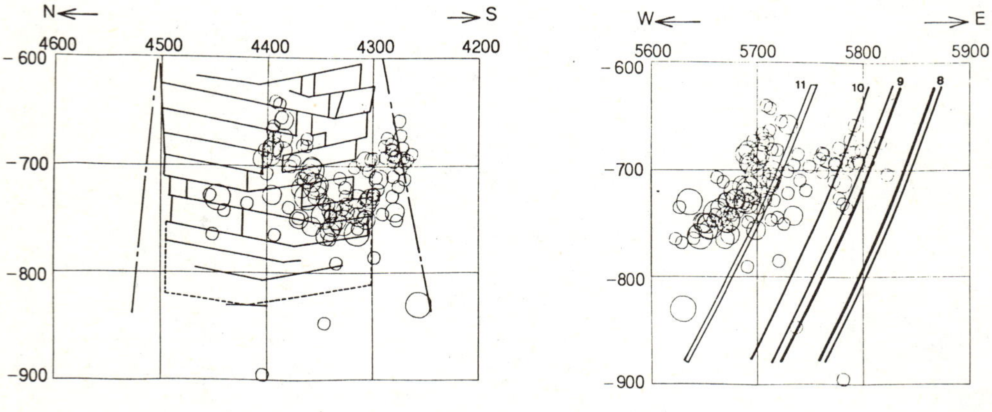

Fig.10 Focal depth distribution of seismic events occurring the alternative mining at the N2 and S2 blocks from Oct.6 to Oct.23,1981.

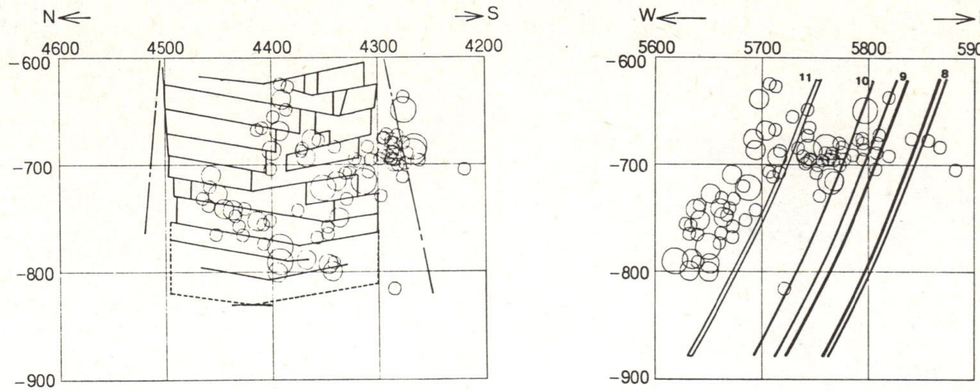

Fig.11 Focal depth distribution of seismic events occurring the mining of S0 and S1 blocks in the Nos. 10 and 9 coal seams from Oct.24 to Dec.1,1981.

dormant period of mining, the scale of microseismic events became smaller than those in the pevious days, and they were distributed over the whole mining panel with an exception of a concentration on the preexisting fault occurred at the south border of the panel.

An alternative mining that was tried at the mining both the N2 and S2 blocks reduced considerably release of seismic energy, although the number of events was not supressed. Fig.10 shows the spatial distribution of foci of microseismic events in this period.

Finally a drastic decrease in release of seismic energy was realized in the mining of the lower seams. The hypocenters of the seismic events occurring in this period were considerably shifted towards the strata below the No.11 coal seam as shown in Fig.11.

DISCUSSION

The seismic energy released during the mining of the N1 and S1 blocks was highest among all the blocks as mentioned in the previous section. A contributor to the great release of seismic energy may be referred to the fracture of the coal pillar left in the 760L by the reason why a large number of microseismic events clustered on the region containing the coal pillar.

A numerical experiment has been carried out to obtain a proper comprehension with regard to the microseismic activity at these mining blocks. In the calculation the displacement discontinuity method was used by assuming that the strata above and below the No.11 coal seam can be represented by a linearly isotropic elastic medium and that seam material is also elastic and is sandwiched between semi-infinite elastic media as if it was a plate-like inclusion, Ishijima et al (1980). The Young's moduli of 5.0 GPa and 1.0 GPa are assigned to the strata and the seam respectively, and the Poisson's ratio is 0.2 for both of them. The rock pressure in the primitive state is assumed to be in a lithostatic pressure due to the gravity of overburden.

As shown in Fig.12, the mining of the N1 and S1 blocks was simulated by dividing the sequence of mining into eleven steps so that the spatial distribution of normal stress to the No.11 seam and closure between roof and floor of the seam were determined as well as the work done by gravity during the mining. Fig.13 illustrates the spatial distribution of normal stress at the final stage of the mining of the N1 block. The normal stress in the coal pillar varies as the mining proceeds step by step as shown in Fig.14. Since the fracture criterion of the seam material has not yet been determined, the result of the present calculation can not immediately be available for delineating fracture zone in the coal pillar. However, the observation of microseismic activity can supplement an insufficient knowledge with respect to the fracture criterion. That

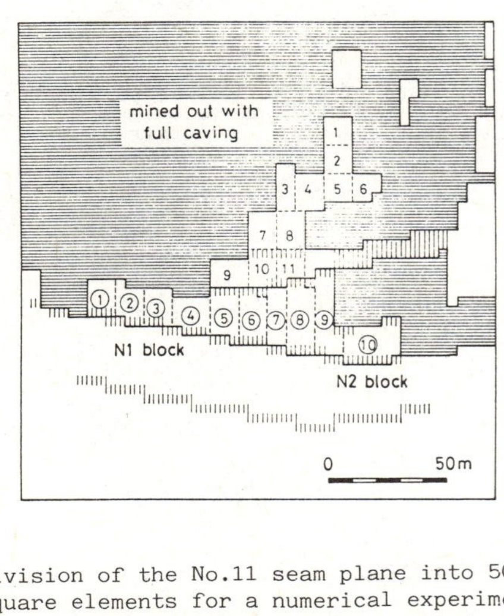

Fig.12 Division of the No.11 seam plane into 50 x 50 square elements for a numerical experiment using the displacement discontinuity method. Encircled number represents the number of mining step, and number which is inserted in the coal pillar corresponds to the partition of the pillar at which the change in pillar stress is observed during mining.

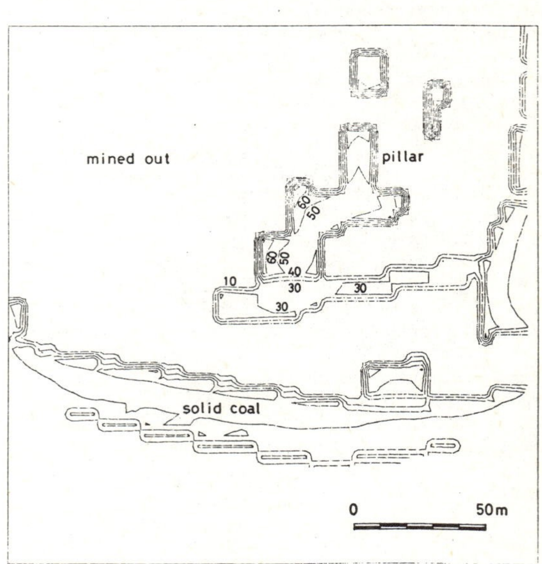

Fig.13 Contour of normal stress to the No.11 seam plane at the No.9 mining step shown in Fig.12, which was determined from the displacement discontinuity method. The interval of contour is 10 MPa.

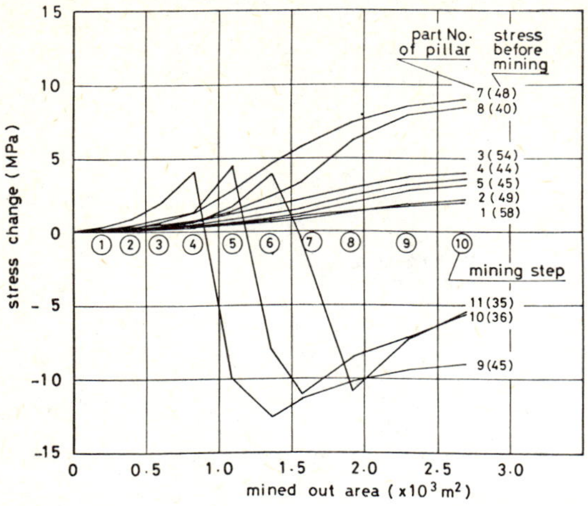

Fig.14 Change of the pillar stress during mining, which was determined from the displacement discontinuity method.

is, the incidence of microseismic event may be referred to the fracture of the coal seam or strata, or at least to the stress drop from a preexisting equilibrium.

In this point of view the spatial rate of energy release during the mining of the N1 and S1 blocks was evaluated for two different situations: the first case assumed that the coal pillar held a sound state, while second one assumed that the coal pillar yielded and its stiffness becames a fraction of that of the sound state. The spatial rates of energy release were determined by the displacement discontinuity method as 6.67 MJ/m^2 and 9.02 MJ/m^2 for the former and later situations respectively. Furthermore the spatial rate of energy release associated with mining of north portion of the N2 block was determined as 7.02MJ/m^2. The ratio of the spatial rate of energy release of N1 plus S1 block to that of the N2 block is, thereby, 9.02/7.02=1.28, if the coal pillar yielded during the mining of the N1 and S1 blocks. On the other hand the ratio of spatial rates of seismic energy release between the same blocks as cited above is estimated as 51.6/21.2=2.43 from the result in Table 2. If the both ratios are close to one another, then the assumption that the pillar might yield during the mining of the N1 and S1 blocks can be accepted, since the spatial rate of seismic energy release is proportional to the spatial rate of energy release, Salamon (1974), Salamon and Wagner (1979). Thus a discrepancy between them implies that yielding of the pillar is, at least, neccessary to the microseismic activity in these blocks, but is not sufficient to it. Other geological conditions that were not included in the present experiment should be taken account for an inerpretation about the microseismic activity.

A drastic reduction in release of seismic energy during the mining of the lower seams has been appreciated in Sunagawa Coal Mine, Isobe et al (1981). The result of the present study has a good agreement with the previous one that has been already explained by the same numerical experiment as that mentioned here.

A significant decrease of release of seismic energy was experienced during the alternative mining of the N2 and S2 blocks. Although a preferable change in microseismic activity might be yielded by a reduction in the rate of face advance, its reasonable explanation can hardly be detailed at the present.

The study presented in this paper is in continuation at the current mining panel and it will be undertaken at the other panels in next year. On the middle of January 1982, a new system will be installed at the mine office for monitoring microsesmic activity. The new system is to make responsible for not only recording automatically seismic data, but also analyzing automatically the source location and seismic energy of event, printing the result and displaying the location of seismic foci relative to mining panel on a colour graphic terminal by means of microcomputers. Besides an improvement of the microseismic study, a rational method for designing support of access roadway and pereventive measures for rockburst should be looked for to alleviate the hazards of rockburst. An intimate linkage of microseismic study and numerical method to preventive measures will enrich the status of strata mechanics so as to spread its potential application to rational design of underground mine layout and neccessary and sufficient support system of underground excavation.

ACKNOWLEDGEMENT

The authors would like to express their thanks for the facilities provided by the Japan Coal Mine Research Center, especially, Mr. S. Fukai and Dr. A. Fukushima who have given many suggestions to the discussion of the microseismic study described in this paper.
The authors are greatful Mr. S. Shiojima, Head of the Department of Machinery and Electrics of Sumagawa Mine who has cooperated in the installation of the microseismic monitoring facility, Mr. K. Watanabe, Mr. S. Yoshinaga and Mr. S. Matsuyama, Mining Engineers of Sunagawa Mine, who have assisted the analyses of the seismic data and provided the mining data.

The source location has been carried out by the COSMO 700 computer system at Muroran Institute of Technology. The numerical experiment discussed in this paper was excuted by using the HITAC M-180 computer system at Hokkaido University.

REFERENCES

Ishijima,Y., Sato,K. and Kinoshita,S., 1980, "Application of the displacement discontinuity method to some crack problems", Theor.Appl.Mech., 28, 167-186.

Isobe,T., Mori,N., Sato,K. and Goto,T., 1979, "Development and application of computer system for monitoring seismicity induced by underground coal mining", Application of Computers and Operations Research in the Mineral Industry, Edited by T.J.O'Neil, A.I.M.E., 513-527.

Isobe,T., Mori,N., Sato,K. and Goto,T., 1981, "Seismicity associated with mining of multiple coal seam - Study of the design of mine layout for working coal seams based on rock mechanics (1st Report) -", J.Min.Metall.Inst.Japan, 97, 1169-1174.

Ohtsuka,T., 1980, "Hydraulic mining at Suangawa Coal Mine", Fourth Joint Meeting MMIJ-AIME,Tokyo, Technical Session B-4, 63-75.

Salamon,M.D.G., 1974, "Rock mechanics of underground excavatios", Proc.3rd Congr.I.S.R.M.,Denver, vol.I, 951-1099.

Salamon,M.D.G. and Wagner,H., 1979, "Role of stabilizing pillars in the alleviation of rock burst hazard in deep mines", Proc.4th Int.Congr. I.S.R.M., 561-566.

INDUCED SEISMICITY RESULTING FROM ROOF CAVING AND PILLAR FAILURE IN LONGWALL MINING

N. J. KUSZNIR*
N. H. AL-SAIGH*
I. W. FARMER **

* Department of Geology, University of Keele, Keele, Staffs., ST5 5BG, England.

** Department of Mining Engineering, University of Newcastle upon Tyne, Newcastle upon Tyne, NE1 7RU, England.

SUMMARY

The causal relationship between seismicity and longwall coal extraction in the North Staffordshire coalfield has been established. The larger seismic events (M_L = 2.5 - 3.5) occur in the pillars of previous workings when the active face passes under or above these pillars. These events have a shear source mechanism and are generated by failure within these pillars. The smaller seismic events, many of which have an implosional mechanism, are interpreted as being generated by strata collapse behind the active face and within the waste of adjacent panels.

INTRODUCTION

Seismicity associated with the mining of coal and ore is a relatively common phenomenon occuring throughout the world. In many of the reported case histories a direct causal relationship between mining and seismicity has been established. Earth tremors have occurred in the North Staffordshire coalfield since the latter part of the last century; however,in the period 1975-1977 and 1980-1981, seismicity levels increased substantially. This seismicity has been shown in most cases to be directly related to the active extraction of coal. The problem of mining induced seismicity is of considerable interest within the North Staffordshire coalfield, since the mined areas underlie the densely populated conurbation of Stoke-on-Trent. Maximum tremor magnitudes of approximately 3.5 have occurred, with hypocentres of the order of 1 km depth, resulting in the worst cases in maximum Mercalli intensities of between V and VI.

The seismicity occurring in the southern part of the conurbation in the Trent-Vale and Hanford area has been studied in greatest detail and has been described previously by Kusznir et al. (1980a) and Westbrook et al. (1980). Other seismicity in the northern part of Stoke-on-Trent (areas A-D, Figure 1) has also been examined. The seismicity within the Trent-Vale area is associated with longwall mining in the Ten Feet seam at a depth of approximately 1000m. The extractions in the Ten Feet seams are over- or underlaid by old workings and major faults are shown in Figure 2.

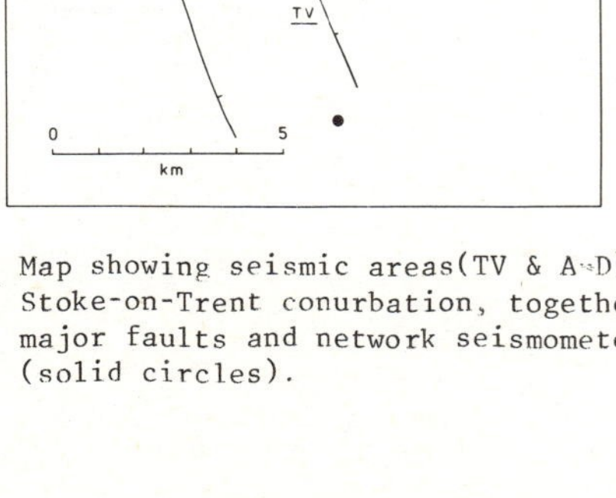

Figure 1: Map showing seismic areas(TV & A-D) in the Stoke-on-Trent conurbation, together with major faults and network seismometers (solid circles).

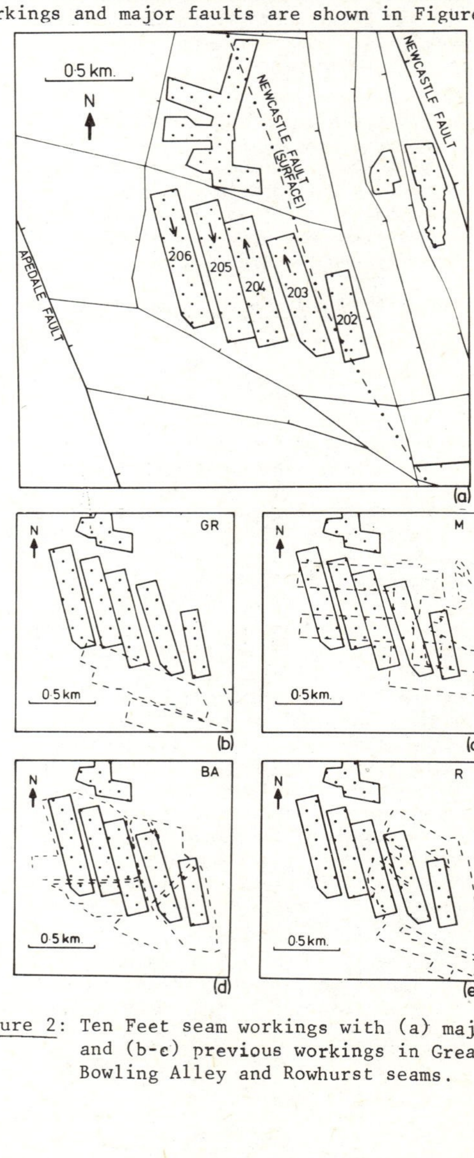

Figure 2: Ten Feet seam workings with (a) major faults and (b-e) previous workings in Great Row, Moss, Bowling Alley and Rowhurst seams.

Mining induced seismicity was studied in greatest detail for panels 205 and 206. A geological section showing the predominant lithologies and the depths of the other seams is shown in Figure 3.

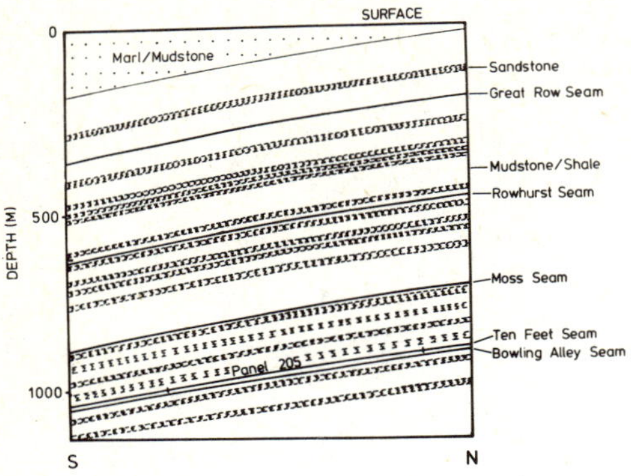

Figure 3: Geological section along the lines of panel 205.

The ground consists of predominantly shale and mudstone with intervening coal seams and lenticular sandstone bodies.

The Ten Feet seam panels were mined by longwall extraction with total caving. Panels 203 and 204 were mined using advance faces, while panels 205 and 206 used retreat faces. Typical advance rates were between 10 and 20m per week with an extraction height of the order of 2.5m. Mining conditions were generally good, although some deterioration of the condition of panel access roads did occur where they passed under or over the Moss and Bowling Alley pillars. No rockbursts were reported to have occurred.

The seismic network used to investigate the mining induced seismicity is shown in Figure 1. Two networks were used: a large, radiolinked network, with aperture approximately 10 km (see Figure 1) and a smaller 5-component network with aperture 1.5 km centred over the panel (not shown in Figure 1). Westbrook et al. (1980) have described the network operations and the technique used for tremor hypocentre determination.

THE OBSERVED RELATIONSHIPS BETWEEN MINING AND SEISMICITY

The seismicity in the Trent-Vale and Hanford area was monitored continually during the mining of panels 205 and 206 using the large aperture network. Tremor hypocentres and magnitudes were determined from the recorded seismic data. The distribution of the tremor hypocentres occurring during the mining of panels 205 and 206 are shown in Figure 4, where they can be related to faults and to the distribution of previous workings, in particular to those in the Moss and Bowling Alley seams. Plan and vertical sections parallel and perpendicular to the face axis are shown. It can be seen that the tremors lie adjacent to the active mine workings at a depth predominently between 600 and 1000m. In addition to the locations shown in Figure 4, the higher resolution small aperture network was used in conjunction with the large aperture network to give tremor hypocentres occurring during one month's mining of panel 206. These locations have been described by Kusznir et al. (1980b) and provide greater hypocentre accuracy than those obtained using the large networks alone. A vertical section parallel to the face advance direction showing these higher accuracy locations is shown in Figure 5.

First motion analysis was also carried out to determine the nature of the seismic source mechanism. Two types of mechanism were identified; one with a shear type mechanism and the other with an implosional source mechanism.

The pillars in previously worked seams and below the Ten Feet seam play an important role in earth tremor generation. In Figure 6 tremor depth is shown against distance advanced by the face and it can be seen that the large events felt by people on the surface occur predominently when the Ten Feet face passes under or over the Moss and Bowling Alley pillars. These large events occur predominently at the level of the Moss seam, as shown by the high resolution data (Figure 5) and have a shear source mechanism.

The analysis of the data to provide the spatial and temporal relationships between tremors and mining have been described previously by Kusznir et al. (1980a) in detail. The conclusions, taken from Kusznir et al. (1980), are summarised below:

1. Tremors were associated with the extraction of faces in the Ten Feet seam.
2. Tremor hypocentres moved in unison with the advancing coal face.
3. The tremors did not lie on either of the major faults bounding the area.
4. The majority of tremors were not felt at the surface. Of 169 tremors monitored while panel 205 was working, 39 were felt. Of 540 tremors associated with the working of panel 206, 16 were felt. None have been felt underground.
5. Felt tremors occurred as the Ten Feet faces passed under or over pillars in the Moss and Bowling Alley seams.
6. The greatest concentration of tremors occurred as the Ten Feet faces passed over a pillar in the underlying Bowling Alley seam.
7. The tremors did not start until the face had progressed about 250m, indicating a critical span for the onset of seismicity
8. The tremor hypocentres appear to be displaced about 200m to the east of the active Ten Feet panel.
9. The larger magnitude events which were felt show a shear type source mechanism, while the remaining smaller, unfelt events have an implosional mechanism.

The relationship between magnitude and tremor occurrence frequency is shown in Figure 7 for tremors occurring during the mining of panels 205 and 206. The frequency-magnitude relationship may generally be expressed in the form:

$$\log_{10} N = a - b M \qquad ..... (1)$$

where N is the frequency of occurrence, M is Magnitude and a and b are constants. For panels 205 and 206 the following relationships were found:

$$\log N = 3.1761 - 0.61 M_L \quad \text{(Panel 205)}$$

$$\log N = 4.6128 - 1.19 M_L \quad \text{(Panel 206)}$$

Both plots show a linear relationship between magnitude and logN in the magnitude range 1.7 to 2.7. The deviation of the experimental data from the straight line for the larger magnitudes suggests that a maximum magnitude exists, indicating that the rocks have an upper limit of strain energy. The seismic events follow Gumbel's third statistical distribution (Gumbel, 1958), and this statistic predicts upper magnitude values for panels 205 and 206 of 3.2 and 3.25 respectively. These values are only a little larger than those actually experienced.

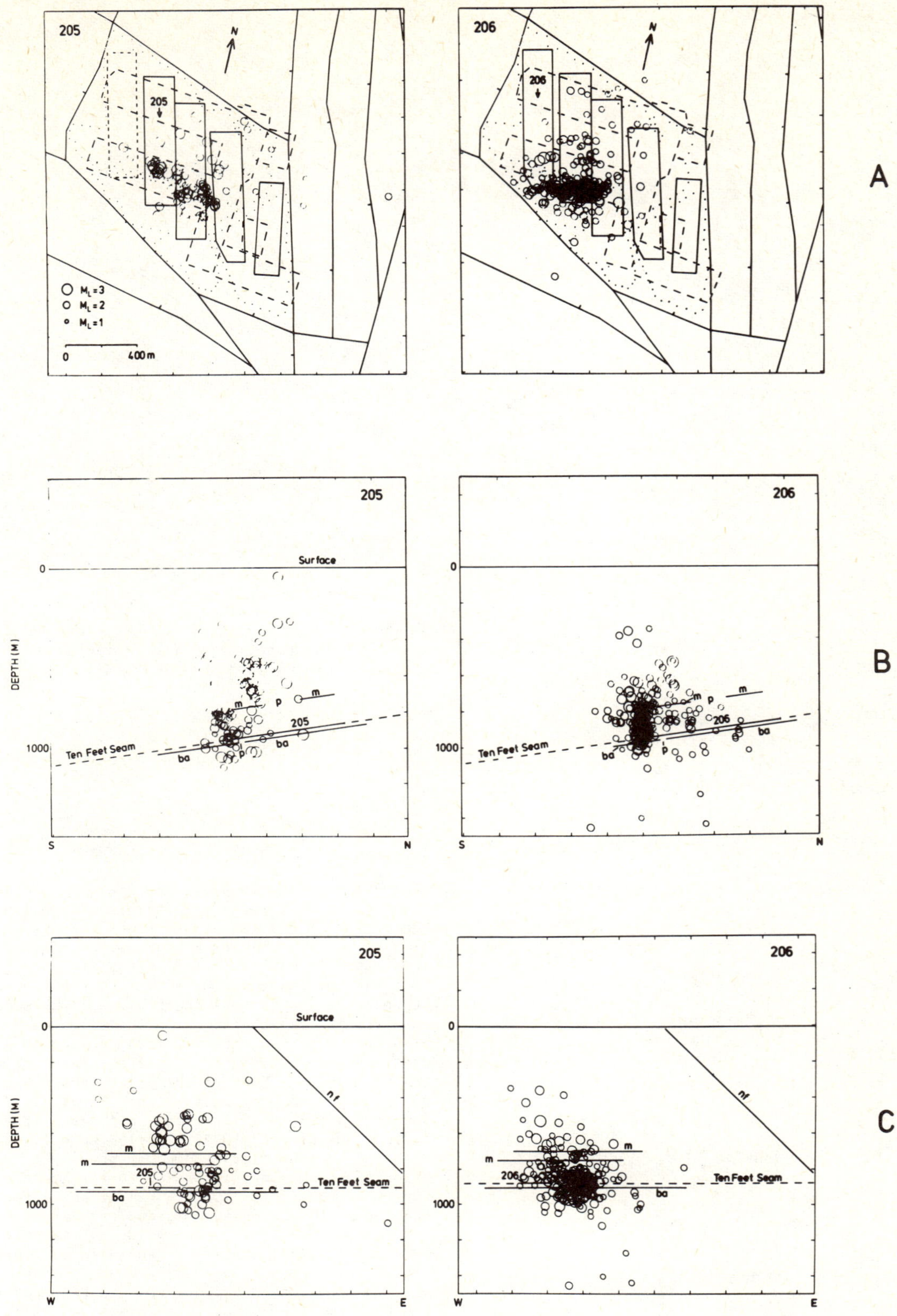

Figure 4: Tremor hypocentres occurring during the mining of panels 205 and 206 (A) plan view, (B) vertical section perpendicular to face, (C) vertical section parallel to face. Faults and previous workings in Moss and Bowling Alley seams are also shown.

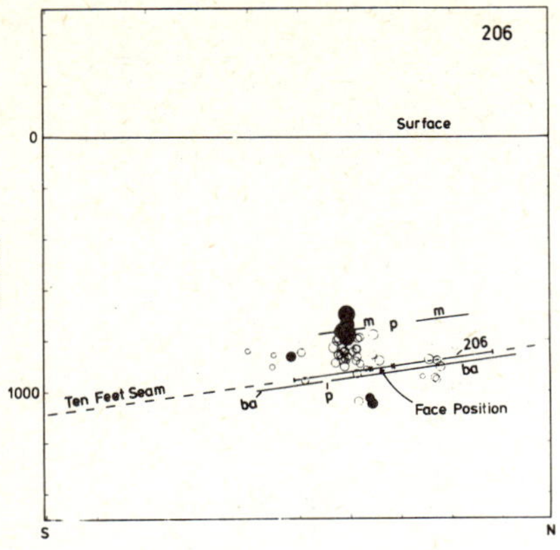

Figure 5: Higher accuracy tremor locations - Vertical section perpendicular to face. Solid circles - shear mechanism, open circles - implosional mechanism.

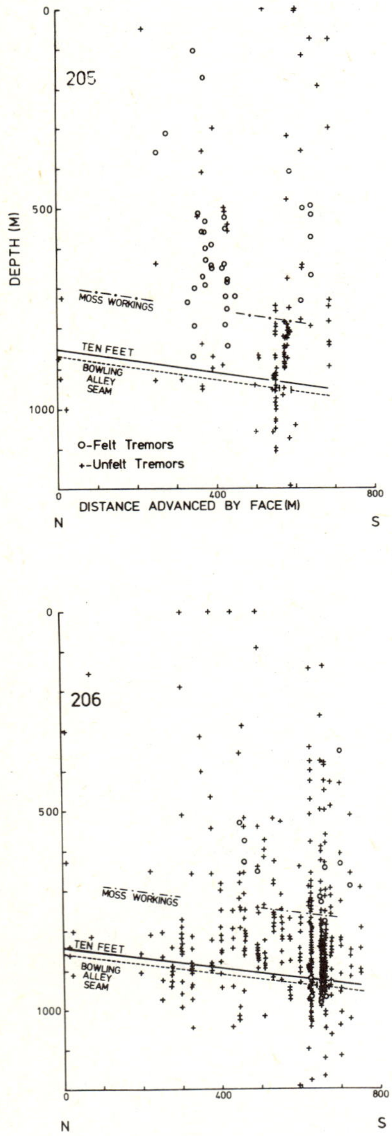

Figure 6: Tremor depth plotted against distance advanced by face. Seam and pillar positions are also shown.

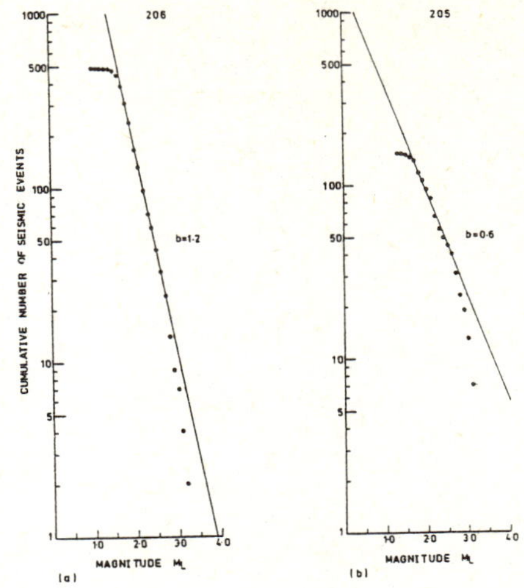

Figure 7: Frequency-magnitude distribution for panels occuring during the mining of panels 205 and 206.

THE ROLE OF PILLARS IN TREMOR GENERATION

Pillars left in previous workings have been shown to be an important factor in controlling the generation of the larger magnitude tremors. These tremors, having a shear source mechanism, occur predominently within these pillars when the Ten Feet face passed under or above these pillars. The pillars in the Moss and Bowling Alley seams, as a result of extracitons in these seams, will have been left with vertical stresses considerably greater than lithostatic and with lateral stresses greatly reduced. Even if the pillar cones had failed during the mining of those seams, large residual vertical stresses will remain in those pillars (Wilson and Ashwin, 1972). While the pillars of the Moss and Bowling Alley seams may have been stable prior to the mining of the Ten Feet seam, the superposition of the front abutment pressure of the Ten Feet face on the residual stress of the old pillars caused renewed failure of these pillars. This failure in compression generated the larger tremors with a shear source mechanism. The vertical separation between the Ten Feet and Moss and Bowling Alley seams is such that considerable interaction can occur between stresses produced by extraction in those seams.

While some large shear events associated with the passage of the Ten Feet face over the Bowling Alley pillar did occur, the majority of the shear events occurred when the face passed under the Moss pillars. Calculations suggest that, while the Moss pillars (width 200m) would have been left with a competent core, the Bowling Alley pillar (width 25m) would have completely failed. The greater number of shear events associated with the Moss pillar is, therefore, related to the greater competence of the Moss pillar and its ability to store more strain energy than that of the Bowling Alley. Although completely fractured the Bowling Alley pillar will still have large residual vertical stress and would fail in compression. The Bowling Alley pillar is, however, more effective in creating a large number of unflet smaller events at the level of the Bowling Alley and Ten Feet seams. These events are of unknown mechanism type, but may be small, shear failures, or more likely implosional events associated with waste collapse modified by the influence of the Bowling Alley pillar.

The a and b parameters of equation(1) may be obtained as a function of face advance. The a parameter provides a measure of the total number of seismic events as M_L tends to zero, while b gives an indication of the ratio of the number of small to large magnitude events. The a parameter is an indication of general seismicity and is related to both the competence of the material and the stress level. The factors controlling the b parameter are not known; however, it has been suggested that it may increase with increase in rock strength heterogeneity (Mogi, 1962), or decrease with increase in the level of stress (Scholz, 1968).

In Figure 8, a and b parameters are shown for panels 205 and 206, for 100m sections of face advance. Lower a and b values occur when the Ten Feet face is under the Moss pillars. The lower a values in the pillar regions are to be expected since this material is more competent and fails less easily than the goaf areas.

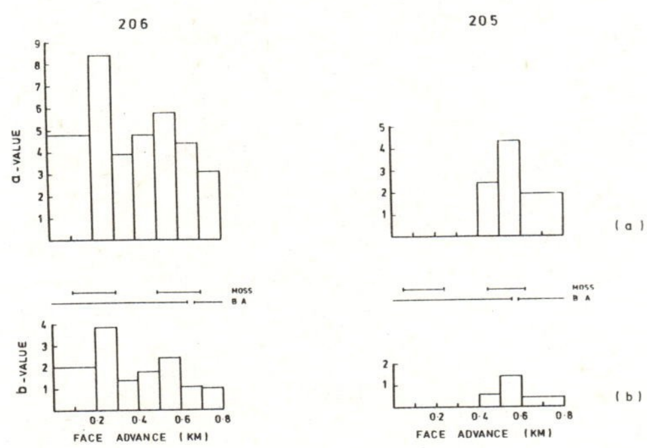

Figure 8: Plots of a and b values for each 100m of face advance for panels 205 and 206. The regions where the Ten Feet face lies above or below the Moss and Bowling Alley goaf is also shown.

The low b values in pillar regions could be attributed to either the greater homogeneity or stress levels within the pillars.

The frequency-magnitude relationship is substantially different for each panel (Figure 7), panel 205 showing lower a and b values than panel 206. The seismicity level per unit volume of extraction, S, may be given by the equation:

$$S = 10^{a} / Vm \qquad \ldots\ldots (2)$$

where Vm is the volume of rock mined (McGarr and Wiebols, 1977). The seismic energy released per unit volume of rock, E_L, can also be calculated using the relationship:

$$E_L = E / Vm$$

where E is the total seismic energy released from the excavation. Calculating these parameters using these relationships gives:

Panel 205

$S = 0.005 \qquad E_L = 1.97 \times 10^4 J.$

Panel 206

$S = 0.155 \qquad E_L = 1.962 \times 10^4 J.$

It can be seen that while the two panels produce nearly identical seismic energy release per unit volume, the seismicity per unit volume of panel 206 is x30 that of panel 205. This substantial difference in seismicity statistics between the two panels can also be illustrated by the fact that 179 tremors were recorded during the life of panel 205, while 500 were recorded during the life of panel 206.

It is suggested that the higher seismicity of panel 205 may be attributed to the barrier pillar left between panels 205 and 206. No similar pillar was left between panels 205 and 204. The effect of the pillar was to greatly increase the number of tremors but also to decrease the number of large events.

SEISMIC EVENTS GENERATED BY WASTE COLLAPSE

The smaller magnitude events with implosional source mechanisms occurred closer in depth to the Ten Feet seam than the shear events. It is suggested that these implosional events were generated by tensile failure in waste collapse behind or to the side of the Ten Feet face. Many of these small magnitude implosional events were shifted eastward relative to the active panel. While this may be due to location error it is more likely that these events were situated over the previously extracted panels.

These events in the Ten Feet goaf adjacent to the active panel may have been caused by the reactivation of collapse by the passage of the abutment pressures of the active face. An alternative explanation may, however, exist. As successive Ten Feet panels are worked, the span of the workings parallel to the face axis increased so increasing the tensile stresses and strains in the strata spanning the workings. A critical span of 250m must be exceeded in the face advance direction for the onset of seismicity. Such a critical span in the direction parallel to the face, of the order of 300m, could explain why many events are displaced over the previously extracted Ten Feet panel.

In contrast to the shear events which were generated by failure in compression and produced large magnitude events, the waste events with tensile failure released much less energy and had smaller magnitudes, since less strain energy can be stored in tension than compression.

DISCUSSION AND CONCLUSION

Two types of seismic event, generated by the longwall mining of the Ten Feet seam, have been recognised, each with different generative mechanisms. The smaller events, with an implosional mechanism and generated by waste collapse, do not appear to form any hazard and are thought to be the inevitable and normal seismic noise generated by waste caving in longwall mining. In contrast, the larger magnitude events had a shear source mechanism and were felt at the surface, although they were not felt at the face. Pillars in previous workings in adjacent seams appear to generate these events by stress interaction effects. Tremors with this type of mechanism may be controlled or moderated by panel layouts which minimise the effects of such interseam interaction. The use of barrier pillars between adjacent active panels may also assist in moderating the effects of such seismicity by increasing the seismicity, but at the same time reducing the number of large events. Gumbel's third statistical distribution predicts upper magnitude values for these events of 3.25; however, the applicability of this statistic to mining induced seismicity problems may be open to criticism. The importance of pillars in adjacent seams in generating mining induced seismicity has been reported elsewhere (Salustowic and Parysiewic, 1963).

Seismicity has also been detected in other parts of the North Staffordshire coalfield and the majority of the seismicity coincides with active longwall mining. During the period 1975-1977 seismicity was also detected in the areas A-D shown in Figure 1. The seismicity in areas A, C and D had similar mechanisms to that of the Trent-Vale seismicity, with again the interaction of current and old workings generating many of the larger events. The seismicity in area B appears to have a different mechanism; the tremors probably being generated by induced slip on an old fault plane. The mechanism of seismicity in area E is unknown, but is not thought to be associated with mining. The seismicity in areas A-D was not felt and had magnitude values much smaller than those of the Trent-Vale seismicity.

REFERENCES

Gumbel E.J., 1958, Statistics of Extremes. New York and London: Columbia University Press.

Kusznir N.J., Ashwin D.P. and Bradley A.G., 1980a, Mining Induced seismicity in the North Staffordshire Coalfield, England. Int. J. Rock Mech. Min.Sci. & Geomech. Abstr., 17, pp.45-55.

Kusznir N.J., Farmer I.W., Ashwin D.P., Bradley A.G. and Al-Saigh N.H., 1980b, Observation and Mechanics of seismicity associated with coal mining in North Staffordshire, England. In: Proc. 21st U.S. Rock Mech. Symp., pp. 163-171.

McGarr A and Wiebols G.A., 1977, Influence of mine geometry and closure volume on seismicity in a deep-level mine. Int. J. Rock Mech.Min.Sci. & Geomech. Abstr., 14, 139-49.

Mogi K, 1962, Study of elastic shocks caused by the fracture of heterogeneous materials and its relationship to earthquake phenomena, Bulletin of the Earthquake Research Institute, 40, pp.125-73.

Salustowic H.A. and Poryriewic W, 1963, The problems of rock bursts in mines, Proc. Third Int.Mining Congress, Salzburg.

Scholz C.H., 1968, The frequency magnitude relation of microfracturing in rock and its relationship to earthquakes. Bull. Seism. Soc. Am., 58, pp.399-415.

Westbrook G.K., Kusznir N.J., Browitt C.W.A. and Holdsworth B.K., 1980, Seismicity induced by coal mining in Stoke-on-Trent (U.K.), Engineering Geology, 16, pp.225-41.

Wilson A and Ashwin D.P., 1972, Research into the determination of pillar size, The Mining Engineer, 131, pp.409-27.

A STUDY OF THE BEHAVIOUR OF OVERLYING STRATA IN LONGWALL MINING AND ITS APPLICATION TO STRATA CONTROL

Chien Ming-Gao

Associate Professor, Head of the Laboratory of Strata Control,
Department of Mining Engineering,
China Institute of Mining,
People's Republic of China.

SUMMARY: The objective of this investigation was to describe the behaviour of strata above a longwall face through a study of the movement of inter-strata plugs in a longwall working area. The investigations were conducted in the Dai-Tun coal mine, Province Jiangsu, China. By analysing the subsidence curves of the overlying strata a structural model was constructed to examine the behaviour of the strata. By using this model, some of the phenomena of ground subsidence and roof pressure in the longwall mining can be explained.

INTRODUCTION

In the Chinese coal mining industry systems of exploitation and of face support are determined by roof conditions and the effects of multi-seam exploitation. In the last 10 years in China hydraulic powered support installations have been widely used in many coalfields and in various roof conditions. In order to define the field of their application and to determine the rock loads which the supports must be capable of resisting, many studies have been undertaken to investigate the interaction between the support and roof pressure.

An important basis for the study of roof control and ground subsidence is the behaviour of strata overlying the working coal seam.

UNDERGROUND INVESTIGATIONS

Conditions and methods of investigation

The general outline of the experimental roadway and investigation boreholes at the Dai-Tun coal mine are shown in Figure 1.

There were 6 boreholes (S_1 .... S_6) placed in the roadway which was over the middle of the working face No. 8111 in the direction of the face advance.

The spacing between the adjacent boreholes along the roadway was 8m. 3-5 plugs were placed in each borehole and these were placed at intervals of 5-10, 10-15 and 15-20m above coal seam No. 8, to intercept beds of interest.

The experiment was essentially designed to observe the behaviour of the floor in the roadway overlying coal face No 8 111 with precise level measurement, and to determine the relative displacement between the plugs and the floor in this roadway through the multi-wire boreholes. The roadway was 178m deep and coal face No. 8111 was 115m long and inclined at 25°. Its extracted height was 2m. The vertical distance between the coal seam No. 8 and the roadway was 24.78m.

In addition to the bor[illegible] on the floor of the roadway there were other su[illegible] points J_1 - J_{15} placed for levelling and traverse surveying.

The strata overlying coal seam No. 8 are mainly sandstones. The lowest sandstones, having a total thickness of 10m, which overlie the coal seam are inherently weak so that they readily fracture during mining operations. They contain frequent natural weakness planes and partings. Above the immediate roof there are four stronger strata of sandstones having thicknesses of 4.05, 2.6, 4.6 and 2.5m.

Vertical displacement (V.D.)

In discussing vertical displacements, the results of measurements at boreholes S_1 are taken as an example. Fig. 2 shows the behaviour of the inter-strata plugs in S_1.

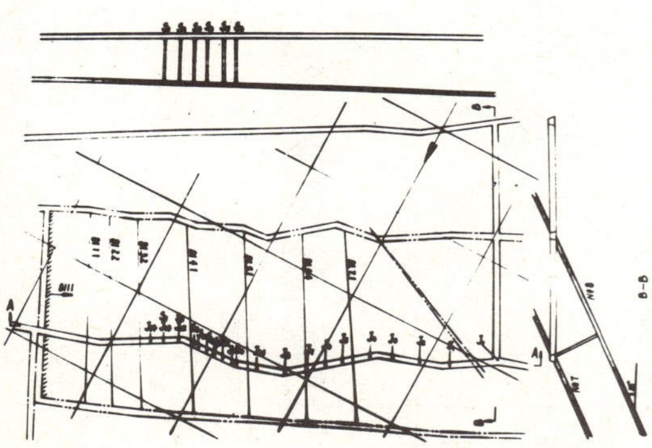

Fig. 1

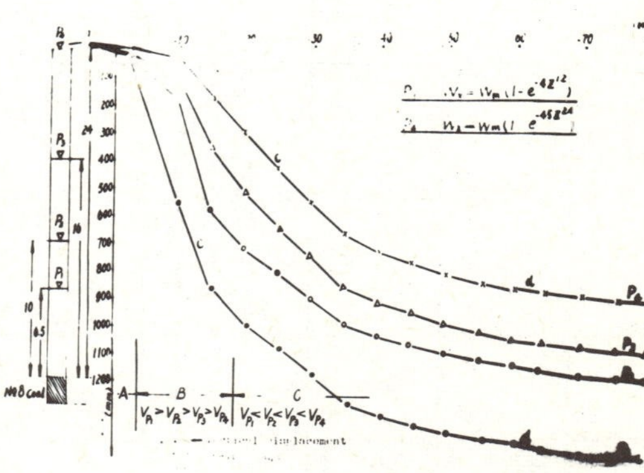

Fig. 2

The form of the curves is similar to a negative exponent curve and can be expressed by:

$$W_x = W_m(1 - e^{-az^b})$$

where W_x is the vertical displacement at distance X from face,
and W_m is the displacement at distance L from face, where the variation of this curve is just stable.

$Z = \frac{X}{L}$, a and b are two coefficients, which are closely related to the mechanical properties of the overlying strata and the interval between the working coal seam and the strata being investigated.

An example of this curve in the 24.78m vertical interval above the working coal seam is shown in the following table:

$Z = \frac{X}{L}$	0.1	0.2	0.3	0.4	0.5	0.6	0.7	0.8	0.9	1.0
measured value (mm)	12	50	200	350	475	600	700	750	800	825
calculated value (mm)	14.6	74.4183	324	473	604	703	765	800	816	

In this example L = 50m, W_m = 825mm.

According to this curve, the distribution of its gradient is an abnormal curve. On the basis of the mechanical characteristics of these curves, the overlying strata can be divided into three zones (see Figure 2) along the direction of face advance.

1. Zone ab-abutment pressure influence (API or A) zone.

The strata in this zone is supported by the influence of the working coal seam and the vertical displacement of the plugs is very slight, and usually, but not always, a slight negative magnitude occurred. This was always less than 40mm.

The vertical displacement of the point b was slight, even of the lower plugs (5-10m from the coal seam), until the face was 4-8m past the borehole.

2. Zone bc-Bed separation (BSor B) zone.

When the influence of the working coal seam is removed the displacement rate of the points increases rapidly. The plug displacement rates in borehole S_1 are shown as follows:

The interval from plug to the working seam (m)	6-10	10-15	15-20	25
The max. displacement rate of the plug (mm/m)	180	100	70	50
The average displacement rate of the plug (mm)	80-100	60-80	55	45

This indicates that the displacement rate of the overlying bed is not as fast as its underlying bed, i.e.

$$V_{P_1} > V_{P_2} > V_{P_3} > V_{P_4}$$

In this zone the strata groups are separated from each other.

3. Zone cd-consolidation (C) zone.

In this zone the separated strata are reconsolidated and the rate of displacement of the plugs in boreholes S_1 is as follows:

The interval from plug to the working coal seam (m)	6-10	10-15	15-20	25
The average rate of displacement (mm/m)	18.5	18.4	22.1	23.5

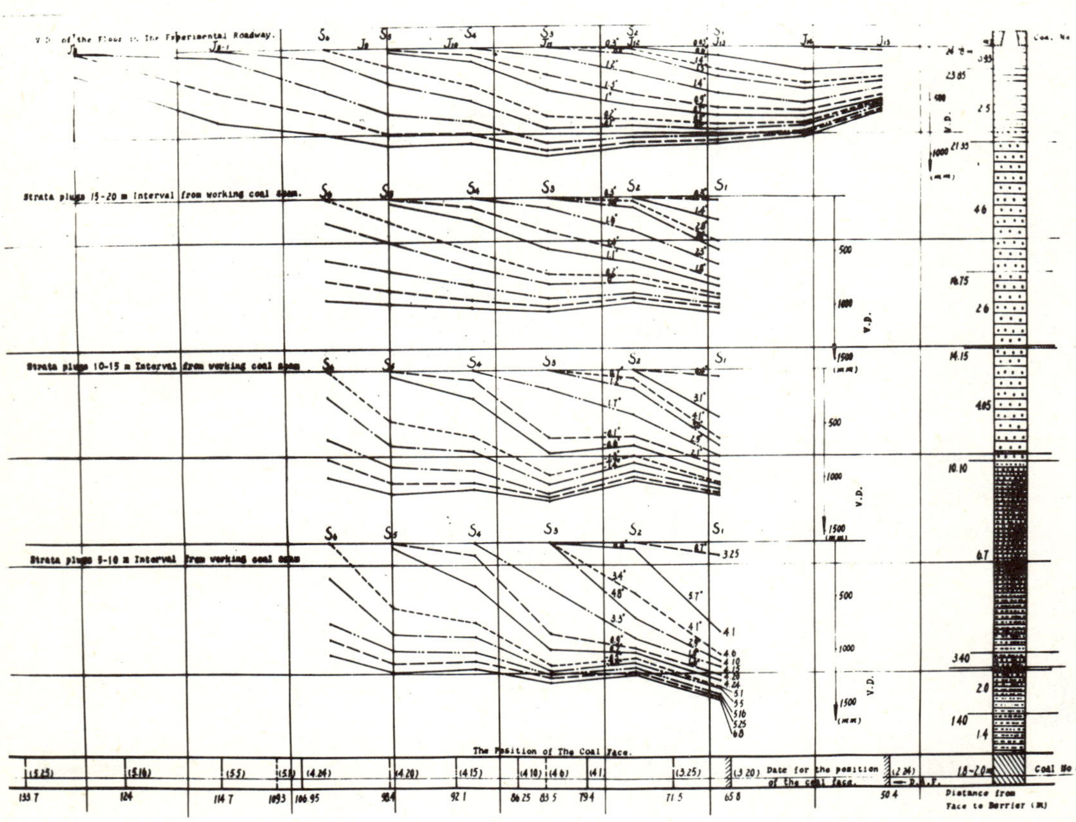

Fig. 3

The phenomenon is the opposite to that in separation zone bc, i.e.

$$v_{P_1} < v_{P_2} < v_{P_3} < v_{P_4}$$

Fig. 3 shows the vertical displacement of all the inter-strata plugs placed in the six boreholes.

The behaviour of the strata can be expressed through the variation of the linear gradient between the pairs of plugs in adjacent boreholes. The example for S_1-S_2 is shown in the following table.

Date of the measurement	25th Mar.	1st Apr.	6th Apr.	15th Apr.	1st May	5th May
The gradient for the level 6-10m (%)	12	100	72	33	30	24

The linear gradient of S_1-S_2 at the same date (1st Apr.) for the different intervals from the working coal seam is expressed below:

The interval from plug to working coal seam (m)	At the roof	5-10	10-15	15-22	25
The gradient of the line S_1-S_2 (%)	123	100	54	24	10

Horizontal displacement (H.D.)

The horizontal displacement was measured only for the points placed in the floor of the roadway by theodolite.

Fig. 4 shows the horizontal displacement path of the points. It illustrates that when the roadway is undermined, the movement at the beginning is in a direction opposite to that of the face advance and then after a time the displacement changes to the direction of face advance. At the same time a horizontal displacement occurs in the direction of rise of the seam inclination.

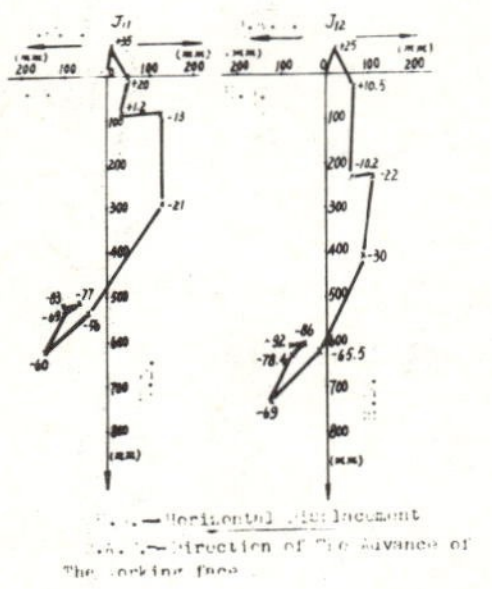

Fig. 4

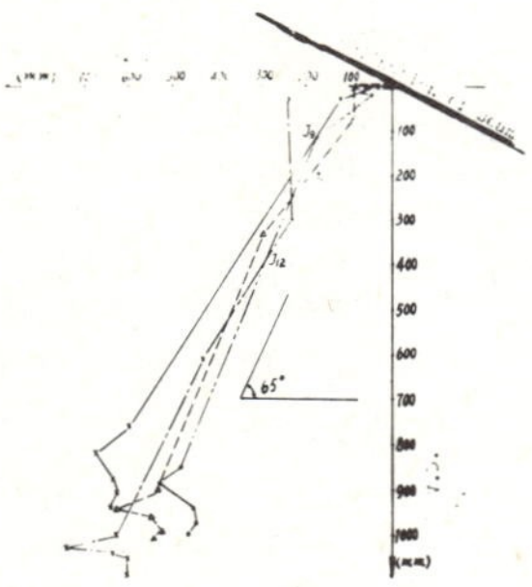

Fig. 5

From this datum, the displacement path in the vertical plane along the seam inclination can be obtained as in Fig. 5. In Fig. 5 it can be seen that the path of the point is just normal to the stratification of the worked coal seam.

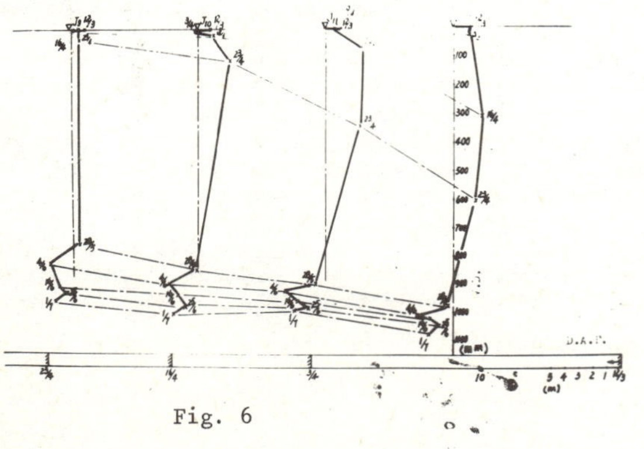

Fig. 6

Fig. 6 shows the displacement path in the vertical plane along the direction of face advance. In this figure it can be seen that the final position of the point always overpasses the original space by 100-200mm.

From this data, it can be seen that the intermediate friction between the strata is insufficient to resist horizontal displacement.

Analysis

From the above results, Fig. 7 can be put forward to illustrate the situation which arises as the overlying strata are undermined. In Fig. 7 the lower immediate roof caves irregularly into the goaf area. Above it are stronger strata (main roof), which are broken into regular blocks with the advance of the coal face. These regular blocks may be interlocked and grade into the subsurface strata.

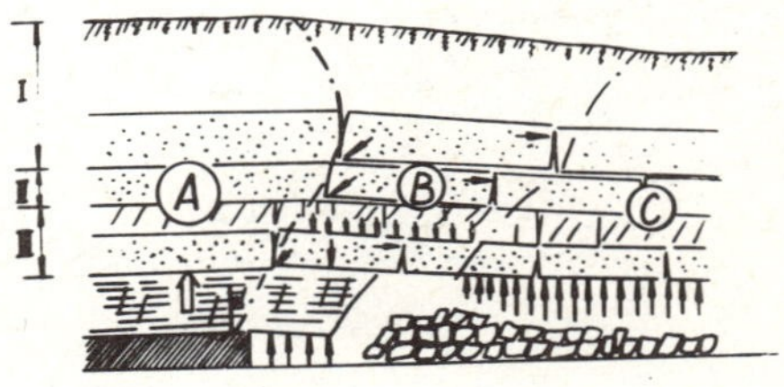

Fig. 7

The weight of the higher layers of overlying strata are supported by a system of "working coal seam - waste caved blocks" and the lower layers are supported by a system of "working coal seam - roof support - waste caved materials".

For this reason, in order to develop a "structural" description for the support of overlying strata, the influence of the roof support resistance on the higher stronger strata should be ignored. From an engineering standpoint, the overlying strata can be divided into several groups.

The lower bed of every group is the strongest and thickest bed (e.g. sandstone or limestone and others). The weak strata which overlie this bed can be considered as a load acting on this stronger bed and as an intermediate supporting-layer for the overlying group strata.

For example, the strata shown in Fig. 7 can be divided into 3 groups (that is I, II and III). The interlocked space between the adjacent blocks of the stronger bed was determined by the characteristic of its vertical displacement curve. For example, when the curve was bending concave downwards, the interlocked space was at the bottom of the blocks and in the opposite case, i.e. bending concave upwards, the interlocked space was at the head of the blocks.

STRUCTURAL MODEL

Assumptions

On the basis of the foregoing statement the following assumptions can be made to build a "structural" model.

1. Every stronger bed in every group stata is assumed to be a "structural" formation.

   Thus, in group strata I, as shown in Fig. 7, the surface deposit is assumed to behave like a uniform boundary load under its own weight, which acts on the stronger bed. For the lower group strata the load acting on the stronger bed will not be uniform.

2. Because the stronger bed undermined by the working coal face has been separated into a series of regular blocks, it is assumed that the regular blocks can be considered as the

elements of the structural formation, and its overlying weak strata as the load acting on this structural formation.

3. Considering the conditions in interbed separation zone bc, there is no resistance from the blocks in this zone to the higher overlying strata.

4. Because the intermediate friction between strata cannot resist the horizontal deformation of the strata, the weak strata between two adjacent stronger beds can be assumed to act as a series of columns to support overlying strata and to load the underlying strata.

Then, considering these conditions, the structural formation of Fig. 7 can be designated as in Fig. 8.

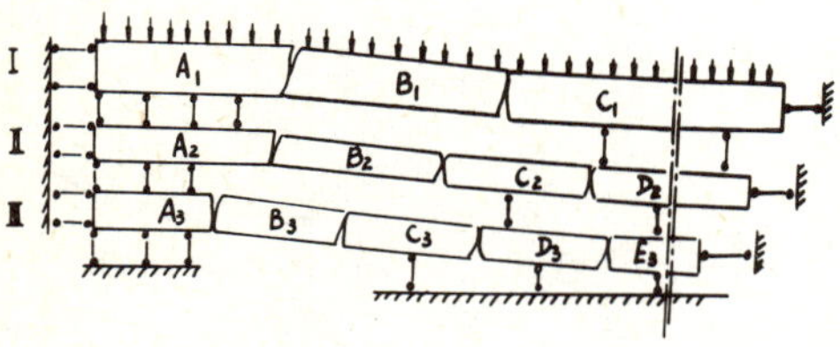

Fig. 8

Model Calculation

It is now possible to analyse the interaction of forces in every structural formation as shown in Fig. 9.

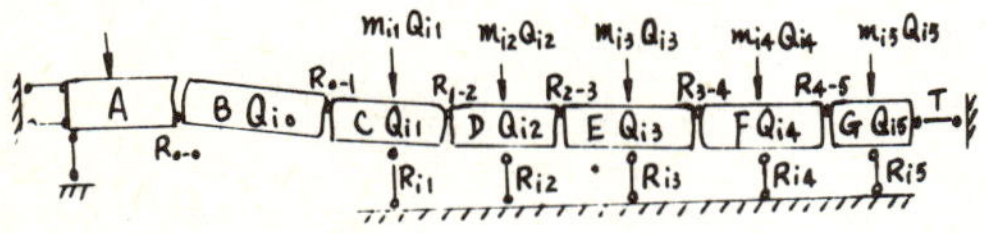

Fig 9

The symbols which appear in the subsequent analysis are given below:

A.B.C.D....	-	The symbols of blocks.
T	-	Lateral thrust of the blocks.
R	-	Resistance of the underlying strata and the shear force between the adjacent blocks.
q	-	Uniform load per unit length, per unit width of the block.
Q	-	The weight of a block in a group strata.
L	-	Length of the block.
S_1 and S_2	-	The relative displacement from one end to another of a block.
h	-	Thickness of the bed.
n	-	Gradient of the block.
m	-	Loaded coefficient of the blocks.

In order to represent and distinguish every bed and every block, to every symbol there is a subscript. For example in R_{11}, '11' means that the first subscript represents the number of the seam and the second one is the number of the block along the direction to the goaf area.

Taking the bed, number i, for the calculation (Fig. 9) the following abbreviated matrix equation can be obtained:

$$\{R_i\} = \{A_i\}\{M_i\}$$

Where $\{R_i\}$ is the column matrix of the force;

$\{M_i\}$ is the column matrix of the moment;

$\{A_i\}$ is the matrix of coefficient.

From this matrix, the horizontal thrust acting on the blocks can be calculated by the following formula:

$$2\left[\frac{h_i - S_{io}}{L_{io}} + 2(n_{i1}-n_{i2}+n_{i3}-n_{i4})\right] T_i = Q_{io}$$

If it is assumed that the gradient of a pair of adjacent blocks is the same, i.e. $n_{i1} = n_{i2}$; $n_{13} = n_{i4}$, then the approximate value of the horizontal thrust can be obtained as follows:

$$T_i = \frac{L_{io} Q_{io}}{2(h_{io} - S_{io})}$$

This means the magntidue of T_i depends only on the physical and geometrical characteristics of the block B.

The approximate value of the other unknown forces of the "structural" model can be calculated as follows:

$$(R_i)_{o-1} = 0 \qquad (R_i)_{o-o} = Q_{io}$$

$$R_{i1} = m_{i1} Q_{i1} - \frac{n_{i1} L_{io} Q_{io}}{h_i - S_{io}}$$

$$R_{i2} = m_{i2} Q_{i2} + \frac{n_{i1} L_{io} Q_{io}}{h_i - S_{io}}$$

$$R_{i3} = m_{i3} Q_{i3} - \frac{n_{i3} L_{io} Q_{io}}{h_i - S_{io}}$$

$$R_{i4} = m_{i4} Q_{i4} + \frac{n_{i3} L_{io} Q_{io}}{h_i - S_{io}}$$

From these formulae, some interesting results can be obtained, as follows:

1. The weight of the strata blocks in the inter-bed separation zone bc is loaded almost on the abutment of the working coal seam.

2. The shear force $(R_i)_{0-1}$ between the block B and C is equal to zero. The interlocked space 0-1 is like the top of the "half arch" for every "structural" formation.

3. The maximum shear force in the structural model occurs in the interlocked space 0-0, i.e. between blocks A and B, and its value is equal to the total weight of the block B.

Conditions required for equilibrium of the structural model

From the standpoint of roof control, it follows that in the inter-bed separation zone bc, degradation between adjacent blocks should not be allowed to occur.

The safety factor can be considered in the light of the relation of the friction force to shear force at the interlocked space. If the shear force exceeds the friction force, sliding between block A and B will occur along the crack and the roof will collapse unless additional support is provided.

A simple method to analyse possible sliding along the cracks is to resolve the resultant of the horizontal thrust T_i and the shear force $(R_i)_{o-o}$ into components, normal and parallel to the crack surface.

Hence, the conditions, required for roof control, can be expressed in the following formula:

$$T_i \cdot t_g(\varphi - \theta) > (R_i)_{o-o}$$

$$\text{i.e. } L_{io} > 2\,\frac{h_i - S_{io}}{tg(\varphi - \theta)}$$

If we take $\theta = 0^o$, $tg\varphi = 1$, $L_{io} > 2h_i$, the structural formation in the strata can be determined.

All the previous analyses are based on the assumption that the coal face is always moving forward. So that the "structural" formation of the strata is always changing with the advance of the face, which can be shown in Fig. 10.

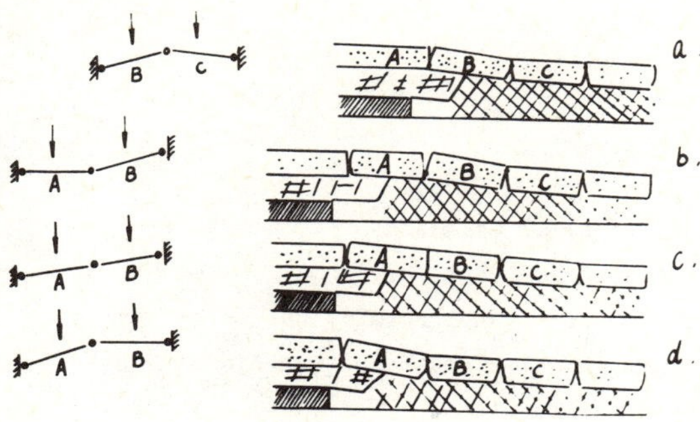

Fig. 10

Fig. 10 shows that block A will be out of balance when it was broken, and under the influence of the moment block A will rotate until the foregoing structural formation appears again.

In order to prevent the appearance of steps and falls in block A, enough resistance must be provided to act on block A. Unless there is enough friction force in the interlocked space, it must be resisted by means of supports in the working area.

APPLICATION

From the previous analysis, the immediate roof is the caving block with lateral expansion. Above the immediate roof the caved material in the goaf provides the overlying strata with support and a thrust force parallel to the strata is created by this buttressing forming a structural unit in the strata.

Now let P_R represent the resistance to be provided by face supports to prevent degradation of the immediate roof. Then it must be equal to:

$$P_R + T_i \cdot tg(\varphi - \theta) \geqslant (R_i)_{o-o} + Q_{io}$$

If $(R_i)_{o-o}$ is equal to the maximum, it will be equal to: Q_{io}, i.e.:

$$P_R + T_i \cdot tg(\varphi - \theta) \geqslant 2Q_{io}$$

then

$$P_R \geqslant 2 - \frac{L_{io}.tg(\ \ - \theta)}{2(h_i - S_{io})}\ Q_{io}$$

If the calculated value of P_R is negative, this means that the conditions required for equilibrium in the stratum are perfect and the P_R may not be needed.

If the supports must have sufficient resistance to prevent the immediate roof from falling, the resistance of the support per unit length of the face can approximately be calculated as follows:

$$P \geqslant \Sigma\, h.r.W + \left[2 - \frac{L_{io}.tg(\varphi - \theta)}{2(h_i - S_{io})}\right] Q_{io} \cdot \text{T/M}$$

Where Σh represents the total thickness of the stratified immediate roof (in the caving block), W is the width of the working area and r is the unit weight of the rock.

Thus, the density of the resistance of the supports is:

$$P \geqslant \Sigma h.r. + \left[2 - \frac{L_{io}.tg\ (\varphi - \theta)}{2(h_i - S_{io})}\right] \frac{Q_{io}}{W}\ ,\ \text{T/M}^2$$

Here P and p are approximate values for designing the resistance of the support in a longwall working area.

Besides the aforesaid application, some other phenomena in ground subsidence and in roof pressure can be explained by this model.

CONCLUSIONS

On the basis of measurement from inter-strata plugs in the working area in Dai-Tun coal mine, the vertical displacement of the plugs, relative to the coal face, can be assumed to be a negative exponent curve. The distribution of the gradient dW_x/dZ of this curve is an abnormal curve.

Depending on the strength of the overlying strata, these can be divided into several groups. The lowest layer of every group is a stronger and thicker bed. With the advance of the face, between the adjacent blocks in the bed, a lateral thrust will occur creating a structural formation in the bed. These structural formations are supported by a system of "working coal seam-roof support-caved material in the goaf" and a system of "working coal seam-caved or broken blocks underlying strata". The rock masses of the overlying strata in the working area can then be divided into three zones.

On the basis of approximate calculations, the value of the force acting on the block B can be shown to be independent of the vertical stress distribution in the goaf area.

According to calculations, the total weight in the inter-bed separation zone is loaded almost on the abutment of the working coal seam. From the formation, the space in which the shear force is maximum, can be obtained and the conditions, required for equilibrium can be analysed.

With the advance of the coal face, the process "equilibrium-dynamic equilibrium-equilibrium again" occurs in every structural formation.

In order to prevent the degradations of block A (steps and falls), there should be enough resistance (which can be approximately calculated) from the face supports.

REFERENCES

Chien Ming-Gao "Conditions Required for Equilibrium of overlying strata at working areas". Journal of China Institute of Mining Technology. 1981.2.

Li hon-zhang, Chien Ming Gao "A study of the system of the exploitation in upward order in Dai-Tun coal mine". 1981.

King, H.I., Whittaker, B.N. and Batchelor, A.S. "The effects of interaction in mine layouts". 5th Inter. Strata Control Conf. 1972, London.

Proceedings of European Congress on Ground Movement. 1957.

Wright, F.D. "Roof control through beam action and arching." SME Mining Engineering Handbook, 1973.

INDUCED CHANGES IN PERMEABILITY OF COAL MEASURE STRATA AS AN INDICATOR OF THE MECHANICS OF ROCK DEFORMATION ABOVE A LONGWALL COAL FACE

Keith R. Whitworth

National Coal Board, Western Area Geological Services,
Chatterley Whitfield Colliery, Tunstall, Stoke-on-Trent, ST6 8UN.

SUMMARY. In the Coal Measures of South Staffordshire below the unconformity at the base of Trias, there is a zone of increased fissure permeability and porosity in which most sandstones and coals give small makes of water. Very occasionally, considerably larger flushes of up to 1000 g.p.m. have occurred on longwall coal faces working in or close to this zone, even though these faces are at levels much further below the Trias unconformity than the minimum 60m. permitted. One such flush at West Cannock Colliery and the resultant investigation boreholes are described. The information gained from these boreholes demonstrated the existence of tensile zones in the form of bed separations which occur in a regular geometric pattern relating to a series of beams of uniform thickness over the edge of a worked panel.

INTRODUCTION.

Changes in the permeability of strata due to the effects of underground mining are well recognised. The problem exists of obtaining accurate quantitative data and relating it to strata deformation. This paper concentrates on the problems and mechanics of induced permeability above longwall faces in the South Staffordshire Coalfield. The present day coal production in South Staffordshire is concentrated in the Cannock Chase district (Figure 1) where the Coal Measures are concealed by an unconformable cover of Triassic rocks. The high permeabilities and water storage characters of these Triassic rocks are well documented, as they form one of the main aquifers in the Midlands.

Water problems associated with mining under this aquifer are accepted as normal, and since 1908 when an "Irruption of Water" occurred at Brereton Colliery (see Figure 1) with the loss of three lives there has been a local rule that workings must not encroach within 60m. of the base of Trias.

WATER PROBLEMS ASSOCIATED WITH WORKINGS CLOSE TO BASE OF TRIAS.

The normal problems associated with working close to the 60m. cover line to the base of Trias are those of nuisance water which relates to a zone of increased fissure permeability and porosity found in an irregular band extending to 250m. below the unconformity where almost all sandstone bands and coal seams give small makes of water. The typical make of water for a longwall face is less than 50 g.p.m. which is consistant with laboratory and borehole determined permeabilities of the Coal Measures in South Staffordshire. Cementation (1976). The flow rates although small remain constant for long periods of time and originate either from the seam being mined, or from a sandstone or coal above the worked panel up to a maximum height of about 30m.

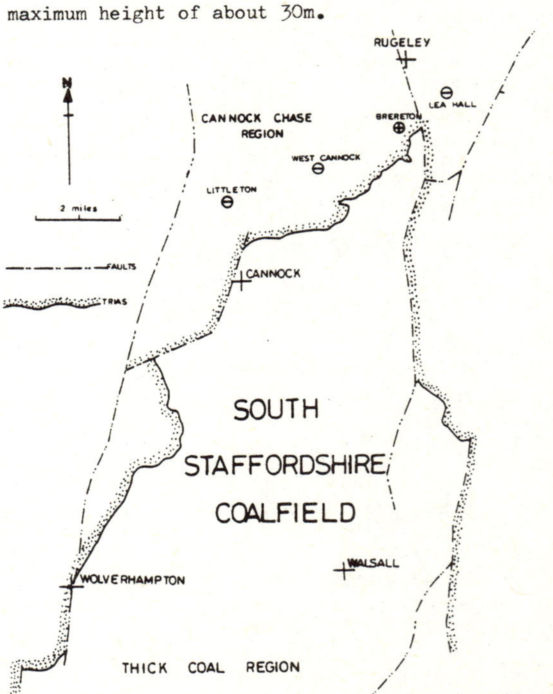

Fig.1 The South Staffordshire Coalfield showing the position of present working collieries.

Very occasionally larger makes of water of up to 1000 g.p.m. have been encountered on longwall faces. These 'flushes' are characterised by a high initial flow gradually reducing over a short period of time to flow rates more consistant with normal Coal Measure permeabilities, (i.e. less than 50 g.p.m.) Chemical analysis of the waters has shown that they originate from within the Coal Measures, from a level higher than that of the workings, but not directly from the Trias.

The circumstances which give rise to these flushes of water are interpreted as being due to a zone of greatly increased storage and permeability above a longwall face, the water from which can by some mechanism be ejected onto the face as a sudden flush,

THE PARK SEAM WEST CANNOCK COLLIERY

On the 18th June 1976 a large inflow of water of about 1000 g.p.m. occurred on 901's Park face; this was later officially classified as an 'inrush' by H.M. Inspector of Mines. 901's Park face was situated on the steep western limb of an anticline running NNE to SSW. and dipping at about 1 in 2. (See Figure 2). The anticline is eroded, with most of the upper seams incropping into the base of Trias. The Park Seam is about 220cms. in total section and was worked by longwall faces advancing on strike, (See Figure 3) taking a 183cm. extraction and leaving roof and floor coal. The face was wet when driven, the water coming from the coal, and the seam continued to make up to 30 g.p.m. as the face was advanced.

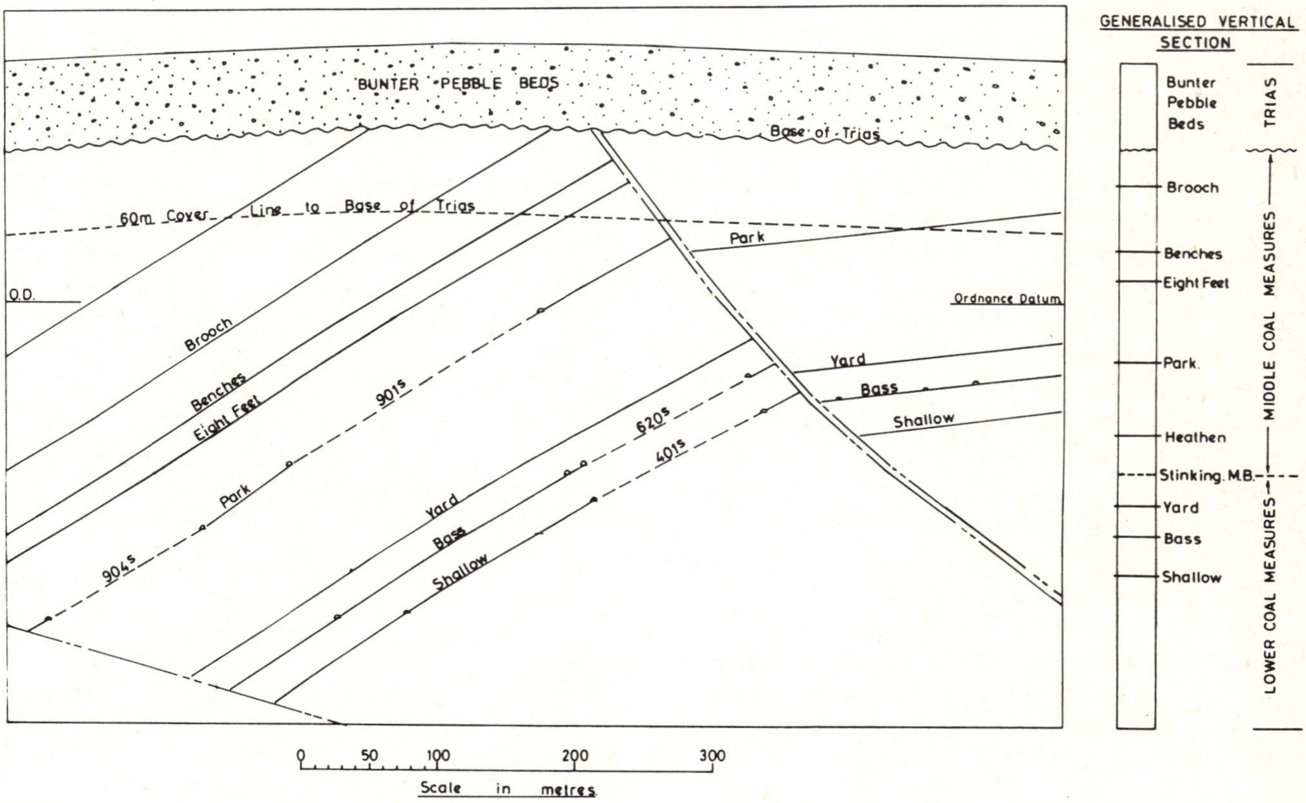

Fig.2 East-West Geological Section through West Cannock No.5 Colliery.

usually associated with a 'weighting' on the face. It was postulated some time ago that this increased storage is the result of bed separation above the coal extraction.

Attempts have been made in the past to de-water the strata above longwall panels suspected as being at risk from flushes. Saul (1970). Limited success was achieved at Lea Hall Colliery in the late sixties due to an imperfect understanding of the circumstances involved.

When the face had advanced 640m. a 3m. fault was encountered which gave small flushes of water of about 20 g.p.m. The face was reheaded inbye of the fault, and small flushes of water occurred from breaks in the roof when the face had advanced 120m.

The face had advanced 200m. when the inrush of water occurred at the main gate end of the face. The initial flow was about 1000 g.p.m. which gradually declined over the next three to four days.

The problem was diagnosed as being due to bed separation water; in fact the face had been recognised as being at risk prior to working, but because of the

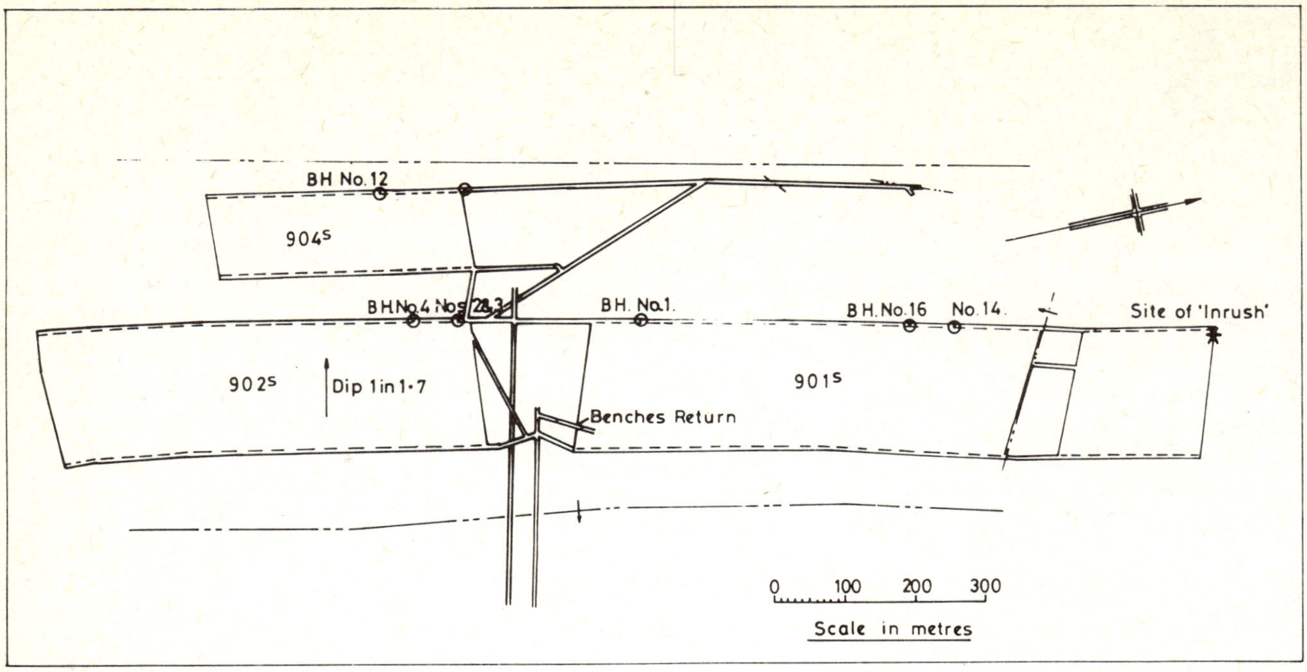

Fig. 3. Plan of Western Extension No.1 Area Park Workings at West Cannock No.5 Colliery

limited success of previous de-watering boreholes no remedial action was taken. Concern was generated over 902's Park panel which had been successfully worked during the reheading of 901's face, and the first 901's Park panel, both of which could still have large bodies of water over them. This water could be ejected into existing mine workings or cause problems to future panels planned in the seams above the Park.

In order to investigate the problem, two boreholes were drilled, one in 901's Park main gate and one in 902's Park main gate (See Figure 3). Borehole No.1 in 901's Park main gate was inclined at 15° to vertical, square to the gate, and towards the solid. As shown in Figure 4, flushes of water of up to 60 g.p.m. occurred at 18m, 38m, 57m. and 73m. The water from each horizon was allowed to drain until a steady flow was attained, then cased to that horizon and drilled up further. All makes of water from the district were being carefully monitored and it was noted that the flushes of water from the lower horizons in No.1 Borehole were coincident with a reduced flow from the Benches return roadway being driven over, and up dip from 901's face start position, showing as was suspected a hydraulic connection between the bed separation horizons and the roadway. (See Figure 4).

Borehole No.2 was drilled in 902's Main Gate aligned as No.1 Borehole, but unlike the first borehole no flush of water occurred at the lowest horizon (18m.). A flush of about 60 g.p.m. occurred at 38m, when a pressure reading of 80 p.s.i. was recorded. This borehole was suspected of deviating above this horizon and was replaced by a third borehole which confirmed the absence of water filled bed separation at 18m.

The pressure head of 80 p.s.i. recorded in No.2 borehole is equivalent to a free head of water of 56m, when the borehole was only at 38m, the remaining 18m. being in bed separation up dip. (See Figure 5). The volume of the bed separation at the various horizons can be calculated by subtracting the base flow rate from the total make for that horizon. Table I lists the calculated volumes in cubic feet for the flushes from the first three boreholes.

Table 1.

Borehole No.	Base flow g.p.m.	Flushes at m.	Max. flow g.p.m.	Calc.Vol. cu.m.
1	10-12	18	60	400
		38	70	1100
		37	60	110
2	5-12	38	60	160
3	5-12	38	52	200
		68	33	310

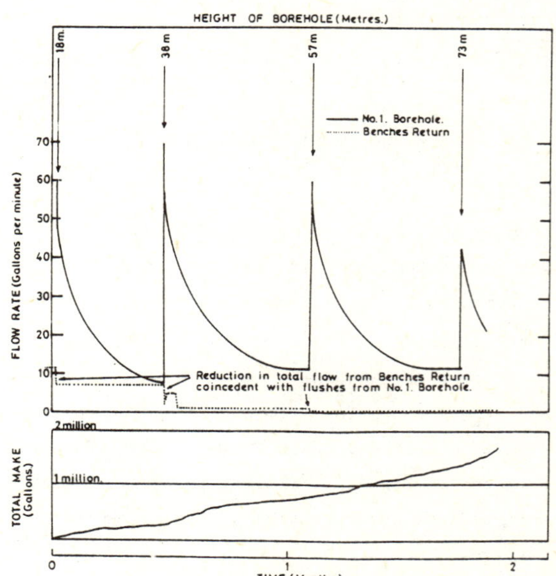

Fig.4 Diagram of flow rates and total make of water from Borehole No.1 West Cannock Park Seam.

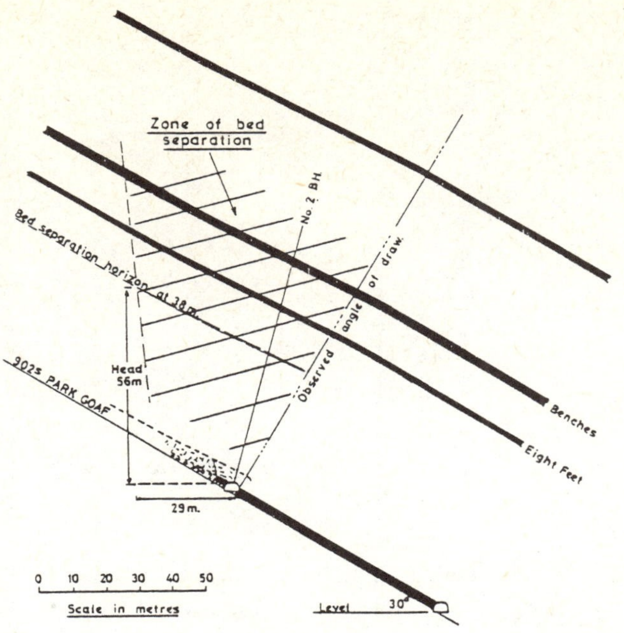

Fig.5 Section through No.2 Borehole West Cannock Park showing head of water in bed separation horizon at 38m.

Several boreholes were drilled in 901's and 902's Park, and in the subsequent panel 904's. In all, up to four bed separation horizons were encountered in the boreholes, up to a maximum height of 76m. The boreholes had been drilled at slightly differing angles and inclinations, and it was only when all the horizons where water had occurred were plotted normal to the seam extracted that it was realised that all the horizons bar two came within four fairly narrow bands, the mean heights of which occurred at 19m, 37m, 55m, and 72m. above the seam extracted. (See Figure 6).

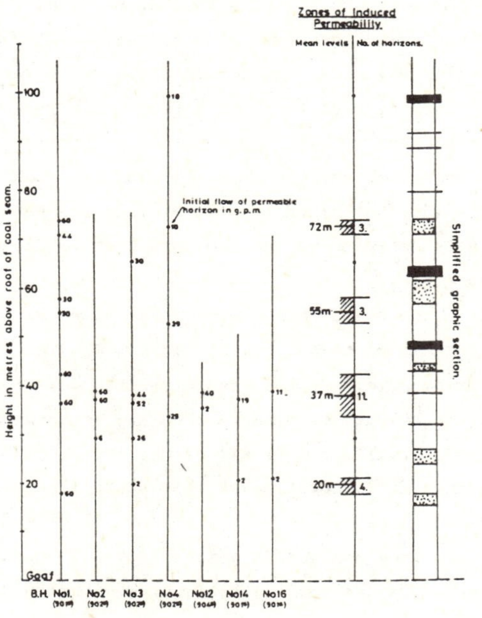

Fig.6 De-watering Boreholes - West Cannock Park Seam.

THE BOTTOM ROBINS SEAM - LITTLETON COLLIERY.

The Bottom Robins Seam to the north of Littleton Colliery occupies a similar geological environment to the Park Seam at West Cannock Colliery and was identified as being at risk from bed separation water. Prior to working advancing longwall faces a borehole programme was implemented to investigate and drain any water from the bed separations above the panels. The greater knowledge of the problem obtained from the boreholes at West Cannock meant that the boreholes in the Bottom Robins workings were concentrated on the lower bed separation horizons, it being felt that induced vertical breaks would not connect the panel with any bed separation horizons above 36m. (c.f. Barrier to old workings).

The mean levels of the bed separations proved in the boreholes from the Bottom Robins Seam were almost identical to those established above the Park Seam. (See Figure 7).

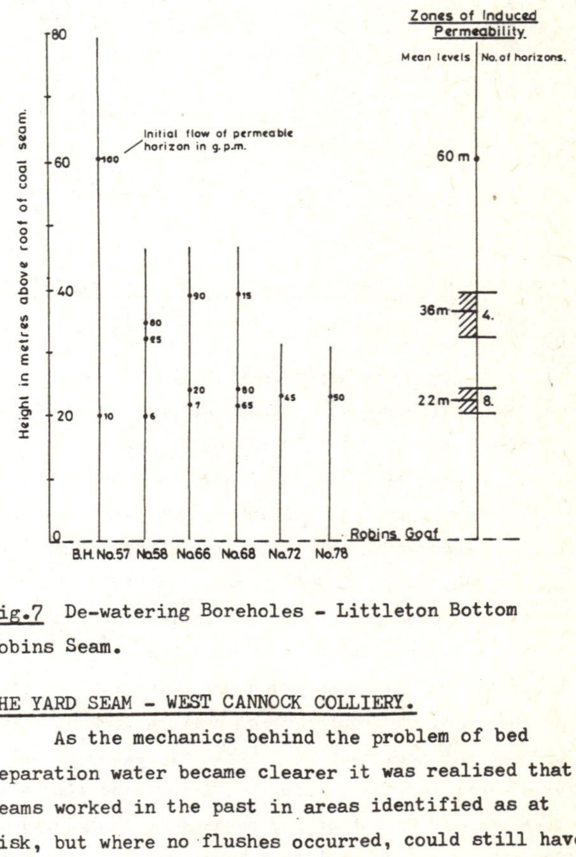

Fig.7 De-watering Boreholes - Littleton Bottom Robins Seam.

THE YARD SEAM - WEST CANNOCK COLLIERY.

As the mechanics behind the problem of bed separation water became clearer it was realised that seams worked in the past in areas identified as at risk, but where no flushes occurred, could still have large bodies of water in bed separation existing above them. Some years ago an attempt to work the Yard above the Bass Seam at West Cannock had encountered large quantities of water from the roof measures which affected the productivity from the panel. Prior to a recent attempt to work the Yard Seam over the Bass an investigation borehole was drilled from an old roadway below the workings to establish if any bed separations still remained with water in them. The first borehole showed that although the bed separation horizon at 21m.

had not filled with water, and the 38m. horizon made little water, there were considerable quantities of water at the 52m. horizon. (See Figure 8). This information was used to design a borehole dewatering programme, following the implementation of which the panel was successfully worked. The bed separation horizons were again shown to occur at the same regular intervals, which had been correctly forecast beforehand. (See Figure 8).

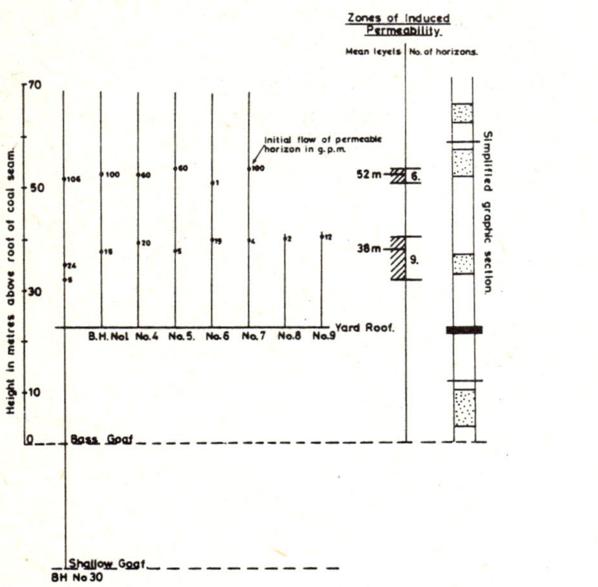

Fig.8 De-watering Boreholes - West Cannock Yard (759's)

DISCUSSION.

Since the early work in the Park Seam at West Cannock Colliery demonstrated the existence of distinct bed separation horizons at regular intervals above a longwall face, there have been several dewatering programmes initiated. In every case but two the bed separation horizons fall within the narrow zones predicted. (See Figure 9). It is felt that sufficient evidence is now available to demonstrate that the tensile stress zones which manifest themselves as bed separation are a function of the mechanics of rock deformation above a longwall face, and are not dependant on lithology. Although not the primary factor controlling bed separation, lithology does have an effect on bed separation in two ways. Firstly, lithology will to a large degree govern the amount of bed separation; strong massively bedded strata may have little or no development of bed separation, whereas a weak rock or thinly bedded strata may very easily form voids. Secondly lithology determines how much water if any will occur, as a normally permeable horizon is required to feed the bed separation; the absence of water does not imply the absence of bed separation.

It is suggested that the tensile zones which give rise to the bed separations occur between a series

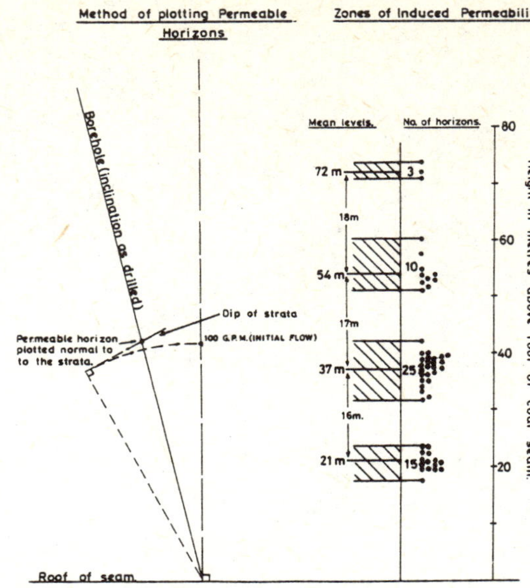

Fig.9 All Permeable Horizons Encountered in De-watering Boreholes.

of beams which form over the margins of any longwall face. The beams are 'fixed' at one end, while becoming progressively supported at the other end as convergence brings the roof down onto the caved roof material. (See Figure 10). Bending of the beams at the 'fixed' end gives a tensile zone around the margins of the extraction, Timoshenko (1955), and may even result in a slight upward flexure of the strata. Wardell (1968).

The thickness of the beams that form are believed to be a function of the length of the beam, which in turn is governed by the thickness of extraction, the strength of the roof strata and the angle of convergence. Figure 10 shows how the stresses are believed to be distributed over a 152cm. extraction using a convergence angle of $2^{o}43'$. N.C.B.(1972). The beam will only form once the excavation has reached a certain width, when the 'first weight' is felt. This represents the change from a 'pressure arch' to a 'fixed end beam', and the fracturing of the strata associated with this allows water from any naturally permeable horizons in the roof measures to flow into the workings.

The beams form over all the edges of a panel providing it is of sufficient width. The face line is however dynamic, and this can alter the way the stresses are distributed, for example the rate of advance of the face, which will in turn affect the length of the beam. It is significant that where flushes of water have occurred they were almost all at a time when the faces were not working, i.e. at weekends and holidays. The mechanism that causes water held in bed separation to be suddenly ejected onto the face is not known, but could be a combination of factors, for example minor faulting, increased joint development, or facies changes in the roof measures. Garritty (1980).

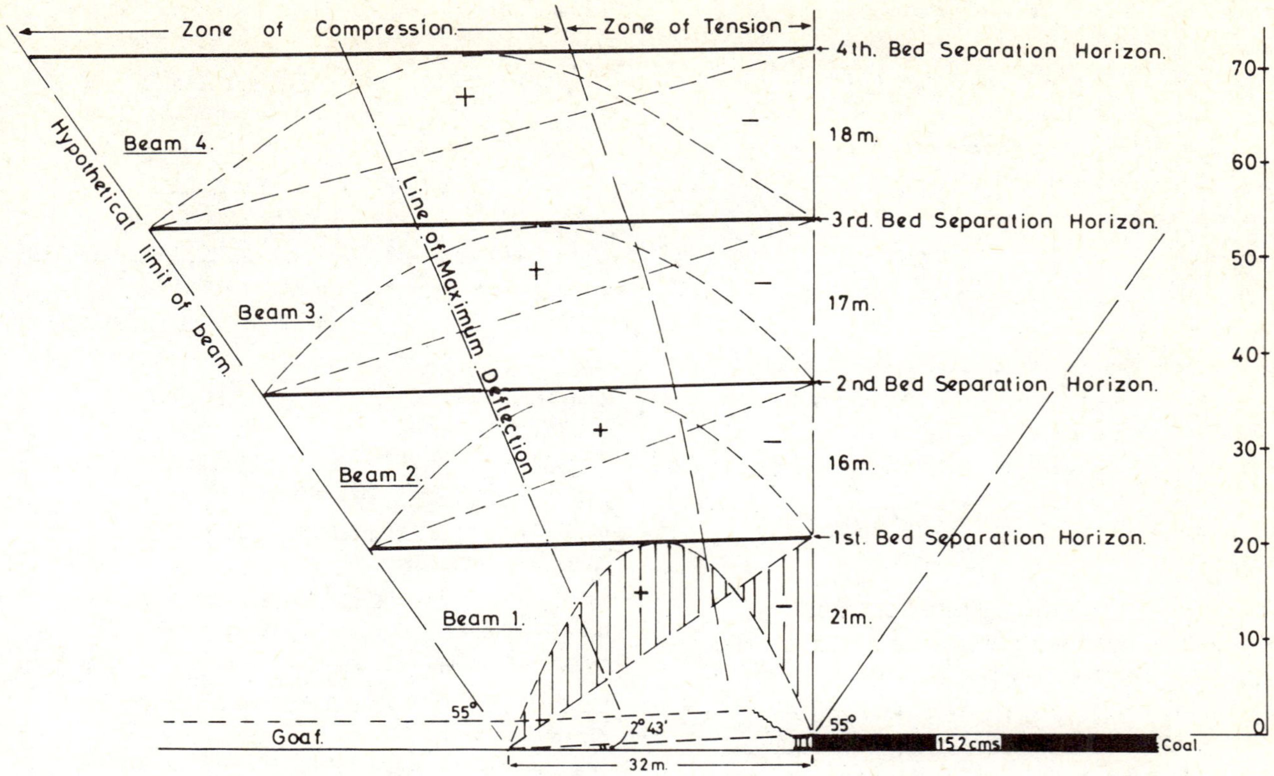

Fig.10 Section through a typical longwall face showing the bed separation horizons between a series of beams and the stress distribution within the beams.

At present the policy in N.C.B. Western Area when dealing with bed separation is to drain as much water as possible by drilling, thereby minimising the risk of flushes of water occurring.

OTHER EXAMPLES OF TENSILE ZONES OR INCREASED PERMEABILITY ABOVE LONGWALL FACES.

There are numerous references in the literature to tensile zones and zones of increased permeability both of gas and water. Recent work by Tubby et al (1981) demonstrates the existence of tensile strain zones above longwall faces in the North East of England, and although attributed by the authors to the effects of lithology, they generally fall into zones described in this paper. Not only does evidence exist of the tensile zones parallel to the excavations, but the tensile zone that extends up from each rib line has been demonstrated by interpretation of subsidence observations. N.C.B. (1982).

Permeable zones affecting both water and gas above a longwall face have been much discussed. Saul (1970) reports examples at Horden and Blackhall collieries of 'hydraulic pressure zones' which conform to the 37m. bed separation horizon in South Staffordshire.

Whittaker (1979) has also shown great variations in permeability above longwall faces, but due to the method used of testing large zones, it would be unlikely to detect the relatively narrow zones proposed in this paper.

The changes of permeability as it affects gas is less well documented. Unpublished examples of gas pockets under pressure have been reported to the author in South Staffordshire which fit the predicted bed separation zones. There are also records of gas emissions from several collieries, M.R.D.E. (1980), but no clear pattern can be seen, although most of the gas makes appear at the correct levels, some do not. This may well be a function of the difference in gas permeability compared to water.

CONCLUSION.

There is now a considerable amount of evidence to demonstrate that above the margins of a longwall panel there exists a series of beams of uniform thickness, separated by tensile zones where bed separation develops, and permeability is greatly increased.

Using remote methods of sensing such as strain gauges or packer tests, it is unlikely that the tensile zones described in this paper would be observed without prior knowledge of their existence, allowing for specific testing of discreet zones. The positions of the bed separation horizons are purely a function of the mechanics of rock deformation, but the degree of bed separation is believed to be related to lithology, and to some extent lithology may also control the amount of stress by concentrating it in an inherently weak band, or by spreading the stress in strong massively bedded strata. The effects of the tensile stress zones on faces above old workings may be too small to have any significant effect on mining by conventional methods, but knowledge of their existence must help in understanding how rock deforms above a longwall face and may have some relevance in the field of underground gasification, as it does in mine water and gas drainage.

ACKNOWLEDGEMENTS.

The paper is presented with the permission of Mr. J.H. Northard, Area Director. The views expressed are the author's and not necessarily those of the National Coal Board.

The author gratefully acknowledges the contributions to this paper of the other members of N.C.B. Western Area Geological Services, and the various members of staff at the collieries involved.

REFERENCES.

Cementation 1977. "Report on laboratory tests on samples of rock core from Park Colliery No.1 Borehole for N.C.B." Cementation Ground Engineering Report. Unpublished.

Cementation 1977. "Report on water make tests at Park Colliery for N.C.B." Cementation Ground Engineering Report. Unpublished.

Garritty P. 1980. "Effects of mining on surface and Sub-surface water bodies." Ph.D. Thesis. University of Newcastle-upon-Tyne.

Hoare R.H. and Whitworth K.R. 1978. "Water around a coal face." National Coal Board, Geological Branch Seminar. Unpublished.

Imperial College, Dept. of Geology. 1967. "Prediction and control of water feeders in mining." Report No. P.12857/13F.

Marr J.E. 1981. "Strata subsidence - Methane Drainage." National Coal Board, Surveying Department Report. Unpublished.

National Coal Board. 1980. "Firedamp prediction and control". M.R.D.E. Progress Report No.4. Research Project 7220-AC/809.

National Coal Board. 1975. "Subsidence Engineers Handbook." National Coal Board, Mining Dept. pp111.

National Coal Board. 1972. "Design of Mine Layouts." Working Party Report. Mining Dept. pp52.

Neale C.J. and Whittaker B.N. 1979. "Influence of the proximity of longwall mining on strata permeability and ground water." Proc. 20th U.S. Symp. on Rock Mechs., Austin pp.217-224.

Orchard R.J. 1969. "The control of Ground Movements in Undersea Workings." Mining Engineer Vol.28, pp.259-273.

Saul H. 1970. "Current Mine Drainage Problems." Mining Engineer, Vol.129, pp.643-657.

Timoshenko S. 1955. "Strength of Materials." 3rd Ed. D. Van Nostrand Co. Inc. New York. pp.422.

Tubby J. and Farmer J.W. 1981. "Stability of Undersea workings at Lynemouth and Ellington Collieries." Mining Engineer.

Wardell K. and Eynon P. 1968. "Structural Concept of Strata Control and Mine Design." Trans. of the Inst. of Min. Engrs. Vol.128, pp.125-150.

Whittaker B.N., Singh R.N. and Neale, C.J. 1979. "Investigation and Evaluation studies of surface and sub-surface drainage pattern changes resulting from longwall mining subsidence." 1st. Inter. Mine Drainage Symp. Denver.

WATER PERCOLATION INTO FULLY CAVED LONGWALL FACES

P. Garritty

Department of Mining Engineering
University of Newcastle upon Tyne
Newcastle upon Tyne, NE1 7RU, England.

SUMMARY. Production at a number of mines operating off the North-East coast has been seriously affected by the percolation of major water feeders onto working coal faces. A detailed study of the factors affecting the incidence and distribution of water feeders was undertaken, using operational geological and hydrological data. Analysis indicates that operations tend to encounter excessive water feeders when:

a) The cover to the Permian aquifer is less than 100m, or major Coal Measures aquifers exist within 45m or so of workings;
b) Tensile face strains calculated at the base of Permian exceed 0.6%;
c) Sandstone strata forms less than 35% of the overburden, implying a lack of competence in the seam roof.
d) Faults or other major discontinuities are present in the roof strata;
e) Major strata 'weights' or 'bumps' occur.

INTRODUCTION

A strata zone of increased porosity and permeability is induced around longwall faces. If a system of interconnected fractures within this zone intersect an aquifer or surface water body, either directly or indirectly, water feeders ranging from the negligible to the catastrophic may be drawn into workings. Major coal reserves occur off the counties of Northumberland and Durham and workings now extend over many km^2 beneath the seabed and the offshore Permian aquifer. 180 M tonnes of classified coal reserves, constituting over 80% of the coalfield total, are now assessed to exist within the royalties of offshore North-East mines. The problem lies in the design of systems of working which maximise extraction while minimising the risk of inundation. It is therefore important to examine the factors affecting water inflow.

WATER INFLOW INTO MINE WORKINGS

Theoretical considerations

Conventional flow analysis has tended to assume that rock en-masse constitutes a natural porous medium. The flow of water through a rock formation can then be assumed to obey approximately an empirical law first discovered by Henri Darcy (1) whose investigations suggested a relationship of the form:

$$Q = AKi$$

Q = Discharge rate
A = Discharge area
K = Coefficient of permeability
i = Hydraulic gradient

Permeability is a measure of the ease with which a fluid will percolate through a porous medium, however k, as defined is also dependent upon fluid properties. Rationalisation of the concept of permeability indicated a relationship of the form:

$$K = Cd^2 \frac{\gamma}{\mu}$$

C = Dimensionless constant
d = Characteristic dimension of the medium
γ = Fluid density
μ = Fluid viscosity

between fluid properties and the Darcian coefficient 'C' is controlled by properties of the medium including porosity, packing, void size distribution, and shape.

Numerous investigators have attempted to relate permeability to the physical properties of porous media. The Kozeny-Carman equation can be considered as typical:

$$K = \frac{1}{k_T k_o S^2} \frac{\gamma}{\mu} \frac{n^3}{(1-n)^2}$$

k_T = Pore tortuosity factor
k_o = Pore shape factor
S = Surface area of voids/unit volume of solids
n = Porosity

This expression clearly illustrates the relative importances of porosity, pore nature and fluid properties in regulating rate of discharge.

As a first approach to the analysis of fluid flow in fissured media, it is customary to approximate the discontinuity geometry to an equivalent simple structure, and then to describe the flow regime by analogy with flow between parallel plates. Flow through a single plane fissure inclined at an angle α to the horizontal is then defined by the equation:

$$Q = \frac{be^3}{12\mu} \left(\frac{P_1-P_2}{1} + \gamma \sin \alpha\right)$$

b = Fissure width
e = Fissure opening
Px= Pressure at point x

Although of necessity a gross simplification, it is evident that the theoretical proportionality of flowrate to the third power of fracture opening will be of extreme significance in a study of the mechanics of increased strata permeability associated with longwall extraction.

Water discharge into mine workings will therefore be affected if any of the following are varied:

Discharge area

Hydraulic gradient - horizontal or vertical

Strata permeability, which will in turn be modified by changes in: fluid density, viscosity and compressibility; Degree of strata saturation; Flow regime - laminar, turbulent or transitional; Fracture porosity; Pore nature.

Inflow into individual operations will also vary as a consequence of:

Strata competence

Strata anisotropy (lithology)
Strata inhomogeneities and geological structure.

OPERATIONAL CONSIDERATIONS

Removal of coal is followed by collapse of the immediate roof layers to a height estimated to be between 2 and 4 times the extraction height. Water within this formation is liberated as the roof caves. As work progresses, beds in the upper roof begin to sag, probably giving rise to voids or cavities along bedding planes. When a critical area of excavation is attained, transition from strata bridging to fully developed upper roof curvature is generally accompanied by a major roof break hading over the goaf in the neighbourhood of the original back abutment. Over ribsides induced tensile strains will initiate fracturing, the principal breaks are thought to hade slightly over the goaf. A system of interconnected stepped-planar fractures, capable of transmitting large quantities of water probably form within the tensile zones over ribsides and the back abutment. Major water feeders would therefore appear to enter workings by percolation in an essentially vertical direction along the fractures, over ribsides into the vicinity of an excavation, before entering horizontal breaks and being distributed over the working area. The area of excavation is therefore only indirectly related to the magnitude of water feeders, as it is only one of the factors governing the magnitude of the induced roof breaks. Figure 1 illustrates qualitatively the probable principal flow paths (2, 4).

coalfield is an irregular syncline with a NW-SE axis passing beneath the sea at Sunderland. The eastern limits dip between 2° and 5° North of East, the net effect being to reduce cover to the South and East, where seams incrop successively against the base of the Permian. Structural features include:

Major faulting in an ENE-WSW direction and minor faulting in a NW-SE direction

A dome structure off Vane Tempest Colliery

The Newbiggin anticline with an axis North of East.

The area can be conveniently divided into two parts by the geology and the nature of the water problems encountered:

(i) Northumberland, where the Coal Measures outcrop directly to the seabed and where water feeders characteristically emanate from local roof sandstones.

(ii) Durham, where the Coal Measures are overlain by the Permian, whose waters have been shown by pressure measurement to be under a static head equivalent to sea level.

Any attempt to predict changes of the flow regime in a mining context must allow for changes in the variables discussed in the previous section. Unfortunately, a number of these parameters must inevitably remain highly uncertain, others prove amenable only to general estimation, while few are determined directly in meaningful detail.

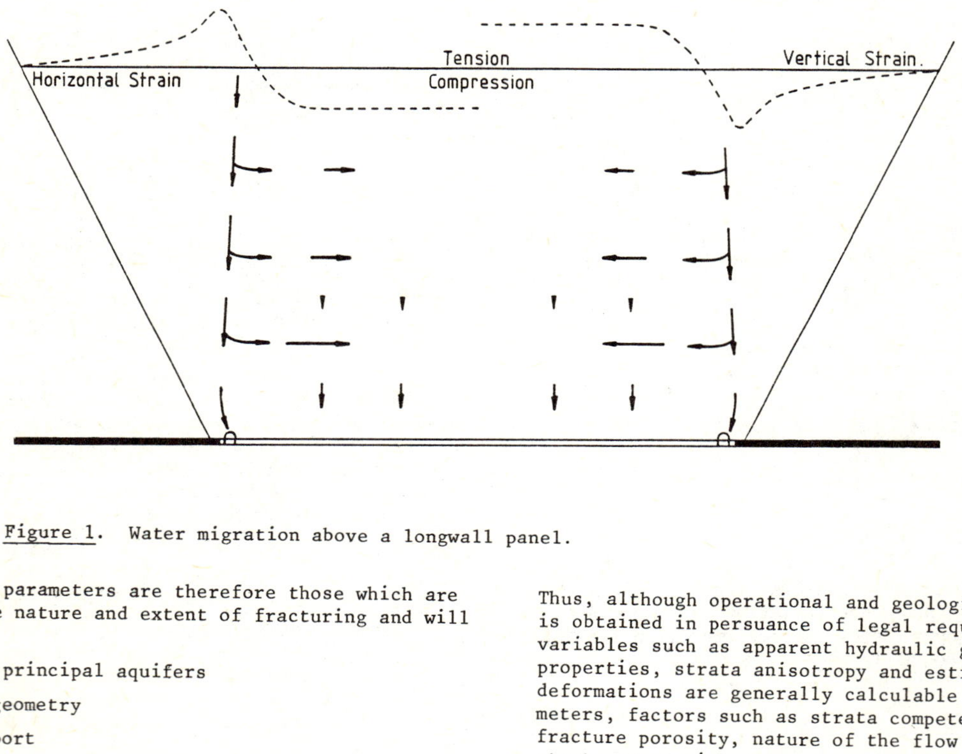

Figure 1. Water migration above a longwall panel.

Critical parameters are therefore those which are related to the nature and extent of fracturing and will include:

Cover to principal aquifers

Working geometry

Roof support

Induced strata deformation - essentially a function of the preceeding factors

Workings in nearby seams

WATER PROBLEMS IN THE NORTH-EAST COALFIELD

The extent of the Coal Measures in the North-East coalfield is shown in Figure 2. The productive Middle Coal Measures outcrop on a southerly line from Amble to Barnard Castle. To the south and east they are overlain unconformably by the Permian. The Permian outcrop crosses the shoreline at South Shields and continues under the North Sea, reaching a position 6 km from the shoreline at Blyth. The principal structure of the

Thus, although operational and geological information is obtained in persuance of legal requirements, and variables such as apparent hydraulic gradients, fluid properties, strata anisotropy and estimates of aquifer deformations are generally calculable from these parameters, factors such as strata competence, pore nature, fracture porosity, nature of the flow regime and strata saturation are never assesed. An additional complication is that a number of factors which individually affect the flow regime are only known in terms of a single determined parameter, e.g. geological bores are the only real source of information relevant to strata lithology, anisotropy and competence. Experience, theoretical considerations, and the overall quality and availability of information indicated that the parameters listed below were the most likely to lend themselves to profitable investigation. For the areas of Northumberland and Durham, data was collected on a historical and actual basis to relate maximum water inflow to:

(i) Cover to the Permian or the seabed;

(ii) Face width/carboniferous cover ratio or face width /seabed cover ratio;

(iii) The maximum tensile strain induced at the base of the Permian or the seabed calculated from the Subsidence Engineer's Handbook (6).

(iv) The aggregate tensile strain induced at the base of the Permian or the seabed including estimates of strains from adjacent workings in the same seam.

(v) The total tensile strain at the base of the Permian or the seabed including estimates of strain from all workings in the same and adjacent seams.

(vi) The ratio between the head of water at the base of the Permian and the carboniferous cover.

(vii) Strata inhomogeneities and geological structure.

(viii) Strata lithology, in particular the percentage of sandstones in the succession beneath the Permian or the seabed.

In the latter case, emphasis was placed on sandstone since it was considered that sandstone strata represented both potentially strong roof rocks which would be less likely to collapse, and water bearing aquifer rocks which could hold and transmit water. It can be seen that there are potentially many combinations of strata which would lead to water ingress, ranging from weak rocks above the seam collapsing to intercept an aquifer, to aquifer rocks close to the seam fracturing and causing increased flow through increased permeability. The percentage of sandstone in the vicinity of workings and above limiting heights of 20 m, 30 m and 45 m, which are thought to be of particular significance (8, 5, 9) was examined as a contributing factor in water inflow.

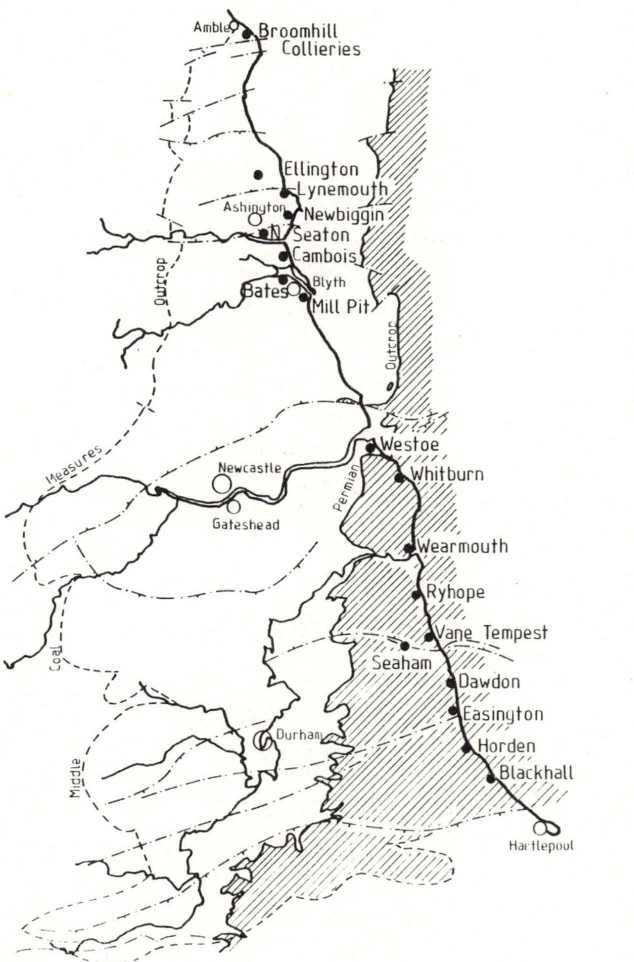

Figure 2. The North-East Coalfield.

WATER FLOW DATA ANALYSIS

Data from the Northumberland part of the coalfield and from the Durham part of the coalfield are summarised in Tables 1 and 2 respectively, where an attempt is made to assess the statistical significance of the relations investigated since the frequency of water influx as a function of any of the variables considered did not follow any form of probability distribution curve a non-parametric rank-sum test was used to estimate the significance of the relationships.

This analysis was used to test the hypothesis that the median of the population of 'wet' faces and the median of the population of 'dry' faces are equal. For purposes of analysis, 'wet' faces are considered in two basic groupings: those which encountered water in quantities in excess of 4 l/s, and those which liberated feeders greater than 20 l/s. Results indicated as correlating at the 5% level are treated as 'significant', while correlation at the 1% level is considered 'highly significant'.

The space available does not permit inclusion of the graphical plots of the data summarised in Tables 1 and 2. Values of parameters mentioned in the text are derived from these plots. Anyone wishing to study the graphical relationships is directed to references (3, 4), where they are discussed at length.

In Northumberland, incidence of feeders in excess of 20 l/s appears to be principally a function of the lithology of the overburden, although the cover to the seabed would also seem to be of some significance. Parameters related to the magnitude of the disruption induced in the strata between seam and seabed have comparatively little correlation with incidence of major feeders.

In this light, the regulations currently controlling offshore extraction appear to guard adequately against inundation of workings directly by the sea, although it should be noted that no individual operations have induced the full 10% seabed strain allowed by the undersea working instruction (7). Operations appear to be at especial risk of encountering feeders in excess of 20 l/s where:

a) Cover to the seabed is less than 140 m. This probably reflects the comparative ease with which overlying Coal Measure aquifers are replenished by the sea, and the increased porosity and permeability of aquifers subjected to erosional stress relief.

b) Major Coal Measure aquifers exist within 45 m or so of workings.

c) The competence of the immediate seam roof is comparatively low; that is where the first 20 to 30 m of overlying strata contains less than about 35% sandstone.

d) Work is prosecuted in the vicinity of faults, particularly those which hade towards workings.

Incidence of feeders of 4 l/s or more seems to be a function of the presence of sandstone aquifers in proximity to workings. Notable influxes of water are associated principally with aquifers within 45 m or so of the seam being worked, which for many Yard, Main, High Main and Ashington seam operations implies the major aquiferous sandstone sequence above the horizon of the Ashington seam. This does not imply, however, that longwall operations cannot tap the waters of aquifers at greater height. It is interesting that feeders have been encountered at depths below the seabed of up to 200 m, which indicates that sandstones are water bearing at depth. Total extraction systems will generally liberate feeders from nearby Coal Measure aquifer , a circumstance which should be borne in mind during layout planning and design. However, feeders encountered at depth and in the absence of faults and dykes will usually amount to no more than nuisance.

Table 1. Significance of relations - Northumberland Coalfield

FEEDERS > 20 l/s		FEEDERS >4 l/s	
"Z" FACTOR	PARAMETER	"Z" FACTOR	PARAMETER
4.72	Sandstone >20m	9.12	Sandstone >20m
2.67	Sandstone >30m	6.80	Sandstone <45m
-2.66	Cover	6.13	Sandstone <20m
1.80	Sandstone >45m	5.68*	W/H Ratio
-1.24	Sandstone <30m	4.79	Sandstone <30m
1.00	Head/Cover Ratio	-3.10*	Individual Strain
-0.79	Sandstone <20m	-2.94*	Aggregate Strain
0.47	Sandstone <45m	2.46*	Head/Cover Ratio
0.38	Individual Strain	2.46	Sandstone >30m
0.20	W/H Ratio	0.88*	Cover
-0.19*	Aggregate Strain	-0.41	Sandstone >45m

*Indicates that the parameter is related to incidence of water feeders in the opposite sense to what would logically be expected.

"Z" values of 2.32 or more indicate a "highly significant" correlation and values greater than 1.64 a "significant" correlation.

Table 2. Significance of relations - Durham Coalfield

FEEDERS > 20 l/s		FEEDERS >4 l/s	
"Z"FACTOR	PARAMETER	"Z" FACTOR	PARAMETER
-5.17	Cover	6.85	Individual Strain
5.12	Individual Strain	6.78	W/H Ratio
4.54	Aggregate Strain	-6.29	Cover
4.17	Head/Cover Ratio	6.20	Head/Cover Ratio
3.49	W/H Ratio	5.44	Aggregate Strain
-2.18	Sandstone <45m	-2.40	Sandstone <20m
-1.37	Sandstone <30m	-1.37	Sandstone >30m
-1.16	Sandstone <20m	-1.28	Sandstone <45m
0.78	Sandstone >45m	-0.75	Sandstone >20m
-0.53	Sandstone >20m	-0.26	Sandstone <30m
-0.44	Sandstone >30m	-0.27	Sandstone >45m

"Z" values of 2.32 or more indicate a "highly significant" correlation and values greater than 1.64 a significant correlation.

In the Durham Coalfield, parameters related to the magnitude of the deformations induced in the strata between workings and the Permian have the most significant correlation with incidence of water feeders, of secondary importance appear lithological considerations, although there is a general correlation between lower proportions of overburden sandstone, especially in the immediate seam roof, and incidence of water feeders.

Consideration of the above and individual case histories suggests that incidence of noteable water feeders is essentially a function of dynamic strata fracturing at the time of major roof breaks, which induces open, effectively continuous, discontinuities within the overburden capable of transmitting water into workings from overlying aquiferous formations, notably the Permian and aquiferous sandstones in proximity to it. It is possible that a calculated tensile strain of 6 or 7 mm/m induced at the base of Permian reflects a "critical" degree of fracturing for "average" conditions.

Workings in Durham appear to run a comparatively high risk of inundation where:

a) The cover to the Permian is less than 100m,

b) Individual tensile strains calculated at the base of Permian exceed 6 mm/m,

c) The total net tensile strain induced at the base of Permian is greater than 12 mm/m, taking into consideration (b) above,

d) The overburden comprises less than 35% of sandstone,

e) Adequate roof control is not maintained and major strata breaks are allowed to form.

The incidence of water has often been associated with strata discontinuities.

Considering the intrusive nature of igneous dyke formation and their association in the North-East with water filled cavities or heavily watered zones, it is not suprising that feeders have been encountered during proving questions, during operations, and whilst working alongside them.

Water in quantity has not generally been associated with faults, especially depositional faults, having throws ranging up to 1 m, although when water is encountered the make tends to be greatest in the vicinity of any fault. Increased risk is apparent where workings approach or encounter:

a) Faulting approximately parallel to the face-line which aggravates strata control difficulties,

b) Faults which hade over workings,

c) Faults of increasing throw,

d) Faults whose nature or siting suggests a significance beyond that indicated by consideration of throw along, for example faults against which smaller faults terminate,

e) Large faults where strata disruption and dislocation are greatest.

Feeders in Durham have also been encountered by operations in the vicinity of the following geological features:

a) Swalley banks,

b) Areas of steeply dipping strata, or dip workings in faulted blocks,

c) Areas of rapidly increasing dip,

d) Monoclinal structures,

e) Lenticular sandstones,

f) The incrop to the Permian of major sandstone aquifers,

g) Massive aquiferous roof sandstones,

h) Major roof jointing parallel to the face-line.

REFERENCES

Darcy, H. (1856), Fontaines Publiques de la Ville de Dijon. Dalmont, Paris.

Elliott, R.E., (1978), Strain Zones and Water Migration above a Working Panel. N.C.B. Internal Report.

Farmer, I.W. (1980), Face and Roadway Stability in Underground Coal Mines: Geotechnical Criteria. Report to the N.C.B.

Garritty, P. (1980), Effects of Mining on Surface and Sub-surface Water Bodies. Ph.D. Thesis, University of Newcastle upon Tyne.

Kapp, W.A., Williams, R.C. (1972), Extraction of Coal in the Sydney Basin from Beneath Large Bodies of Water. Conf. Australian Inst. Mining & Metallurgy, pp 77-87.

National Coal Board (1975), Subsidence Engineers Handbook, 2nd Ed., N.C.B. Mining Dept., London.

National Coal Board (1968), Working Under the Sea. Mining Department Instruction, P.I. (Rev. 1971)

Orchard, R.J. (1975), Working Under Bodies of Water. The Mining Engineer, Vol. 134, pp 261-270.

Von Schonfeldt et al (1979). Subsidence and its Effect on Longwall Mine Design. Coal Convention of the American Mining Congress, St. Louis

GROUND DEFORMATION ABOVE LONGWALL PANELS GEOMETRY OF THE THREE DIMENSIONAL SYSTEM

Donald Burton

Assistant Mining Engineer
Mining Engineering Department
British Railways Board
Midland Station, Derby, DE1 2SA.

SUMMARY. The paper describes the purpose of a three dimensional variable geometry begining with an illustration involving the strains measured in a vertical section across a longwall face, and showing that certain features of rock movement which could be related to geology, and the particular case, are features of geometry. The paper continues with a description of the general application of the variable geometry.

INTRODUCTION

If the observer measures the difference in displacement of rock, and describes deformation using strain values, he will recognise some peculiar features in the patterns of rock movement and may be tempted to relate these features to particular rocks in layers, with structure-related properties. It may be demonstrated with the aid of a digital computer that some of the important features are in fact geometrical properties of the three dimensional system. This is usually shown with the aid of networks of survey stations on the surface, but measurements of vertical strains across the edges of panels provides an opportunity to illustrate the same thing below the surface, and to show the value of a three dimensional system outside the field of surface predictions.

At the second International Conference on Ground Movements and Structures held at the University of South Wales Institute of Science and Technology in April 1980, I explained that the engineer could be spared the complexity of a three dimensional geometry because a constructive investigator can replace a geometric mode of thought by a collection of functions and program a computer to do most of the work.

The paper was based upon the methods which were employed by the Railways Board to introduce their early computer programs in subsidence prediction, and it was closely followed in the Proceedings of the Conference by a different subject, a text by I.W.Farmer and P.F.R.Altounyan which deals with vertical strains across the edges of panels. Figures 2 and 3 have been extracted from the paper and they show contours of vertical strain in sections across longwall faces, the width of the panels being 140 and 180 metres. A glance at the patterns of strain is sufficient to create the idea that the movements involved are closely related to the geology, and contain few if any essential components that are related to other facts, something remote such as the differential displacement of two bridge abutments on the surface.

Figure 1 is a diagram indicating some of the main features of the strain contours that could be picked out by an observer. There is nothing obvious about these contours of vertical strain near the source of the disturbance and so a simple question arises. Could they be anticipated prior to the event given a geometry in which a lamina or surface is subject to curvature, or an analytical structure which forces points to travel along curved paths in a space of three dimensions? Would this be the same geometry which I have been using for many years to compute surface displacements, horizontally and vertically? It seems unlikely but to find the answers it is a simple matter of removing one of the three dimensional systems from the computer program I use in practice, leaving only a three dimensional geometry, and instruct the machine to contour a section across a face with vertical strains. There is no way in which the computer could detect features A to E if they related to the particular case and to geology, because the computer does not live in a world of experience, and the control system which deals with reality has been taken away from the system.

VERTICAL STRAINS

Selecting a convenient grid size, the computer calculates vertical strains, and interpolates the contours as seen in Figure 4. The same situation is illustrated by the machine in Figure 5, the computer using the magnitude of the strain values to fix the size of the character 'o' which it plots and shades the section accordingly.

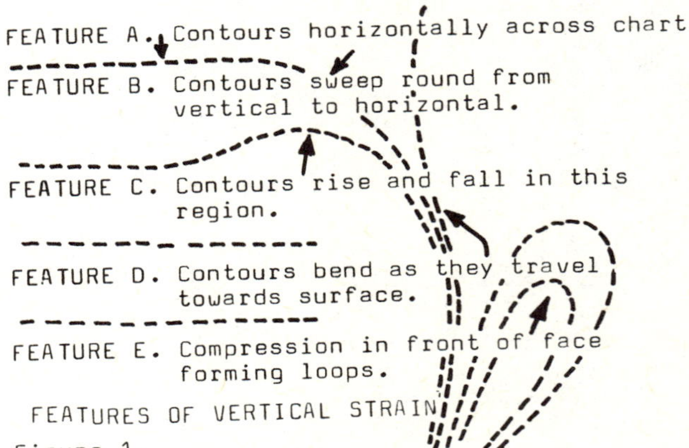

FEATURES OF VERTICAL STRAIN

Figure 1.

It is evident from Figures 4 and 5 that geometry alone predicts a correct pattern of rock movement across the face, even to the extent of the detail in Feature C. This is not accidental, but what follows is just as important, because we could have made the basic mistake of forgetting that the ground moves in a space of three dimensions, and if we did not inform the computer that the width of working was limited to 140 or 180 metres, then with no ribsides to the panel the computer would have drawn Figures 6 and 7. We could then conclude that the geometry did not apply, or arrive at the wrong conclusions using the facts outside a conceptual framework.

By pressing a few keys on the computer **keyboard one** can alter the grid size for the interpolation, and the curvature (of space), together with other details one might consider relevant to the geometry of the physical situation, and thereby create detail to match the type of detail obtained by the independent observer, so it is only necessary to produce the one pattern as seen in Figures 4 and 5 to conclude this argument. Whilst it is obvious that the different layers of rock have structure related properties covering a wide spectrum, no investigator would willingly choose to confuse the particular with the general, and talk about the contours of vertical strains as if they were controlled by geology, or the peculiar way in which certain beds of rock fracture in the space above the mine working, when some of the main features under discussion are known to belong to the subject of geometry, and recognised immediately by thinking in three dimensions.

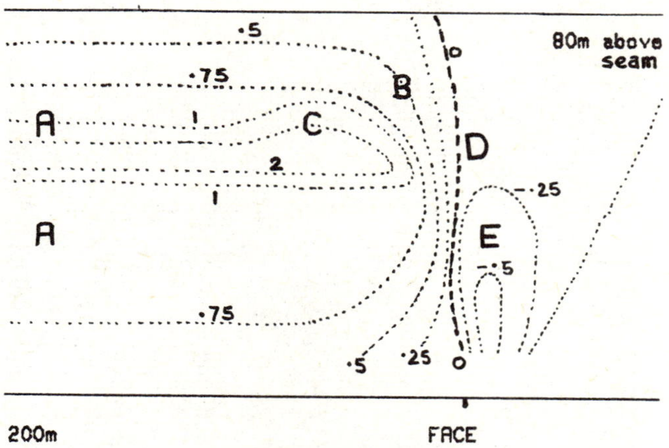

CONTOURS OF VERTICAL STRAIN ACROSS LONGWALL FACES. FARMER AND ALTOUNYAN, 1980.

Figure 2.

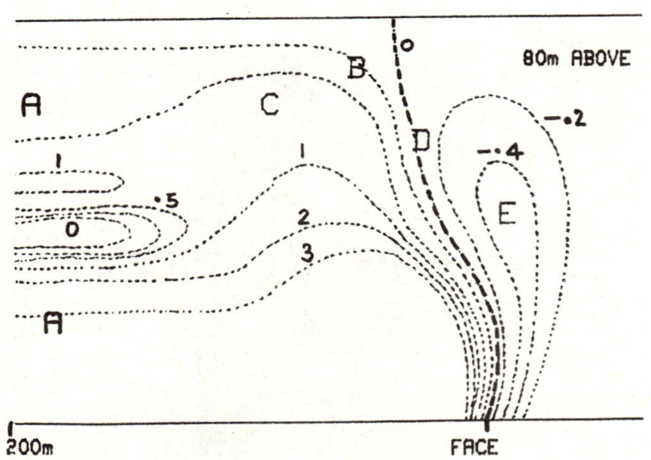

Figure 3.

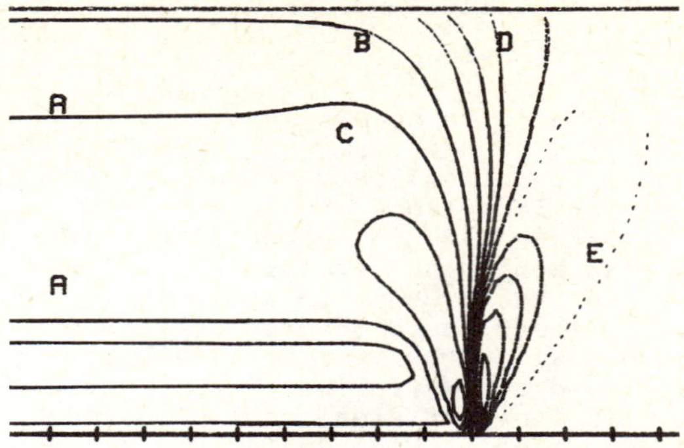

Figure 4. CONTOURS OF VERTICAL STRAIN GENERATED BY A COMPUTER.

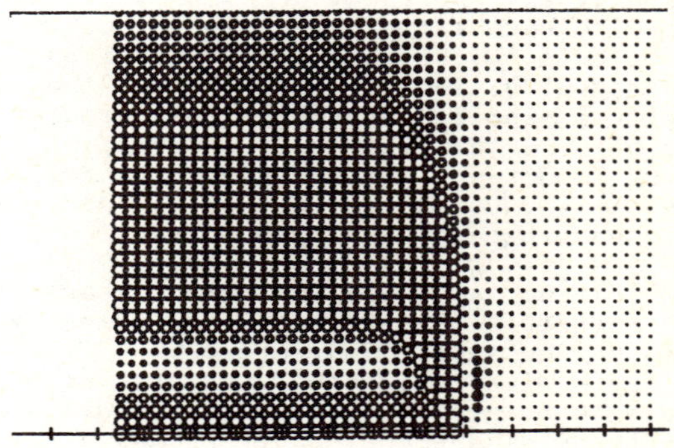

Figure 5.

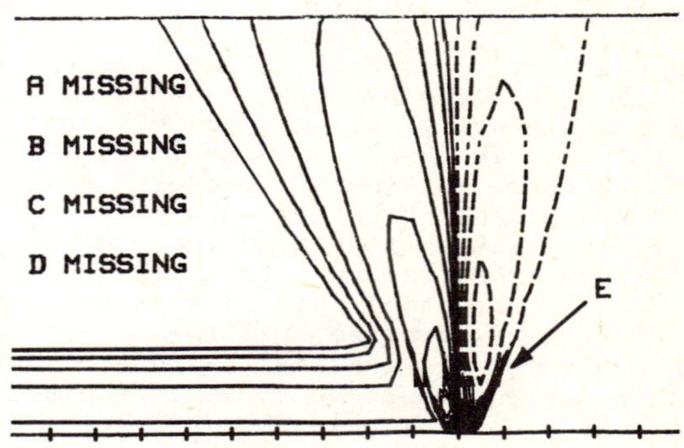

CONTOURS OF VERTICAL STRAIN - EXTENSIVE LONGWALL FACE.

Figure 6.

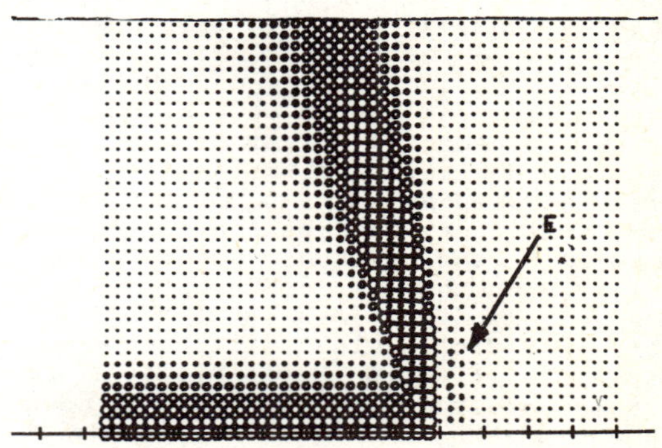

Figure 7.

THE THREE DIMENSIONAL GEOMETRY

Generally speaking, the surface movement above a modern panel is highly predictable, but to see this it is necessary to think of the deformation of the rock and the surface in a space of three dimensions. If you do not appreciate the importance of the direction of a line with respect to the edges of the mine working, then subsidence immediately becomes unpredictable. There would be no point in thinking that that one could predict differential displacements along the line of a railway, or a bridge, given a description of subsidence and strain in lines at right angles to the edges of panels of varying depth, width and length. Unfortunately, this fact means that railway engineers either have to spend a good deal of time in calculation, to give the civil engineer a lively appreciation of what he will be faced with each time a panel is worked, every case being different, or the civil engineer must adopt a wait-and-see policy. At one time vertical displacements were most important, and being a scalar quantity, direction was not so important, but today our standard track is continuous welded rail which means we are most concerned with changes in surface length.

The writing was on the wall many years ago, and the first three dimensional geometry I created in September 1964, after many false starts, was a direct attack on the problem of being able to think about ground movement in a space of three dimensions, as if it had always been a natural thing to do. As soon as I had devised this solution however, I realised that few observers or engineers would want to know about it. Having constructed something at all complex, it may seem to be concrete, a given reality to the designer, but a rather odd product of the imagination to anyone else, and I was proved correct in thinking it would not catch on. I decided there and then to plan a computer program to do all the work, but to do it in such a way that members of an organisation with more than a hundred years experience in the investigation of mine workings behind them, would not be offended by the idea of a machine being smarter than they were, which included myself amongst the human observers.

The product of this work is a very fast and efficient program, willing to take advice, and work out all the implications. Figure 8 gives the interpolated strains from 380 measurements along 20m bays, at 128° to the rib of a panel, and Figure 9 is the prediction made by the computer. In this case the Coal Measures dip fairly steeply below the Permo-Trias in the East Midlands, and the relationship between the curvature of the surface and the strains is far from average, as it would also be above panels worked at depths of say 100 to 200m near Coalville in the South Midlands. The depth in this particular case is 670 metres, but the computer makes the necessary distinctions between one case and another despite the geology and despite the previous mining history of each site. Most important, despite the particular configuration of the reference points being used.

Taking the many different subjects into account which are involved in making a quantitative prediction that will be checked by measurements, it seems impossible to think of the process of mining subsidence being dealt with by a computer, and equally unlikely that it could extend its influence into the field of rock movement below the surface. Every observer has his own concept of ground movement, which depends upon the information he has access to, what experiences have impressed him in the past, and what he expects of the future. This has never been a subject which lent itself to

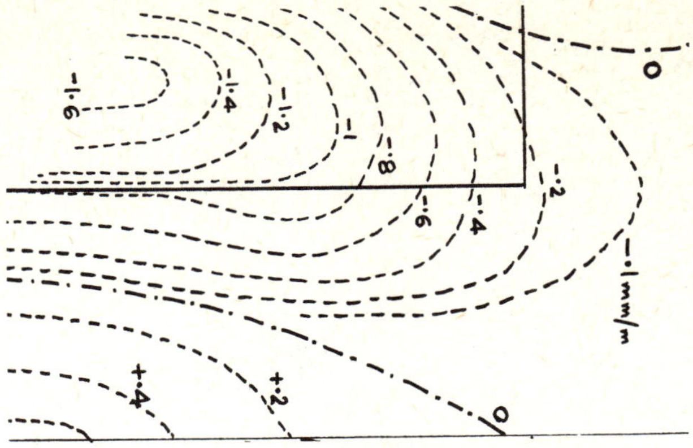

Figure 8. HORIZONTAL STRAINS ABOVE A PANEL. 128° TO RIBSIDE

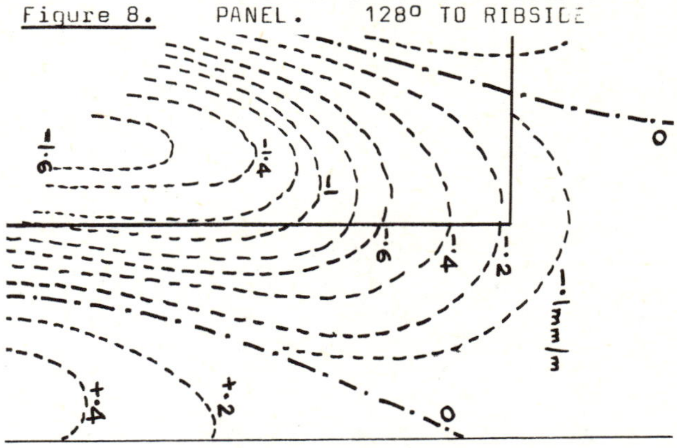

Figure 9. PREDICTED PATTERN

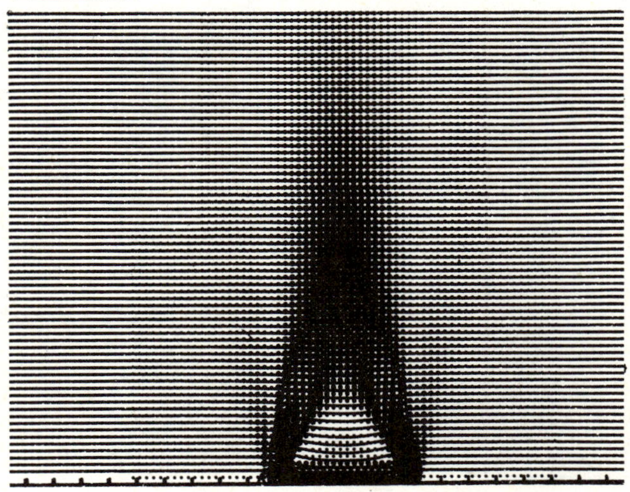

Figure 10. VERTICAL STRAIN ACROSS PANEL.

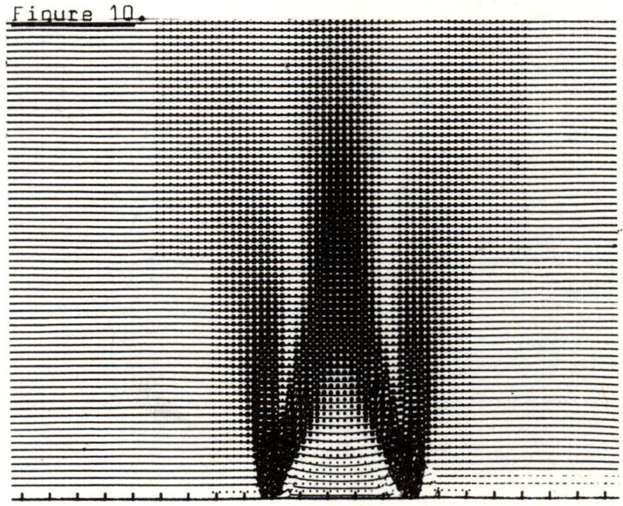

Figure 11. HORIZONTAL STRAINS.

a clinical gaze. I was introduced to ground movements by investigators who had obtained their skills from the previous generation, and to think of a unified system, a common language and a common conceptual framework, shared with a machine, must have appeared to be nonsense. However, the subject of geometry is concerned with organisation and the means to ask the right questions, and it was this subject I chose in 1964 to design the computer program.

A number of drawings as seen in Figures 10 to 13 were used to illustrate a simple problem and its obvious solution. Each pattern of ground movement is connected to the process of measurement. Figure 10 is a simple pattern of vertical strain in a section across a mine working that results from measuring vertical displacements along various surfaces. In the centre there is tension vertically, observers seeing a reduction in maximum subsidence as depth increased. To each side there are small amounts of compression. Selecting any block from the centre, the top and bottom must be moving apart to result in the observed reduction in vertical displacement, which is exactly what one would expect if the disurbance spread out from the source, and the sides of the block therefore moved closer. So measurements of vertical displacement entailed measurements of horizontal changes in length, resulting in the second basic pattern seen in Figure 11.

In this pattern, compression in the centre was flanked by the two outer areas of tension. Above the working, within a limited area, one enters the fairly chaotic regime where there is no lateral compression to hold the structure together. Given these two patterns, the third basic pattern is entailed, the difference between the two sets of quantities as in Figure 12. The early railway engineers knew they were dealing with a dissipative system, and the third pattern illustrated the dissipation of energy in the two outer areas, where vertical compression fails to offset tension laterally. The quantities must always balance, and the loss in the original space exactly matches the reduction in the cross-sectional area of the subsidence profiles as width/depth ratio decreases.

Measurements made it obvious to the early railway engineer that it was insufficient to work coal to cause rock movement and subsidence, it was first necessary to add enough gravitational potential energy to overwhelm the force of friction. On the other hand, the familiar result of friction in a dissipative system was that moving objects grind to a halt unless energy is pumped in, and because longwall faces continued to advance, energy was pumped in and then the system headed for some form of steady state. Since rock movement liberated the gravitational potential energy, and deviations were damped out, it appeared to some observers that working coal was directly related to subsidence, but the intrinsic properties of the rocks were always evident on the long railway levelling sections, because the centre of the additional subsidence was never located centrally above the strip of coal to which it was supposed to be related. Consequently they never created concepts like the Notion of Zone Values, used on the Continent which said nothing about the loss in space. They could explain why the same patterns appeared at different depths, by pointing out that the quantities in the equation, gravitational potential energy and friction, cancelled out to leave a dimensionless ratio, a pure number which applied irrespective of the units of measurement employed. So they could say, as a rule of thumb, a face advance of say 0.2 x depth was the energy normally required to cross the threshold which continuity demanded between the two extremes, when surface movement was first noticed.

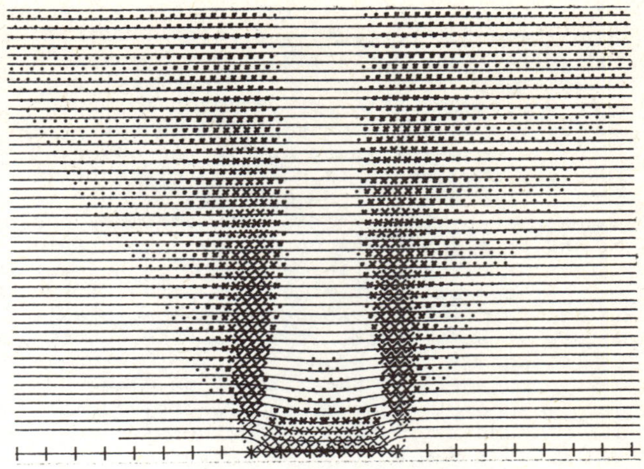

DISSIPATION OF ENERGY

Figure 12.

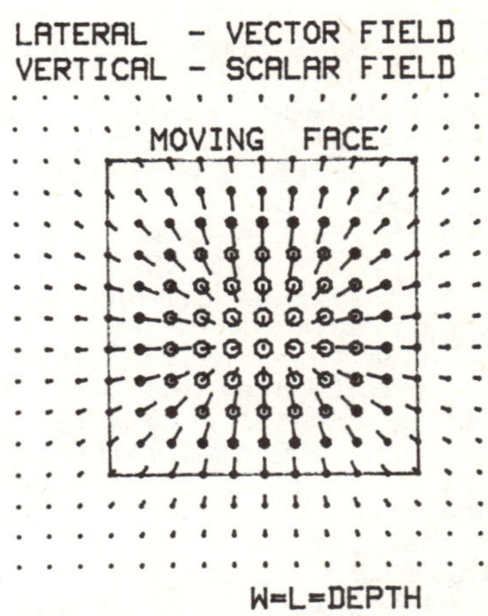

Figure 13.

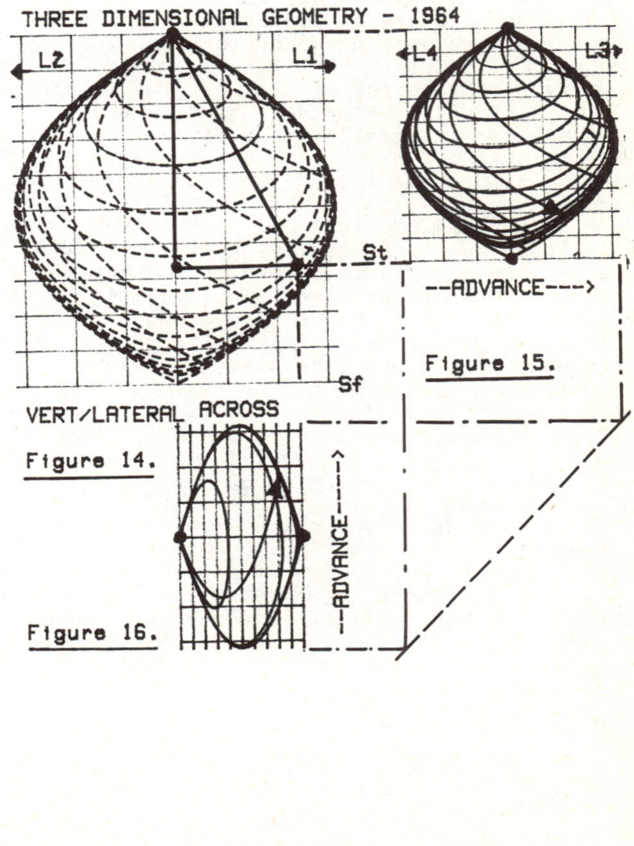

Figure 14.

Figure 15.

Figure 16.

Figure 13, showing the scalar and vector quantities for vertical and lateral displacement across any surface horizon, at any instant, could be added to the three basic patterns of ground movement that were related to measurements. The simple problem all such patterns illustrated was that to obtain the observed results, one had to think of every particle of rock having a unique path. Before one could compare a particular case with another, one had to envisage the means of comparing one set of unique paths, an infinite number of them, with another complete set of unique paths. There was no point in talking about the unpredictable nature of subsidence, if it was only unpredictable because one had no idea what path a point was supposed to be on at the time. That was merely a combination of complexity and ignorance, not unpredictability. So the simple solution I used, was to take advantage of the fact that every path was different. If ground movement had been vertical, all the points would plot along one line or scale, but if every path was different, then simply by labelling each point in space with co-ordinates, we would find the path in its proper place. To do this one only required two drawings on a flat sheet as seen in Figures 14 and 15. This geometry has been explained many times in the past, and by selecting any co-ordinate position, using say the left-hand edge and start of a panel for the origin, one can only obtain one path for one point, and one position on that path for one instant in time.

It was the next step which attracted me. By giving Figures 14 and 15 average dimensions, for example on average the maximum lateral displacement of a point in a set is 0.16 x the full subsidence, one created a model of 'average' mining subsidence, and then it was possible to go to an area like Coalville, where the average maximum lateral displacement is 0.28Sf, and predict a large number of changes in surface length as faces advanced, knowing full well that the predicted quantities would be too small. The chart one obtained was as shown in Figure 17 on the left. The lines could be of any length and point in any direction. To bring the quantities into line, obviously it was going to be necessary to expand the model laterally by 75%, producing the chart on the right of Figure 17. What I wanted to demonstrate was that this change was a geometric deformation, or transformation, which would be demanded in every case, such that one needed a three dimensional model of infinite variety and shape, vertically and laterally. This applied whatever the source of the relevant information used by the investigator, be it an explanation of the characteristics of certain events, or a description. The reason I wanted to illustrate this was because the computer program was going to make a very subtle use of geometry.

To explain what I had in mind required an analogy, but first it was necessary to know that when the displacements of surface objects are arranged three dimensionally as in Figures 14 to 16, whilst they do not form the neat patterns of the geometry itself, they repeatedly turn up in the expected place. Moreover, the geometrical patterns can be reproduced or replaced by a collection of functions. This answers the question about a computer, which cannot have a geometric mode of thought, and why it is possible to talk about geometry and spaces when one is going to deal with an analytical structure. The computer will possess one three dimensional system in which a collection of functions is capable of representing a three dimensional geometry. Now it is apparent that if these functions would produce the curv-paths of points given co-ordinate positions, as represented by Figure 19 for example using a

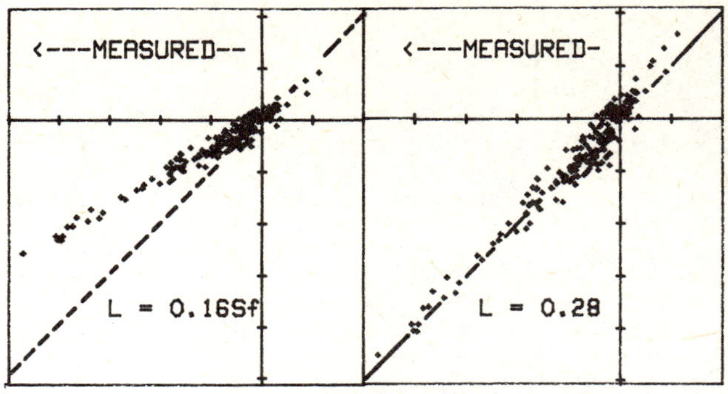

GEOMETRIC TRANSFORMATION

Figure 17.

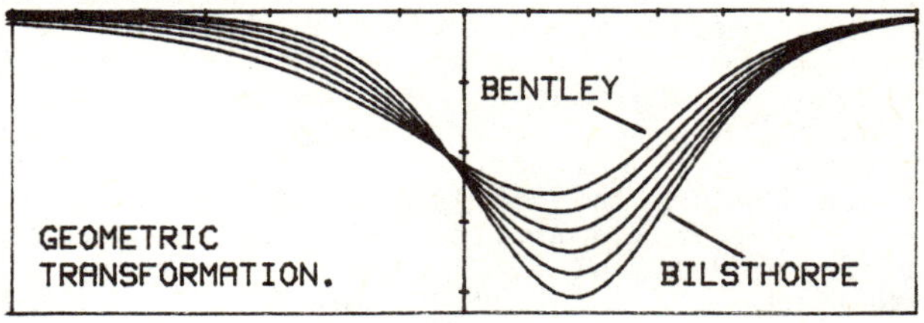

Figure 18.

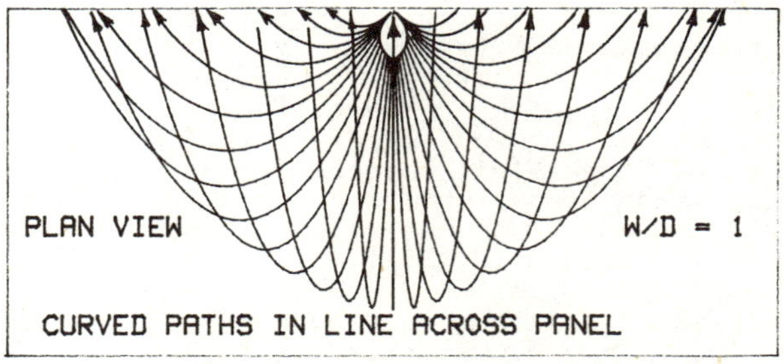

Figure 19.

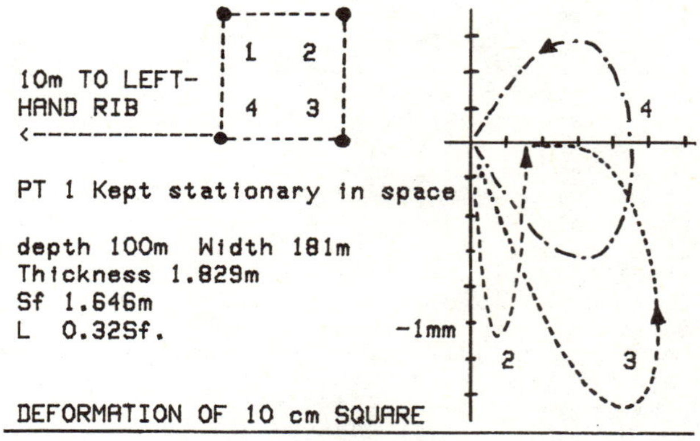

Figure 20.

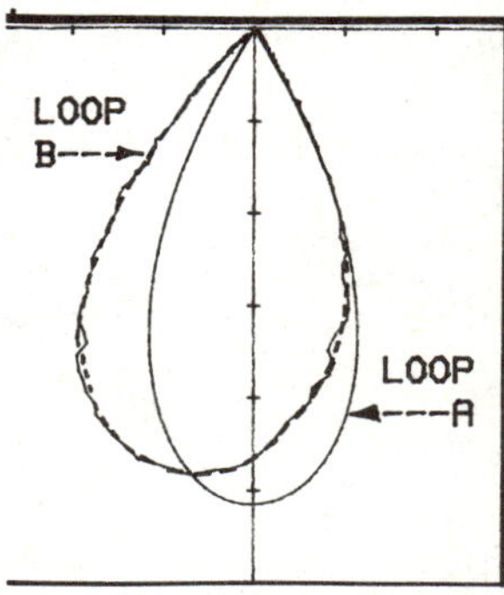

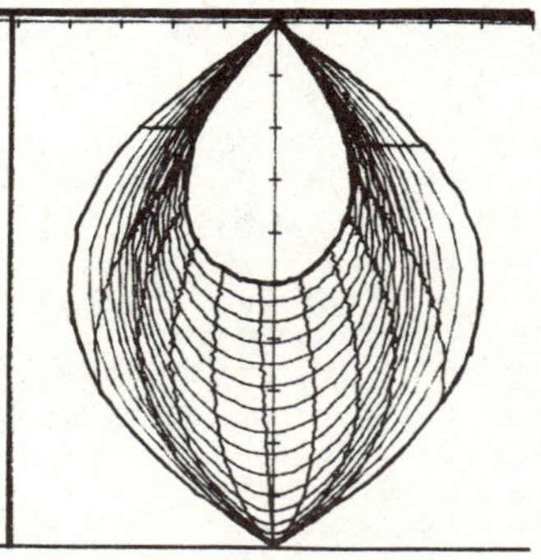

POTENTIALLY UNLIMITED NUMBER OF DEFORMATIONS.

Figure 21. Figure 22.

line of points spaced out across a panel, seen in plan view, then these functions connected every point in a set. One had an intrinsic surface theory such that given just a few bits of information, one could determine the complete state of the system. The computer can use its collection of functions to move from an actual series of events at say Bilsthorpe in Figure 18 to a series of events at Bentley in Yorkshire, where an old working 47 metres outside the left-hand edge of the current panel changes the geometry point by point across the basin, and in the process it would contain in its memory the meaning of this transformation in terms of any point in the system. So an observer who wanted to 'talk' to the computer in a sensible way, might decide that the only points he was going to refer to were two points placed at say 0.2depth inside and outside the edges of the panel, these edges treated in isolation one by one as if the others were not there. He might in fact be dealing with a surface with the length/ and width/depth ratios being 0.3, but he could talk to the computer in terms of isolated edges. If he did this he would find that to match actual events, the language he used made sense, the numbers did not change in a random way, they were easy to follow. Think of the particular choice of points as being a distance formula.

The analogy then comes naturally. If one selects three points on a sphere and joins them by ordinary straight-line segments, the result is a Euclidean triangle, and all that entails. However, choose the shortest distance to be along great circles along the surface, and the result is a spherical triangle, a different geometry altogether. So designing a geometry for a mountainous region, the distance formula which determines the geometry, changes from place to place, even from point to point. One has proposed a geometry for nonhomogeneous spaces, spaces of varying curvature. Because of this feature of geometry, Einstein being too clever by half, upset many philosophers by allowing the different masses in space to determine the distance formula, and the distance formula determines the geometry, so matter determines the geometry, which then goes on to explain all the phenomena previously ascribed to gravitation. The gravitational effect of gross matter was reduced to geometry, he geometrised gravity. On that basis, the properties of solid rock determine the distance formula, which determines the geometry, which goes on to explain the phenomena previously ascribed to the properties of rock. In one jump, geometry does not deal with models of physical space, it deals with anything whose properties and concepts fit the geometric framework. So when two railway mining engineers talk about changes in geology, the effect of the edge of an old working in one specific location, or the size of an underground roadway, any subject which affects the differential displacement of two or more surface objects, they only require one language, the variation in a distance formula which determines the local curvature of space, which determines the geometry known to the computer. Since they can look in four directions at once, across each edge, and dimension a problem if only one or two edges are involved, cutting the problem down to size, and as the computer works out all the implications at high speed, they have an infinitely variable geometry to work with that does not require a great deal of effort, one might say none at all. The only criterion is the pragmatic one where predictions are made and tested by observation.

Clearly it is always necessary to keep a sound sense of proportion. The computer will calculate the differential displacement of any points, as in Figure 20 for example using a 10cm square. It is known for a fact that if all 10cm cubes at this surface obeyed the rules then the surface points which were observed followed exactly the same geometry. In fact points up to 80 metres apart provided changes in observed length that were extremely accurate, but when the distances between points are reduced beyond a certain **limit**, the observer comes face to face with the aggregate properties of the material. Accurate predictions have been made by the computer on a surface only 39 metres above a panel, but it requires no imagination to see what would have happened if the reference points were placed ten centimetres apart, with a fissure half a metre wide eventually separating them. However the geometry itself does not fail.

The computer can describe 'average' ground movement, represented by loop A in Figure 21, or approximate a series of events as represented by loop B, but given say 500 points in a line, it can make 500 separate geometric transformations, backwards and forwards, to match the events in detail. It is not as if such things as mathematical models cannot deal with reality, only the observer fails to obtain the perfect information to make perfect predictions. Similarly, the computer can take 5000 points in columns and rows in a section across a panel, and compute each separate displacement as in Figure 22. Depending upon the relevant information, the computer will be asked to make some major changes in geometry as depth increases, but it could as easily change the geometry 5000 times, say within a prescribed range, to deal with supposed detail. This has no relevance to the railway mining engineer who only deals with the gross properties of the system, but it makes the point that the geometry is never stopped.

One chooses certain mathematical materials to build a unified system, with intuition rather than knowing rationally that they will complete the task, and only when the design is complete will rational criticism and observational tests show if it is scientifically sound. Twelve years were spent testing the geometry before the first program was written and then the speed of the computer revealed some non-trivial consequences of apparently trivial advice, which affected the next design. I now use the third generation program which gives an extremely polished performance.

Recognising that a three dimensional system links information which previously seemed remote and unconnected, investigators in the field of rock mechanics may decide that a variable geometry is an ideal way to remove certain difficulties, and may in fact already employ similar techniques. It would certainly be a waste of modern computing facilities, desktop machines, not to think in terms of models that provide infinite variety. They are surprisingly effective and easy to construct.

REFERENCES

Farmer, I.W. and Altounyan, P.F.R.,1980, "The mechanics of ground deformation above a longwall face", Proc. Conf. Ground Movements and structures, UWIST.Cardiff.

THE APPLICATION OF AN INTEGRATED INSTRUMENTATION SYSTEM

FOR COALMINE ROCK MECHANICS INVESTIGATIONS

A. K. Isaac, B. G. D. Smart, M. A. Price, and J. D. Thornton

Department of Mineral Exploitation
University College Cardiff
Cardiff, CF1 1XL, Wales.

SUMMARY. From a fundamental understanding of strata loading and deformation behaviour, support system design may improve performance at longwall units considered reasonably satisfactory. The pre-requisite of such an understanding is the ability to accurately quantify and interpret in situ observations.

This Paper briefly describes the design features of a new instrumentation system designed to integrate the separate processes of data acquisition, storage and processing. The emphasis in the Paper is placed upon successful field trials of the new system at a high output, rapid advance coalface in the anthracite belt of the South Wales Coalfield.

INTRODUCTION

The British Coalmining Industry has since its nationalisation in 1947, given priority to comprehensive mechanisation and high level technology as the means of achieving higher profitability, increased efficiency, improved safety and reduced manning. Examination of statistical data reveals the success of this policy, but in productivity terms, viz., output per manshift, the rate of improvement from 1.2 tonnes in 1947 to 2.15 tonnes in 1960, has not been maintained. Although various contributory factors have combined to produce this plateau effect, the question remains regarding the efficiency levels achieved at longwall coalfaces. To effect further productivity improvement and to meet the costs of equipping modern coalfaces, current levels of machine utilisation time require to be raised above the present average level of around 30-40%.

A major factor affecting machine utilisation time is that of strata control along the faceline and at the junctions of the faceline with its gateroads. The inability of support systems to control strata deformation in these circumstances adversely affects output performance, and hence the design and application of coalface powered support systems assumes significance. The generally successful mechanisation policy of the National Coal Board has to some extent obscured the relatively low efficiency levels of most coalface installations. In the last 10-15 years, renewed attention has been given to understanding the fundamental behaviour of strata around longwall coalfaces. Such knowledge undoubtedly assists support system design and therefore attention has been devoted to examination of the relevance and accuracy of methods of assessing support performance. A system that successfully integrates data acquisition and processing of support system behaviour is highly desirable because of the increased volume of results that may be generated and analysed, and of more importance, the additional time made available to the research worker for their interpretation. In the context of modern transducerised instrumentation and sophisticated recording techniques, and continuous developments in computer technology, the Strata Mechanics Research Group at Cardiff have produced a system that may be applied to a variety of coalmine rock mechanics investigations, including possibly the most hazardous, that of the coalface.

DATA ACQUISITION, ANALYSIS AND PRESENTATION

The instrumentation system developed by the Mining Research Establishment, N.C.B., for the first major phase of coalface investigations covering the decade 1954-64 and discussed by Shepherd (1964), was used in the field investigations summarised by Wilson (1964). With some exceptions the literature reveals that between 1964-78, data acquisition continued to employ the early range of devices and recording methods, with data processing being essentially traditional in nature.

Because of the problems associated with existing instrumentation systems, viz., operational difficulties, low order accuracy, and discontinuous monitoring, research at Cardiff from 1971 was directed towards the development of a complete system for data generation and treatment.

Significant stages in development are summarised in Table 1 which also indicates the desired parameter and the continuing nature of the research programme. The period 1971-74 witnessed the use of manual methods of measurement and recording. From 1974-79, emphasis was placed upon the development of a prototype intrinsically safe range of transducers, and clockwork driven chart recorders employing existing signal conditioning units for continuous recording of relevant parameters. The instrumentation system and results obtained from this stage of development were presented by Platt et al. (1977). Subsequent evaluation of the system indicated the general unsuitability of the chart recorders and the need to refine transducer design. The result of work conducted from 1977-81 and presented by Smart et al. (1981) was the production of an integrated instrumentation system that employed: -

(i) a range of transducer devices for measurement of pressure, linear and angular displacement.

(ii) a twelve-channel, portable, battery-operated data logger modified to meet intrinsic safety requirements, using standard C60 or C90 magnetic cassette tapes, and with automatic scanning facility from 2 second to hourly intervals.

(iii) computing facilities for translation, processing and presentation of data, viz., translation unit, floppy discs and minicomputer or minicomputer/computer.

Figure 1 illustrates diagrammatically the general arrangement of the above system, while Figure 2 shows the junction box, interface unit (Type M 200 U, Microdata Ltd) and modified data logger (Type M 200 L, Microdata Ltd), with the convergence, compaction and pressure transducers. Figure 3 illustrates the data processing arrangement of translator unit (Type M 200 TRZ) and a 32K memory minicomputer (Type PDP 11/03, Declab). Presentation of data in the system is achieved via software that combines 'raw' data and ancillary information to provide output on a visual display unit in digital or graphical form, with 'hard copy' an available option. This arrangement facilitates rapid interpretation of data and is further enhanced by the linking of the minicomputer to a larger computer (Type 4070, G.E.C.) with multi-channel and multi-colour graph plotter facility (Type 1302, Benson). This alternative of mini-computer or minicomputer/computer processing gives for the former the option of rapid preliminary interpretation despite certain limitations in accuracy and resolution, and for the latter lineprinter, typewriter, visual plot displays and greater processing capacity.

Rock mechanics investigations using this data logging system may record up to twelve channels of data in which a typical coalface investigation may use the following: -

Channel No.	Function
1	Scan Counter. Visual display available on LED allowing correlation of visual observations with scan number.
2-8	Seven pressure transducers monitoring a bank of three supports to measure front and rear leg pressures, together with middle leg pressure of the central support.
9-10	Strata convergence measured by two linear displacement transducers, mounted adjacent to the central support.
11-12	Support or roof beam inclination measured by angular displacement transducer.

This combination reflects the variety of options available; in addition a facility exists to electrically connect transducers in series or parallel to the same channel, further maximising data acquisition.

After normal development programme and certification process delays, the system was taken for field trials in July 1981 upon the successful completion of laboratory tests.

Table 1 Stages in Transducer Development for Coalface Investigations

Parameter	Stages in Development		
	1. Manual (1971-74)	2. Semi-automatic (1974-1979)	3. Automatic (1979-1985)
Powered Support Unit			
Leg Hydraulic Circuit Pressure	Pressure Gauges (Conflow) Chart Recorders (Bristol)	Pressure Transducer (Mk.1) (Univ. Coll., Cardiff)	Pressure Transducer (Univ. Coll., Cardiff)
Leg Closure	Dial Gauges (Mercer)	Leg Closure Transducer (Mk.1) (Univ. Coll., Cardiff)	Leg Closure Transducer (Mk.1,2) (Univ. Coll., Cardiff)
Leg Attitude	Clinometer (Hilger & Watts)	Inclinometer Transducer (Mk.2) (Univ. Coll., Cardiff)	Inclinometer Transducer (Mk.1,2) (Univ. Coll., Cardiff)
Base/Canopy Attitude	-	" " "	" " "
Strata			
Roof-Floor Convergence	Convergence Recorder (Davis-Derby)	Convergence Transducer (Mk.1) (Univ. Coll., Cardiff)	Convergence Transducer (Mk.2) Univ. Coll., Cardiff)
Roof-Floor lateral movement	Romometer (N.C.B.)	-	-
Roof and floor	-	Dial Gauges (Mercer)	Compaction Transducer (Mk.1) (Univ. Coll., Cardiff)

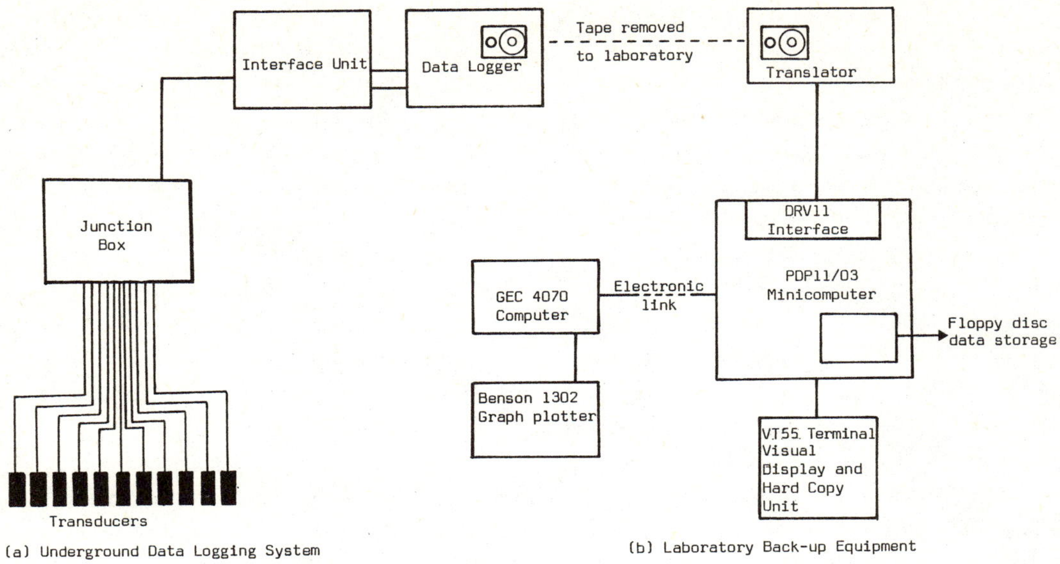

Figure 1 Diagrammatic Representation of the new Integrated Instrumentation System

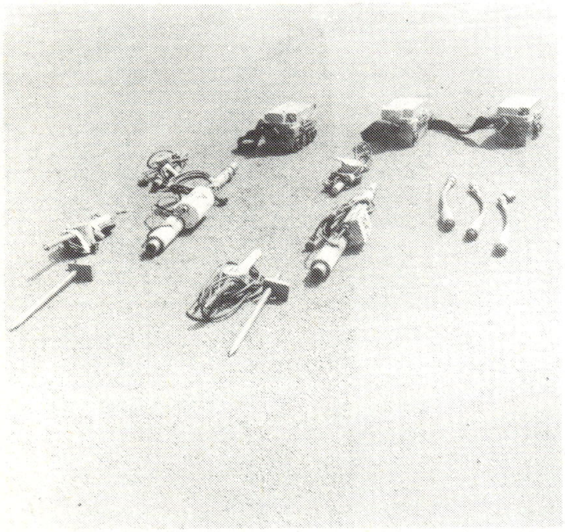

Figure 2 Intrinsically Safe Data Logger with Associated Transducers in Foreground

Figure 3 Minicomputer and Translator Units

FIELD TRIALS OF NEW INSTRUMENTATION

Location

At Betws Colliery, Ammanford, a new mine situated in the north west corner of the South Wales Coalfield, the Red Vein is extracted for its high grade anthracite coal. W3 Coalface at Betws was selected for the field trials because of associated gateroad investigations and a previous coalface investigation at W1 Coalface. In addition to the rapid face advance of W3 and its favourable proximate geology, colliery personnel consistently demonstrated keen interest in all areas of research by their high level of co-operation.

The layout of W3 District is given in Figure 4 together with details of seam section and physical properties of roof and floor. At the time of the investigation, the face had retreated almost 100 m from face entry position up a 1 in 10 gradient, to a point where cover depth was approximately 650 m. Two ranging drum shearers operated along the 185 m faceline which dipped at 1 in 13 toward the Main-gate (left hand roadway), producing an average of 6000 tonnes/week from extraction of the full seam section of 1.4 m, and for a face advance of 20 m. Gullick Dobson 6/240 powered supports possessing articulated canopies and rigid bases, were set at 14 MPa pressure and yielded at 44 MPa pressure.

Geology

In the general geology surrounding W3 District, the seam dipped in a south westerly manner as indicated above. Although no large scale structural features were present, the seam roof was prone to small washouts,roof rolls and mudstone lenses. Although undulating in nature, the seam roof

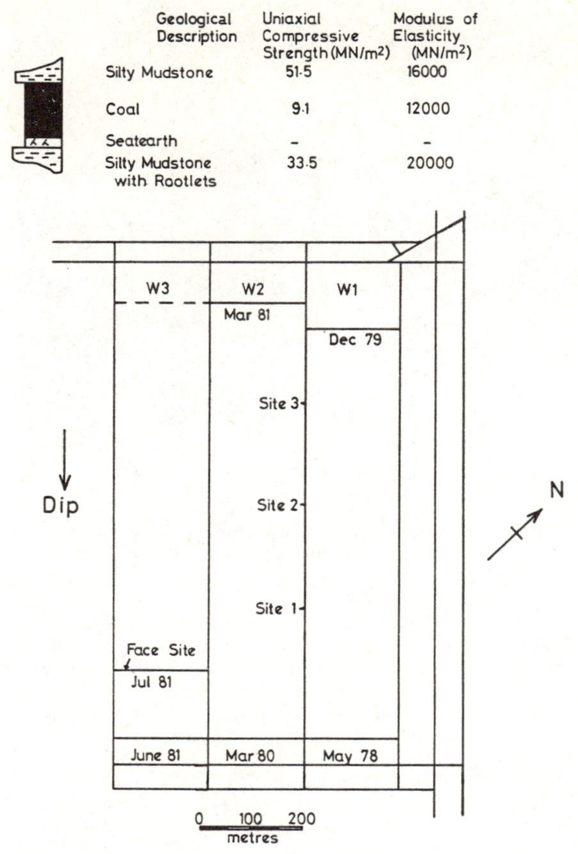

Figure 4 Layout and Seam Section, W3

generally appeared competent except in the vicinity of washouts where extensive low angled slickensided faulting was present. The immediate seam floor was typical of the weak underclays found in South Wales.

The Red Vein displayed a consistent coal thickness and quality over the general area of extraction in which the major slip was found to have a strike of about 335°, and the average dip of all planes was 50°.

An intensive core drilling programme revealed the proximate geology of the Red Vein. The immediate 1.5 m of roof comprised laminated silty mudstone, which was overlain by a further 6.5 m of mudstone containing silty bands up to 0.5 m thick. The seam was underlain by a weak underclay band approximately 0.5 m thick, which in turn lay on a succession of silty mudstones approximately 5 m in thickness. Average uniaxial compressive strength of roof rocks was 51.5 MPa, that of the main floor rocks was 33.5 MPa, and for the coal was 9.1 MPa.

Installation

A 'health' survey of the support system conducted prior to installation indicated that the area from Support No. 25-40 offered a representative situation. Subsequently the complete instrumentation system which was installed during the pre-production shift of 16th July 1981, on Powered Support No. 32, comprised: -

Pressure transducers (3) on front, middle and rear leg circuits;

Roof debris compaction transducers (2) each side of front canopy;

Floor debris compaction transducers (2) loosely mounted on anchor stakes each side of support adjacent to front legs;

Convergence transducers (2) between competent roof and floor each side of support adjacent to front legs;

Junction box mounted in rear of support structure to accept armoured cables from transducers;

Interface Unit and Logger mounted in rear of Support No. 33, connected by armoured cable to junction box.

The pressure transducers were connected by Conflow adaptors to the pressure testing ports of the support, and the debris compaction transducers were affixed securely to the front canopy and base units. All transducers were connected to the junction box via a cable 'highway', which was given maximum protection by a carefully selected route within the support structure.

Immediately following installation, the compaction and convergence transducers were tested by manual adjustment with observation of logger values, after which a five minute scan period, at two second intervals, was recorded to establish the initial values of all transducers. The logger unit was disconnected and taken to surface where the battery and cassette were replaced by fresh equipment, the unit then being returned to site and reconnected.

Investigation

The duration of study was from Thursday 16th - Tuesday 21st July 1981 with 2 research personnel serving each of three shifts, of which two were production periods. Although the system was arranged to provide automatic monitoring and recording of data, the presence of research personnel was considered desirable at this stage of system development to record details such as cutting operations, to advance transducers with support advance, and to control the logging system.

Scan frequency was set at 2 minute intervals, except for periods when production operations were in the 'zone of influence' (2 second intervals) i.e., 10 supports either side of monitored support (Number 32), and over the weekend (10 minute intervals). Figures 5-8 illustrate the sequence of events associated with passage of shearer and advance of supports, and Table 2 indicates general investigation details.

Eight complete production cycles and the weekend were monitored in the study period during which W3 face retreated 5.3 m. The cycles encompassed some 1380 minutes, the total investigation period lasted over 100 hours, and a total of 7656 scans were recorded. A production cycle is defined as the completion of all operations associated with the removal of one machine cut off the faceline.

Figure 5 Data Logger and Pressure Transducers Installed on W3 Coalface, Betws

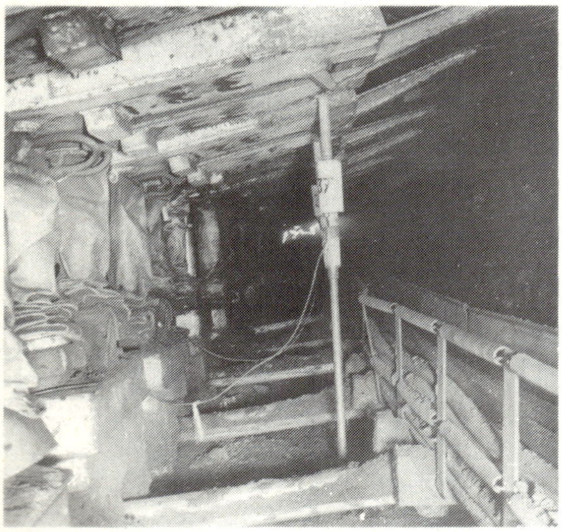

Figure 6 Convergence Transducers Advanced Prior to Powered Support Advance

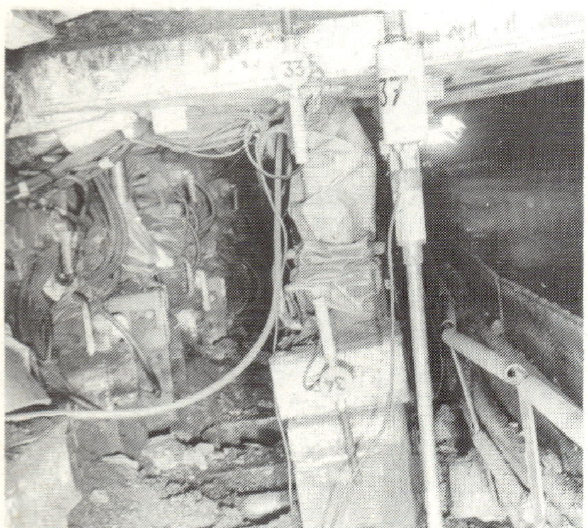

Figure 7 Powered Supports Lowered and Advanced

Figure 8 Powered Supports re-set - Roof and Floor Debris Compaction Sites Re-established

Table 2 W3 Coalface Investigation Details

Date	Period Monitored	Tape No.
Thurs. 16th July 1981	Initial Readings	1
Fri. 17th July	Production Shift 1	2
	Production Shift 2) Maintenance Shift)	3
Sat. 18th July - Mon. 20th July	Weekend and Production Shift 1	4
	Production Shift 2	5
	Maintenance Shift	6
Tues. 21st July	Production Shift 1	7

Data Treatment

Data on each cassette was transferred to the minicomputer via the translation unit using appropriate software (Program DRV11), a fully recorded C60 tape taking 22 minutes to transfer. The completion of transfer enabled data to be written into direct access files on a 250K byte capacity floppy disc thus providing a data bank facility.

Computer program TAPEIN enabled the entry of control data from each tape to access files, e.g. FTN50.DAT, a file which identified missed readings, transducer positions and anomalous results. These files were subsequently computer processed (programs TAPCON and TAPRES) to analyse convergence/closure and pressure results respectively. Each operational cycle was split into sections to provide values of rates of convergence, pressure increase and Mean Load Density. Example output files shown in Table 3 and Figure 9 indicate calculated pressure values and front leg pressure development during the period logged on Tape No. 7.

Table 3 Pressure Values Recorded by Data Logging System

TABLE OF SCAN NUMBERS : ELAPSED TIME : PRESSURES
****

	SCAN NUMBER	ELAPSED TIME (MINUTES)	PRESSURE VALUES (MPa)		
1259	9/106	154.427	26.88	23.30	27.06
1260	9/107	154.461	26.88	23.06	27.06
1261	9/108	154.494	26.14	22.83	24.35
1262	9/109	154.527	25.64	22.59	23.12
1263	9/110	154.561	25.64	22.59	22.63
1264	9/111	154.594	25.40	22.59	22.39
1265	9/112	154.627	25.64	22.59	22.14
1266	9/113	154.661	25.64	22.59	21.89
1267	9/114	154.694	24.90	21.65	21.16
1268	9/115	154.727	24.16	21.18	20.42
1269	9/116	154.760	23.91	20.95	19.43
1270	9/117	154.794	22.92	20.48	16.97
1271	9/118	154.827	22.92	20.48	16.73
1272	9/119	154.860	22.92	20.48	16.48
1273	9/120	154.894	22.92	20.48	16.48
1274	9/121	154.927	22.67	20.24	15.99
1275	9/122	154.960	22.67	20.48	15.99
1276	9/123	154.993	22.67	20.48	15.99
1277	9/124	155.027	22.67	20.24	15.99
1278	9/125	155.060	22.67	20.24	15.99
1279	9/126	155.093	2.33	2.38	2.46
1280	9/127	155.127	2.33	2.15	2.46
1281	10/ 0	155.160	2.33	2.15	2.21
1282	10/ 1	155.193	0.35	0.50	0.49
1283	10/ 2	155.227	0.35	0.74	0.98
1284	10/ 3	155.260	0.35	0.74	1.23
1285	10/ 4	155.293	0.35	0.74	1.23
1286	10/ 5	155.326	0.35	0.74	0.98
1287	10/ 6	155.360	0.35	0.74	0.74
1288	10/ 7	155.393	0.35	0.74	0.74
1289	10/ 8	155.426	0.35	0.74	0.74
1290	10/ 9	155.460	0.35	0.74	0.74

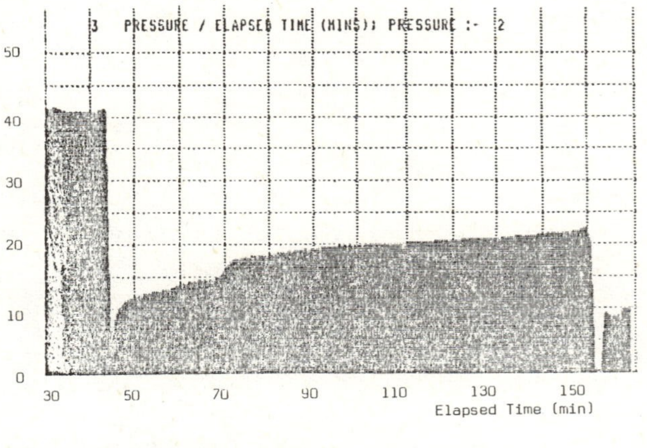

Figure 9 Front Leg Circuit Pressure Recorded on Tape 7

Results

The analyses of data automatically processed as indicated above, provided a comprehensive picture of support performance and strata behaviour at W3 Coalface. From summaries of results by Price (1981) and Thornton (1981) and given in Table 4, calculation may be made of the pressure/convergence ratio, an indicator of support efficiency, viz: -

$$\frac{\text{Average Rate of Pressure Increase (MPa)}}{\text{Average Rate of Roof-Floor Convergence (mm)}}$$

In the case of Betws W3, this value was calculated to be 0.5 MPa/mm.

Because an intensive analysis of results is beyond the scope of this Paper, a few pertinent comments only are made in this section.

Pressure transducers were completely automatic in operation compared with convergence and compaction measurement, hence the monitoring period was longer. Adjusting the observed convergence rate over this period revealed a value of 42.7 mm/m of face advance, which is very close to that predicted by Wilson (1964) and Ashwin et al. (1970). This emphasises the need for as near continuous recording as possible for greatest accuracy of data.

Compared with previous coalface investigations in South Wales, Mean Load Density values were the highest recorded, as also the total rate of pressure change. These results confirm visual observation and production results of W3 District regarding high operational efficiency and good strata control.

Pressure/convergence ratios were relatively high for each element of the operational cycle, indicating the efficient conversion of convergence to increased leg pressures. Further examination of the data identifies an increasing 'stiffness' of support system through the cycle (0.4 MPa/mm-1.0 MPa/mm) except for the period immediately prior to support advance.

CONCLUSIONS

The 35 coalface investigations conducted in South Wales during the development period 1971-81 provided considerable fundamental information about the interaction of strata and powered support. With the exception of the investigation reported in this Paper, almost all investigations were prone to incomplete data acquisition due to inadequate instrumentation system design or accidental displacement and/or damage of instruments. The field trials of the new integrated instrumentation system provided the most complete results to date, with the only missed values occurring during re-establishment of floor compaction stations.

The introduction of a continuous monitoring system has provided measurements of greater accuracy and of a more representative nature. Identification of the pressure/convergence ratio as a method of assessing powered support efficiency, requires the accuracy of measurement of pressure and convergence that is provided by the new system. Although the system may be considered to provide continuous monitoring, it was designed to operate with one or more observers relating production operations and other influences to recorded values. This ensured accurate and safe operation of transducers and logging system. Therefore it may not be considered fully automatic at this stage of development. However, a comparison of the scan frequency of the system (2 seconds) with that of manual methods (1 minute) reveals that at least 30 times as much data was recorded at periods of high face activity.

Table 4 Summary of Investigation Results at W3 Coalface, Betws

Pressure and Load			Convergence and Compaction		
No. of Monitored Cycles				:	8
Period of Monitoring	:	1380 min	Period of Monitoring	:	1100 min
Lowest Pressure	:	7.09	Face Advance	:	5.3 m
Proportion of Yield	:	16%	Roof-Floor Convergence Rate	:	0.164 mm/min
Highest Pressure	:	44	Observed Convergence Rate/metre Face Advance	:	34 mm/m
Proportion of Yield	:	100%			
Rate of Pressure Change			Rate of Debris Compaction		
Front Leg Circuit	:	0.064 MPa/min	Front Roof	:	0.061 mm/min
Middle Let Circuit	:	0.091 " "	Rear Roof	:	0.054 mm/min
Rear Leg Circuit	:	0.073 " "	Proportion of Convergence	:	70%
Total	:	0.233 " "			
Setting Load Density	:	0.210 MPa			
Yield Load Density	:	0.660 "			
Overall Mean Load Density	:	0.275 "			

Automatic data processing by computer has improved accuracy and speed of results analysis, presentation and interpretation. Data storage in computer files as well as on magnetic tape cassettes and results sheets allowed for rapid data re-call and transfer, as well as providing a data bank facility.

The system described in the Paper was designed to provide compatibility of modern methods of data acquisition, storage and processing. Laboratory and field trials have shown that it answers the requirements of a modern investigative technique at coalmines. Its robust, uncomplicated and intrinsically safe design provides a compact and portable system that may be applied to a range of mining environments to secure the basis for support system design, sound strata control, and improved coalface performance.

ACKNOWLEDGEMENTS

The authors gratefully acknowledge the research sponsorship of Gullick Dobson Ltd., Wigan and Microdata Ltd., Bristol.

Thanks are due to the Director and Staff of South Wales Area, N.C.B., for permission to investigate coalface behaviour at collieries of the Coalfield. Especial thanks are due to the General Manager and personnel at Betws Colliery.

Particular appreciation is expressed to H.M. Senior District Inspector of Mines and Quarries, South Wales, for advice and guidance towards the certification of the Data Logger, and for general encouragement in strata mechanics work.

Information in this Paper has resulted from the contributions of all members of the Strata Mechanics Research Group, and sincere thanks are extended to them.

REFERENCES

Ashwin, D.P. et al., 1970, "Some Fundamental Aspects of Face Powered Support Design", Mining Engineer, 129, 659-670.

Platt, J., Isaac, A.K., Smart, B.G.D., 1977, "Investigations of Design Criteria for Longwall Powered Support Systems", Proceedings 6th Int. Strata Control Conf. Banff, Paper 20.

Price, M.A., 1981 "The Prediction of Powered Support Behaviour on Longwall Coalfaces", Ph.D. Thesis, University of Wales, Cardiff.

Shepherd, R., 1964, "Study of Strata Control on Mechanised Coalfaces", Proc. 4th Int. Conf. on Strata Control, N. York, 230-247.

Smart, B.G.D., Isaac, A.K., Price, M.A., Thornton, J.D.., 1981, "An Intrinsically Safe Data Logger and Associated Transducers for Monitoring Displacement, Strain and Pressure", Proc. Int. Conf. British Soc. Strain Measurement, Edinburgh.

Thornton, J.D., 1981, "The Qualitative Assessment of Hydraulic Powered Support Behaviour on Longwall Coalfaces in Relation to Proximate Geology", Ph.D. Thesis, University of Wales, Cardiff.

Wilson, A.H., 1964, "Conclusions from Recent Strata Control Measurements made by the Mining Research Establishment", Mining Engineer, 43, 367-380.

MECHANISM FOR OUTBURSTS OF COAL AND GAS IN THE SOUTH WALES COALFIELD

S.J. GODDEN

Department of Mining Engineering
University of Newcastle upon Tyne
Newcastle upon Tyne, NE1 7RU, England.

SUMMARY

A hypothesis for the formation of outburst zones is described and is based on research carried out in the South Wales Coalfield. Outburst zones are seen as syntectonic features formed in response to the prevalent geological and mechanical conditions. A hypothesis for the spontaneous initiation of outbursts is described, based on conclusions from part one and various sytems for identification, prediction and control are briefly discussed.

INTRODUCTION

The phenomenon of outbursts; the explosive release of coal (mostly fine powder) and gas (or gases) from within a seam into a working area, is an old, dangerous and global problem. They have occurred in all types of excavation and over 14,000 events have been recorded from twelve countries. They vary in size from a few tonnes of extruded material to events of about 6000 tonnes and up to 5 x $10^5 m^3$ of evolved gas and vary in ferocity from gentle pushes of coal to events where structural damage was incurred or parts of a mine lost.

In this country the problem is largely restricted to the Gwendraeth Valley in the anthracite (west) part of the South Wales Coalfield. Since 1913, some 250 events have occurred (up to 1000 tonnes), mostly in two main seams (Big Vein and Pumpchwart) with the loss of 28 lives.

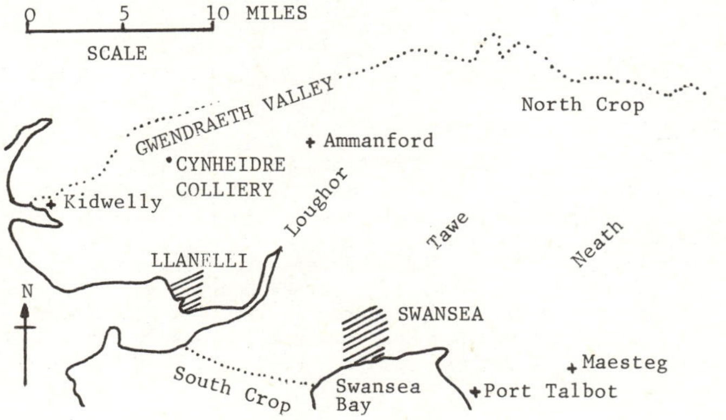

Figure 1. Location Plan

A considerable amount of research has been carried out worldwide, looking into problems of identification, prediction and control but with limited success. To the author's knowledge, no serious attempt has been made to collate and critically evaluate the wealth of information, then form a hypothesis to account for all stages of outburst formation and initiation: An understanding of outbursts would go a long way to resolving the associated problems.

Recent events at Cynheidre Colliery led to a special investigation into outbursts which form the basis of this paper. The investigation showed there to be certain characteristics common to outbursts which when combined lead to outburst formation, in-situ.

THE FORMATION OF OUTBURST ZONES

Tectonically Induced Shear Zones

Outbursts are always associated with tectonically induced shear zones and can form whole or part, or come from several isolated areas within such a zone. Shear zones are very common in the heavily disturbed semi-brittle to brittle anthracite (75% vitrain) seams. However, the fracture patterns can be different. Evans and Brown (1973) defined various coal groups according to their microstructure, namely:

a) Normal anthracite - typically hard, strong, clean and bright coal which breaks with perfect to sub-conchoidal fractures.

b) Abnormal anthracite - semi fiable, some shearing, typically showing conchoidal fractures. Some loss of bright lustre.

c) Outburst prone anthracite - soft, weak and highly sheared coal commonly displaying feather type fracturing. Dull with almost complete loss of recognisable structure.

d) Outburst type anthracite - very soft and weak, sometimes fine unconsolidated powder (30 x 20 x 10 microns in size). Dull with no recognisable structure.

Careful examination of coal sections and outburst locations revealed a progressive transition from one coal type to another towards outburst type coal with increased intensity of shearing. Barker-Read (1980) showed there is a complete difference in, and mode of formation of fractures associated with outburst type coal and those associated with the normal breakdown of anthracite. It is proposed that:

1) The formation of outburst type coal is an integral part of outburst formation.

2) It is generated by the breakdown of normal anthracite, by shear, in a compressive stress field.

and 3) The shear zones are syntectonic features (Hercynian) and formed late in the anthracitization of the coal.

There is no evidence to suggest a different mode of origin of the various coal types. Further reduction in particle size and additional shearing of outburst type and outburst prone coal, respectively, may occur during an event by the explosive release of strain energy and shock loading.

The size of a shear zone depends on the in-seam geological dimension of a disturbance. The nature of a disturbance observed in surrounding strata often belies its overall size in a seam and this can lead to a misinterpretation of the extent of shearing. The in-seam geological dimension is a function of the nature and mode of formation of the disturbance, the boundary conditions and seam thickness.

The Distribution of Events

The most meaningful classification is according to the type of associated structure:

Structure Type	% of Total Distribution
Folds	13%
Faults	35%
Overthrusts	5%
Low-angle in-seam disturbances	31%
Other causes	16%
Total	100%

Table 1: Distribution of Causes

This distribution is not suprising considering that the greatest number of disturbances are faults and low-angle in-seam structures. The section 'other causes' refers to those events where accurate data are so sparse that a proper classification is not possible.

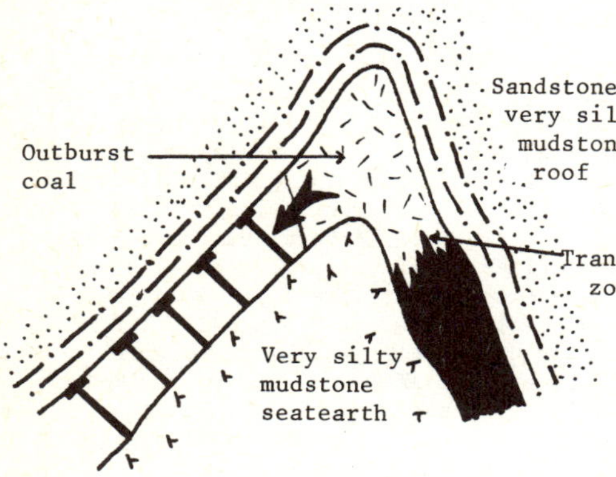

Figure 2 An Outburst Associated with a Fold. Type Example One. Pumpchwart Seam, Great Mountain Colliery 26/10/1925.

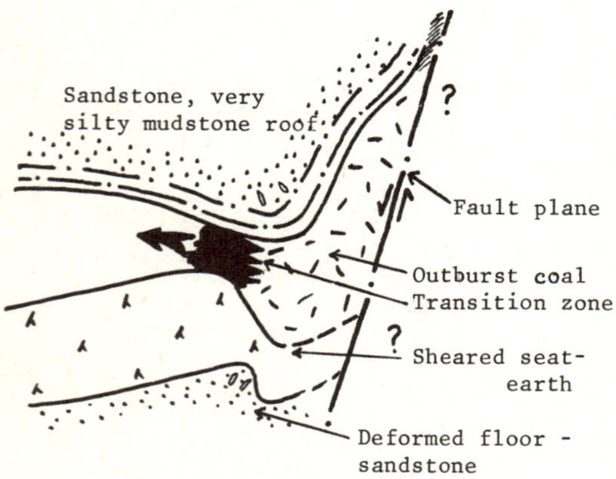

Figure 3 An Outburst Associated with a Normal Fault. Type Example Two. Big Vein, BV21 District, Cynheidre Colliery 25/3/1981.

Shear zones are always found where it is normally accepted that concentrations of stress and shear deformation occur (Figures 2 - 5). However, only a small proportion give rise to outbursts. If, as is sometimes suggested, mining induced stresses were solely responsible for outbursts, the inference is made that all shear zones would display some degree of outburst proneness. This is not the case. The feature that separates outburst zones is seen as the presence of exceptional and residual tectonic stress fields. Their importance cannot be over emphasised. It is considered that no plausible or valid hypothesis excluding such stresses at any stage can reasonably be brought forward.

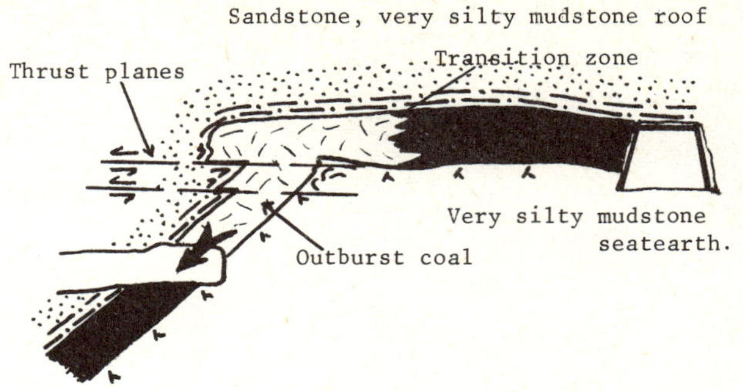

Figure 4 An Outburst Associated with a Confined Thrust. Type Example 3. Pumpchwart Seam. Pontyberem Colliery 11/10/1944.

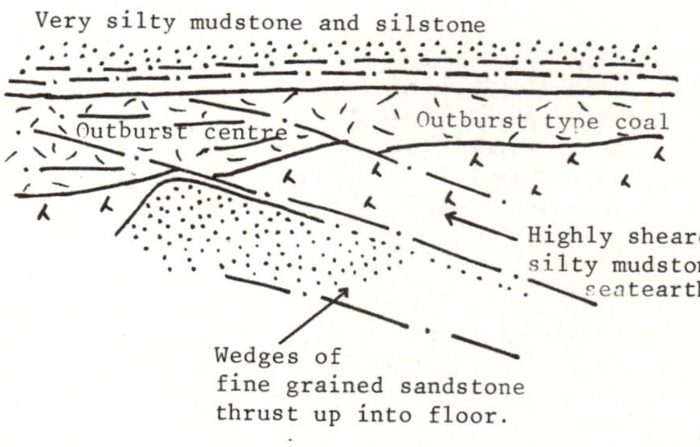

Figure 5 An Outburst Associated with a Confined In-seam Disturbance. Type Example 4. Big Vein BU99 District. Cynheidre Colliery 6/1/1981

Residual Tectonic Stresses

Although the stress/strain characteristics of outburst zones have not been directly measured; due mainly to problems inherent in weak and sheared coal, it is not difficult to infer the presence of exceptional stress conditions. The association of stress has been known for a long time (e.g. Farmer and Pooley, 1967; Evans and Brown, 1973; Hargraves, 1980) and Kidybinshi, (1980) has shown that there is a relationship between depth and the size of recorded events at Nowa Ruda Mine, Poland.

Exceptional stress conditions are characterised by:

a) A very marked increase in micro-seismic activity immediately prior to and during an event.
b) An increase in volume and decrease in density of coal samples taken directly from an outburst zone.
and c) Bad spalling and general relaxation of a coal face around an outburst zone.

Furthermore, exceptional stress conditions can be inferred from:

a) The geological, mechanical and structural environment.
b) The behaviour and nature of the outburst type coal and gas in-situ.

As is known, a geological disturbance will only continue to develop so long as the active stresses exceed the total confining stresses and/or the effective frictional forces in an active shear plane. If a disturbance becomes confined then any residual active stress will also become confined within the geometry of that disturbance. Residual (shear) stress zones are known to exist in the South Wales Coalfield (e.g. N.J. Price, 1975) and are manifested by increased rates and amounts of roadway deformation. Only a small percentage of such zones (when associated with a seam) give rise to outbursts. It is where exceptional levels of residual stress exist that outburst conditions prevail, the difference is in the nature and disposition of the boundary conditions.

The Boundary Conditions

The boundary conditions normally associated with the confinement of structures are the effective overburden pressure and the in-situ strength of the associated strata. This can be complicated by the effects of evacuating connate water during diagenesis and orogenic events, of which there is some evidence in South Wales but is not commented on further here.

The maximum thickness of coal measure strata in the area is about 2450 metres. N.J. Price (1975) notes that the total and initial overburden could have been in excess of 6,500 metres. In either case, the potential for confining active structures is clear. Furthermore, the tectonic forces necessary to propagate the complex, extensive and frequent disturbances, well known in this part of the coalfield, must also have been substantial. It is suggested that under these conditions it is possible for exceptional residual stress fields to form. However, at present mining depths of around 1,000 metres, it is inconceivable that exceptional residual stresses could be maintained. Regional relaxation of residual stresses would occur unless other factors were involved.

The investigation showed that strong strata in the seam environment had a marked influence on the proneness to outbursts as is best shown by the Big Vein Rock. This sandstone bed lies at a variable height 0 to 15 metres plus) above the Big Vein and varies in lithology across the take. In the west it is an off-white, structurally continuous orthoquartzite (>170 MPa, cone indentor test), progressively changing to a grey, argillaceous quartzite with discrete mica partings (approx 100 MPa, cone indentor test) in the east.

The variation in in-seam structural complexity, shear deformation characteristics and outburst potential of the Big Vein has been related to this sandstone unit and varies with:

1) the vertical proximity of the bed of the seam.
2) the thickness.
3) the structural continuity.
and 4) the mechanical behaviour of the bed.

That is, an increase in stiffness and in-situ strength of the associated strata results in an increase in proneness to outbursts. The presence of very strong seatearths (e.g. the Pumpchwart Seam) or tectonically emplaced sandstone beds associated with low angled, in-seam structures may also have an influence on outburst proneness.

The Effect of Stress on the Shear Zone

The effect of compressing sheared coal is to induce gas adsorption, decrease permeability, increase internal pore pressures and increase density. Adsorption of gas; the concentration of gas molecules near a solid surface by long range intermolecular forces, is a function of pressure, solid surface area and composition of the (seam) gas.

An increase in pressure results in an increase of adsorption to a maximum level which for Big Vein Anthracite is reached between 50-150 atmospheres (Griffin, 1978), where an assumed monomolecular structure is attained with a saturation level of 35cc/gramme. An increase in available surface area would result in higher saturation levels but Barker-Read (1980) has shown there is an insignificant difference between the porosity of the various coal types and is about 22·4m^2/gramme for Big Vein Coal.

The seam gas almost definitely represents that gas which could not escape during diagenesis and consists of:

Methane	CH_4	90%	
Carbon Dioxide	CO_2	4%	
Nitrogen	N_2	4%	
Oxygen	O_2	2%	(N.C.B.)

Carbon dioxide has a greater sorbtive capacity than methane (approx. twice) and in general carbon dioxide outbursts tend to be far more aggressive, a tendancy demonstrated in the laboratory by Famin (1959). The maintenance of saturated levels of adsorbed gas relies on an equivalent pressure in the free gas phase. Realistically, this can only be achieved with very low rates of permeability. Gunter (1965) showed that the permeability of outburst type anthracite reduces dramatically with little increase in pressure.

It is proposed that the imposition of exceptional compressive stress on a shear zone results in an outburst zone of finely sheared and relatively dense coal, very high internal gas pressures saturated adsorbed gas and a permeability which approximates to zero. If the conditions which initially gave rise to an outburst zone prevail and are not significantly altered, then there is no reason to suppose that the potential for outburst will change either, even in geological time.

THE SPONTANEOUS INITIATION OF OUTBURST

As a face advances towards an outburst zone, the retaining wall of coal (the distance between a face and outburst zone) is actively reduced. A stage will be reached where the combined effects of thickness and intact strength of this 'wall' will no longer be able to contain the exceptional strain energy and gas pressure in the outburst zone. An event is initiated by their simultaneous and explosive reslease, characterized by a sudden and dramatic increase in micro-seismic activity. This point need not necessarily be where the exceptional residual and mining induced stress fields combine. It is more a function of the size, nature and relative disposition of the two stress fields.

The release of energy probably results in further shearing of the coal combined with fracturing and dilation. The gas desorbs into these fractures and their combined effects results in the relaxation and dilation of the wall of coal. This may only take a few seconds and is what the miners describe as the coal 'shaking and heaving'. The coal relaxes and settles under gravity and the rapidly desorbing gas forces its way out into the working area resulting in 'jets of hissing gas'. When the desorbing gas can exert enough pressure to blow the face coal out, a mixture of coal and gas explodes into the working area. If the coal relaxes sufficiently, a continuous void is formed on top of the seam along which the outburst issues.

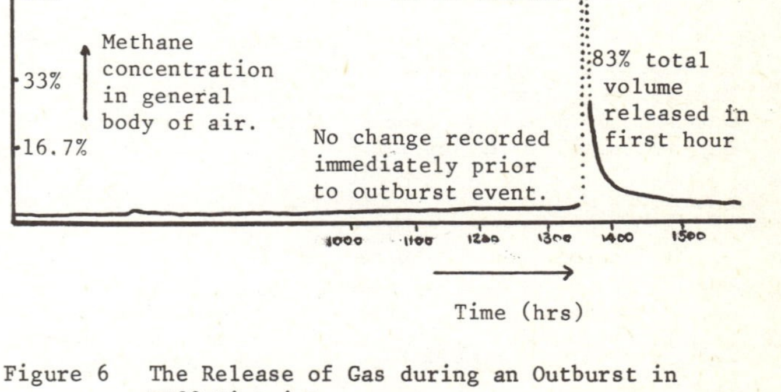

Figure 6 The Release of Gas during an Outburst in BV99 District.

The energy released by the gas mobilizes the coal as a fluidized system. At first this can be very violent and large lumps of coal and rock can be thrown out. With time this becomes progressively more passive (the shape of the curve depends on the desorption characteristics of the gas) until the energy released insufficient to continue the process, and the outburst ends. The outburst product always 'flows' in the direction of least resistance while fluidized, the accumulating coal product eventually forming a characteristic outburst pile.

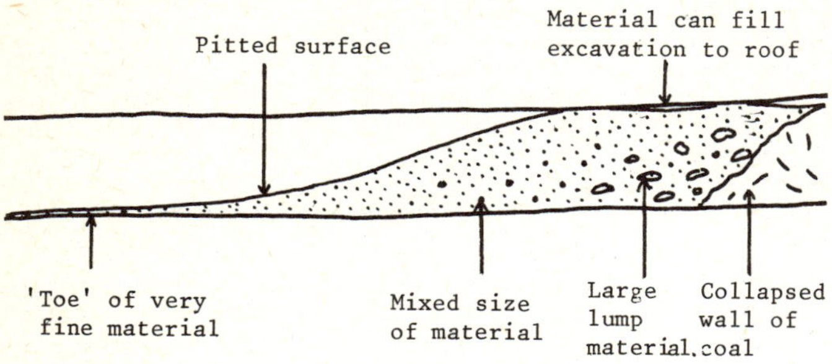

Figure 7 A Typical Outburst Pile.

Passive desorption can continue for long periods after an event, the gas bubbling up through the outburst pile resulting in a characteristic pitted surface. Relaxation of stress by shear deformation usually closes any cavity left during an event but this depends on the level of shear stress, shear strength of the surrounding coal or rock and the size of the event.

IDENTIFICATION, PREDICTION AND CONTROL

The most important part of any investigation concerning outbursts is to isolate methods for identification and prediction of outburst zones and to control their worst effects.

The structural environment and boundary conditions are the most readily recognisable features; given careful and detailed analysis. It is possible that the stress/strain characteristics may alter significantly with advance towards an outburst zone, but this needs to be investigated.

The explosive nature of an event is such that little or no warning, over any appreciable length of time, is given. For this reason, little future is seen in measuring any changes in physical or chemical conditions immediately prior to an event. However, an early warning system is highly desireable. An analysis of slip/strain and dilation characteristics of coal faces is normal and exceptional circumstances may yield results.

The only effective methods of remotely controlling outburst potential are degassation by boreholes or destressing by under or overmining an outburst prone zone. Such methods have had some success elsewhere in the world (Hargraves, 1980) with careful planning and design. However, practical mining considerations can preclude their use.

Volley firing techniques have been used extensively for short term control, with varying degrees of success. An adequate volley firing system which simmulates spontaneous initiation should induce outbursts in safe and controlled conditions.

ACKNOWLEDGEMENTS:

The author would like to thank Mr R.H. Price, retired Head of Geological Services and Mr C.J. Davies, Chief Mining Engineer, N.C.B. South Wales, whose assistance made this paper possible. Any views expressed are those of the author and are not necessarily those of the National Coal Board.

REFERENCES:

BARKER-READ, G.R., 1980, "The Geology and Related Aspects of Coal and Gas Outbursts in the Gwendraeth Valley", MSc Thesis, University of Wales.

EVANS, H., and BROWN, K.M., 1973, "Coal Structures in Outbursts of Coal and Firedamp Conditions", Mining Engineer, 148, 171-179.

FAMIN, L.B., 1959, "Instantaneous Outbursts of Coal and Gas in a Laboratory Experiment". Proc. Conf. Problems of Mine Ventilation, Min. Inst. Acad. Sci. U.S.S.R. 219-224, Cerchar Trans. 471-60.

FARMER, I.W. and POOLEY, F.D., 1967, "A Hypothesis to Explain the Occurrence of Outbursts of Coal, based on a study of West Wales Outburst Coal", Int. J. Rock Mech. 4, 189-193.

Geol. Survey of Great Britain. "Geology of the Gwendraeth Valley and adjoining areas". H.M.S.O.

GODDEN, S.J., 1981 "A Special Geological Report on the Problems of Outbursts of Coal and Gas" N.C.B. Internal Report.

GUNTER, J., 1965, "Etude de la liaison Gaz-Charban". Revue de L'Industrie Minerale, 47/9, 693-708.

HARGRAVES, A.J., 1980, "A Review of Instantaneous Outburst Data". Proc. Aus. IMM Symp. The Prediction and and Control of Outbursts in Coal Mines.

KIDYBINSKI, A., 1980, "Significance of in-situ Strength Measurements for Prediction of Outburst Hazard in Coal Mines in Lower Silesia". Proc. Aus. I.M.M. Symp. The Prediction and Control of Outbursts in Coal Mines.

MORGAN, G.D., 1981, "Gas Emission in Outburst Conditions", N.C.B. Internal Report.

PRICE, N.J., 1975, "Fault and Joint Development in Brittle and Semi Brittle Rock", 2nd Edition, London.

VOIGHT, C., Editor, "Thrust Faults and Decollement", Benchmark Papers in Geology, Vol. 32.

LAY-OUT AND OPERATION OF MINE WORKINGS WITHIN A ROCK MASS PRONE TO ROCK BURSTS

Ing. Bohumil Zamarski, CSc.,

Scientific Coal Research Institute, Ostrava-Radvanice,

Prof. Ing. Lubomir Siska, DrSc.,

Corresponding Member of Czechoslovak Academy of Sciences,

Ing. Petr Konecny, CSc.,

Institute of Geology and Geotechnics CSAV, Ostrava, Czechoslovakia.

SUMMARY. The paper describes the rock burst prevention system applied in the complex geological and mining conditions which are characteristic of the Czechoslovak part of the Upper Silesian Coal Basin which is the most important coal deposit in Czechoslovakia.

A rock burst may be defined as a "sudden failure of the rock mass with resultant release of elastic energy and its transformation into work which leads to sudden transfer or ejection of structural parts of the rock mass into a mine working so that the latter ceases to fulfull its functions from operational and safety viewpoints. Gaseous and liquid components of the rock mass do not participate in this sudden transfer of energy".

The failed structural part of the rock mass which is ejected into the mine opening is on most occasions the coal seam. In certain circumstances the rocks in the overlying or underlying strata may also be affected. Figure 1 illustrates a typical example of the destruction of a mine working as a result of rock burst.

Figure 1. Section of a longwall face following a rock burst.

The mechanism of a rock burst may be described in the following way. The "primary state of stress" which exists within the virgin rock mass may be characterized as a vertical principal stress:

$$\sigma_z = \gamma_o \cdot H \qquad (1)$$

and horizontal principal stresses:

$$\sigma_x = \sigma_y = \frac{\sigma_z}{m - 1} \qquad (2)$$

where:

γ_o is the density of rocks

H is the depth below the surface

m is the Poisson Number.

Depending on the strength of the rock mass and prevailing stresses, two eventualities can occur in the vicinity of the respective mine working as schematically presented in Figure 2.

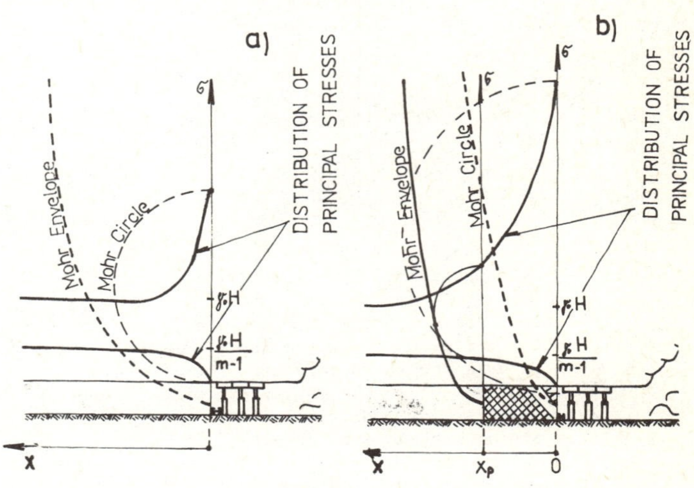

Figure 2. Principal scheme of conditions from which failure of coal and/or rocks near the mine opening may occur.

The left side of Figure 2 represents a longwall face where the resultant stresses do not exceed the strength of the adjoining rock mass. Occurence of a rock burst is, therefore, not possible. The right side of the figure represents a case where the resultant stresses exceed of the strength of the adjoining rock mass. If the properties of the coal seam and adjoining rocks are such as to allow sufficiently large amounts of energy to be released during subsequent failure, conditions for rockburst initiation will occur.

In order to prevent rockbursts, the following steps, based on the principal conditions of occurrence, may be considered:

1. The stress in the vicinity of mine workings must be prevented from exceeding the strength of the respective part of the rock mass.

2. The properties of the rock mass around mine workings may be altered with the objective of preventing elastic energy release during failure.

3. A "zone of protection" may be created around the mine workings.

When practical situations are considered it is recommended that the possibilities offered by the first approach are used to the maximum since experience indicates that they are usually cost advantageous.

Generally, the zones of stress illustrated in Figure 3 may be encountered in a mined rock mass:

- a zone of original stress which can include zones of favourable or unfavourable residual stresses.

- a zone affected by the existence of mine workings which can include zones of reduced (favourable) or additional stresses.

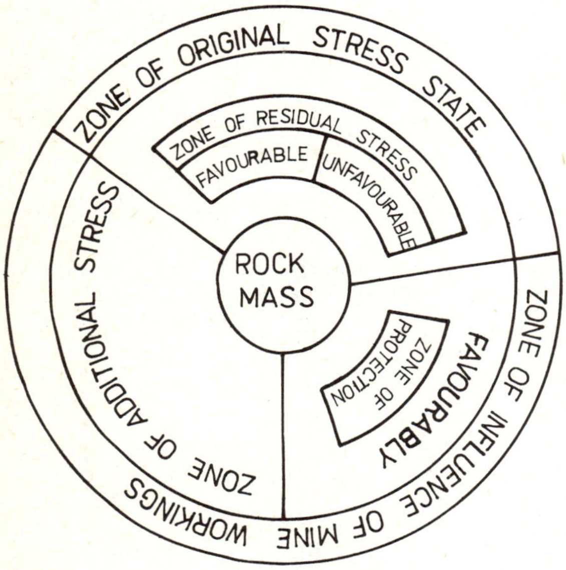

Figure 3. Demarcation of the rock mass into zones from viewpoints of resultant stress.

The favourably affected zone can comprise a "zone of protection", i.e. a zone where the stress does not reach values high enough to release a rock burst. In conditions prevailing in the Ostrava-Karviná Coal Field, this value corresponds approximately to the level of geostatic pressure at a depth of 300m.

The first approach to the prevention of rock bursts requires that mine workings be concentrated in areas where favourable residual stresses prevail, e.g. close to tectonic faults, and in areas favourably affected by the presence of other mine workings. Mine workings should not be planned in areas where additional stresses exist.

The use of this approach in the complex geological and operational conditions characteristic of the Ostrava-Karviná Coal Field requires careful planning of lay-outs of mine workings both in space and time and this must be done at an early stage. An optimisation scheme for lay-out planning is illustrated in Figure 4. This system designed for optimization of operation of mine workings both in time and space comprises analytical elements for the evaluation of existing stress situations in the deposit from the viewpoint of geological and operational conditions. It also synthesises elements of practical programs for mining activities in areas at the boundary between two adjacent mines. An appropriate design for the operation of individual longwall faces, drivage of access roadways and preparation roadways in time and space are suggested. Mine workings are classified according to the level of risk of rock burst occurence and operational programmes - including other measures for the rock burst prevention system - are prepared.

ANALYTICAL PART

1. Regional prediction of rock burst occurence; demarcation of seams prone to rock bursts.
2. Demarcation of mined, unmined, and unminable parts of deposits.
3. Assessment of possibilities for mining of protected seams.
4. Determination of optimum sequence of working for individual seams.
5. Demarcation of zones affected by previous and future operation of mine workings.
6. Assessment of the situation near the boundary between two adjacent mines.

SYNTHETICAL PART

1. Solution of situation near the boundary between the takes of adjoining mines.
2. Lay-out of longwalls both in time and space.
3. Lay-out of access and preparation roadways both in time and space.
4. Classification of mine workings according to the level of the rock burst risk.
5. Proposals of technologies for the operation of mine workings.

Figure 4. Optimization scheme for the operation of mine workings in space and time.

Several partial solutions can be obtained by means of computers. Particular advantages for the definition of individual zones within a rock mass, as illustrated in Figure 3, are offered by methods of mathematical modelling. Special programs developed for these purposes allow increased flexibility of calculation.

Figure 5 illustrates the result of one simple and practical calculation. In this case, it was necessary to define the zones of additional stresses around an unmined pillar 50m wide at a depth of approximately 600m. Calculations indicated that unfavourable effects would be expected at a vertical distance of 200-250m in the overlying strata. Figure 5 also indicates where beneficial effects of protection might be expected. It should be noted that the beneficial effects of mining out a protective seam are time dependent. In conditions prevailing in the Ostrava-Karviná Coal Field, the effect is lost after about five years.

The second principal approach to prevention of rock burst occurence is through alteration of properties of the rock mass so as to prevent the release of large amounts of elastic energy during failure. The principal method used to alter the properties of Carboniferous rocks in the Ostrava-Karviná Coal Field is water infusion. Water infusion is currently applied by means of high pressure pumps with working pressure approx. 30 MPa. In addition to this, long term water infusion is used at normal pressures available in the mine water mains.

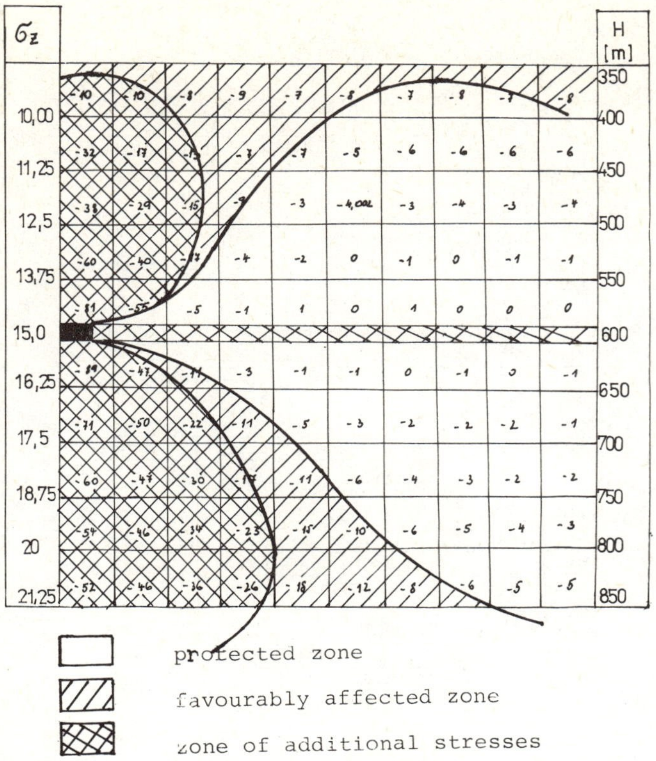

Figure 5. Example of calculation of zones near an unmined pillar.

Figure 6 shows an example of the influence of water infusion on the Mohr Envelope in siltstone.

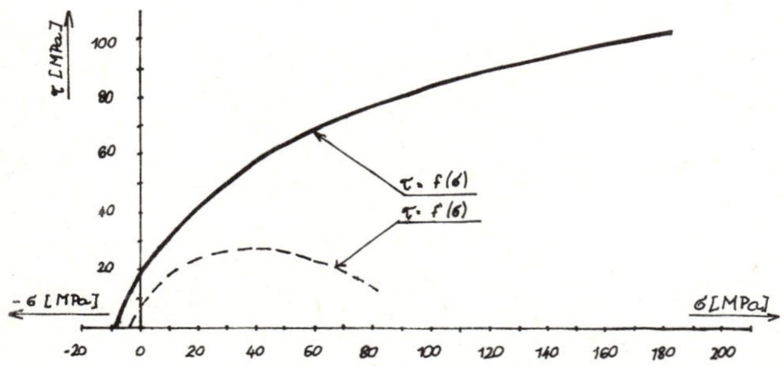

Figure 6. Strength of rock affected by water.

Effects of the water infusion were confirmed on numerous occasions as illustrated in Figure 7, which shows the increase of convergence in a roadway caused by water infusion which was carried out in order to protect against imminent rock burst risk. It should be emphasised that water infusion has not always given satisfactory results since some rocks are not affected by pressurized water. This approach to the rock burst prevention can therefore be applied only after thorough laboratory tests.

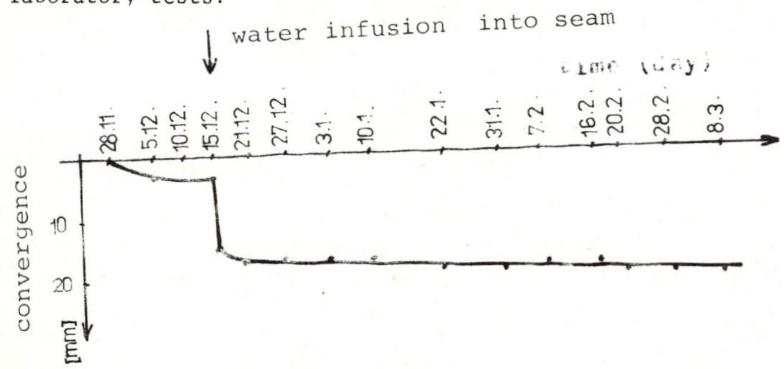

Figure 7. Convergence in a roadway due to water infusion.

The third approach to rock burst prevention is through formation of a protective zone around mine workings. This is explained on Figure 8.

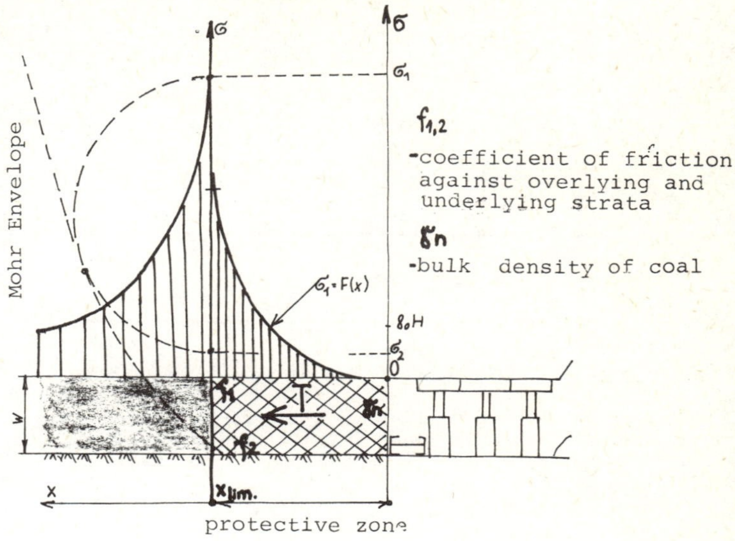

Figure 8. Calculation of the equilibrium state of stress around a mine working due to the effect of the protective zone.

The protective zone, represented on the illustration by hatching, comprises artificially failed rock. The required width of the protective zone can be deduced from the relation for the definition of the resistance against the movement into the working

$$T = f\left[\left(2\int_0^{x_{lim}} F(x)\,dx + \gamma_n \cdot w \cdot x_{lim}\right)\right] \qquad (3)$$

where the symbols are explained in Figure 8. For appropriate routine definition of the width of the protective zone in practical situations underground, a nomograph has been developed, the use of which is obligatory in the Ostrava-Karviná Coal Field.

The means most often used for the preparation of the protective zone are special shotfiring techniques and destressing drilling.

Figure 9 shows a practical example of a drilling pattern used for volley shotfiring and Figure 10 shows a pattern for destressing drilling. The routine application of both methods is usual in the Ostrava-Karviná Coal Field at present.

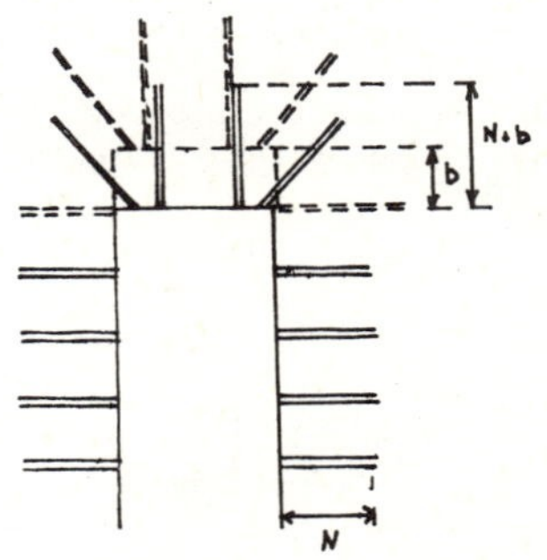

Figure 9. Drilling scheme for destressing shotfiring in a heading.

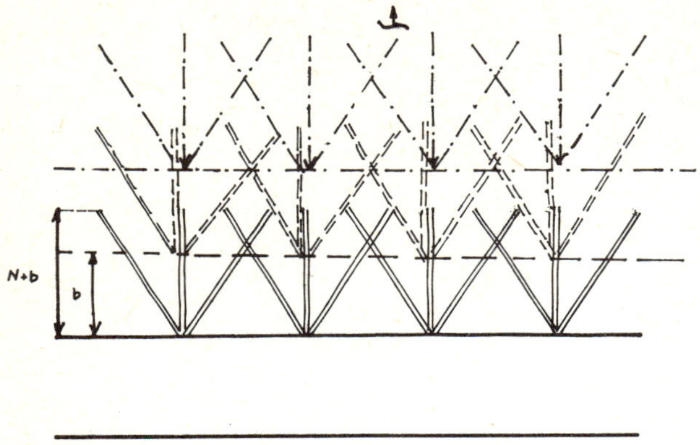

Figure 10. Scheme for destressing drilling on face.

Both the shotfiring and the destressing drilling are applied either ahead of actual mining faces or as an inherent part of the mining cycle. It should be emphasized that both of these methods are relatively expensive but they may nevertheless be necessary to secure the required level of safety at work.

In the Ostrava-Karviná Coal Field, the rock burst risk is the highest in Czechoslovakia. Difficulties encountered in the prevention of rock bursts result both from complex geological conditions and from the fact that the deposit is exploited intensively. The lay-out and operation of mine workings in seams prone to rock bursts is, therefore, carried out with particular care. As result of this, research into related problems is carried out at almost every scientific and research institution.

Rock burst prevention in the Ostrava-Karviná Coal Field is controlled and enforced through safety regulations and "Instructions" issued from the Scientific Coal Research Institute at Ostrava-Radvanice. The application of the latest knowledge from the scientific and research organizations underground is the task of a special division of the General Direction of the Ostrava-Karviná Mines. At individual collieries, a position of Mine Geomechanic has been established to carry out both the prediction and prevention of rock bursts. Complicated questions related to rock burst prevention are decided by the Committee for the Rock Burst Prevention, of the General Direction of the Ostrava-Karviná Coal Mines, which includes the best Czechoslovak experts in the field.

Due to the attention given to this particular mining problem, the rock burst risk has been reduced. There remain however, several areas which require further investigation. These include the comparatively high cost of rock burst prevention. In the future, the main objective will be to achieve improved effectiveness of the necessary preventative measures. In addition the question of the resistance of mine supports to dynamic loadings resulting from rock bursts has not been answered satisfactorily. This question is particularly delicate when longwall supports are involved because dynamic loading has caused severe damage on several occasions.

REFERENCES

Dopita M., 1968, "Geological and mining problems of working the saddle seams in the Ostrava-Karviná Coal Field", Přírodovědecký sborník, 1968, Ostrava.

Konečný P., 1978, "Simple method to determine the zones of increased stress in the overlying and underlying strata of unmined pillars and its application in the Ostrava-Karviná Coal Field", Proc. of the Conf. Nové poznatky V geomechanice, Tatr.Lomnice.

Straube R. et al, 1972, "Rock bursts in the Carboniferous ", SNTL Praha.

Šiška L., 1967, "Die Konvergenz als ein Mass für Gebirgsschlaggefahr/Convergence as a measure for the rock burst risk", Glückauf-Forschungshefte, 1967, Nr. 2.

Šiška L., Zamarski B., 1972, "Problems related to rock bursts in the Ostrava-Karviná Coal Field", Proc. Conf.: Nejnovější poznatky v geomechanics, Košice 1972.

Vavro M., 1967, "Analyses of the natural conditions of the rock burst occurence in the Ostrava Karviná Coal Field", Annual Report VŠB Ostrava, 1967, No 3 .

Zamarski B., 1971, "Contribution to the theory of the rock burst occurence and principles of their prevention", Uhlí, 1971, no. 8.

Zamarski B., Franěk J., 1977, "Theoretical principles, organization, and practice of ground control in subsurface mines in Czechoslovakia", Proc. 6th Internat. Conf. on Strata Control, Banff, Canada.

Zamarski B., 1973, "Theory of the occurence of rock bursts and principles of their prevention", Proc. Conf.: "Hornická Ostrava", DT GŘ OKD Ostrava.

DEVELOPMENT OF ROOF CONTROL CRITERIA FOR LONGWALL MINING - PARAMETRIC MODELLING

by

Syd S. Peng, Professor and Chairman

and

S. M. Hsiung, Graduate Student

Department of Mining Engineering
College of Mineral and Energy Resources
West Virginia University
Morgantown, WV 26506
U.S.A.

SUMMARY. Three-dimensional finite element analyses were performed to assess the effects of physical and mechanical properties of the immediate roof before and after longwall mining begins, the minimum principal stress distributions in the roof over the gobs as mining progresses, and the first caving and periodic weighting. Comparison of the first caving and periodic weighting lengths under various types of roof conditions are presented. The structural stability of three special types of model are compared.

INTRODUCTION

The adoption of longwall mining in the U.S. has been slow during the past two decades. Apart from its high capital investment, the major reasons that the industry is reluctant to accept the longwall system are the following unanswered questions: what is the longwallability of a coal seam, e.g. under what conditions the longwall system is suitable and if so, how to determine the panel layouts? What will be the optimum size of the chain pillars? When will the first caving occur and what effect will it have on the front and side abutments? Does the periodic weighting really exist and how will it affect the face area? And will the tailgate remain stable for the next panel?

The work presented here is the first phase of a parametric structural analyses of a three-entry retreating longwall system. It was carried out by three-dimensional modellings using NASTRAN program which was developed by NASA (McCormick, 1977). The major objectives of this study are; (1) to evaluate the effect of Young's modulus, Possion's ratio, thickness of the immediate roof before and after panel mining begins; (2) to estimate the effects of caving under various types of the immediate and main roofs; (3) to determine the length of the first caving and periodic weighting; and (4) to evaluate the stability of chain pillars and overall structure. Throughout the analyses, the artificial support was excluded for simplification since it has a negligible effect on the front abutment pressure (Peng, 1980a).

FINITE ELEMENT MODELS

Since the sequential development of a longwall mining results in a three-dimensional structure, three dimensional models rather than the popularly used two-dimensional models were employed. Two typical three-entry longwall panels and their corresponding finite element mesh layouts are shown in Figs. 1-3. Figs. 1 and 2 were used for modelling before and during the beginning of the panel extraction while Fig. 3 was used to explore the effects of periodic weighting when panel mining has been well advanced.

Each finite element mesh is laid out such that its edges coincide with the pillar edges and bedding planes except for the area of major concern. All the crosscuts in the coal seam 90m away from the area of major concern were neglected (Peng 1980b) to avoid creating an excessive number of elements. Because of symmetry, only a quarter of the longwall panel was used in both models. Hexahedron elements with double precision were used and the total number of nodal points were 2160 in Fig. 2 and 2520 in Fig. 3.

The width and length of the panel used in the models were 150m and 1200m, respectively. Both the widths of entries and crosscuts were 6m. The size of the chain and bleeder pillars were 30m x 30m. The height of coal seam was 2.4m and the model overburden depth was 236m. The model overburden consisted of four strata while the floor was made of two strata. The geologic sequences are shown in Fig. 4. An uniform pressure of 375 kg/cm² was applied to the top surface of the model to represent the remaining overburden weight. The mechancial behavior of each stratum was assumed to be isotropic, homogeneous and perfectly elastic. Their physical properties are given in Table 1. The physical properties of the immediate roof, and the thickness of the main roof for various models are listed in Table 2.

VERTICAL STRESS DISTRIBUTION BEFORE AND AFTER LONGWALL MINING BEGINS

The average pillar stresses under a certain depth, h, can be calculated, before longwall mining begins, as follows:

$$\sigma_v = c_1 \cdot c_2 \cdot h$$

where c_1 is the unit stress for a square prism 2.54cm wide by 30cm long, for which the upper bound value is 0.4 kg/cm². While c_2 is the ratio of the supported roof to that of the cross-sectional area of the pillar. As longwall mining begins, the increase of the exposed open area or the caving of the roof makes it difficult to find a simple equation to estimate the stress distribution in the pillars or around the face area. Besides the stress distributions in the pillars and front abutment are highly influenced by the physical properties of the roof. The effects of those parameters are discussed as follows:

Effect of Young's Modulus of the Immediate Roof

The effect of the Young's modulus of the immediate roof on the vertical stress of the pillars is illustrated in Fig. 5. Since the overall trend is the same, only one element in the chain and bleeder pillars is illustrated. R is the ratio of the Young's modulus for the immediate roof to that for the coal. As R increases, the vertical stress in both pillars decreases. The decrease levels off when R exceeds 10 for element A and 6 for the bleeder pillars.

The vertical stress in element D is the average stress and its concentration factor with respect to the average overburden stress is within the range of 1.35-1.4 while the stress concentration factor for element A ranges from 1.45 to 1.7. As R exceeds 6, the stress concentration factor for element A approaches that of the upper bound for element D. That is, no high stress concentration exists when the Young's modulus of the immediate roof is large. Since a larger Young's modulus of the immediate roof induces a much less stress concentration in the pillar, the pillar will be stabler which in turn improves the stability of the overall structure.

As longwall mining begins and if no roof caving occurs, the smaller is the Young's modulus of the immediate roof the larger is the vertical stress in the chain pillar. But for the bleeder pillars it is different. Fig. 6 shows that shortly after 6m of extraction, a larger Young's modulus of the immediate roof causes a higher vertical stress in the bleeder pillars.

Effect of Thickness of the Immediate Roof

Fig. 7 shows the effect of the thickness of the immediate roof on the vertical stress in the pillars before the longwall mining begins. Model B1, B2 and B3 were used (Table 2). It indicates that the thickness of the immediate roof has some effect on the vertical stress in element A and C at low values of roof thickness. The increase of vertical stress with the thickness of the immediate roof levels off when the thickness is larger than 12m. Comparison between Fig. 5 and Fig. 7 shows that the effect of the thickness is much smaller than that of the Young's modulus. Thus a weak but thick immediate roof will be the worst case.

On the other hand, the increase of vertical stress in element D due to increase in thickness of the immediate roof is larger than that due to the decrease of the Young's modulus and the increase remains well beyond the range shown in the figure.

As a whole, the pressure responses under different roof conditions are not the same for bleeder and chain pillars. Firstly under the same R, stress concentration in the chain pillar will level off as the thickness of the immediate roof increases, but not in the bleeder pillars, and secondly, a larger R induces a smaller stress concentration in the chain pillar before/when panel mining begins, but the trend in the bleeder pillar is just the opposite.

Figs. 8 and 9 show the effect of Young's modulus of the immediate roof on front abutment pressure when the immediate roofs were 6m and 12m thick, respectively. As indicated, the front abutment pressure decreases with increase in Young's modulus in both cases. But the rate of decrease is much larger for thicker immediate roof.

CAVING OVER THE GOB AREA

As the longwall face advances away from the bleeder pillars, the gob area and thus the exposed roof increases. When the span of the exposed roof becomes sufficiently large and satisfies some sort of failure criteria, the immediate roof begins to collapse. Thereafter, the magnitude of the front abutment depends very much on the behavior of the main roof. The main roof overlying immediately above the immediate roof is the first major strong stratum that with the weight of the overlying strata, can self-support for a certain roof span measured between the rear and front abutment supports. The front abutment support is the solid coal in front of the faceline while the rear abutment support is however not fixed. As the roof caving reaches a sufficient height and allows the caved fragments to be sufficiently compacted, it will then start to support the unbroken overlying roof and act as the rear abutment support. The pressure on the front abutment is therefore cyclic, the period of which depends on the maximum span the main roof is capable of maintaining. It follows that the main roof may become the immediate roof and vice versa if the immediate roof is strong and self-supporting.

It is therefore important to know the causes and characteristics of caving, the effect of caving under different roof conditions, the first caving length of the immediate roof and its effect, the period of the periodic weighting and its effect, and the location of the face when the caved gob starts to act.

Failure Criterion

Although the process of rock failure is very complex, the following failure criteria proposed by Paul (1961) was used for practical purpose.

$$\sigma_1 = C_0 + q\,\sigma_3 \quad \text{when } \sigma_1 > C_0[1 - C_0T_0/4S_0{}^2] \qquad (2)$$

$$\sigma_3 = -T_0 \quad \text{when } \sigma_1 < C_0[1 - C_0T_0/4S_0{}^2] \qquad (3)$$

where σ_1 and σ_3 are the maximum and minimum principal stresses, respectively. S_0 is cohesive strength, C_0 is uniaxial compressive strength, T_0 is tensile strength and q is triaxial stress factor (Wilson, 1977). The stress distribution in each finite element in each computer run is checked against Eqs. 2 and 3 for failure. Table 3 lists various parameters for each stratum as required by Eqs. 2 and 3. The strength values were obtained by reducing five times from the laboratory measured values.

Minimum Principal Stress Distributions in the Roof Over the Gob

Throughout the analyses on the first caving, the failure of roof rock elements is tensile strength dominated regardless of the locations and types of models. In this aspect, contour figures of the minimum principal stress distribution in the roof over the gob area can offer a good idea as to the location and extent of the potential areas of caving.

Fig. 10 shows two minimum principal stress distributions in the roof over the gob areas before the first caving occurs. Model A1 and D1 were used. Basically, tensile stress zone occurs at all cases and higher tensile stress concentrations were induced by a larger Young's modulus of the roof. Furthermore, the maximum tensile stress shifts gradually from the center of the exposed roof to the edge as the Young's modulus of the roof increases.

Fig. 11 shows the variations of the minimum principle stress as the face advances. Model A1, B2, C1 and D1 were used. For each model two curves are shown: One for tensile stress in the exposed roof near the edge of the chain pillar and the other in the center of the panel roof. For model A1, the highest tensile stress is always in the center area of the exposed roof. One can roughly determine the most critical length at which the failure of the immediate roof initates provided the tensile strength is known. For example, if the tensile strength of the roof for model A1 is 10 kg/cm², the approximate critical length is 21m, which is measured from the edge of the bleeder pillar to the faceline. The same procedure can be applied to model B2. The conditions for model C1 and D1 are more complex. Tensile stress near the edges of the exposed roof will exceed that at the center when the face advances 33m in model C1. If the tensile strength is less than 48 kg/cm², failure of the roof will start at the center which will then generate large roof caving. If the roof fails first near the chain pillar, it can only add to the tensile stress in the nearby roof and reduces slightly the approximate length of caving. The same thing is true for model D1 except the magnitudes are different.

Fig. 12 illustrates the variations of the minimum principal stress due to face advance when the immediate roof is 12m thick. Model C2 and E1 were used. In contrast to the cases described previously, tensile stress zone occurs around the edges of the exposed roof as a result of high shear stress. The tensile stress with immediate roof of 12m thick is smaller than that with immediate roof of 6m thick (Figs. 11 and 12). Thus a longer distance is needed to induce and generate roof caving for thicker roof. Comparison between model C1 (Fig. 11) and model C2 (Fig. 12) by using an assigned tensile strength of 17 kg/cm² shows that the roof span for model E1 to induce the first roof caving is around 24m while it will be at least 60m for model C2 depending on the extent of local roof failure along the chain pillar. The location of the maximum tnesile stress in the roof is at the face area for model C2 while it occurs near the bleeder pillars for model E1.

In summary, the first roof caving induced by the face moving away from the bleeder pillars can be roughly estimated once the characteristics of the tensile stress development and tensile strength of the roof rock are known.

Length of First Caving

First caving is referred to the very first large area of roof caving after panel mining begins. Thus the first caving length is the distance between the face and the edge of the bleeder pillars when the first caving occurs. First caving length can be used as an index for evaluating the immediate roof and to some extent for selecting the face powered support because it is related to the physical and mechanical properties of the immediate roof.

Table 4 lists the lengths of the first caving and the projected lengths that roof caving initiates for six different models. The projected lengths that the roof caving initiated were estimated from Figs. 11 and 12. The difference between the projected and the first caving length is less than 6m which represents the delay distance that the longwall face needs to move ahead in order to cause the first caving. Table 4 also shows that the first caving length is highly influenced by the Poisson's ratio of the immediate roof (see Model E1 and E2).

Length of Periodic Weighting

Length of periodic weighting is defined as the maximum span the roof is capable of self-supporting. Therefore it is the distance between two adjacent rapid drops of the front abutment pressure.

Table 5 lists the length of periodic weighting. Seven cases were examined. For model A2 and A4, periodic weighting simply does not exist. Because the roof constantly fractures ahead of the faceline and cave immediately after it is exposed. Furthermore, because of the bulking effect, the caved fragments can fill up the open area and starts to act as a rear abutment support. Therefore, the thick and strong main roof in model A4 does not have any effect on the periodic weighting. The periodic lengths for model C2 and C4 are about the same. Once again, the thick and strong main roof in model C4 does not show any effect on the periodic weighting. The geologic conditions for model A3 and C3 are basically similar except the Young's modulus of the immediate roof for model C3 is four times that for model A3. Yet the periodic lengths for those two models are essentially the same.

In summary, the immediate roof for model C2, C4 and E1 does serve as a main roof with various periods as controlled by the Young's modulus. Comparisons between the periodic and first caving length for the corresponding models show that the first caving length is always longer than periodic length.

Effect of Caving on the Chain and Bleeder Pillars and the Front Abutment

Figs. 13 and 14 show the variations of pressures in side abutment and bleeder pillars, respectively, for Model E1. Pressure in the chain pillar drops sharply when local roof failure occurs along the edge of the chain pillar while pressure in the bleeder pillars shows a slight increase. When the first caving occurs, pressures in both chain and bleeder pillars drops rapidly and considerably.

Figs. 15 and 16 also show the variations of side abutment and bleeder pillars pressures, respectively for Model E2. They indicate that the pressure drops sharply in element A and both bleeder pillars when the first caving occurs while a slight pressure increase is induced for element C. As the face continues to advance, the pressure in bleeder pillars and element A builds up gradually while element C experiences a sharp jump of pressure. As the face advances further, the cantilevered roof with the fixed end on the front abutment caves and causes a sharp release of pressure for element C. But it has a negligible effect on the stress condition in the bleeder pillars.

Fig. 17 shows the variations of side abutment pressure as the face advances for model E1. As the longwall face moves from one end of pillar A to the center, it brings about a large increase in side abutment pressure for pillar A and B. As the longwall face approaches the other end of pillar A, the side abutment pressure of pillar A experiences a stress redistribution while pillar B has a considerable amount of pressure decrease due to entry roof failure behind the longwall face, though the effect is restricted to pillar A and B only. As the longwall face moves away from pillar A and reaches the location for first periodic weighting, the side abutment pressure maintains the increasing trend and shows no sign of being affected.

Fig. 18 shows the variation of the side abutment pressure as the face advances for model A3. As longwall face moves from one end of pillar A to the center of pillar E, the side abutment pressure increases constantly independent of the periodic weighting. The sharp decrease in the pressure of pillar D which is two pillars behind the longwall face is purely the effect of the roof failure immediately above the abutment.

Comparison of Fig. 17 and Fig. 18 indicates that model A3 with relatively thin and weak immediate roof creates a much higher side abutment pressure than that of model E1 with thick and strong immediate roof.

Fig. 19 shows the variations of the front abutment pressure as the face advances for model E1. Each curve in the figure represents the front abutment pressures at four different locations. The front abutment pressure steadily increases as the periodic weighting approaches. It drops sharply at the end of the periodic weighting when the roof caves. The total pressure increases ranged from 340 to 610 kg/cm^2. Fig. 20 shows the variations of the front abutment pressure as the face increase advances for model A3. It also shows the pressure increase during the period of periodic weighting. The normal front abutment pressure for model E1 is essentially lower than that for model A3. Although the peak pressure for both cases is about the same, model E1 experiences a much higher pressure variation.

STRUCTURAL STABILITY OF LONGWALL PANEL

The structural stabilities of the entry system and the face are the major concerns for panel design. The stability of overall structural would invariably contribute to higher production rates. To determine the optimum size of chain pillars, width of entry and panel in order to acquire better performance and optimum recovery has long been a challenge for the coal mining industry. This cannot be solved without incorporating the geologic conditions with mine layouts. In this section, three cases which result in relatively unstable conditions will be discussed.

Figs. 21 and 22 illustrate the maximum principal stress distributions in the pillars when the longwall face is 7.5m and 22.5m into a pillar for model A2, respectively. The shaded areas shown in both figures represent the failed portion of the pillar and the longwall panel. As the face advances, the failed area in the panel extends consistently outby for a distance around 15m. Similarly the failed areas in the pillars also grow in the direction of face advance. Those failed regions together with the broken roof in the face area constitute the longwall stability problems. Moreover, the failed zones in the pillars also grow toward the center of the pillars. This will contribute to the instability of the 2nd entry.

Table 6 lists the percentage of failed zone in the gob-side pillar and the length of failed zone in the longwall panel. It indicates that the conditions for model C4 with Young's modulus of the immediate roof four times that of the model A2 and A4 are much stable, especially in the face area.

CONCLUSIONS

The length of the first caving is usually longer than that of the periodic weighting, especially for the thick but weak roof. Therefore, the impact of the first caving is more intimately tied to the stability of the working area. Accordingly, the prediction of the first caving length is necessary for selecting a proper type of powered support. By knowing the tensile strength of the roof and its characteristics of the tensile stress induced during face advance, the length at which roof failure initiates can be estimated.

The effect of periodic weighting is found to be restricted to the longwall face. It has a negligible effect on the side abutment.

Although a thick and strong immediate roof can serve as a main roof which brings about relatively high pressure in the front abutment, it induces considerably less pressure in the chain pillars. Therefore, the longwall panel is more stable from the overall structure point of view, though a stronger and well designed powered support will be required. A thick but very weak immediate roof causes failure in a large portion of the pillars on the gob side and in the solid coal in front of the faceline, both of which tend to create less stable overall structure. Under such condition, it will be necessary to increase the pillar size.

REFERENCES

McCormick, C. W. (ed.), 1977, "The NASTRAN User's Manual (Level 17.5)", National Aeronautics and Space Administration, Washington, DC. Sec 2.1-1-2.7-33.

Paul, B., 1961, "Modification of the Coulomb-Mohr Theory of Fracture", J. Appl. Mech., p. 259-268.

Peng, S. S., Matsuki, K. and Su, W. H., 1980a, "3-D Structure Analysis of Longwall Panels", Proc. 21st U.S. Sym. of Rock Mech. University of Missouri, Rolla, MO. p. 44-56.

Peng, S. S. and Su, W. H., 1980b, "3-D Structural Analysis of Retreating Longwall Panel", Proc. of 4th Joint Meeting of MMIJ/AIME, Tokyo, Japan. Vol. T.S. B-4, p. 1-16.

Unrug, K. F. and Szwilski, A. B., 1980, "Influence of Strata Control Parameters on Longwall Mining Design", Proc. 21st U.S. Sym. of Rock Mech., University of Missouri, Rolla, MO. p. 285-298.

Wilson, A. H., 1977, "The Effect of Yield Zones on the Control of Ground", 6th Int'l. Strata Control Conference. Banff, Canada.

Table 1
Physical Properties of the Strata

Material	Young's Modulus (kg/m^2)	Poisson's Ratio	Density (kg/m^2)
Coal	17	0.20	2.28×10^{-6}
Sandstone	170	0.17	3.69×10^{-6}
Shale	68	0.25	3.83×10^{-6}
Clay Shale	92	0.30	3.69×10^{-6}

Table 2
Physical Properties of the Immediate and Main Roofs for Various Models Employed

Model No.	Young's Modulus (kg/m^2)		Poisson's Ratio		Thickness (m)	
	Immed. Roof	Main Roof	Immed. Roof	Main Roof	Immed. Roof	Main Roof
A1	17	170	0.25	0.17	6	3
A2	17	170	0.25	0.17	12	3
A3	17	170	0.25	0.17	3	12
A4	17	170	0.25	0.17	12	12
B1	34	170	0.25	0.17	3	3
B2	34	170	0.25	0.17	6	3
B3	34	170	0.25	0.17	12	3
C1	68	170	0.25	0.17	6	3
C2	68	170	0.25	0.17	12	3
C3	68	170	0.25	0.17	3	12
C4	68	170	0.25	0.17	12	12
D1	102	170	0.25	0.17	6	3
E1	170	170	0.25	0.17	12	3
E2	170	170	0.2	0.17	12	3

Table 3
Uniaxial Compressive Strength, Uniaxial Tensile Strength and Triaxial Stress Factor for Each Stratum

Strata	Young's Modulus (kg/m^2)	Uniaxial Compressive Strength (kg/cm^2)	Uniaxial Tensile Strength (kg/cm^2)	Triaxial Stress Factor
Coal	17	307	11.6	4
Shale	68	1009	15.7	3.8
Sandstone	170	1232	54.5	5

Table 4
Length of First Caving

Model No.	First Caving Length (m)	Projected Length for Initiation of Roof Caving (m)	Remark
A1	28.5	≃ 24.5	$\sigma_t = 0$
A1	36	≃ 30.5	$\sigma_t = 11.6\ kg/cm^2$
C1	28.5	≃ 24	$\sigma_t = 15.7\ kg/cm^2$
E1	42	≃ 40.5	$\sigma_t = 54.5\ kg/m^2$, $\upsilon = 0.25$
E2	28.5	–	$\sigma_t = 54.5\ kg/cm^2$, $\upsilon = 0.2$
C2	> 36	> 60	$\sigma_t = 15.7\ kg/cm^2$

Table 5
Length of Periodic Weighting

Model No.	Periodic Weighting Length (m)	Model No.	Periodic Weighting Length (m)
A2	no	C3	36
A3	36	C4	15
A4	no	E1	36
C2	15-21		

Table 6
Percentage of Failed Zone in Pillar on the Gob Side and Length of Failed Zone from the Face

Model No.	Failed Zone in the Pillar	Length of Failed Zone from the Face
A2	50%	15 m
A4	50%	7.5 m
C4	25%	NONE

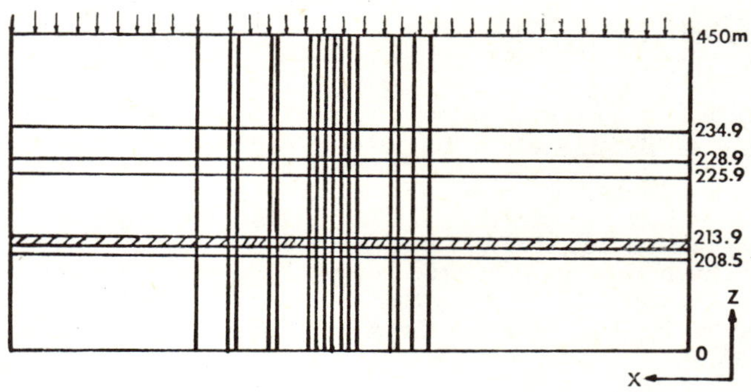

Fig. 1. FINITE ELEMENT MODEL, CROSS SECTIONAL VIEW

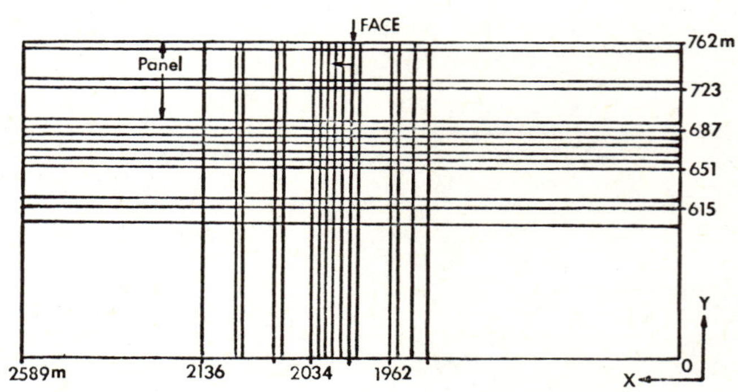

Fig. 2. FINITE ELEMENT MODEL, PLANE VIEW

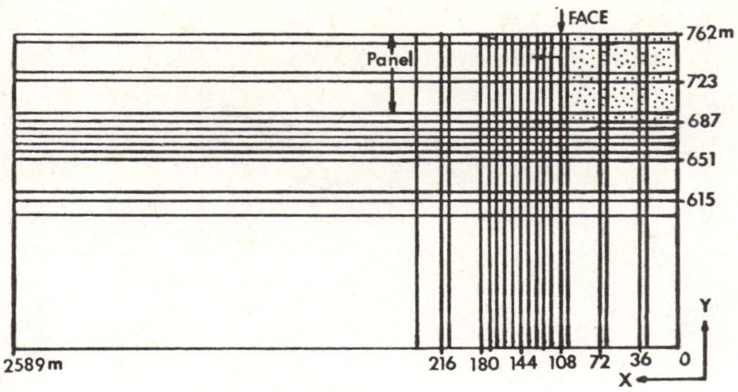

Fig. 3. FINITE ELEMENT MODEL AWAY FROM THE BLEEDER PILLAR, PLANE VIEW

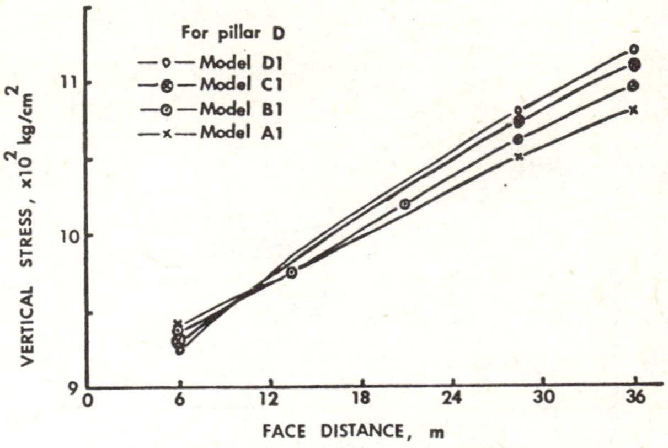

Fig. 6. VERTICAL STRESS IN THE BLEEDER PILLAR AS THE FACE ADVANCES

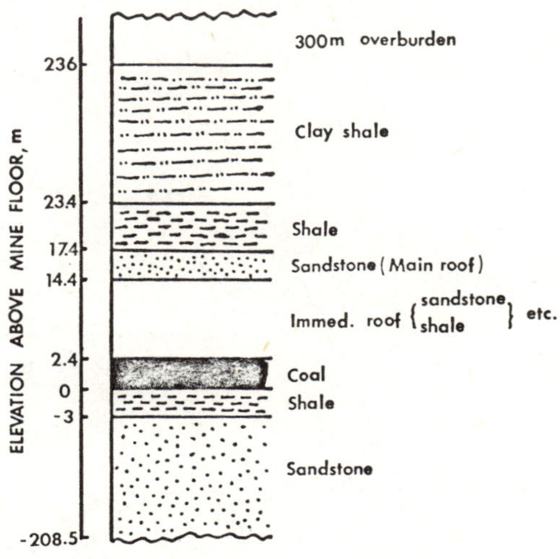

Fig. 4. IDEALIZED STRATIGRAPHIC SEQUENCES

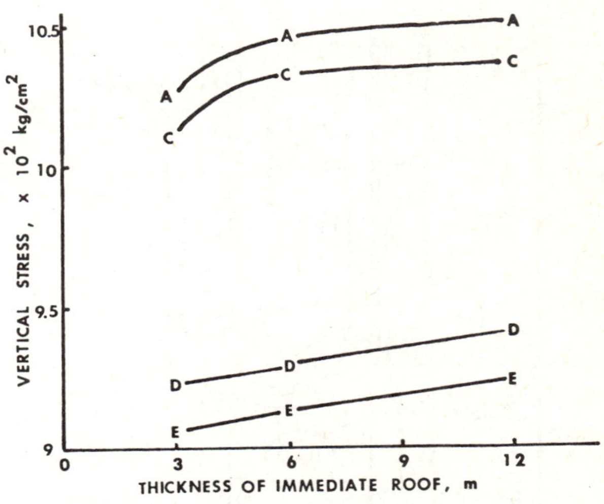

Fig. 7. EFFECT OF THE THICKNESS OF IMMEDIATE ROOF ON VERTICAL STRESS IN THE PILLARS BEFORE MINING BEGINS

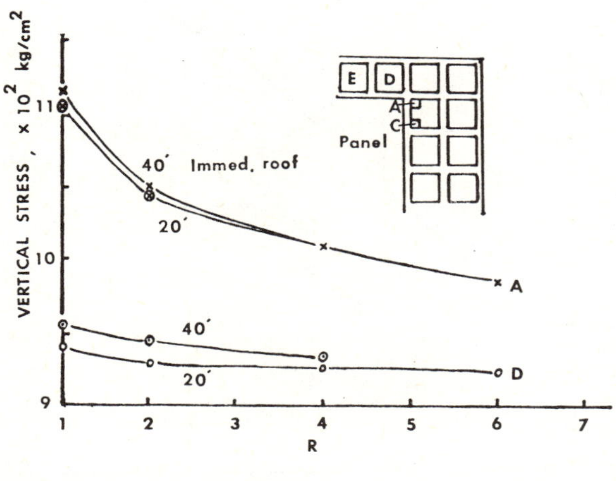

Fig. 5. EFFECT OF YOUNG'S MODULUS OF IMMEDIATE ROOF ON VERTICAL STRESS IN THE PILLARS BEFORE MINING BEGINS

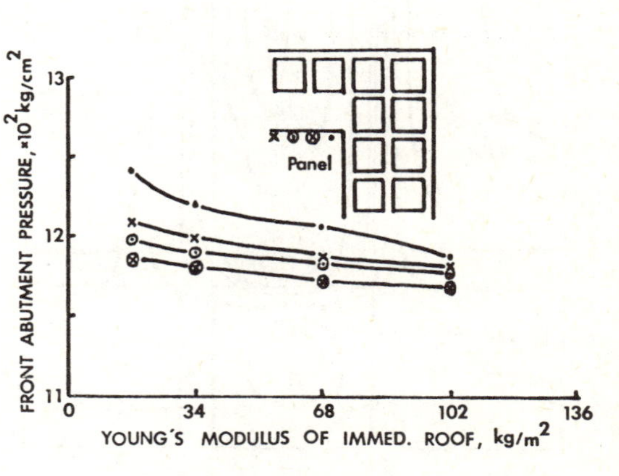

Fig. 8. EFFECT OF YOUNG'S MODULUS ON FRONT ABUTMENT WITH 6m THICK OF IMMEDIATE ROOF

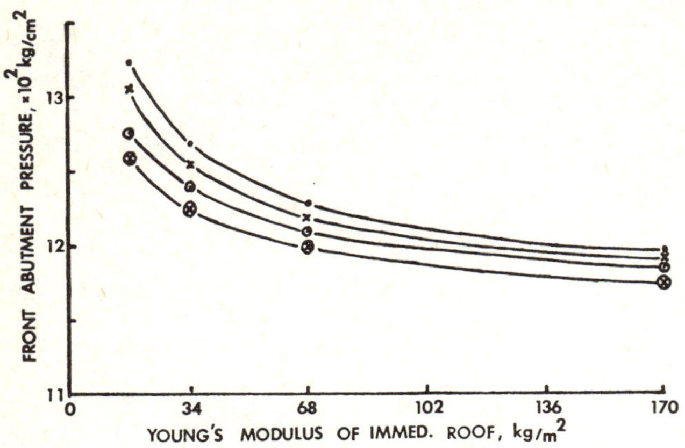

Fig. 9. EFFECT OF YOUNG'S MODULUS ON FRONT ABUTMENT WITH 12m THICK OF IMMEDIATE ROOF

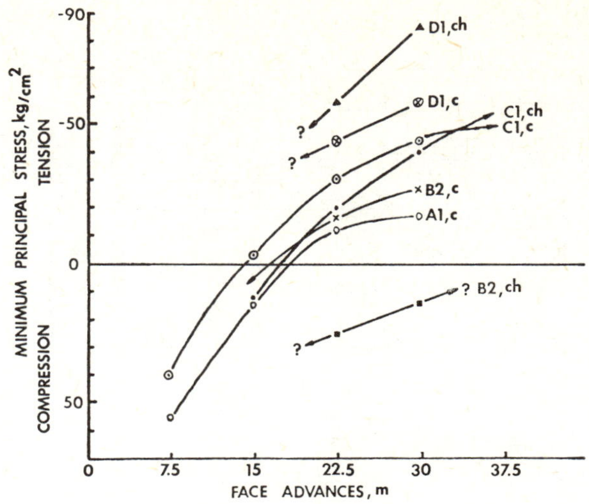

Fig. 11. VARIATIONS OF MINIMUM PRINCIPAL STRESS FOR MODEL A1, B2, C1 AND D1 AS THE FACE ADVANCES

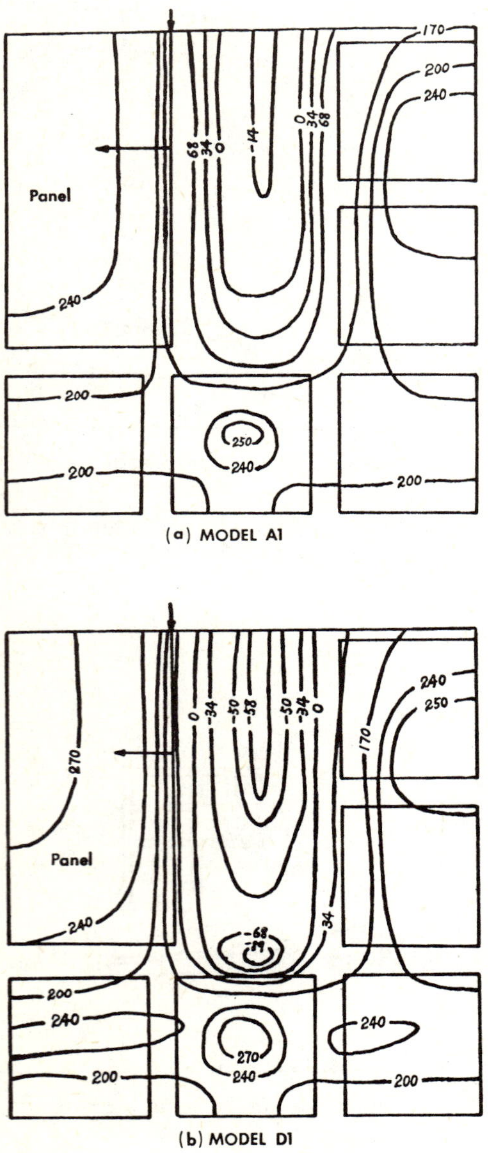

Fig. 10. MINIMUM PRINCIPAL STRESS DISTRIBUTIONS IN THE ROOF OVER THE GOB BEFORE THE FIRST CAVING OCCURS (kg/cm²)

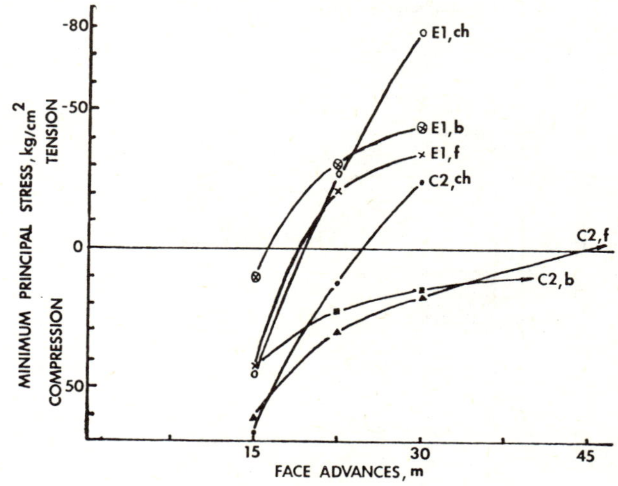

Fig. 12. VARIATIONS OF MINIMUM PRINCIPAL STRESS FOR MODEL C2 AND E1 AS FACE ADVANCES

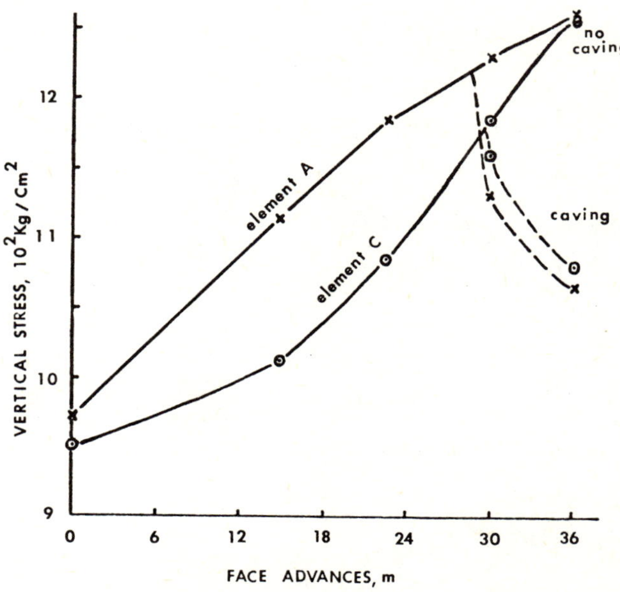

Fig. 13. VARIATIONS OF SIDE ABUTMENT PRESSURE FOR MODEL E1 AS THE FACE ADVANCES

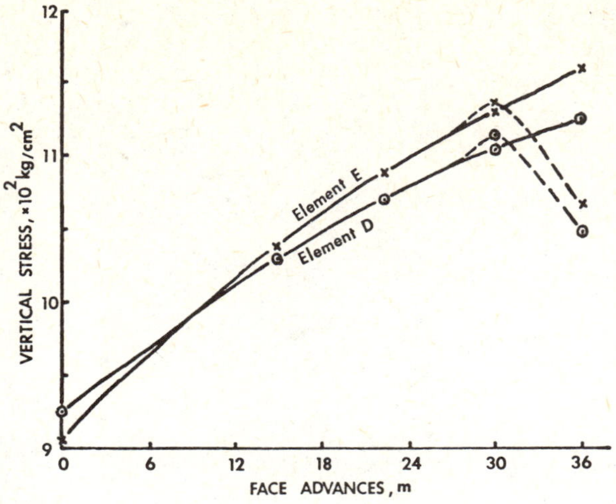

Fig. 14. VARIATIONS OF BLEEDER PILLAR PRESSURE FOR MODEL E1 AS THE FACE ADVANCES

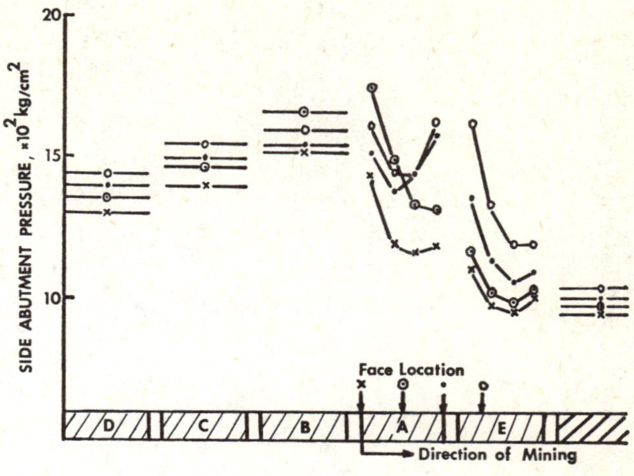

Fig. 17. VARIATIONS OF SIDE ABUTMENT PRESSURE FOR MODEL E1 AS THE FACE ADVANCES

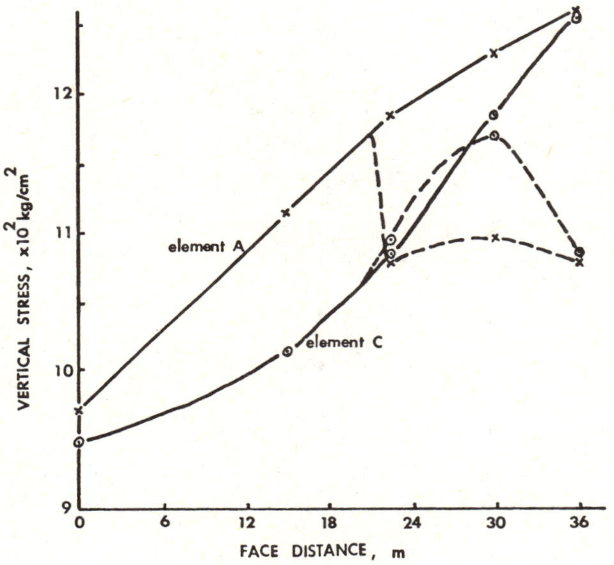

Fig. 15. VARIATIONS OF SIDE ABUTMENT PRESSURE FOR MODEL E2 AS THE FACE ADVANCES

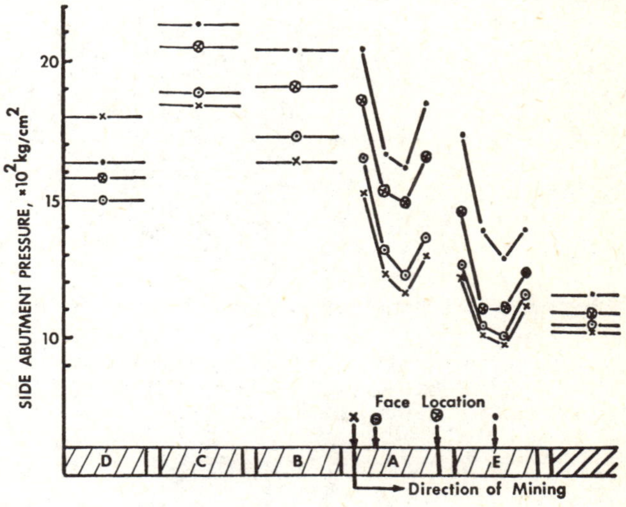

Fig. 18. VARIATIONS OF SIDE ABUTMENT PRESSURE FOR MODEL A3 AS THE FACE ADVANCES

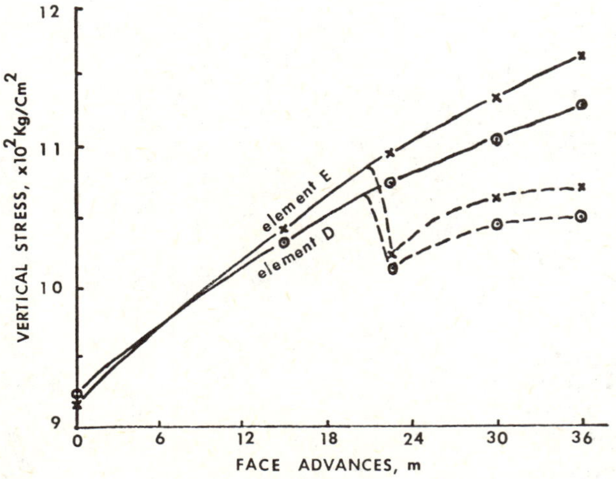

Fig. 16. VARIATIONS OF BLEEDER PILLAR PRESSURE FOR MODEL E2 AS THE FACE ADVANCES

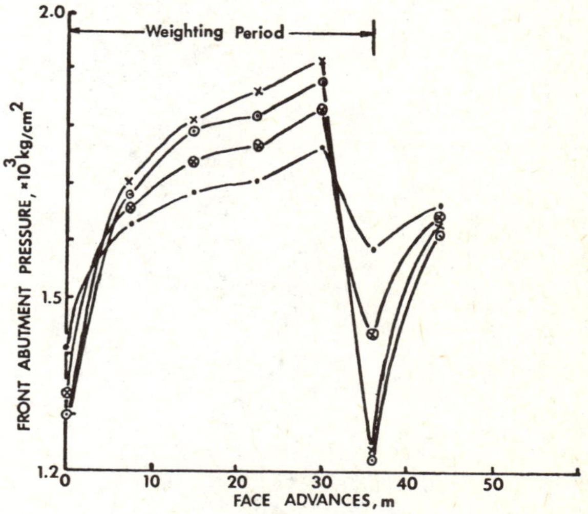

Fig. 19. VARIATIONS OF FRONT ABUTMENT PRESSURE FOR MODEL E1 AS THE FACE ADVANCES

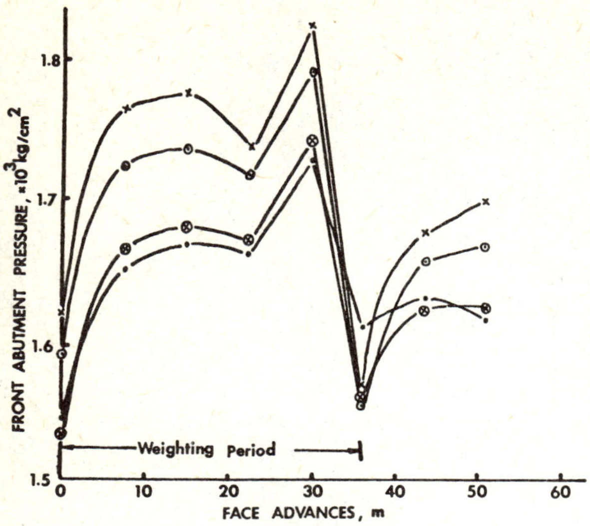

Fig. 20. VARIATIONS OF FRONT ABUTMENT PRESSURE FOR MODEL A3 AS THE FACE ADVANCES

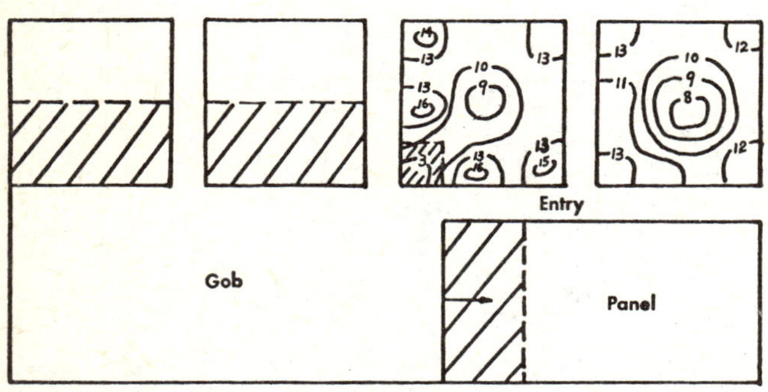

Fig. 21. MAXIMUM PRINCIPAL STRESS DISTRIBUTION (kg/cm²) IN THE PILLARS WHEN THE LONGWALL FACE IS 7.5m INTO A PILLAR FOR MODEL A2

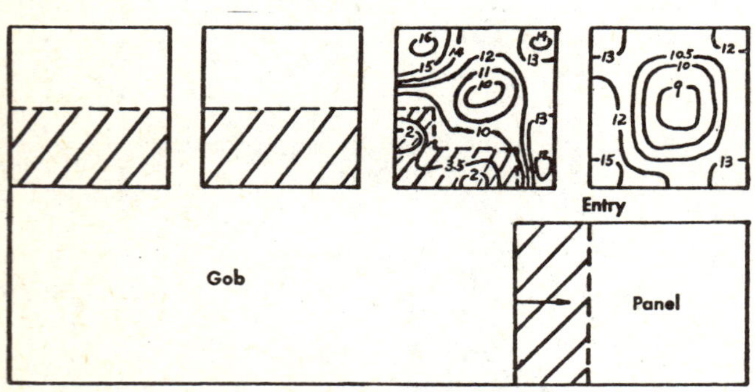

Fig. 22. MAXIMUM PRINCIPAL STRESS DISTRIBUTION (kg/cm²) IN THE PILLARS WHEN THE FACE IS 22.5m INTO A PILLAR FOR MODEL A2

SOME OBSERVATIONS ON THE USE OF STOCHASTIC METHODS FOR THE PREDICTION OF SUBSIDENCE OVER LONGWALL PANELS

N. H. GLOSSOP

Department of Mining Engineering
University of Newcastle upon Tyne
Newcastle upon Tyne NE1 7RU., U.K.

SUMMARY

The paper describes briefly the stochastic method of subsidence prediction. Volume and trough width parameters are derived from the Subsidence Engineers Handbook to fit this model and thus enable a direct comparison to be made between the Handbook's empirical troughs and those predicted stochastically. This comparison indicates that the stochastic method does not accurately predict the shape of the settlement trough.

INTRODUCTION

Several authors (Litwinitzyn, 1964; Sweet and Bogdanoff, 1965) have discussed the use of stochastic modelling for the prediction of subsidence over openings of various shapes in various types of ground. An adaptation of this method has been used with some success to predict the ground movements caused by tunnelling in cohesive soils (Glossop, 1978; Glossop et al., 1979), and computer programs have been written by the author to handle numerically the problem of differently shaped excavations. Clearly there are some similarities in principle between subsidence over tunnels and subsidence over mine excavations and it is reasonable to hope that stochastic modelling may be successful in the latter case, provided that suitable coefficients for the constituent equations can be found.

Currently the Subsidence Engineers Handbook (1975) is widely used for the estimation of ground movements caused by longwall face coal extraction. It contains several tables and nomograms derived from a large number of empirical observations of subsidence troughs in the United Kingdom. It therefore provides a suitable base for the assessment of any alternative method of subsidence prediction.

This paper briefly summarises the principles of the stochastic method in this context, and compares the results of the application of this method with the predictions of the Subsidence Engineers Handbook for a specific case. From this comparison, some conclusions are drawn about the realism of the stochastic model and about the volumetric changes which would appear to take place during ground subsidence.

The Stochastic Model

For many years the prediction of subsidence caused by mining excavations has been carried out on an empirical basis, usually with the aid of the Subsidence Engineers Handbook. This method has proved quite reliable but has the limitation that it is difficult to apply to seams or excavations of complicated geometry. Clearly, there is a need for a better theoretical understanding of the subsidence phenomenon and, in particular, for a more general method for subsidence prediction, applicable to all geometries and all types of ground. Stochastic modelling addresses the latter problem.

The basic stochastic theory, as applied to problems of ground subsidence was originally discussed by Litwinitzyn in a series of papers from 1955 onwards and has been developed by Baoshen et al. (1979) but the papers by Sweet and Bogdanoff (1965) and Sweet (1965) present the theory in its most graphical form. A stochastic medium is one whose behaviour is governed, at least in part, by probabilistic rather than deterministic laws. Sweet and Bogdanoff consider the ground to be made up of an array of rock or soil particles whose movements are controlled by their size, the geometry of their packing, and the laws of probability, under the influence of the gravitational field. If one of these particles is removed, then an adjacent, higher particle will fall down to fill the resulting void, a process which will recur until the void reaches the surface, resulting in an element of subsidence. The subsidence process can thus be visualised as the migration of a large number of elemental voids from the source of ground loss to the surface, the path of each migration being analogous to a one-dimensional random walk, with the vertical space dimension in place of time. This process will, of course, take place in three dimensions but, for the modelling of the fully developed subsidence over a longwall face, it is adequate to consider the simpler two-dimensional case.

The shape of the resulting subsidence trough is the same as the probability distribution of the voids' arrival at the surface and is an inverted Gaussian distribution, described by the equation:

$$S_x = \frac{V_s}{\sqrt{2\pi}\, i} \exp\left(\frac{-x^2}{2i^2}\right) \qquad \ldots\ldots (1)$$

where S = vertical settlement at distance x from centreline.

V_s = total settlement volume.

i = horizontal distance to point of inflection of trough.

It is necessary, therefore, to estimate the settlement volume and the position of the inflection point in order to calculate subsidence from this equation. The stochastic model as described assumes that all voids ultimately reach the surface, resulting in zero volumetric strain. This assumption appears reasonable for shallow, soft ground tunnels, where volume changes can be calculated separately, but for mining subsidence it will be necessary to estimate the expected ground dilation.

Sweet and Bogdanoff show how the point of inflection of the settlement curve can be related to mean size and variance of the particle size, as well as the packing or angle of slope of the particle array. It may be possible to equate these to certain aspects of the rock jointing pattern. In cohesive ground the point of inflection can be shown empirically to be equal to half the depth to the source of ground loss. A similar relationship may be derived from the Subsidence Engineers Handbook, as described in the following section. This relationship can be considered to be a basic ground property, although it is unclear how, or if, it would relate to any other geotechnical properties of the rock in question.

It is possible to calculate the tilt (T) and curvation (C) of the ground directly from equation 2 by differentiation:

$$T = \partial S/\partial x = (-4x/z^2)\, S \quad \ldots\ldots (2)$$

$$C \approx \partial^2 S/\partial x^2 = (4x^2/z^2 - 1)(4/z^2)\, S \quad \ldots\ldots (3)$$

Horizontal strain (ε_h) and horizontal displacement (S_h) are dependent upon the amount of volumetric strain experienced by the ground. If there is no volumetric strain then horizontal strain is equal and opposite to vertical strain ($\varepsilon_h = -\varepsilon_z$ for plane strain) and:

$$\varepsilon_h = -\varepsilon_z = -\partial S/\partial z = (1 - 4x^2/z^2)S/z \ldots\ldots (4)$$

$$S_h = \int_{-\infty}^{x} \varepsilon_h \, dx = (x/z)\, S \quad \ldots\ldots (5)$$

The above equations (1) to (5) apply to point sources of ground loss in two dimensions and can be considered to be 'source functions'. To estimate the settlement parameters due to a real excavation, these source functions must be integrated over the extent of the excavation. This is most easily accomplished numerically by computer. (Effectively, the computer sums the source functions for a large number of point sources evenly distributed through the excavation.) In this way a tabular source of ground loss equivalent to a caved longwall excavation can be simulated.

It should be noted that the above arguments can be translated into three dimensions with minimal modification if required, although this will result in a major increase both in computer storage and computing time required. In the case of fully developed subsidence over a reasonably long panel the two dimensional model is adequate.

The estimation of V_s and i

As discussed in the previous section, in order to use the stochastic model for subsidence prediction, it is necessary to find the value of two parameters, the volume of the subsidence trough (V_s) and the position of the point of inflection (i). It seems likely that both of these parameters will be related to the depth of the excavation and it should be possible to estimate the relationships from the Subsidence Engineers Handbook.

Prediction of the Settlement Trough Volume

It is quite straightforward to measure the volume of the fully developed subsidence trough for various values of depth-to-width ratio (h/w) by plotting the subsidence profiles using the nomograms presented on pages 9 and 13 of the Subsidence Engineers Handbook. For a hypothetical seam of width 200m and seam thickness 1m, we can construct Figure 1 relating the settlement volume, normalised with respect to the total seam volume (V_s/V_t), to the depth-to-width ratio. The curve reaches a maximum V_s/V_t of 0.9, showing a maximum settlement volume of 90% seam volume (for caved or strip-packed workings). This probably indicates a 10% bulking factor in the caved seam itself. The volume decreases rapidly as h/w increases, with an apparent decrease in gradient at h/w = 1.6, indicating ground dilation. The final point on the

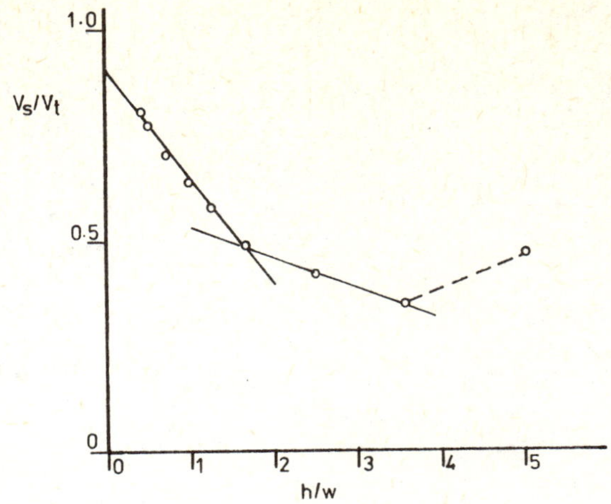

Figure 1 Ground dilation

graph, at h/w = 5 would seem to show a rise in volume. It is very difficult to account for this volume increase. It occurs at the extreme end of the range of the nomograms where settlements are relatively small and difficult to measure. For h/w ratio of less than 5 it would appear that the curve may be described as two straight lines described by the following equations:

$$V_s/V_t = 0.9 - 0.25\, h/w \quad (0 < h/w < 1.6) \quad \ldots\ldots (6)$$

$$V_s/V_t = 0.61 - 0.073\, h/w \quad (1.6 < h/w < 3.6) \quad \ldots\ldots (7)$$

It is possible that these two relationships reflect two distinct mechanisms of ground dilation. This may reflect the difference between the sub-critical condition, where the seam can be regarded as a line source (or a point source in two dimensions), and the super-critical condition, where the shape of the source takes on more importance, with a greater dilation rate in the latter case.

Alternatively, it is possible to obtain a reasonable straight line by plotting log (V_s/V_t) against log (h/w), as in Figure 2.

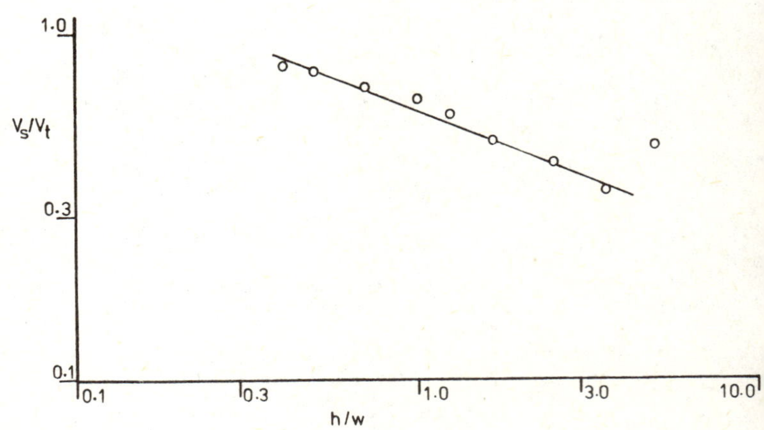

Figure 2 Ground dilation plotted logarithmically

This gives the relationship:

$$V_s/V_t = 0.61\ (h/w)^{-0.38} \qquad \text{..... (8)}$$

This equation can be used directly to modify the stochastic model though it should be noted that the excavation width (w) is not otherwise used in the basic source functions. It should also be noted that the mechanism of ground dilation is not explained by the stochastic model as described here. The volume relationships are purely empirically derived, independently of the model, and are used to modify its results.

Prediction of the inflection point (i)

The position of the point of inflection (i) is dependent on some 'ground property' which, according to the stochastic model, may itself depend upon some measure of 'angle of repose' and 'particle size' (Sweet and Bogdanoff, 1965). Neither of these properties can easily be interpreted in this particular context but it may be possible to make a direct estimate of i from the Subsidence Engineers Handbook. In view of the fact that the stochastic model operates for point sources, with i controlled by the above ground properties, it should be possible to relate i directly to depth (h) rather than depth-to-width ratio, in a similar way to soft ground tunnel settlement (Glossop, 1978). The effect of the width of the excavation on the trough width, namely the transition from super-critical to sub-critical, will be taken into account by the numerical integration. Strictly speaking, this means that i must be estimated from cases where the depth-to-width ratio is sufficiently high that the seam acts virtually as a point source.

Figure 3 shows the position of the point of inflection for a 200m wide, 1m thick seam at various depths, both as plotted from the Subsidence Engineers Handbook and as predicted theoretically using the relationship:

$$i = 0.002h + 0.00026h^2 \qquad \text{..... (9)}$$

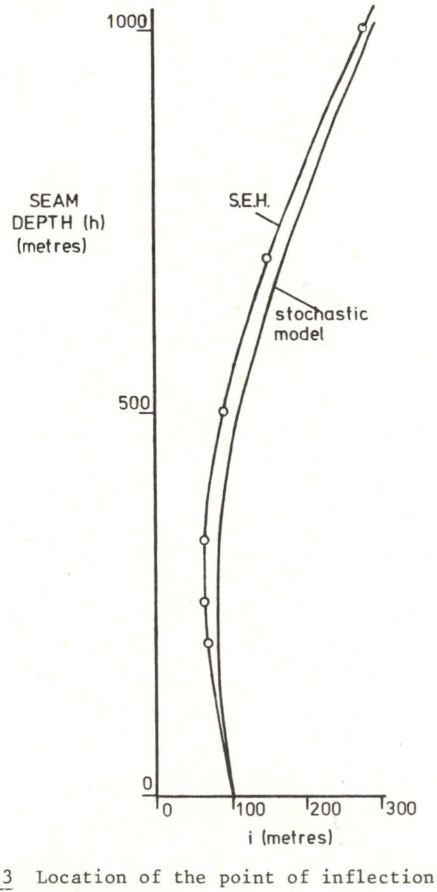

Figure 3 Location of the point of inflection above the seam.

The coefficients of equation (9) are estimated from that portion of the empirical curve where h/w is greater than 2.5 using polynomial regression. The fit of these two curves, whilst by no means perfect, is adequate to enable a reasonable comparison to be made between the stochastic model and the Subsidence Engineers Handbook, although it is possible that a better fit could be found by further refinement, should that be necessary.

Comparison of the Stochastic Model with Empirical Subsidence Profiles

Having found equations for the estimation of subsidence trough volume (equation (8)) and the point of inflection (equation (9)), it is possible to compare the shapes of the settlement troughs predicted by the stochastic model with those actually observed on the basis of the Subsidence Engineers Handbook. The profiles are shown super-imposed in Figure 4.

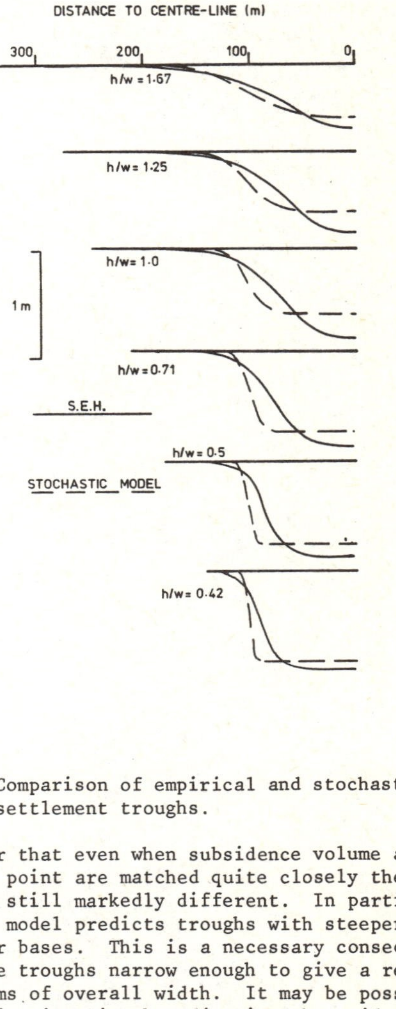

Figure 4 Comparison of empirical and stochastic settlement troughs.

It is clear that even when subsidence volume and inflection point are matched quite closely the trough shapes are still markedly different. In particular, the stochastic model predicts troughs with steeper sides and flatter bases. This is a necessary consequence of keeping the troughs narrow enough to give a reasonable fit in terms of overall width. It may be possible to correct this situation by adjusting the void distribution through the seam but this is difficult to justify on theoretical grounds.

CONCLUSIONS

It is clearly possible to derive relationships for subsidence volume and inflection point to give reasonable agreement on these parameters between the Subsidence Engineers Handbook and the stochastic model. Nonetheless, when these relationships are used in the stochastic model there still remains a major discrepancy in shape between the troughs predicted by the model and those observed in reality. It would appear, therefore, that the stochastic method, as it stands, does not adequately describe the observed process of subsidence.

REFERENCES

Baoshen L., Kuohua L. and Roungui Y., 1981, "Research in the surface ground movement due to mining", Proc. 4th. Cong. I.S.R.M., Montreux, Band 3.

Glossop N.H., 1978, "Ground movements caused by tunnelling in soft soils", Ph.D. Thesis, University of Durham.

Glossop N.H., Saville D.R., Moore J.S., Benson A.P. and Farmer I.W., 1979, "Geotechnical aspects of shallow tunnel construction in Belfast estuarine deposits", Tunnelling '79, I.M.M., London.

Litwiniszyn J., 1964, "On certain linear and non-linear strata theoretical models", Proc. 1964 Int. Conf. Strata Control and Rock Mech.

National Coal Board Mining Department, 1975,"Subsidence Engineers Handbook" National Coal Board, London.

Sweet A.L. and Bogdanoff J.L., 1965, "Stochastic model for predicting subsidence", J. Eng. Mech. Div.,A.S.C.E.

Sweet A.L., 1965, "Validity of a stochastic model for predicting subsidence", J.Eng. Mech. Div., A.S.C.E.

ROCK SURFACE ROUGHNESS AND DISCONTINUITY CLOSURE AT DEPTH

A. Robin Westerman
Malcolm J. Reeves
Peter B. Attewell

Engineering Geology Laboratories,
University of Durham, South Road,
Durham DH1 3LE, England.

SUMMARY. Preliminary consideration is given to two related matters: characterization of rock discontinuity surface roughness, and the normal pressure (as a depth-related parameter) necessary to achieve discontinuity wall closure through crushing of asperities in a type-rock. Experimental evidence is presented for effective acoustic fracture closure at normal stresses equivalent to about 15 percent of the intact unconfined compressive strength of the wall-rock. Fracture openings are implicitly predicted as a function of the combined rock surface roughness across the fissure and overburden pressure.

INTRODUCTION

An important question in rock engineering concerns the contribution of pre-existing fractures to the analysis of mass movement promoted by excavation. It is often assumed that at mining depths the normal pressures on such discontinuitiies are sufficient to maintain closure, and that the shear strength parameters of the intact rock material are most appropriate for the analysis of instability under confined compression conditions.

Closer to ground surface, in a 'civil engineering' setting, the normal pressures on such discontinuities are lower. Depending upon the intrinsic material strength, any shear displacement will either involve the shear strength parameters of both the discontinuity surface and the rock material, as shearing occurs through the roughness asperities and without dilation normal to the shear direction, or it will involve only the shear strength parameters of the discontinuity surface, with an apparent friction angle equal to the summation of the discontinuity friction angle and the asperity angle, and dilation normal to that surface.

ROCK DISCONTINUITY SURFACE DESCRIPTION

Field observation and qualitative description is the classical geological approach to the appraisal and classification of rock discontinuity surfaces. Fecker (1978) subdivides joints into tension fractures, displacement fractures and shear fractures based on gross surface morphology.

A semi-quantitative technique is discussed by Fecker and Rengers (1971) to describe 3D surfaces by plotting large numbers of observations of orientation for circular plates of various sizes placed on the surface at random. Different plate sizes allow roughness to be recorded on various scales. Profile gauges and primitive 'stylus-type' instruments have been used by Barton (1977) and by Fecker and Rengers (1971) to obtain coarse roughness profiles.

Comparison of roughness from one rock to another has relied upon descriptive schemes such as that of Barton (1978), comparison with standard profiles as suggested by Barton (1977) or dimensionless wavelength amplitude ratios as proposed by Mogilevskaya (1974). A more rigorous statistical description of rock surface roughness has been recently presented by Tien and Elfatih (1978).

STATISTICAL DESCRIPTIONS OF SURFACE ROUGHNESS

The statistical description of surfaces for engineering purposes is generally ascribed to Longuet-Higgins (1957) and there has followed an extensive literature on engineering surface metrology. Natural surfaces have received relatively little attention, although Williamson et al. (1969) point out that most naturally-occurring surfaces are likely to have Gaussian roughness as a consequence of the central limit theorem and that non-Gaussian distributions arise only in transitional surface topographies.

Nayak (1971) has shown that the statistical geometry of an isotropically-rough surface can be defined in terms of the moments of the power spectrum of any single profile. Up to five radially-disposed profiles are required for the satisfactory characterization of Gaussian anisotropic surfaces.

Patir (1978) has demonstrated that satisfactory models of rough surfaces can be generated from two parameters. These are the standard deviation of the height profile, a measure of the surface-normal roughness, and the autocorrelation length, a measure of the surface-parallel waviness. This surface description is only possible if the defining parameters are independent of the sampling interval, that is, if the surface roughness may be represented by a stationary series.

It has been demonstrated by Thomas and

Sayles (1978) that for many natural surfaces roughness is non-stationary, that is, the autocorrelation length depends upon scale. This could present problems for the description of natural surfaces, although the geostatistical concepts of semivariograms, which are especially designed to cope with non-stationary series, may be of considerable value in this respect.

STATISTICAL DESCRIPTION OF ROCK SURFACES

Rock surfaces, by the standards of engineering metrology, are extremely rough, so much so that the sliding stylus instruments with resolutions of 2 microns cannot be used. Weissbach (1978) has developed a profile-measuring device with a stepping stylus, and this type of instrument is required for rock surfaces.

Unlike many engineering surfaces, the surfaces of rock discontinuities cannot be assumed to be independently random, since they were, at the moment of fracture, perfectly matched. Subsequent shear and erosion may have modified the surfaces and the resultant match will be a significant factor in determining the magnitude of natural fracture openings.

Applications of the techniques of engineering metrology, based on independently random surfaces, may require adaptation with respect to rock discontinuity surfaces for which roughness will probably be :

1. scale dependent (non-stationary);

2. transitional (non-Gaussian);

3. partially matched (non-independent).

As a virgin rock fracture surface, with initial Gaussian roughness, is sheared, a transition surface develops which comprises :

1. areas of sheared asperities and asperity 'tracks';

2. unmodified areas of fracture surface.

Relatively fresh surfaces may remain in the transitional state but others will be heavily modified, mainly by the action of groundwater, and become Gaussian worn surfaces.

ROCK DISCONTINUITY OPENINGS

For independent Gaussian rough surfaces, the variance of the separation of two surfaces in contact can be obtained by adding the variances of the two surface roughnesses. For perfectly-matched Gaussian rough surfaces the variance of the separation of the two surfaces is zero.

Numerical simulations of non-independent, Gaussian surfaces in shear, which represent a much more realistic model of geological surfaces, have been investigated by Westerman (1981). For shear displacement with no dilation, the mean separation of the initially matched surfaces depends on the magnitude of the displacement relative to the autocorrelation length of the surface. If dilation is allowed, for a surface with a linear autocorrelation function, separation is a linear function of displacement.

Thus the magnitude of natural fracture openings will depend upon the extent of shear displacement and the degree to which dilation was permitted during shear. If no dilation is permitted then crushed material is likely to remain filling any opening that has been created.

EXPERIMENTAL EVIDENCE FOR CRACK CLOSURE

New (1976) and Toombs (1978) describe experiments using a PUNDIT (Portable Ultrasonic Non-destructive Digital Indicating Tester) to monitor the acoustic closure of artificial cracks. The cracks consisted of two prepared, lapped surfaces, and closures were observed for trivial normal stresses of less than 1 MPa.

For the present work, the experiments of New (1976) and Toombs (1978) were extended by Jonasson (1980) to provide a more realistic model of crack closure.

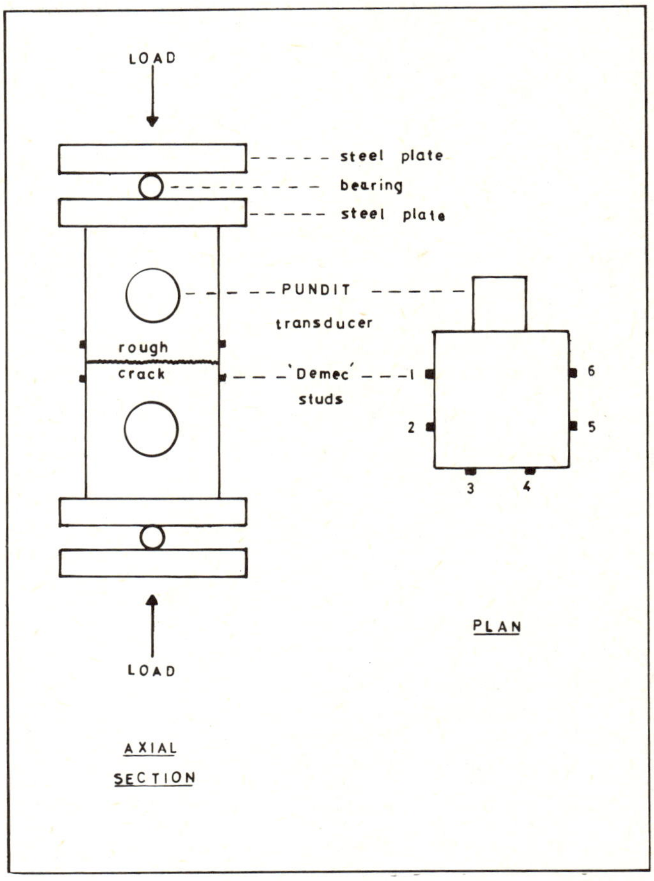

Figure 1 Experimental layout

The experimental layout is illustrated in Fig. 1. The rock used in the experiment was micritic Corallian Limestone from Spaunton Moor Quarry, North Yorkshire. The experimental surfaces were prepared by sawing and grinding followed by acid etching for various lengths of time.

The principal experimental advances were:

1. monitoring the intact rock as a control before sawing and etching the experimental surface,

2. monitoring surface roughness both

before and after the loading to a level sufficient for acoustic closure,

3. comparing the load required for closure on surfaces of varying roughness characteristics,

4. using a 'Demec' gauge to monitor strain for both the intact and fractured rock throughout loading.

A typical etched surface profile and correlogram are shown in Fig. 2. The periodicity in the correlogram illustrates a non-Gaussian transition topography resulting from the saw cut. After further etching a Gaussian surface height distribution is re-established.

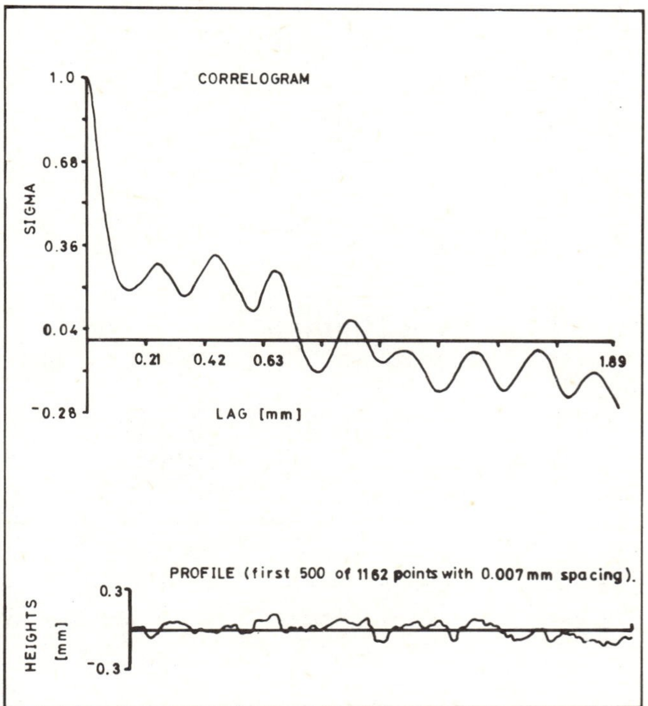

PROFILE, CORRELOGRAM AND HISTOGRAMS after Halli Jónasson (1980)
Sample no. S1A : Micritic Corallian Limestone from Spaunton Moor Quarry ; sawn surface etched for 60 seconds.

Figure 2 Typical profile and correlogram

Table 1 gives the surface statistics for a typical etched surface. The period of etching was sufficient to obliterate any transition topography in this case.

Table 1 Typical surface profile statistics

		Total	Peaks
Number of points	(unit)	1042	73
Sampling interval	(mm)	0.01	0.14 (av)
Roughness range	(mm)	0.8	0.7
Standard deviation	(mm)	0.15	0.12
Variance	(sq mm)	0.022	0.014
Skew	(unit)	-0.2	-0.05
Kurtosis (less 3)	(unit)	-0.004	0.3

The variations in axial strain during the loading of two limestone samples, S1 and S3, are shown in Fig. 3. Stress-strain curves for the intact rock, sawn and ground rock, and etched rock for each sample are presented. Fracture closure can is indicated when the gradients of the stress-strain curves approach that for intact rock. Tangents to the curves for the fractured rocks have been drawn with the characteristic gradient for intact behaviour. Visual inspection of Fig. 3 suggests that the intact gradient is achieved at a normal stress of approximately 18 MPa.

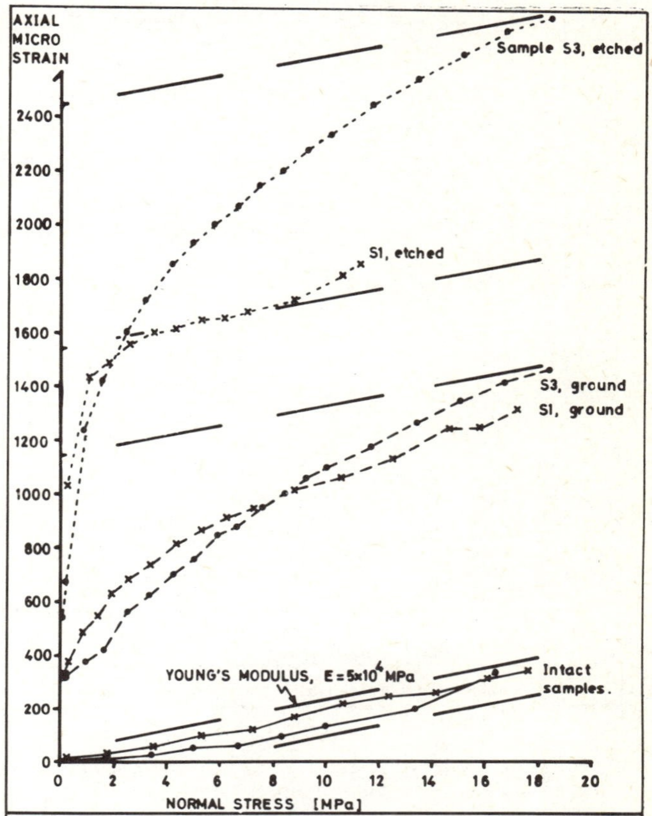

Strain across rough cracks under stress, after Jónasson (1980).

Figure 3 Variation of axial strain with normal stress for two limestone samples

Variations of acoustic velocity and attenuation with normal stress for intact and etched states of sample S3 are illustrated in Fig. 4. The intact acoustic velocity and attenuation values are re-established in the etched, fractured state at a normal stress of about 18 MPa. The experimental measurements of concurrent velocity and attenuation were made by Dowlen (1979), who also gives details of the techniques employed.

The experimental results shown in Fig. 3 suggest that fracture closure, as indicated by the re-establishment of the intact deformation modulus, occurs at a normal stress of around 18 MPa. The sonic velocity and attenuation measurements, summarised in Fig. 4, imply acoustic closure was achieved for a similar normal stress of 18 MPa. In both cases it is perhaps of interest to observe that 50-60% closure occurs for trivial normal stresses of less than 1 MPa, but is is clear that closure is not complete until much greater normal stesses are applied. For example, at 10 MPa normal stress for sample S3 in the etched state in Fig. 3, the average crack opening is still about 10% of the initial opening.

Subsequent testing to failure of cores drilled from sample S3 gave an unconfined compressive strength value of 137 MPa.

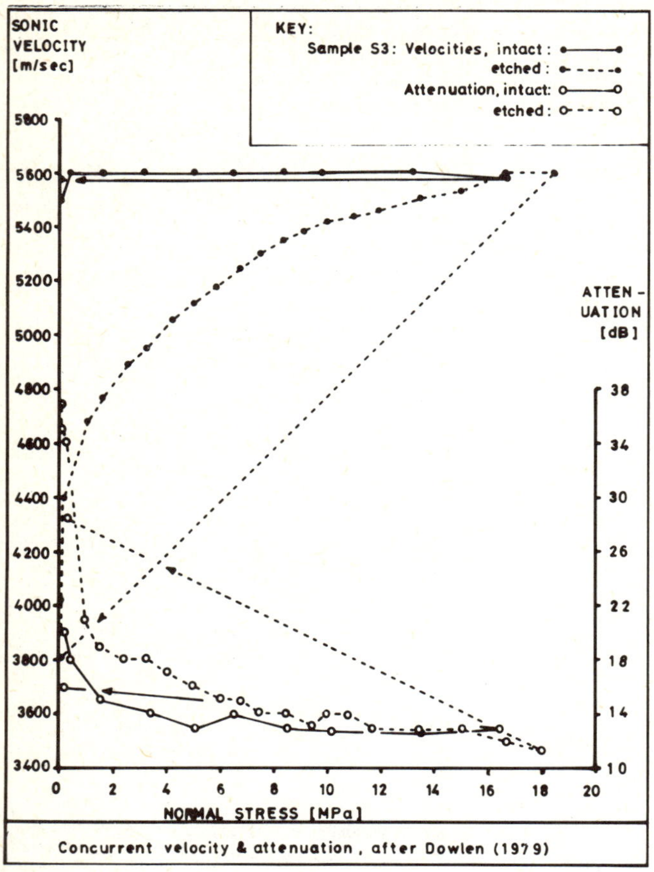

Figure 4 Variations sonic velocity and attenuation with normal stress for one limestone sample

INTERPRETATION OF EXPERIMENTAL DATA

The bearing area equation describes the surface contact area as a Gaussian surface is 'planed away', and so corresponds to the cummulative Gaussian height distribution function:

$$A(h) = 0.5 \text{ erfc } (h / \sqrt{2}.s)$$

where A(h) is the bearing area for separation h,
and s is the standard deviation of the surface roughness.

For the etched sample used in the experiment, acoustic closure occurs for a bearing area of 46%. This can be explained by crushing of asperities and movement of crushed material to 'fill' the remaining roughness.

Comparison of the surface profiles before and after loading, shown in Fig. 5, illustrates that some 'intermeshing' of the surfaces must also have occurred, since some 'high' asperities are preserved.

Acoustic closure is expected to develop when roughly 50% of the asperities have been crushed to fill the fracture opening with the small scale equivalent of fault 'gouge'. The normal stress at which this is likely to occur will be half (because of the 50% bearing area) the point hardness of the intact rock, that is, about one sixth of the unconfined compressive strength. This theoretical expectation can be tested easily for the present experiment.

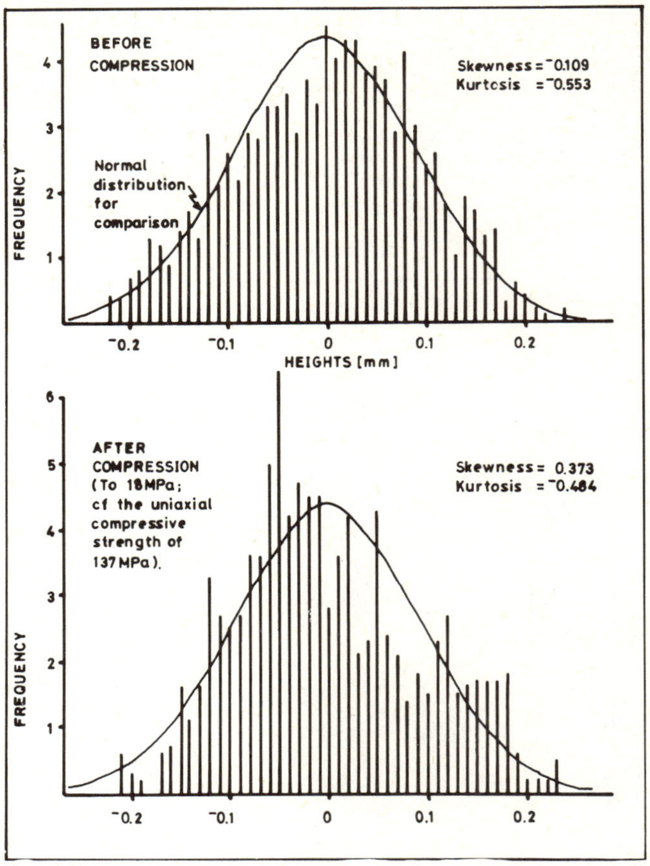

Figure 5 Surface profile statistics before and after loading

For ideal plastic materials the point indentation hardness can be shown both empirically and theoretically to be about one third of the yield stress. It could therefore be argued that for predominantly plastic asperity contact, an unconfined compressive strength of 137 MPa implies a point hardness of about one third, or 46 MPa. Hence, closure would be expected at about half this value, that is 23 MPa at a bearing area of 50%. The observed closure at 18 MPa with a bearing area of 46% is thus in reasonable agreement with a simple 'crush and fill' model.

The experimental results are also consistent with the mixed elastic-plastic asperity yield model, proposed by Francis (1977) which predicts the stiffness characteristics of a rough crack under normal loading.

INFERENCES BASED ON EXPERIMENTAL EVIDENCE

The results of the experimental work described by New (1976) and Toombs (1978) giving smooth-crack closure at 1 MPa was used by New and West (1980) and Hudson et al. (1980) to argue that, since normal stresses of 1 MPa are equivalent to a depth of only 40m, natural fractures will be closed at such a relatively shallow depth.

The experimental evidence presented here, of acoustic closure at about 18 MPa for surfaces having low surface roughness, could be interpreted as implying fracture closure only at considerable ('mining') depths amounting to about 700m. In the presence of normal pore pressures this leads to the conclusion that horizontal fractures would be unlikely to close, to a degree necessary for acoustic transmission, at depths less than 1 km below ground surface and vertical fractures would remain open to depths of 3 km at least.

The experimental work has considered only small-scale roughness, but demonstrates a major effect on the relationship between normal stress and fracture closure.

Fissure openings and the disposition of asperity contacts are predictable functions of rock surface roughnesses and associated shear displacements. If the discontinuity pattern can be described and the surface roughnesses are known, it has been suggested by Westerman (1981) that the work of Patir and Cheng (1978) can be applied to predict the rock mass permeability.

CONCLUSIONS

In the classification proposed by Fecker (1978), rock discontinuities are considered as tensional fractures, displacement fractures and shear fractures. In the first case, the opposed discontinuity surfaces will 'match' and the experimental evidence of New (1976) and Toombs (1978) is likely to be valid when applied to field situations.

For other types of displacement and shear discontinuities, opposing surfaces will not match. The experimental work presented here indicates that fractures then require very significant overburden pressures for complete acoustic closure. Both a simple 'crush and fill' model and a mixed elastic-plastic asperity contact model imply that fractures are likely to be open over the range of depths of major strata mechanics interest in civil and mining engineering.

ACKNOWLEDGEMENTS

The experimental work was carried out by H. Jonasson, with the assistance of G. Dowlen under the supervision of the authors, as part of project dissertation studies for the Advanced Course in Engineering Geology at the University of Durham. The facilities for measuring small-scale roughness were made available by Dr T.R. Thomas of Teesside Polytechnic who also gave considerable advice on the 'state of the art' in engineering metrology.

REFERENCES

Barton, N. 1977. "Shear strength of rock joints in theory and practice", Rock Mech., 10, 1-54.

Barton, N. 1978. "Suggested methods for the quantitative description of discontinuities in rock masses", Internat. J. Rock Mech., 15, 319-368.

Dowlen, G.R. 1979. "Seismic velocity and attenuation anisotropy as studied under laboratory and field conditions", Unpublished MSc dissertation, University of Durham.

Fecker, E. 1978. "Geotechnical description and classification of joint surfaces", Bull. Internat. Ass. eng. geol., 18, 111-120.

Fecker, E. & Rengers, N. 1971. "Measure of large-scale roughness of rock planes by means of profilograph and geological compass", Proc. Internat. Symp. Rock Mech., Nancy, France, 1-18.

Francis, H.A. 1977. "Applications of spherical indentation mechanics to reversible and irreversible contact between rough surfaces", Wear, 45, 221-269.

Hudson, J.A., Jones, E.J.W. & New, B.M. 1980. "P-wave velocity measurements in a machine bored tunnel", Q. J. eng. Geol., 13, 33-44.

Jonasson, H. 1980. "A preliminary study of small-scale roughness of rock surfaces", Unpublished MSc dissertation, University of Durham.

Longuet-Higgins, M.S. 1957. "Statistical properties of an isotropic surface", Phil. Trans. Roy. Soc., A250, 157-171.

Mogilevskaya, S.E. 1974. "Morphology of joint surfaces in rock and its importance in engineering geological examination of dam foundations", Proc. 2nd Internat. Congr. Internat. Ass. eng. geol., Sao Paulo, Brazil, 17.1-17.8.

Nayak, P.R. 1971. "Random process model of rough surfaces", Trans. Amer. Soc. Mech. Engrs J. Lub. Technol., F93, 398-407.

New, B.M. 1976. "Ultrasonic wave propagation in discontinuous rock", Transport and Road Research Laboratory Report, LR720.

New, B.M. & West, G. 1980. "The transmission of compressional waves in jointed rocks", Eng. Geol., 15, 151-161.

Patir, N. 1978. "A numerical procedure for random generation of rough surfaces", Wear, 47, 263-277.

Patir, N. & Cheng, H.S. 1978. "An average flow model for determining effects of 3-dimensional roughness on partial hydrodynamic lubrication", Trans. Amer. Soc. Civ. Engrs J. Fluids Engng, 100, 12-17.

Thomas T.R. & Sayles, R.S. 1978."Some problems in the tribology of rough surfaces", Tribol. Internat., 11, 163-168.

Tien, H.W. & Elfatih, M.A. 1978. "Statistical representation of the roughness of rock joints in the laboratory", Internat. J. Rock Mech., 15, 259-262.

Toombs, A.F. 1978."Ultrasonic wave propagation in strong rocks", Transport and Road Research Laboratory Report, SR384.

Weissbach, G. 1978. "A new method to determine the roughness of rock joints in the laboratory", Internat. J. Rock Mech., 15, 131-133.

Westerman, A.R. 1981. "The development of secondary porosity and permeability in fractured rock", Unpublished PhD thesis, University of Durham.

Williamson, J.B.P., Pullen, J. & Hunt, R.T. 1969. "The shape of solid surfaces", Amer. Soc. Mech. Engrs, New York.

OPTIMISATION OF SUPPORT PERFORMANCE THROUGH CONSIDERATION OF THE PRESSURE/CONVERGENCE RATIO

B. G. D. Smart

Department of Mining and Petroleum Engineering, University of Strathclyde, Glasgow G1 1XW, Scotland.

A. K. Isaac

Department of Mineral Exploitation, University College, Cardiff CF1 1XL, Wales.

SUMMARY. The Paper considers the ratio of pressure increase per unit of convergence in the legs of a powered support to be a measure of its sensitivity to convergence. This sensitivity, which should be maximised in the interests of strata control, is shown to depend on the load bearing properties of the foundation underneath the support base. Methods, including the manipulation of setting and yield pressures, are suggested for modifying these properties in order to generate high sensitivities.

INTRODUCTION

The powered support is an active device while it is being set, developing a reaction between roof and floor of the longwall face as a setting pressure is applied from a remote power pack to the hydraulic legs of the support. The setting pressure has a nominal value controlled by an unloading valve on the power pack, but the actual value can be lower, depending on the demand on the hydraulic supply system at the time of setting and operator practice, although recent developments in hydraulics, e.g. positive set valves, can overcome these deficiencies.

After setting, the support becomes a passive device, accommodating convergence between roof and floor of the longwall face through closure of the hydraulic legs. Thus pressure within the legs increases toward a maximum or yield value determined by relief valves incorporated in the hydraulic circuitry of the legs.

The pressure range over which the hydraulic legs of a support function during a production cycle therefore extends from a setting pressure towards a yield pressure, the driving mechanism causing the increase in pressure and consequently an increase in resistance to convergence being the ongoing convergence between roof and floor. Furthermore, the rating of setting and yield pressures are the only control parameters available for influencing the performance of the powered support during this period.

Recommended values of setting pressure have been determined either from a roof stability consideration, i.e. by determining the minimum force which must be applied to the roof in order to support the largest block liable to separate from the overlying bridging beds (Wilson, 1975), or from an analysis of the relationship between observed average rates of convergence and Mean Load Densities (Ashwin et al., 1970). The concept of Mean Load Density (M.L.D.) was introduced to enable the resistance developed by a support system to be quantified, allowing for the variations which occur during a production cycle in leg pressures and supported roof area. The recommended setting pressure in this case is that which generates a M.L.D. found to limit the average rate of convergence to a minimum but inevitable value.

Alternative empirical approaches to determination of setting pressure values have been based on the identification of their effect on rates of leg pressure increase (Bates, 1978) and on the examination of relationship between rates of leg pressure increase and convergence (Smart et al., 1981).

Yield pressure values are determined by the structural strength of the powered support, which in turn is related to the maximum loads the support is expected to tolerate in service.

In general, conventional chock type supports have performed well in British mining conditions. The authors believe that it is possible to enhance the performance of these supports in certain conditions, based on an appreciation of the optimum relationship between the rates of pressure increase in the support legs and convergence during the production cycle.

The appreciation is inherently simple. If a support is to exert maximum control over convergence while acting as a passive device, then it is essential that all or nearly all of the convergence is absorbed as leg closure resulting in increased leg pressures, i.e. the sensitivity of the support to convergence is high. However, investigations have shown that a substantial proportion of convergence is absorbed by compaction or penetration of the foundation on which the support rests. An examination of the behaviour of the foundation under load suggests how manipulation of setting and yield pressures can be used to ensure that the support retains maximum control over convergence, thus extending the approach suggested previously (Smart et al., 1981) by introducing the quantity:

$$\frac{\text{sum of pressure changes in all support legs per unit time}}{\text{convergence occurring between competent roof and floor per unit time}}$$

as an index of support sensitivity to convergence. (Referred to below as the pressure/convergence ratio.)

THE INFLUENCE OF THE FOUNDATION PROPERTIES UPON THE SENSITIVITY OF THE SUPPORT TO CONVERGENCE

The foundation on which the support sits may be considered to consist of three layers as shown in Figure 1, viz., a bottom and infinitely thick layer of competent rock, an intermediate layer of failed or very weak rock such as seatearth, and a top layer of debris (fines) produced by mining operations. For purposes of this analysis the layer of rock debris which often accumulates on the support canopy may be included in the top layer. The thickness of the intermediate layer will depend on geological conditions, while the thickness of the top layer will depend on mining practice, in particular the efficiency of the shearer and face conveyor in cleaning the floor prior to face equipment advance.

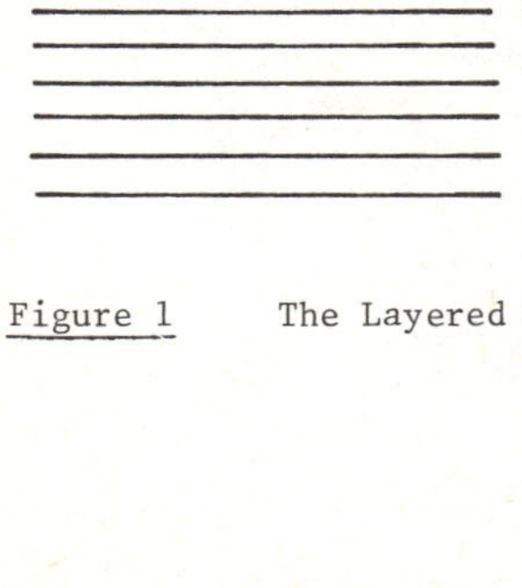

Figure 1 The Layered Foundation

The total penetration of a support base into the foundation will depend on the base shape, contact area and loading, and the load/deformation characteristics of each layer. The characteristics of each layer are presented in Figure 2, as may be obtained from bearing plate tests, and these characteristics depend on the physical properties of the material which constitutes the layer, and layer thickness.

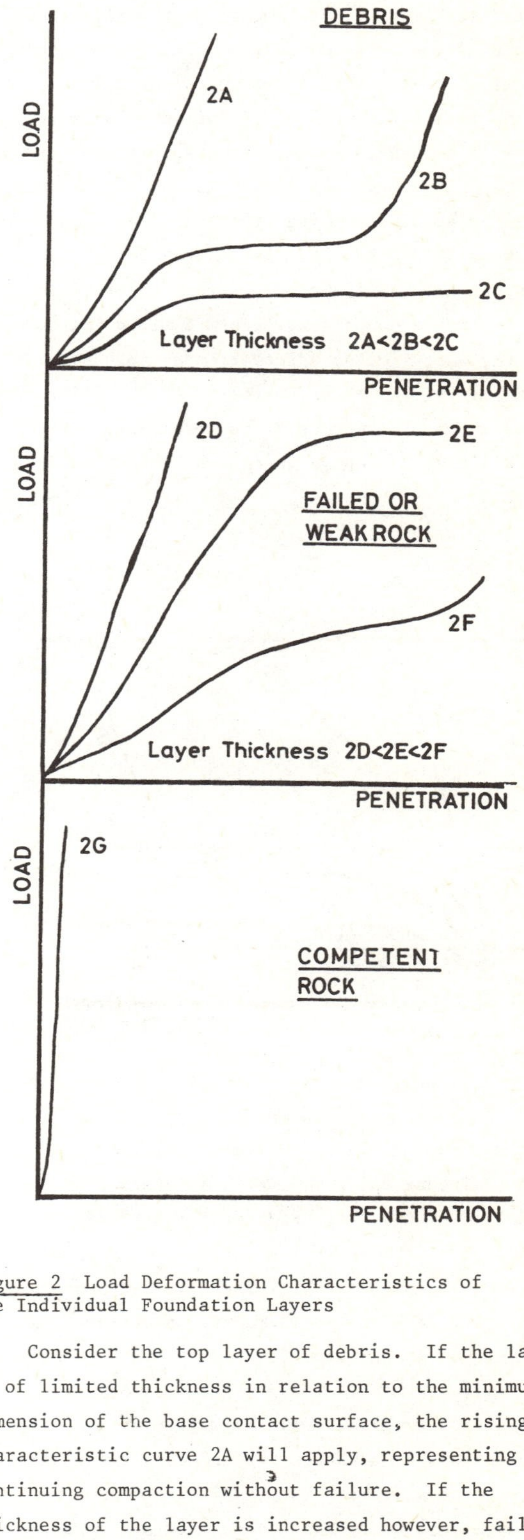

Figure 2 Load Deformation Characteristics of the Individual Foundation Layers

Consider the top layer of debris. If the layer is of limited thickness in relation to the minimum dimension of the base contact surface, the rising characteristic curve 2A will apply, representing continuing compaction without failure. If the thickness of the layer is increased however, failure will occur after some initial compaction. Thereafter the base will be pushed into the debris extruding it sideways until the thickness of the layer trapped underneath the base has been reduced sufficiently to enable a rising characteristic to be generated again as shown in curve 2B. The thickness of the trapped layer which will generate a rising characteristic will be related to the minimum dimension of the base contact area.

If the thickness of the debris layer is excessive, it is possible that the trapped layer will not be produced and after some initial compaction deformation will be excessive as shown in curve 2C.

The intermediate layer of failed or very weak rock will behave in a similar manner to the top layer of debris, but with the transition points on the characteristics occurring at higher loads, as shown in curves 2D, 2E and 2F. The bottom layer of competent rock will, by definition, not fail and will exhibit a comparatively stiff rising characteristic with limited deformation as in curve 2C.

Combining the characteristics of the individual layers in several permutations results in three possibilities, as shown in curves 3A, 3B and 3C of Figure 3.

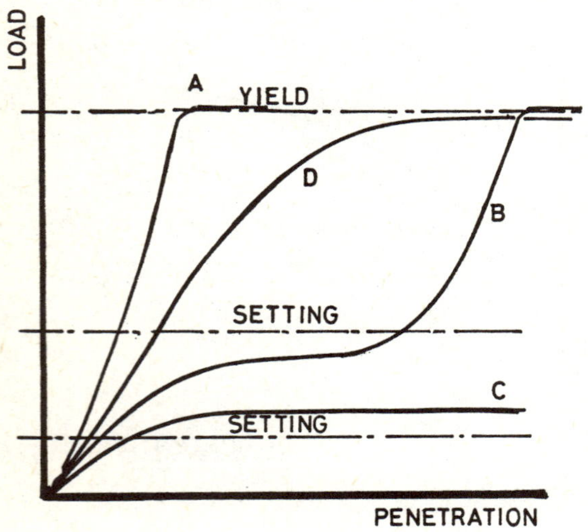

Figure 3 Load Deformation Characteristics of the Combined Layers

Curve 3A will be generated if all the layers compact without failure, i.e. the layers of debris and failed or very weak rock are either absent or of limited thickness. This represents the ideal situation where the sensitivity of the support to convergence is high and increasing resistance will be developed to an upper limit determined by the yield valve setting. The pressure/convergence ratio in this case would be comparatively high and constant, approaching the characteristic of the legs themselves.

Curve 3B will be generated if one or both of the two top layers fails after some initial compaction and then undergoes sufficient reduction in thickness to re-stiffen as a trapped layer. Thus the sensitivity of the support to convergence is reduced for a significant part of the cycle. The pressure/convergence ratio will increase initially, then fall off before increasing again towards the end of the cycle. An example of this type of cycle is illustrated in Figure 4, showing the pressure/convergence ratio about to increase.

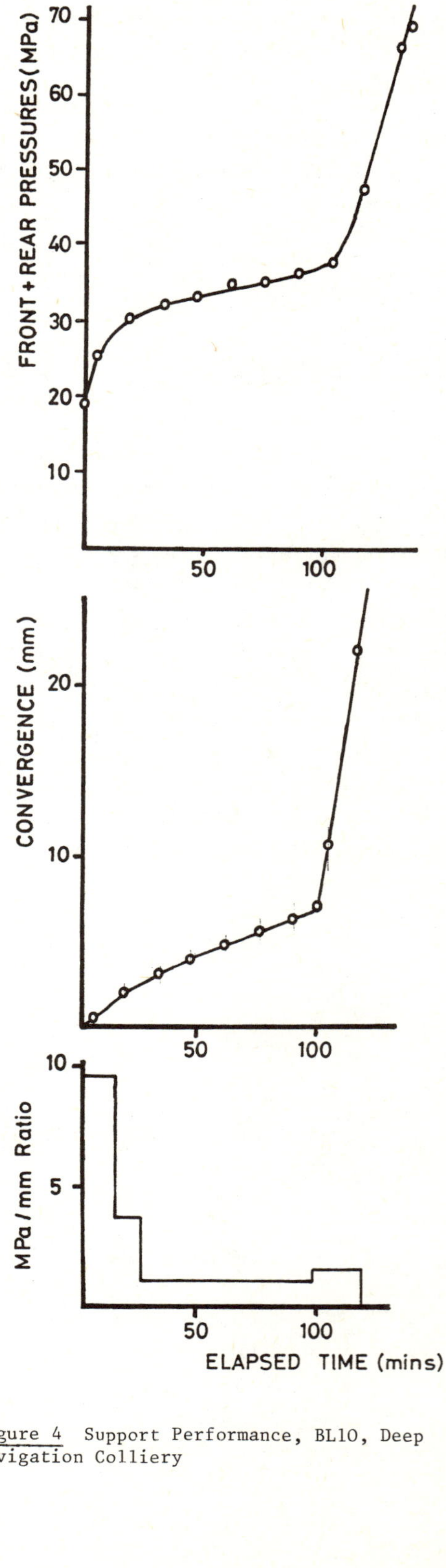

Figure 4 Support Performance, BL10, Deep Navigation Colliery

Curve 3C will be generated if one or both of the two top layers fail after some initial compaction, with the layer thickness being such that no "trapping" effect occurs, and sensitivity of the support is lost. In this case the pressure/convergence ratio will rise initially and then fall off and remain low. Examples are presented in Figure 7.

Curve 3D is similar to 3C but with the failure of the foundation occurring at higher loading level, and would be generated by the failure of a floor rock stratum. Thus the pressure/convergence ratio will be high for the majority of the cycle, decreasing suddenly towards the end before yield pressure is reached. Thus it is postulated that the sensitivity of a support to convergence and therefore the controlling action of the support over convergence is governed by the reaction of the foundation to the loading applied to it through the support base. Furthermore, support sensitivity may be assessed in-situ, at least for chock type supports, by examining the ratio of pressure increase within the hydraulic legs to convergence occurring between competent roof and floor in the same time.

METHODS OF INFLUENCING SUPPORT SENSITIVITY

Accepting that the sensitivity represented by characteristic 3A of Figure 3 is the ideal case; the following recommendations can be made: -

(i) Where a characteristic of type 3B is found to exist at current setting pressures, then the setting pressure should be increased to that sufficient to create a trapped layer immediately on setting and so produce a characteristic similar to type 3A. This approach has been simulated in the laboratory as shown in Figure 5, with a bearing plate being depressed into typical debris. The load on the plate was controlled and both load and plate penetration into the debris monitored continually. Two loading cycles were investigated, the results being presented in Figures 6(a) and (b). In the first cycle a load was applied to the plate to create an average pressure in the debris equivalent to a 6/240 support being set at 9.1 MN/m^2, the actual value. This corresponds to Point A in Figure 6(a). Thereafter the load on the plate was increased, simulating an actual cycle which had been monitored underground. Key points in the cycle are identified in the figure. It can be seen that on Points 1-2 of the curve, the debris layer accepted increasing load, the average stiffness of the foundation being 23 MN/m^3. At Point 2 the layer failed and the plate was pushed into the debris until a trapped layer was produced at Point 3. The average stiffness for Path 2-3 was effectively zero. On production of the trapped layer, however, the stiffness of the foundation increased, the average stiffness of Path 3-4 being 160 MN/m^3.

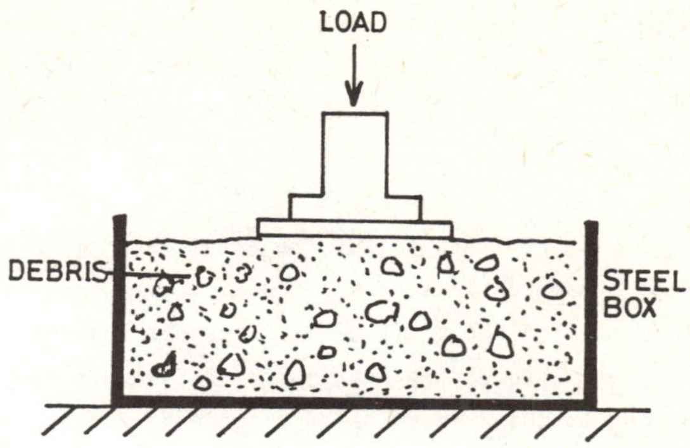

Figure 5 Laboratory Simulation of Loading of Support on Debris Layer

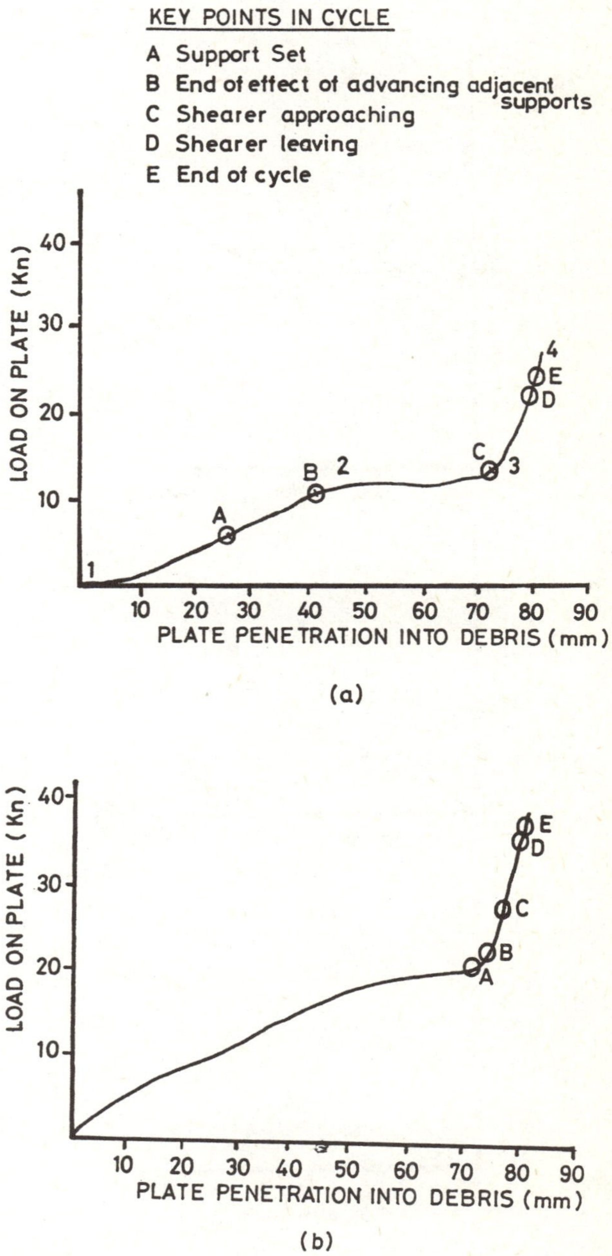

Figure 6 Simulation Results, Actual and Raised Setting Pressures

Between Point A on the graph, representing support setting, and Point E, representing the load carried by the support at the end of the cycle, 53 mm of "convergence" was absorbed as debris compaction, 43 mm of which occurred during the comparatively soft phase of foundation behaviour, i.e. Path 1-3.

In order to investigate the effect of higher setting pressure the test was repeated with the initial load on the plate being increased sufficiently to progress through the soft phase of foundation behaviour to the production of a trapped layer. The load required was equivalent to a setting pressure of 27.3 MN/m^2. Taking the loading on the plate up through the same increments as in the

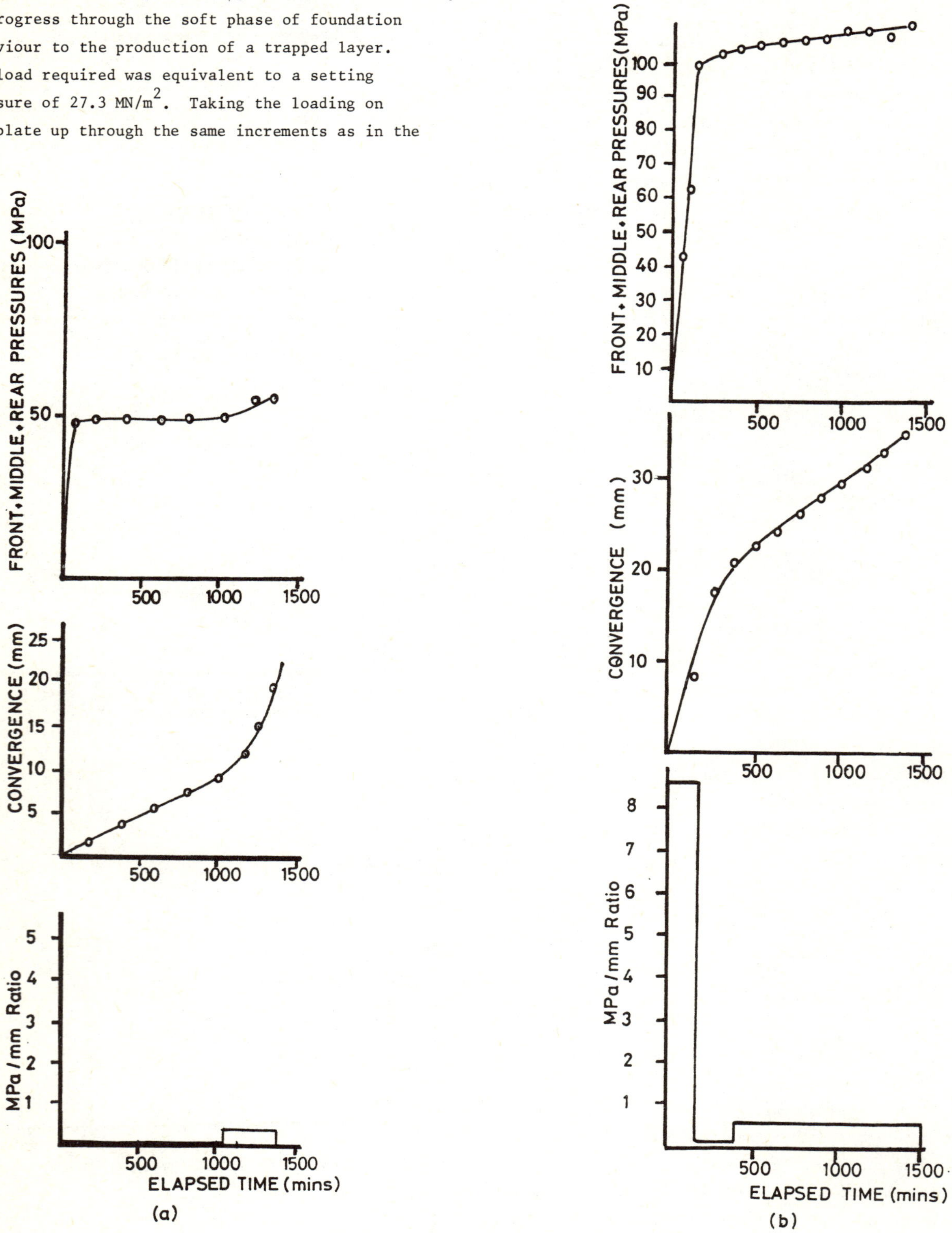

Figure 7   Support Performance, N1 Treforgan Colliery

previous test produced only 10 mm of debris compaction, shown in Figure 6(b); i.e. 43 mm of "convergence" previously absorbed as debris compaction after the simulated setting of the support has been eliminated.

(ii) Where a characteristic of type 3C is found to exist, it will probably be due to excessive thicknesses of either debris or weak floor, or a combination of both. In this situation the foundation compaction required to produce trapped layers would be excessive, resulting in subsequent problems with support advance. Accordingly, if excessive thickness of debris is the cause, then mining practice should be improved to produce a cleaner floor and/or canopy. Alternatively, if an excessive thickness of weak floor is present, then consideration should be given to increasing the contact area of the support base. The effect of increasing base area is shown in Figures 7(a) and (b). Figure 7(a) shows typical results for pressure (sum of all three leg circuits) and convergence for a 6/240 on an excessive thickness of very weak fireclay and rashings. It can be seen that the pressure/convergence ratio was extremely low, with leg pressures barely rising above setting pressure as the support was pushed into the floor.

Results obtained for the same support after the base area had been increased by 46% are shown in Figure 7(b). In this case leg pressures increased briefly after setting, reflecting the improved base area. Soon, however, the foundation failed and again the support was pushed into the floor, but at a higher resistance. This implies that while the increased base area was sufficient to improve the resistance developed by the support, the increase was insufficient to prevent failure of the floor and bring convergence entirely under the control of the support as shown by the low pressure/convergence ratio.

(iii) Where a characteristic of type 3D is found, floor loading could be reduced by increasing the base contact area and/or reducing yield pressure.

CONCLUSIONS AND RECOMMENDATIONS

(1) It is suggested that in order to offer acceptable control over convergence, a support should absorb a high proportion of convergence, i.e. a support should exhibit sensitivity to convergence throughout the production cycle.

(2) The sensitivity of a support to convergence is influenced by the properties of the foundation on which the support rests.

(3) The properties of the foundation and hence the sensitivity of the support can be influenced by setting pressure, yield pressure, base contact area, and mining practice.

(4) It is proposed that the setting pressure for a given powered support installation be determined as follows: -

(a) The setting pressure is set initially at the value determined by current operating practice and/or regulations, whichever is the higher.

(b) After production has commenced, the pressure/convergence ratio should be evaluated for a number of production cycles in order to determine support sensitivity to convergence.

(c) If required the appropriate action as described in the previous section should be taken to maximise sensitivity.

(5) Yield pressure should be set at a level which will enable acceptable pressure/convergence rates to be developed without overloading the foundation to failure.

ACKNOWLEDGEMENTS

The authors gratefully acknowledge the financial support provided by Gullick Dobson Ltd., Wigan, for the work presented in this Paper, and the cooperation of the Staff of the South Wales Area, National Coal Board.

Particular thanks are due to Dr. M. A. Price, Dr. J. D. Thornton and Mr. H. Denning for providing data for analysis.

REFERENCES

Ashwin, D.P. et al., 1970, "Some Fundamental Aspects of Face Powered Support Design", Mining Engineer, 129, 659-670.

Bates, J.J., 1978, "An Analysis of Powered Support Behaviour", Mining Engineer, 203, 681-694.

Smart, B.G.D., Davies, D.O., Isaac, A.K., 1981, "Conclusions from Strata Mechanics Investigations Conducted by Cardiff and Strathclyde Universities at Longwall Coalfaces", Paper presented to the Scottish Branch of the Institute of Mining Engineers, and to be published in Mining Engineer.

Wilson, A.H., 1975, "Support Load Requirements on Longwall Faces", Mining Engineer, 178, 479-488.

RELATIONS BETWEEN STRATA DEFORMATION AND SUPPORT PERFORMANCE ON LONGWALL FACES

by R.N. Gupta and I.W. Farmer

Department of Mining Engineering
University of Newcastle upon Tyne.

SYNOPSIS: An extensive underground experimental programme was designed and carried out on three longwall faces at Westoe Colliery, Tyne and Wear. The objective of the programme was to improve the stability of mechanised longwall faces through investigating the relations between support setting pressure and various other support parameters including convergence and the deformation of roof and face strata. Some of the results and preliminary conclusions are presented showing a strong relation between increased support setting pressure and improved face stability.

INTRODUCTION

Recent developments in powered support design have been concerned with improved mechanical and hydraulic performance. In particular there has been emphasis on higher yield loads and hence higher load densities. There has at the same time tended to be a lack of knowledge of the mechanics of support-roof interaction and of the behaviour of the immediate roof during production operations. A research programme at the University of Newcastle upon Tyne was designed to investigate the relation between strata deformation and support performance through a detailed study of three faces at Westoe Colliery, Tyne and Wear. The results are presented in full by Gupta (1982).

EXPERIMENTAL DESIGN

The major factor determining the efficient performance of a longwall support system is the stability of the immediate or nether roof strata. If this remains intact and unbroken, supports will maintain good contact with the roof and cavities will not develop ahead of and above the supports.

In order to design a satisfactory experimental programme, the basic mechanics of strata deformation at a typical caved longwall coal face were considered. These are described in some detail by Farmer (1980) among others. The face line forms the front abutment of the working area. Stresses which cannot be transmitted through the fractured and partially consolidated rock above the worked area are redistributed over a zone of several metres ahead of the face. The stresses in this area are predominantly vertical with reduced lateral restraint from the caved strata behind the face. The concentrated vertical stress tends to induce vertical tensile fractures above, below and through the coal seam and parallel to the coal face. These may be imposed on or exacerbate existing structural discontinuities such as cleat. The effect is to create fractures at the face line which intersect bedding plane features in the roof. These features - particularly shear zones can create conditions for cavity formation if any yielding or lack of support occurs at the face. Yield resistance of supports is less important therefore than setting resistance and in the experimental design, setting resistance was chosen as the major independent variable. The dependent variables chosen for investigation included convergence (including leg closure and debris compaction), roof flaking and face spalling and roof and face strata deformation.

EXPERIMENTAL SITE

Westoe Colliery workings extend under the North Sea for a distance of about 8m. The current workings (Figure 1) are in the Main (F) and Brass Thill (K) seams on either side of the 90 Fathom Fault. The faces chosen for investigation were F-25 and F-27 at a depth of 277m below sea level and 232m below sea bed level working a 2.13m thick section of the Main Seam and K-36 at a depth of 345m below sea level and about 305m below sea bed level working a 1.8 - 2.0m section of the Brass Thill seam.

The supports used on F-27 and K-36 face were Gullick-Dobson 6/300Ton rigid base fully shielded chock supports with positive set facilities at 1.2m centres. The supports used on F-25 face were Gullick Dobson 4/450 Ton chock shield supports with lemniscate linkages with positive set facilities, at 1.5m centres. All the faces were retreating longwall with lengths of 219m (F-25) 220m (F-27) and 215m (K-36). The face directions were roughly normal to the NW-SE major cleat directions and the face coals were therefore 'on end'.

EXPERIMENTAL PROCEDURE

A similar experimental procedure was used on each face. The setting pressure in the supports was maintained for 8 - 10 days at each of six selected setting pressures itemised in Table 1.

Table 1 Setting pressure and setting load densities

Setting pressure		Setting load density	
MPa	p.s.i.	F-25 MN/m^2	K-36/F-27 MN/m^2
13.8	2000	0.208	0.223
17.2	2500	0.260	0.279
20.7	3000	0.312	0.334
24.1	3500	0.364	0.390
27.5	4000	0.416	0.446
31.0	4500	0.468	0.502

Some variation in these values occurred during the experiment. In Table 1 setting load density is defined as the load or thrust exerted by the support per unit area on the exposed roof supported by it from the rear of the canopy to the face, at the time of setting. It is apparently lower in the case of F-25 face because

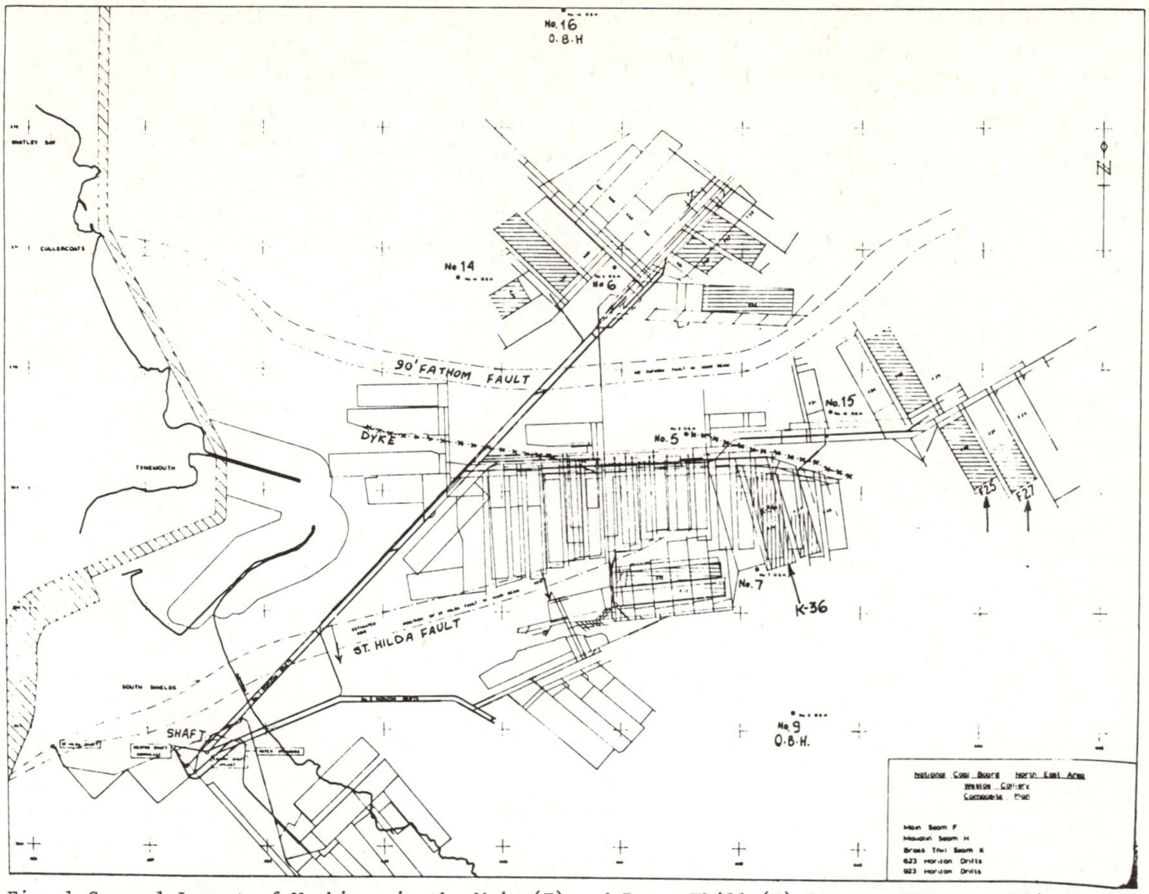

Fig. 1 General Layout of Workings in the Main (F) and Brass Thill (K) Seams at Westoe Colliery, Showing the Location of the Faces F25, F27 and K36 Chosen for Experimentation.

the prop-free-front distance is greater.

At each setting pressure, observations were obtained throughout the face at each support at the time of setting and immediately before move-over for four cutting cycles and for two weekend breaks. The cutting cycles represented an average duration of 2 hrs 30 mins during which the face was advanced 0.45m on F-25 and K-36 face and 0.65m on F-27 face. The weekend break represented a stand or idle period of average duration 60 hours from setting to first move-over. The observations and methodology were:

a) Front (middle) and rear leg circuit pressures at every support on the face, using Conflow linear pressure indicators installed as standard fittings on each support. These were calibrated on K-36 face against a standard dial type pressure indicator.

b) Leg closures at approximately 10% of the supports, using clamps and verniers or dial gauges attached to the cylinder and telescopic tube of the support.

c) Support inclinations at 10% of the supports using a Hilger and Watts clinometer and NCB designed liquid tube type inclinometer.

d) Convergence between roof and floor at approximately the same locations as leg closure measurements, using a telescopic rod convergence meter located between steel pins driven into the roof and floor. Convergence was measured between the face and the middle leg and between the goaf side of the armoured face conveyor and the middle leg.

e) Debris compaction above the support by dial gauges at the same positions as leg closure and convergence measurements.

f) Roof flaking or cavity formation within 1m of the face. Cavities greater than 100mm high and 300mm wide only were monitored.

g) Face spalling throughout the face. This was monitored as a distance greater than 500mm between the tip of the support canopy and the face.

h) Extrusion of the face wall towards the goaf.

i) Lateral movement of roof relative to the floor measured by plumb bob suspended from a roof pin, at the same locations as leg closure and convergence measurements.

In addition at one location on F-25 and K-36 face and three locations on F-27 face roof, floor and face wall deformations were investigated using an anchor wire or magnetic ring-reed switch extensometer system. This has been described in Gupta and Farmer (1981). Frequent observations were also obtained of overhang in the goaf. Details of the geology, lithology and structure of proximate rocks were logged together with relevant progress and production data.

In order to monitor any time related effect six Conflow clockwork driven continuous linear chart recorders were used to monitor circuit pressure on two supports - one at the centre and one at a third of the face length.

RESULTS

The vast amount of data collected during the experimental programme cannot be presented in full, but a sample of the data from Gupta (1982) is reproduced to illustrate some of the major conclusions from the work.

Support setting and final mean load densities for F-27 face during an extraction cycle at lowest nominal setting pressure (15.2 MPa) and highest nominal setting

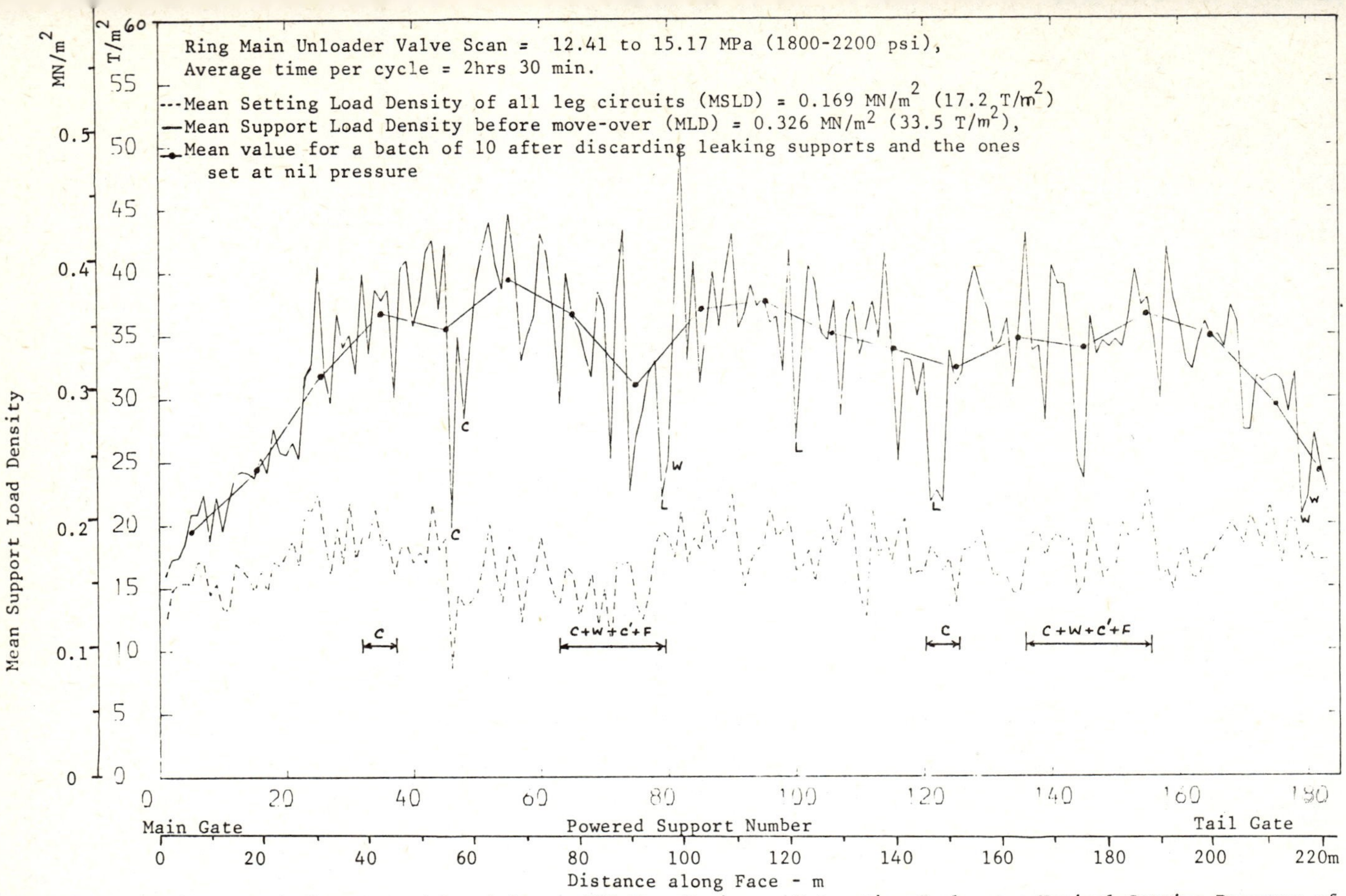

Fig. 2 Setting and Final Mean Load Densities on F27 Face During an Extraction Cycle at a Nominal Setting Pressure of 15.2

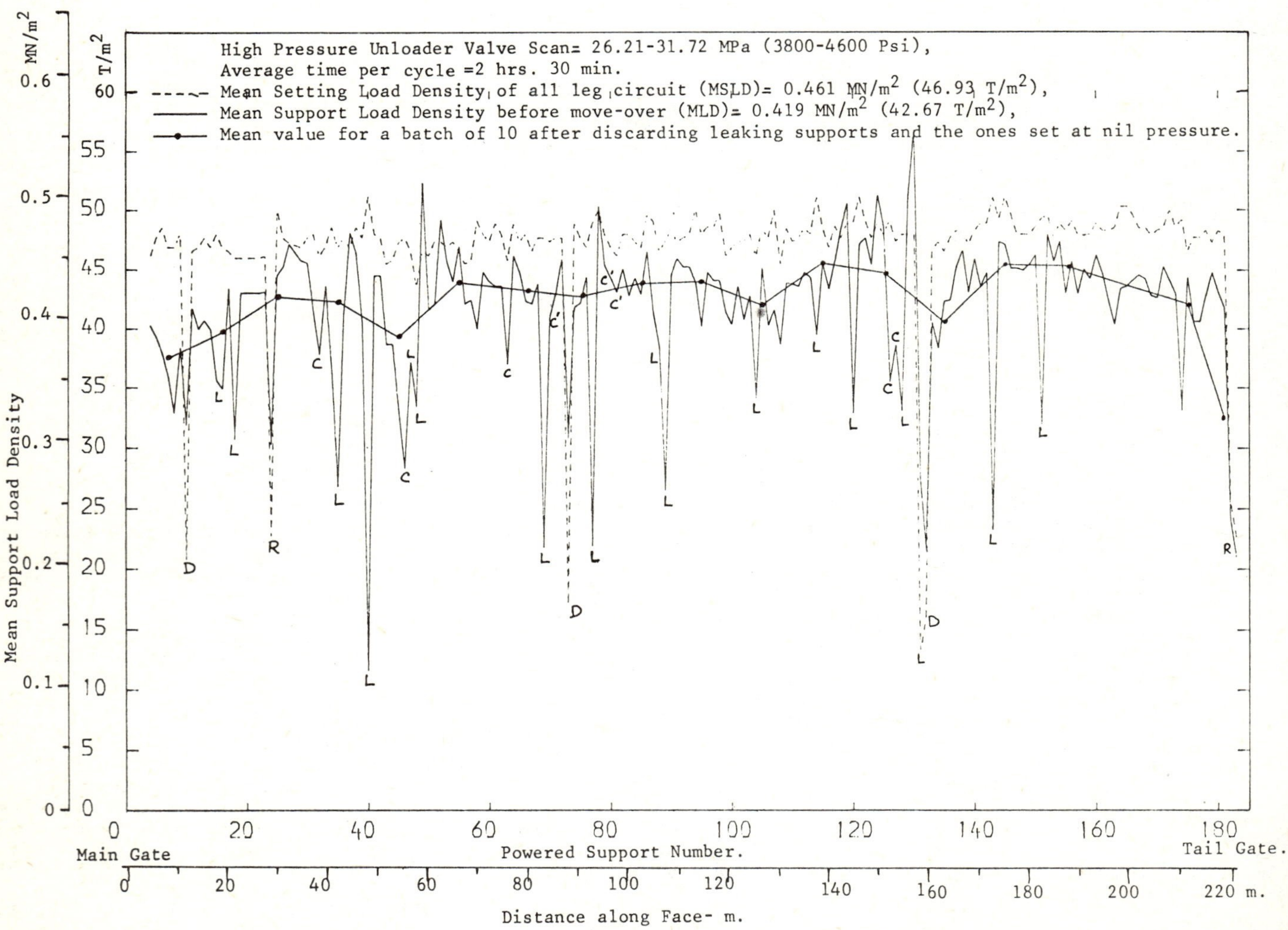

Fig.3 Setting and Final Mean Load Densities on F27 Face During an Extraction Cycle at a Nominal Setting Pressure of 31.7 MPa.

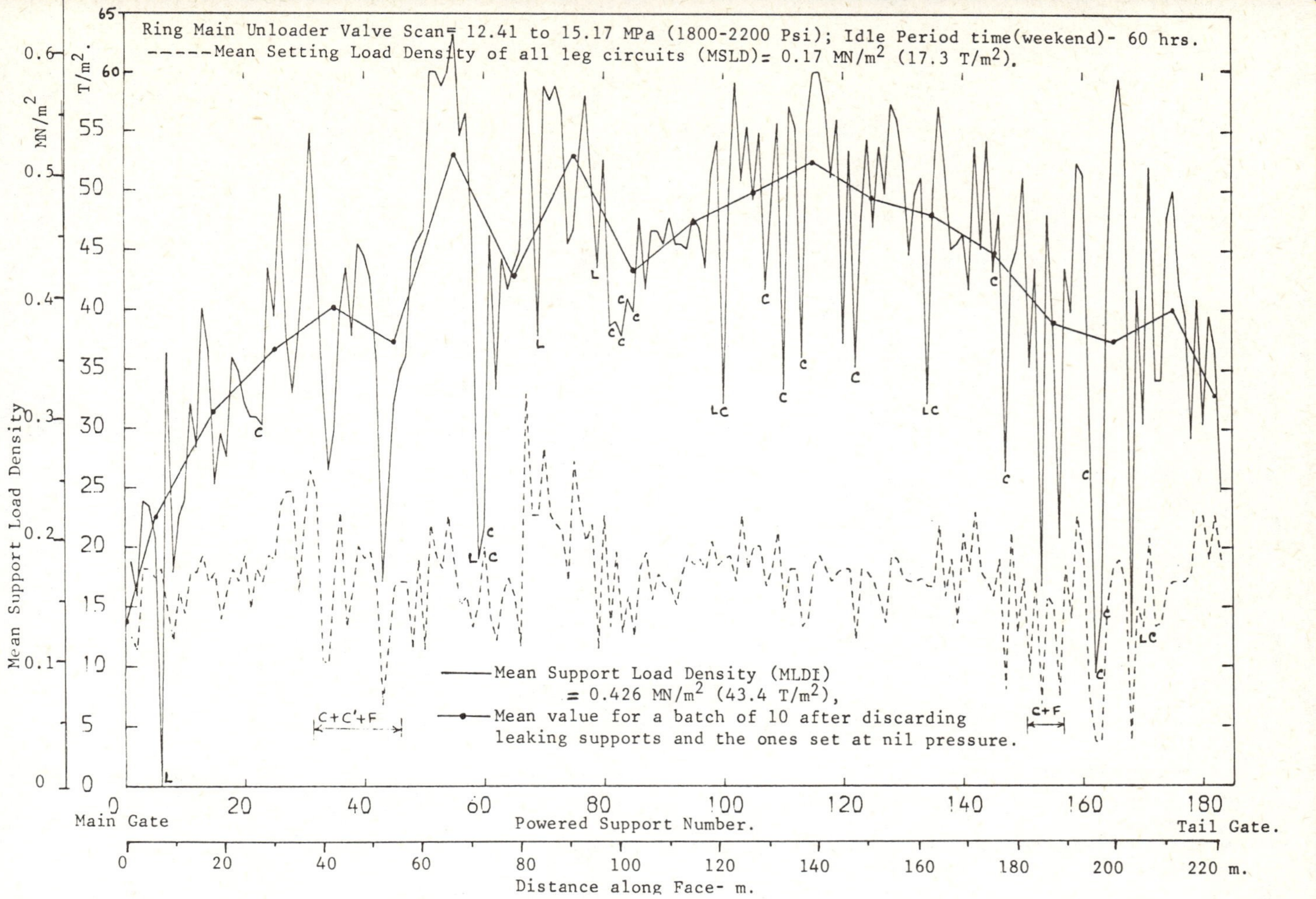

Fig.4 Setting and Final Mean Load Densities on F27 Face During a Weekend Idle Period at a Nominal Setting Pressure of 15.2 MPa.

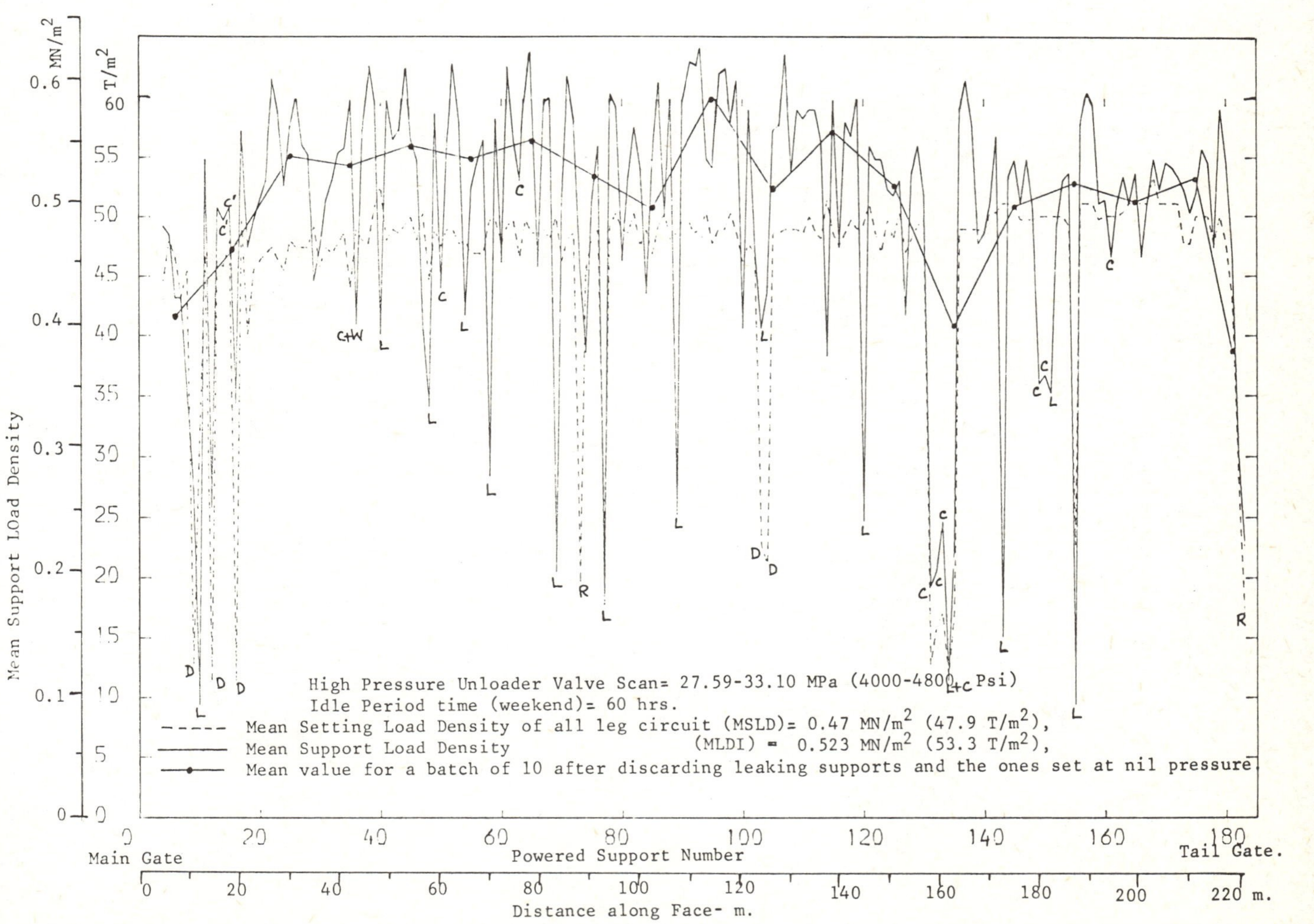

Fig.5 Setting and Final Mean Load Densities on F27 Face During a Weekend Idle Period at a Nominal Setting Pressure of 31.7MPa.

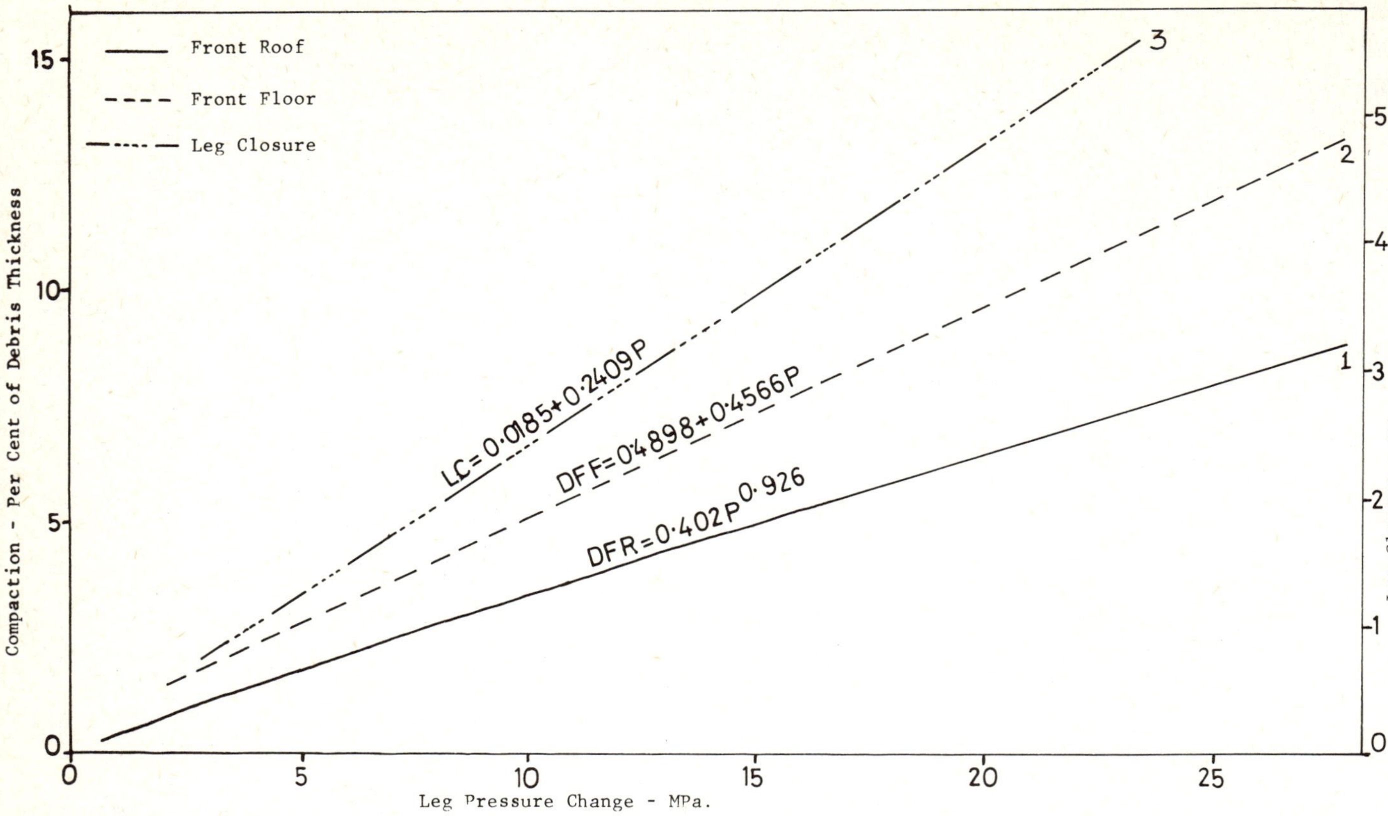

Fig 6 Debris Compaction at Front Roof and Front Floor Above and Below Gullick Dobson 6/300 T Chock Supports on F27 Face, and Leg Closure, Related to Leg Pressure Increase.

pressure (31.7 MPa) are illustrated in Figures 2 and 3. The same data over a weekend stand period are illustrated in Figures 4 and 5. These data have been chosen because they illustrate the important effect on average support pressures of leakages in the support system (designated L in the figures), cavities above the support (designated C) and damaged positive set valves (designated D).

The following points can be made from the figures:

a) Because of multiple demands on the ring main hydraulic power supply and distribution system during support advance the average observed setting pressures were 70-80% of the nominal value and fluctuations in setting pressures are high. Positive set valves enabled face supports to achieve a higher average observed setting pressure of about 90% of the observed value and to reduce fluctuation of setting pressures.

b) A support set at a low setting pressure tends to develop a low final pressure when compared with adjacent supports. This is because the resistance offered by a support is a function of debris compaction and leg closure and these in turn are related to setting pressure and support dimensions. These relations are illustrated in Figure 6, where data for roof and floor debris compaction and leg closure as a function of leg pressure increase are illustrated. The implication of fluctuations in final pressure is that differential strata movements may be caused. The use of positive set valves reduce these fluctuations.

c) Improved face mean load density values are obtained by increasing the mean setting load density. At the higher setting load densities above 0.4 MN/m^2 there is practically no increase in load density during normal extraction cycles. The resistance offered to the roof is therefore maintained throughout the cycle, limiting strata movement in general and in the period immediately after face advance in particular.

d) During the weekend stand period the mean load density increased above the mean setting load density for all setting pressures. The ratio however was 2.5 for the lower setting and 1.1 for the higher setting pressure, again indicating increased resistance to roof deformation.

e) At all pressures and on all faces there is some increase in support pressure together with some convergence which appears inevitable. This is, however, significantly reduced at higher setting pressures.

f) At low setting pressures, final front leg pressures over a weekend period increased until they were greater than rear leg pressures. This was associated with increased flaking and cavity formation, spalling and face extrusion. At higher setting pressures the rear leg pressures remained higher than the front and flaking was reduced.

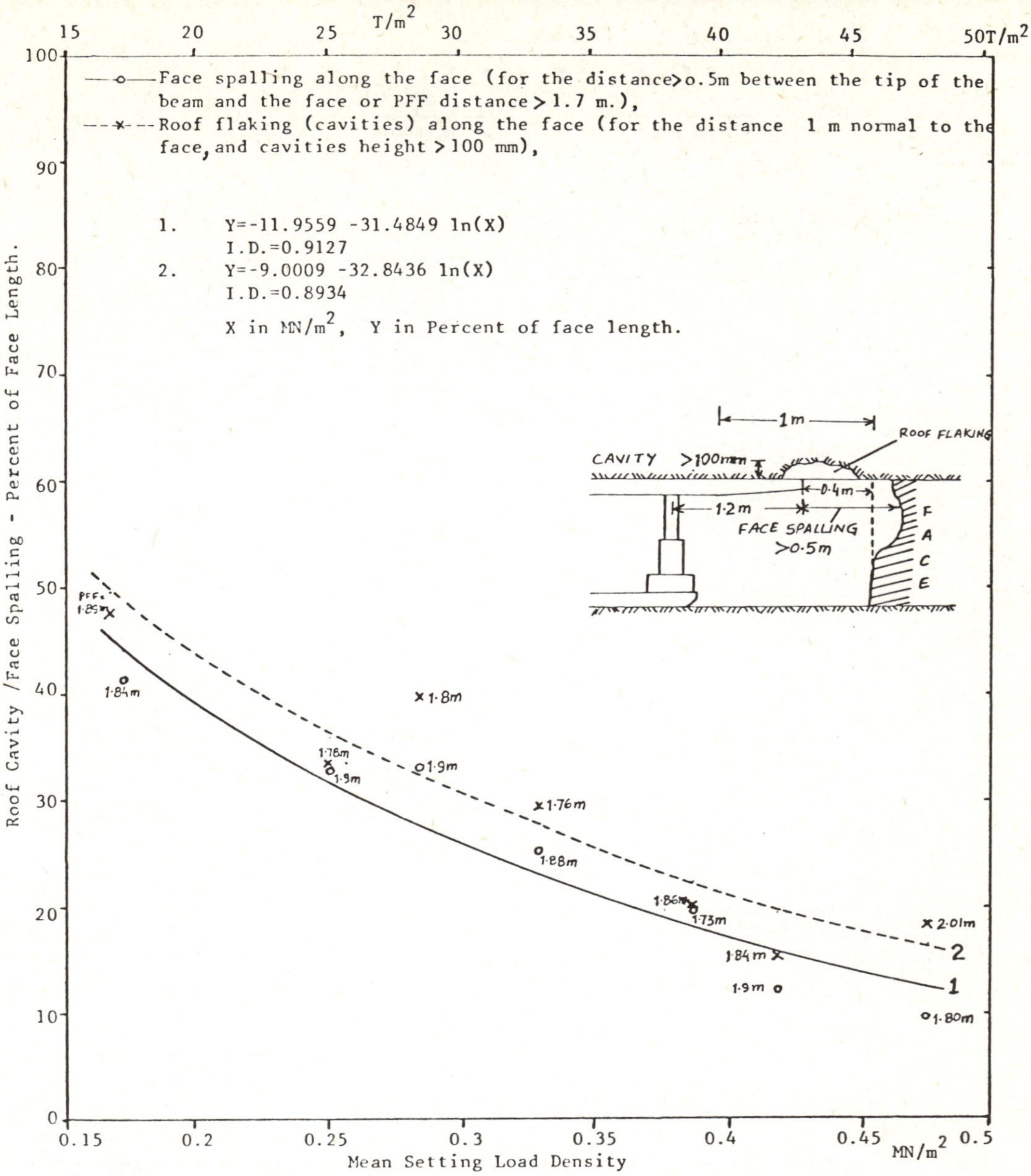

Fig. 7 Influence of Setting Load Density on Roof Flaking and Face Spalling Expressed as a Percentage of Face Length.

g) Roof flaking was always reduced with increased mean setting load density. This is illustrated in Figure 7 for F-27 face.

h) The frequency of leg circuit pressures reaching a pressure equivalent to 85% of rated yield pressure increased with increasing setting pressure. The frequency of leakage also increased and this depended strongly on the quality of support maintenance.

i) Face end zone pressures increased by a smaller amount than face pressures at lower setting pressures. This effect was reduced at higher setting pressures. The result of lower face end zone pressure increases was to reduce convergence in the face end zone with consequent improvement in roof conditions. The face end zone was about 40m from the main gate and 20m from the tail gate.

j) Convergence reduces significantly with increasing setting pressure. This can be explained partly by increased initial debris compaction due to high setting pressures (Figure 6) and partly by recompaction of bed separation in the strata immediately above the supports loosened by lowering and advancing of the support system.

k) Recompaction of separated beds can be illustrated from Figures 8 and 9.

These show that if the principal strata control function of a powered support is to maintain a competent roof by reducing roof convergence to a minimum this can best be achieved by increasing and maintaining a consistant setting load density. At the lower pressure (Figure 8) the roof compression zone is limited to 3m high above the rear and middle legs indicated by the position of the mouth of hole station

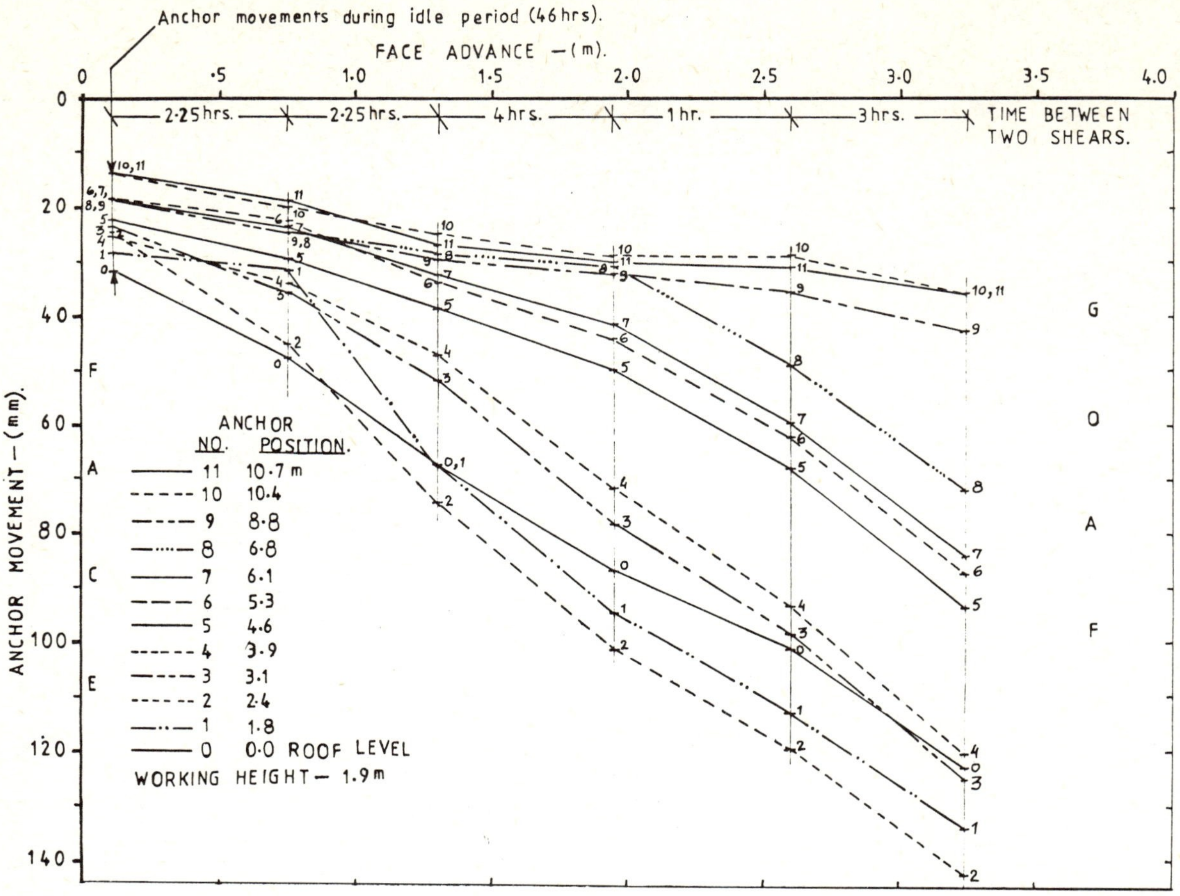

Fig. 8 Absolute Anchor Movements at Heights between 0 and 10.7 m in the Roof above Gullick Dobson 6/300 T Chock Supports on F27 Face During a Face Advance of 3.25 m at a Nominal Setting Pressure of 13.8 MPa.

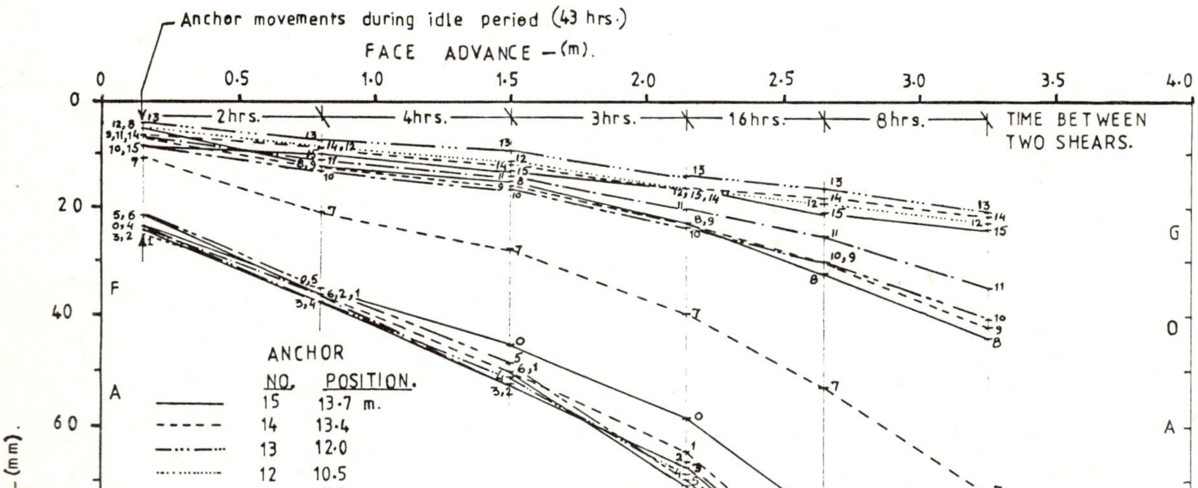

Fig. 9 Absolute Anchor Movements at Heights between 0 and 13.7 m in the Roof above Gullick Dobson 6/300 T Chock Supports on F27 Face During a Face Advance of 3.25 m at a Nominal Setting Pressure of 27.6 MPa.

relative to anchor No. 3. At the higher pressure (Figure 9) the compression above the rear support is increased to between 4 and 5m (between anchor Nos 6 and 7). More important however is the fact that this compression zone is not limited to the rear of the support but spreads over the whole area from face to waste edge, thus forming a competent roof beam.

It can be seen that the even compression from the higher pressure provides an adequate explanation for the reduced flaking and cavity formation.

ACKNOWLEDGEMENTS

The work was supported by a grant from the National Coal Board extramural research fund, and the E.C.S.C. research fund (Contract 7220-AC/806). The authors wish to acknowledge the assistance of Mr N.H. McLeod and Mr D. Kelly of the North East Area Headquarters and Mr I.W. Day and Mr T.F. Burns of Westoe Colliery. The views expressed are the personal views of the authors.

REFERENCES

Farmer I.W. (1980) Face and roadway stability underground in coal mines; geotechnical criteria. Report No. EUR7298 EN to E.C.S.C., Luxembourg.

Gupta, R.N. (1982) The influence of setting pressure and strata behaviour on longwall faces. Submitted, Ph.D. Thesis, University of Newcastle upon Tyne.

Gupta, R.N. and Farmer I.W. (1981) A magnetic ring extensometer system for strata deformation measurement in coal mines. Mining Engineer, 141, 303-305.

STRATA REINFORCEMENT IN THE BARNSLEY AREA

G. V. Silvester, B.Sc., C.Eng., M.I.M.E.

Area Strata Control Engineer,
National Coal Board,
Barnsley Area Headquarters,
Grimethorpe,
Nr. Barnsley,
South Yorkshire.

SUMMARY The paper describes the different types of strata reinforcement techniques used in the Barnsley Area of the National Coal Board, together with their technical classifications, and descriptions of their various applications underground for differing circumstances. The initial section deals with the more conventional types of reinforcement, such as bolting and dowelling and introduces some new developments in the fields of inorganic capsules. This is followed by the topics of injection dowelling, strata infusion and strata injection using cement grouts. This latter technique, the equipment, the types of cement and examples of practical installations are described fully. The new and important role of cavity filling is outlined in its four forms of applications, namely inflatables, cementitious materials, Aerostem and Isoschaum. Finally, the beneficial effects of truss bolting are discussed.

INTRODUCTION

Strata reinforcement is a process whereby the strata, weakened either by natural means i.e. faulting or by man-made factors, such as high stresses set up by interaction, is strengthened in order to minimise yielding in the abutment area, reduce convergence and avoid the collapse of strata around mining excavations. At places where roof fall has already occurred, the aim of strata reinforcement is to secure the roof and sides of the fall by filling in or supporting the cavity and by strengthening the ground around the edges of the fall area, so safe working can recommence with minimum delay. Carr and Silvester (1972)

Coal measures in the Barnsley Area have a high coal seam density, 24 of which have been worked. Consequently difficult mining conditions, caused by interaction, are very common. There are collieries where up to 12 coal seams have been worked creating many pillars and ribsides at different horizons but often in the same vertical plane that affect mining operations simultaneously. Since approximately 40 per cent of all our face delays are being caused by interaction or geology, strata reinforcement plays a very important part in maintaining continuous production.

Reinforcement can be classified technically into the following groups:

1. Strata Bolting (a) mechanical anchorage
   (b) resin anchorage.
2. Strata Dowelling (a) steel rebars
   (b) composite dowels with steel or fibreglass core & wood shells.
   (c) cuttable dowels, i.e. wood or fibreglass.
   (d) tandem dowelling
3. Injection Dowelling, as for strata dowelling.
4. Strata Injection (a) steel pipes
   (b) cuttable pipes i.e. wood & hollow bamboo
   (c) steel RSJ girders
5. Strata Infusion
6. Cavity Filling (a) inflatables
   (b) cementitious material
   (c) aerostem
   (d) chemical foam
7. Truss Bolting

Ideally reinforcement should be:

Preventive Reinforcement

Only good planning by the Strata Control Engineer and Management will help to introduce this method of strata reinforcement.

Curative Reinforcement

Unfortunately most strata reinforcements are in this category since it usually involves face recovery operations after a roof fall.

The locations of reinforcement can be classified into the following groups:

1. Longwall Faces (a) face line heading-out (prior to face operation)
   (b) during face production
   (c) preparation for salvage
2. Roadheads of an advancing face.
   (a) ripping
   (b) roof of face ends
   (c) pack voids
3. T-junctions of retreating faces or advanced headings.
4. Face headings and Advanced headings.
5. Roadways (a) roof
   (b) sides
   (c) floor

STRATA BOLTING

In this process a suitable size hole is drilled through the weak strata into a more competent rock and reinforcement is obtained by tensioning a rod between the anchorage placed to the back of the hole and the plate end, thus allowing the weak beds to be suspended from a strong stratum. The anchorage could be achieved mechanically by expanding shells or by bonding resin capsules. The latter type of anchorage is superior, especially in weak strata, since it spreads over a length of 203mm of the hole. Mechanical anchors induce very high point contact stresses in the strata around the hole which may cause the failure of the rock resulting in the loss of bolt tension. Plates should not be smaller than 203 x 203mm. The most popular bolt length is 1.829m but it can be in lengths between 0.914 - 2.438m.

The diameters of the bolts and the required boreholes and resin capsule sizes are given below:

Dia. of Bolts	Dia. of Boreholes Required	Dia. of Resin capsules
19-20mm steel round rod	28mm	25mm
20-22mm rebar (steel)	28mm	25mm
25mm rebar (steel)	35mm	30/32mm

Resin is used as a bonding agent because of its strong adhesive property, short setting time and non-shrinkage characteristic. Carr (1976). Dry drilling and good cleaning out of the holes is important to achieve efficient bonding. For good mixing action between the resin and the catalyst, both enclosed separately in the capsule, the diameter difference between the mixing end of the rod and the borehole should not be more than 10mm. Normally two 152mm resin capsules are used per bolt and the mixing time should last for 10 seconds after rotating through both capsules. The tension plate can be attached to the bolt some 3-4 minutes after mixing.

The anchorage strength of the bolt varies with the strength of the rock media. The average pull-out force required for a weak - 31 MPa, medium 31-68.9 MPa and hard rock + 68.9 MPa is 29.9-59.8 kN, 59.8-169.4 kN and 169.4 kN respectively.

STRATA DOWELLING

This reinforcement is often described as full column resin bolting. It bonds the strata along the whole length of the rod, therefore, providing a more effective support and making the use of end plates unnecessary.

The maximum length of a single dowel is 2.438m and with tandem wood dowelling a reinforcement length of up to 4.876m has been achieved.

Dowels should always be installed at right angles to the main weakness planes or fracture lines and the setting distance between dowels should not exceed 1.0m.

The diameters of the steel dowels and the required boreholes and capsule sizes are the same as for the bolts described earlier. In addition to the steel dowels, combination dowels and the so called 'cuttable dowels' are also used. The combination dowel is made up of a 19mm inner steel bolt surrounded by a wood sleeve giving an overall diameter of 36mm. It is used in standard 43mm boreholes with 40mm diameter resin capsules.

Cuttable dowels, wood or fibre glass, are used in areas where subsequent mining operations are expected i.e. shearing or dinting. The most popular wooden dowel is 1.83m long, has a 36mm diameter so it can be used in 43mm standard boreholes and it is slant cut at both ends to facilitate the application of tandem dowelling. The majority of wood dowels are made of Keruing-type wood due to its straight grain and high tensile strength characteristics.

Fibre glass dowels are relatively new and less popular than wood dowels due to their higher cost. A new combination dowel developed recently has the steel core replaced by fibre glass. The shear strength of this product is increased by 4 to 5 times over that of the normal wood dowel and it can be used in heavy conditions and where cutting operation of the reinforced ground is to follow.

Tandem dowelling is introduced when long forward reinforcement is required in the coal seam. The individual wood dowels are not coupled together physically in this system but this weakness does not appear to jeopardise their effectiveness.

The resin used for dowelling is different from that used for bolting by having lower viscosity and longer gel time for quicker and easier installation. For tandem dowelling the use of EC Phix resin is recommended which has even lower viscosity, therefore, less force is required in the spinning operation. It is most important to fill completely the annulus between the hole and the dowel and to achieve this, normally, four 330mm long capsules are required for a 1.83m long dowel.

New cement and gypsum capsules have been tried recently giving satisfactory results. They are inert, therefore, completely safe for use underground, cost less than resin type capsules and their bonding strength is adequate in most cases. However, the handling of some of the cementitious capsules, such as the Fosroc Combextra and the Celtite Selfix LH10, which need to be submerged in water prior to installation, cause some inconvenience.

In the LH20, Celtite Selfix Ltd., provides water in a separate compartment of the capsule. Cement and water are mixed prior to the installation of these capsules. There is a critical time limit on the soaking and the installation of the above products which is a constraint in the practical applications.

There is no special preparation or soaking required for the Cemicron 2000 gypsum based capsule. It incorporates water in the form of thousands of tiny spherical micro-capsules captivated in wax shells. In the process of installing the dowels the crushing of micro-capsules in advance of the bolt tip releases water and makes an easily penetratable grout substance which hardens rapidly. This capsule is manufactured by Commercial Plastics Special Products Ltd. A satisfactory field trial has been conducted on this product recently in the Area.

A unique roof bolting experiment was tried in S10's return gate at Bullcliffe Wood Colliery. Here five full column resin bonded steel bolts were installed in every bay of the arched supports behind the ripping. A Victor Pegassus type boring machine was used for the quick drilling of the holes and installation of the bolts. It is intended to withdraw every alternate arch when this advancing face stops production. Finally, the remaining arches are also to be salvaged so the gate will be solely supported by the bolts. M.R.D.E., Bretby has instrumented and now monitors the strata behaviour of the roadway.

The reinforcement strength of a dowel equals to the tensional strength of the rod. Such values are given below for the various types of dowels.

20mm steel rebar	169.4 kN
25mm steel rebar	299 kN
25mm fibreglass	290 kN
36mm combination dowel with steel core	169.4 kN
36mm combination dowel with fibreglass core	160 kN
36mm wood dowel	99.6 - 120 kN

INJECTION DOWELLING

In this reinforcement, normally 3.05 - 5.49 m long holes are drilled into the fractured strata and after the insertion of suitable sized wood or steel dowels special grout is injected around the rod into the hole under pressure. The purpose is, not only to bond the dowel into the hole by filling the annulus, but also to consolidate the weak strata around the borehole by injecting grout into the fractures. The system combines the added advantages of both dowelling and infusion although it is not considered to be as effective as strata injection.

The reinforcing dowels in the holes could be either steel rebars or solid double slant cut wood depending on the location of application. The roof would normally be strengthened by the steel dowels whereas in the coal and floor holes cuttable dowels would be used.

The standard 43mm dia. holes are made by using dry drilling methods for which the Haskel Energy type drill rods are specially recommended. Good cleaning out of the holes is essential otherwise a plug of drillings could easily be formed and stop the passage of the grout.

In most injection dowelling the mouth of the hole would not be sealed off permanently prior to injection but grout pressure would be retained by the use of an injection gun. With this technique longer holes are drilled, so after the insertion of the dowels the mouth of the hole remains free to receive the expandable end of the injection gun.

Organic resin can no longer be used for injection due to its low flashpoint temperature and because it gives off poisonous cyanide fumes if heated or burnt. There are two fully tried and approved gypsum based inorganic injection grouts on the market, Celtite Selfix M.100 and the I.C.I. Rotaset SR1/Y. Both have relatively high adhesive qualities butare rather expensive to use in quantities. The G.P.1 - type cement grout supplied by Pozament Ltd is approximately 16 times cheaper and has prθved highly successful in reinforcement work in Barnsley Area.

STRATA INJECTION

In this most successful method of reinforcement the grout, under pressure, is delivered to the back of the borehole through a pipe line system which is sealed off around its perimeter at the mouth of the hole. The grout spreads outbye in the borehole filling the annulus around the pipes and entering into the fissures and breaks in the rock. Since the highest pressure is away from the seal, air is not trapped in the borehole and a more effective bonding and better infusion is obtained than with injection dowelling.

The boreholes are normally 3.05 - 5.49m long and the individual pipes inserted and coupled together can be made of steel or wood. The diameter of the steel pipes can vary and will be described later. The wood pipes, which have an outside diameter of 36mm could be either the split or the grooved types, both have a 19mm diameter plastic tube inside to act as a passage for the injected grout.

The same type of grouts are used as described for injection dowelling.

In heavy conditions where extra forepoling strength is required the steel pipes are replaced by 76 x 76mm cross section RSJ girders. 114mm diameter holes are drilled and 1.5m long sections of girders are fishplated together and inserted into the borehole. They are bonded into the roof strata by cement injection through a 19mm diameter plastic tube extending to the back of the hole. A short tube is used for 'breathing' purposes and the mouth of the hole is sealed off with Hardstop or Packbind. N.C.B. Broadsheet No.44 (1980)

HISTORY OF STRATA INJECTION WITH CEMENT GROUT

The first cement injection in Barnsley Area was introduced on F.19's retreat face in the Kent Seam at a depth of 174m at North Gawber Colliery in May, 1976. This 101m long face had difficult strata conditions from the start of working, due to saturated, broken roof and soft wet floor. Full column bolting and resin injection was tried for 7.3m of face advance, without much success, when a 21m long 1.83m high roof fall occurred at the middle of the face.

It was decided to drill 25 x 3.66m long 89mm dia. horizontal boreholes 457mm above the seam. See figure 1 for the pattern of reinforcement.

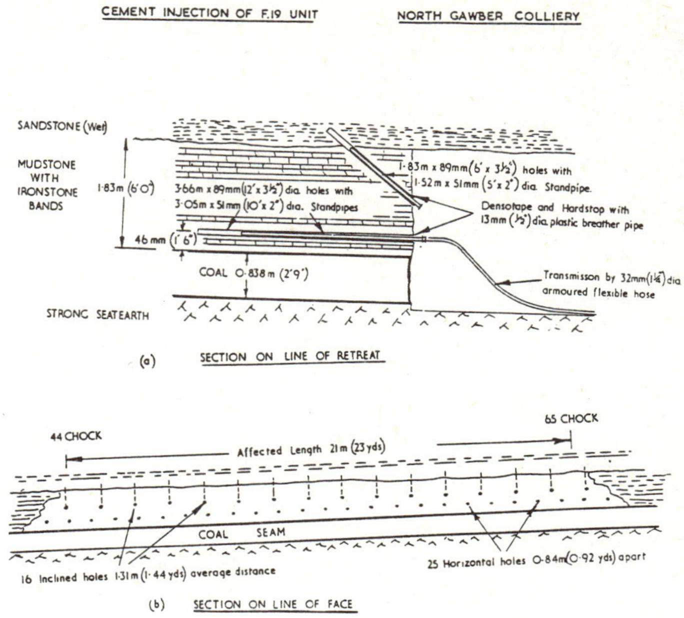

FIGURE I.

3.05m long 51mm dia. standpipes were inserted into each hole and the mouth of the hole was sealed off by a bandage of Densotape and Hardstop allowing a 13mm dia. plastic breather pipe to get through. In addition 16 x 1.83m long, 89mm diameter inclined holes were drilled at an angle of 45°, started 0.91m above the seam and reaching into the sandstone which formed the roof of the fall area. Here 1.52m long 51mm diameter standpipes were used for injections.

The holes were drilled by hand held Victor electric borers. The cement grout pumps, first an Edeco Mini Hydrack drilling machine's sludge pump replaced by a D4 Mono pump, were positioned in the tail gate. The cement and water were mixed in a 182 litres capacity home made mixer tank and the grout was pumped to the fall area through a 91.4m long 32mm diameter armoured flexible hose at a pressure up to 1.38 MPa. The grout was made up of Portland Cement, Bentonite and water in the proportions of 203 kg 4.1 kg and 90.7 kg respectively. These represent 0.45 water/cement and 1/50 bentonite/cement ratio by weight. The total quantity of cement used was 2.54 tonnes amounting to an average of 62.1 kg cement per hole injected. The drilling and injection operations were completed in three days (9 shifts) and the roof was held up successfully by the reinforcement when production recommenced.

Following the success of this operation other collieries in the Area had also started using cement grout for strata injection. Increased practical experience led to several improvements of the described technique.

In the majority of cases, cement injection was applied in fall areas on longwall faces, but it was used effectively also for reinforcing face headings and for stopping air leakage at a booster fan at Dearne Valley Colliery. Successful application of cement injection of the roof were made at the following collieries in Barnsley Area:-

North Gawber, Houghton Main, Woolley and

South Kirkby.

In the majority of cases, adverse roof conditions were caused by interaction resulting in up to 5.0m high roof falls of the fractured strata. As a result the face had to be stopped and 2.7m long cross bars and wooden chocks were packed above the powered supports. When the fall area was cleared and safely supported, the boring operation could commence.

BORING EQUIPMENT AND HOLE SIZES

The recommended type of drilling machine is the Torque Tension TMT 20 which operates hydraulically and held in position by a double acting hydraulic leg.

The hand held electric or hydraulic borers can also be used for holes not in excess of 52mm diameter. The Edeco Mini Hydrack drilling machine was used in a number of cases for boring 76 - 114mm diameter holes. Being a big and heavy machine, it is difficult to manouvre on faces with limited height. Where compressed air is available the Silver 3 Rotary Percussive borer is also used with success.

In a roadway steel dowelling exercise the Victor Pegassus portable drill rig was found to give good results. It is important to use dry drilling whenever possible.

INJECTION PIPES

The combination of the diameters of the holes and that of the inserted pipes for injection can vary as follows:-

Diameter of hole	43mm	52mm	76 - 89mm
Diameter of pipes to be used	25mm	38mm	51mm

The individual pipes are normally 1.52 long and couplers are used to connect up the pipe range. The length and the diameter of the borehole is determined by a number of factors i.e. the number and type of boring equipment, the time and manpower available, the size of the fall, the degree of fracturing of the strata and the weight on the face. In heavy conditions or with big falls the longer and greater diameter holes are recommended.

The pipe at the outbye end incorporates a stop valve and a mechanical injection seal which is compressed and expanded against the perimeter of the hole by operating a sleeve from the outside end of the pipe.

One hour after the injection both the valve and the compression sleeve can be recovered, cleaned and re-used at another borehole. This type of seal is easy and cheap to make and provides better results than Hardstop which often blows out at higher pressures and it is not effective for the first 24 hours.

The injection is continued until:-

Leakage of the grout occurs at the sealing or crevices of the strata, or

The pre-determined grout pressure or pump stalling pressure is reached, or

Despite the high quantity of grout injected no increase of pressure is obtained.

PUMPS AND MIXING TANKS

The mixing tank and the pump were sited, normally in the roadway. The most suitable sizes of the mixing tanks were between 136 - 205 litres capacity. The most successful cement injection pump was the Hany MK V - type marketed by Mindev Ltd. and powered hydraulically. The costs of a Mindev GM.30 grout mixer and Hany MK V injection pump are £1,330 and £4,842 respectively. Other pumps were also used with varying degrees of success, i.e. Edeco Mini Hydrac Sludge Pump, D3, to D8 Mono Pumps, but these were not designed to handle cement grout hence their shortcomings and early breakdowns. The Reader D7/4 grout pump, manufactured by Soil Mechanics Ltd. in this country, may become a good competitor of the Swiss made Hany pump. It is hydraulically operated and can deliver grout at a rate of up to 136 litres per minute at 4.125 MPa pressure and costs some £1,000 less than the Hany pump.

CEMENT GROUT

A complication with ordinary Portland cement is that it may have to be mixed with other substances i.e. pulverised fuel ash, bentonite, polymer, plaster and accelerator in order to be suitable for injection. Such requirements are:-

Thixotropy, i.e. the grout does not segregate, but stays in suspension whilst being mixed or pumped.

Small grain size and spherical in shape so less water is required to mix the grout.

Low resistance against pumping, i.e. low viscosity.

Good penetration in broken ground.

Strong initial setting and good bonding qualities.

It is not practical to do such mixing underground. Pozament Cement Ltd. of Hargreaves Group has already put on market some 35 different brands of cement mixes to choose from. Barnsley Area has decided to use the following three types.

GP1 for normal strata injection.

NCB NUMBER 2 MIX for cavity filling and pack reinforcement.

TGT for injection in wet conditions.

The GP1 - type has the highest compressive strength of all currently available Pozament products which varies with the water to cement ratio and with the time of maturity. Using a ratio of 0.4, the compressive strength of the cement will be 5.52 MPa one day after and 13.1 MPa 7 days after application. It consists of 50% ordinary Portland cement and 50% pulverised fuel ash.

For easier pumping and better strata penetration, the recommended water over cement ratio by weight is between 0.5 - 0.6.

When injecting 'in the solid' it is considered to be advantageous to use pressures up to 4.125 MPa in order to get good rate of strata penetration. With the previously described mechanical sealing such pressure is a practical proposition. However, when an exposed roof is injected or infused, on no account should the pressure be increased over 0.69 MPa because it could result in a roof fall. The intensity of 1.38 MPa roof pressure is capable of raising most roof supports to yield load. Each pump is provided with a pressure gauge and in such reinforcement operation the pumping pressure should be strictly monitored and limited to the value mentioned above.

The Pozament NCB No. 2 Mix is a low free-flow type grout which was developed jointly by Pozament Ltd. and MRDE Bretby, it is more viscous and has a higher setting speed than GP1. It is self supporting in 30 minutes after application and has a compressive strength of 0.41 MPa after two hours.

The TGT Pozament grout was specially designed for

wet conditions. It is claimed that the grout is not diluted and its compressive strength is unaltered by the water present in the strata. This grout was used only once, at M.04's face, Woolley Colliery, therefore, its reinforcing qualities could not be assessed due to limited experience. The same water over cement mixing ratio and injection techniques were used as with GP1 and the wet, fractured roof stayed up in the area of reinforcement.

It is advisable to wait for 24 hours after the completion of all cement injections in order to allow a good setting of the cement before the cutting of the face is recommenced.

Figures 2 and 3 show typical roof reinforcements by using a combination of cement injection and steel dowelling in a fall area and in a heading-out of a longwall face respectively.

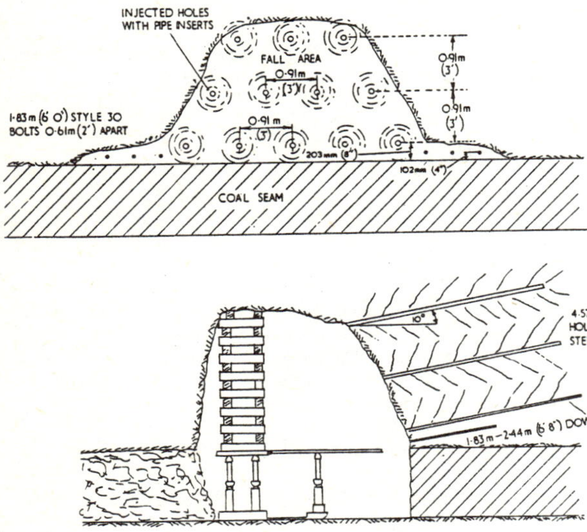

TYPICAL FALL AREA AT A FACE

FIGURE 2.

REINFORCEMENT OF THE FACE HEADING-OUT BY INJECTION AND DOWELLING
X.12 FACE HOUGHTON MAIN COLLIERY

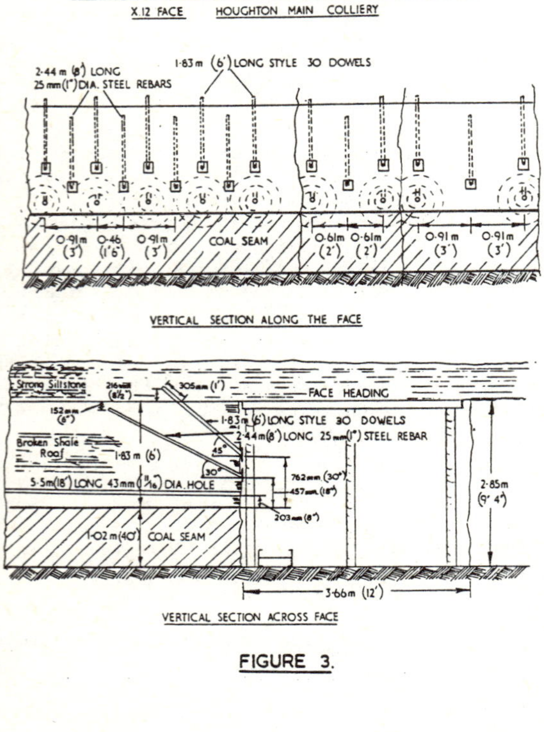

FIGURE 3.

STRATA INFUSION

This technique is used in soft, completely shattered strata, i.e. in goaf area or in the fractured and heaved floor of a roadway, where dowelling would be ineffective and the use of pipes is impracticable. Lower viscosity grout with decreased compressive strength is used in order to achieve a higher degree of penetration.

There is no insert in the borehole for reinforcement or for conducting the grout to the back end of the hole. As a rule this is not as effective as injection but it is quicker and cheaper. The seal is the only device which is introduced into the mouth of the hole in the form of an infusion gun.

The application of strata injection and infusion is described and illustrated in the next chapter.

FLOOR REINFORCEMENT IN E.06's MAIN GATE, NEWHILL SEAM SOUTH KIRKBY COLLIERY.

This seam lies at a depth of 406m and the coal thickness varies between 1.83 and 2.18m. Floor injection was introducted in E.06's main gate in order to reduce floor lift and the penetration of arch legs of the 3.66 x 2.13m size Hollybank supports. The face and panel lengths of the unit were 128m and 640m respectively.

91m length of the roadway was affected by interaction causing wet and heavy conditions, the deflections of support beams, floor lift and arch leg penetration of the floor, resulted gate convergence of between 600 - 750mm.

First a row of five holes were drilled in the floor across the gate in every bay of the supports i.e. 0.91m apart. See figure 4.

FLOOR INJECTION AND INFUSION. E.06 MAIN GATE. SOUTH KIRKBY COLLIERY

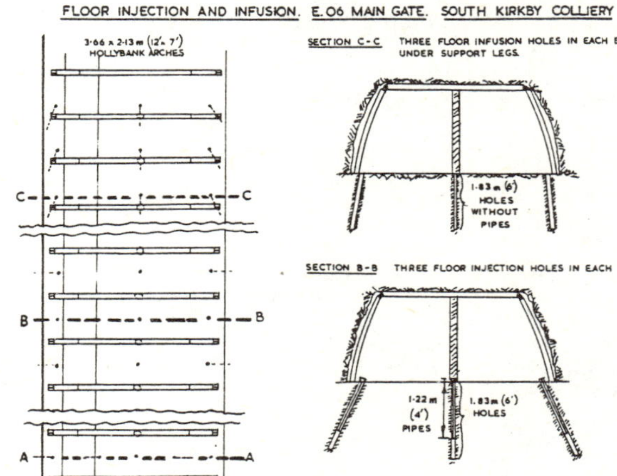

FIGURE 4.

The 1.83m long, 43mm diameter holes were drilled by an electric hand borer. The two outside holes and the one at the middle were to reinforce the floor under the arch legs and the middle props respectively. The 1.22m long, 25mm diameter Unicone pipes were sealed into the floor strata at the top of the hole by Hardstop and injected by a D3 Mono pump at a pressure of 0.83MPa. The pump and the paddle in the mixer were powered by a hand held boring machine.

The cement used was the Pozament N.C.B. No. 2 type. Normally, one bag of cement, i.e. 25 kgs + 10 litres of water was mixed in a 22.7 litres capacity mixing tank and pumped into each hole. Some holes, however, accepted twice the above quantities. The water over cement ratio by weight was 0.4.

Due to the holes being 0.91 apart along the width of the gate the cement grout was often fed through the crevices from one hole to the other without reaching the required pressure of injection. As a result, the 2nd and 4th holes in the row were left out of the boring pattern, injecting only the floor area under the support legs.

Since not all sealings of the pipes were satisfactory and in an effort to save time and money, it was decided to try the strata infusion method. In this reinforcement the hole was left empty and the grout was introduced by an infusion gun which sealed off the hole at the top. The cement grout under pressure filled the borehole and entered into the crevices around it. Up to 20 holes were drilled and infused in each shift by 2 men.

In the reinforced stretch of roadway, the degree of convergence was negligible. This was true even in the part of the roadway which stood for a longer time, during the Christmas holiday, under the influence of the face front abutment pressure. All three types of reinforcement were successful for the following two reasons.

It was found that water, introduced by interaction into the breaks of the floor in the roadway, was squeezed out by the cement grout under pressure. This action not only stopped any further softening of the immediate floor measures but consolidated and strengthened it therebye making it more resistant against forces causing floor lift.

The cement grout infused into the fractures of the floor increased its compressive strength and reduced the rate of floor penetration of the support legs.

PACK REINFORCEMENT IN E.11's TAIL GATE, SOUTH KIRKBY COLLIERY.

Due to interaction, some very difficult roadway conditions were experienced with the previous six retreating faces and in an effort to reduce this problem it was decided to advance E.11's face on an experimental basis.

The 4.27 x 3.05m arched roadway was formed as the face advanced supported by 114 x 114mm section three piece arches set on box stilts 0.91m apart. The 3.66m long, 1.68m high face side packs were put on manually at 1.37m increments. The strength of this pack was lower than that of the solid ribside causing uneven loading of the arches and consequent roadway deformation. In order to strengthen the face side of the roadway, cement grout was injected into each of the last but one pack to fill the voids between the roof and the top of the pack and also inside the debris. See Figure 5.

1.83m long, 51mm diameter Unicone pipes were placed on the top of the packs in the centre position. The protruding ends of the pipes were flanged enabling a quick attachment to the 32mm diameter flexible injection hose leading from the pump, and provided with a Unicone adaptor. The pack was sealed off prior to injection by lagging boards or corrugated sheets and Hardstem on the roadside, by the last pack on the face side, and by a row of wooden buttress chocks and hessian cloth on the goaf side.

PACK REINFORCEMENT E.II TAILGATE SOUTH KIRKBY COLLIERY

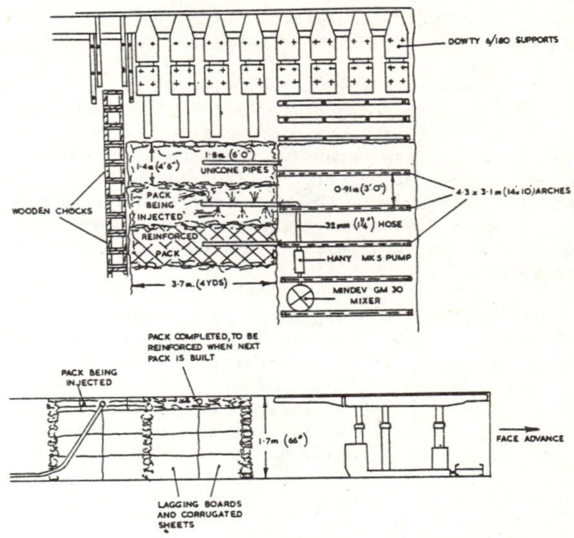

FIGURE 5.

The cement grout consisted of a mixture of POZAMENT, N.C.B. No. 2 type cement and water, using a water ratio of 0.4.
cement

The hydraulically operated Hany MK. V cement injection pump, was used for the delivery of the grout from the mixing tank to the pack. The operating pressure of the pack injection was up to 1.8 MPa . Injection was continued until larger volume of the cement grout started escaping from the packing or the quoted maximum operating pressure was attained.

Two men were employed on each day shift to reinforce up to three packs. The estimated cost of this operation for each metre of advance was £45.

Problems encountered with pack reinforcement were:

It was difficult to contain cement grout on the top of the pack without premature discharge through the pack sides. A more efficient shattering system would be required, especially on the goaf side if this method is to be made more successful.

Pack reinforcement can fall behind schedule, causing it to lose its effectiveness. Therefore, it is important to grout always the last but one completed pack systematically.

The beneficial effects of pack reinforcement were due to the voids left in the pack being filled by the grout resulting in the compressive strength of the pack debris being increased.

Along the length of roadway where effective pack reinforcement was practiced, convergence was decreased by 300mm, on average with very little arch deformation.

CAVITY FILLING

Although cavity filling could not be described in the same category as strata reinforcement, it will improve roof conditions. As a result roof falls will be stopped, bed separation minimised and vertical and lateral movements reduced. It is believed that a material with low specific gravity and compressive strength would be acceptable if it would set quickly after application and it would be safe, simple and cheap to use. One of the greatest problems is how to hold the liquidous substance in position on top of the supports, without escaping before it solidifies. Perhaps the answer is in using various size plastic bags, tailor made for the roof cavities. If these containers could be blown up by air or by the grout to take up the shape of the cavity it could serve a very useful purpose.

This method, when fully developed, could be used above the roof supports not only to fill up cavities caused by roof falls, but also on face-out headings where some roof was excavated with the seam making the overall height bigger than the coal section. Timbering up above the powered supports is an expensive, time consuming and often dangerous operation and it would be very helpful to render this work unnecessary.

The need for developing suitable methods for cavity filling has increased in recent years due to roof problems caused by the ever increasing effects of interaction and the more widespread use of shield supports. Four types of cavity filling have been tried in the Barnsley Area. Inflatables, Cementitious materials, Aerostem and Chemical Foam.

INFLATABLES

The Inflatables were constructed of neoprene or P.V.C. coated nylon, made to cylindrical shape and fitted with a non-return inlet valve and two safety valves. They were pressurised to 0.0069 MPa after being placed into a cavity and intended to keep loose material in position in the roof and sides of the fall. The initial trial was made in a roadway between the arches and the ribside, near the pit bottom at South Kirkby Colliery. The required pressure was obtained by the use of a small foot pump. Even this very low pressure was adequate to displace an arch laterally. Subsequent face trials caused a temporary setback due to the vulnerability of the inflatables to puncturing by sharp objects. It is thought that filling the inflatables with a suitable quick setting, low density material would improve their reliability. As a result aerated cement and Hardstem are being developed in the Area for this purpose.

CEMENTITIOUS MATERIALS

Liquid cement can be pumped into a cavity and retained by the use of shuttering or by pumping it into bags. Surface trials are being held in collaboration with Fosroc C.B.P. Ltd. and H.Q. Strata Control Engineers for aerating liquid cement by the application of a foam therebye expanding its set volume, by approximately 4 or 5 times. The product has many desirable qualities for mining applications. It is safe and much lighter to handle than normal cement. Cost savings in both price and transport coupled with the yielding qualities of the product, make it an attractive proposition.

It is intended to try this substance for filling bags in roof cavities and packholes and for inflating special small envelopes fitted to the top of arches. This latter system fastens arches into the profile of roadway excavations, prevents point loading and increases their load bearing capacity.

AEROSTEM

The first underground application of this material was at Houghton Main Colliery for filling roadway cavities and sealing off the perimeter of a booster fan site. It was produced by mixing sodium bicarbonate with water and Hardstem (gypsium plaster) which make bubbles of carbon dioxide. These are trapped within the body of the material as it sets and form a honeycombed structure. When mixing 100 parts of Aerostem to 60 parts of water the bulk density is reduced from 1200 kg/m^3 to 720 kg/m^3 which represents 60% expansion.

Aerostem has been applied pneumatically by a Markham Spraying Machine and also by pumping it into the roadside cavities using a Monopump. This safe and relatively cheap product is marketed by British Gypsium Ltd.

CHEMICAL FOAM

This foam product is known as Isoschaum. It is made in Germany and has been used in that country for 25 years for cavity filling.

A typical installation consists of the pumps, stop valves, gun, hoses and two solutions and is available as a complete working package which can be purchased in Britain from Rota Engineering Ltd.

Foam Characteristics

Isochaum, a Urea-formaldehyde foam is made up of two liquid components:-

Resin solution B.K. (This is Urea-formaldehyde in an aqueous solution).

Acid foam solution B.(It contains a foaming agent and hardener).

Four parts of resin and three parts of foam solution, are brought together in a gun, by two pumps operating on compressed air at 0.55 MPa. In order to produce 1 m^3 of Isoschaum approximately 20 litres of resin and 15 litres of foam solution are required. The foam expands 30 times its liquidous state by volume. The densities of the wet and dry foam are 42 kg/m^3 and 12 kg/m^3 respectively. The compressive strength of the foam, after 48 hours, or after 24 hours in warm conditions, will be 5 tonnes/m^2. The foam will harden in approximately one hour's time, depending on catalist quantity and ambient conditions. It is water resistant and excellent thermal insulator, similar to that used in the building industry for cavity wall fillings. Isoschaum can be applied remotely into the roof cavities, therefore, nobody needs to enter unsupported or unsafe ground. The material is compressible and will only lose its elasticity after 49% of shrinkage. There is no adhesive bonding between Isoschaum and the surrounding strata.

During the hardening of the Isoschaum foam, small quantities of Urea-formaldehyde are given off. These fumes can irritate the eyes and throat, and could cause headaches. Therefore, a Threshold Limit Value of 2 parts per million is the maximum concentration that is allowed in the general body of air on the return side of the cavity filling. This can be detected by the use of Drager tubes.

The Use of Isoschaum

Because of the fire and the consequent gas hazards of this organic foam product, the Hazards Assessment Committee of the N.C.B. have allowed only a limited number of controlled applications of Isoschaum in this country. It was for the purpose of filling roof cavities on face lines under close supervision. It is not allowed to use this foam in gate roads or in packhole areas.

Isoschaum has been used in Barnsley Area at the following Collieries: South Kirkby, Bullcliffe Wood, Park Mill, North Gawber, and Barrow. It was first introduced in August 1980 and since that time (in 16 months) 4060m^3 of foam have been used in the Area.

Isoschaum Equipment

The equipment consists of a compressed air powered twin piston pump, one draws the resin the other the foam solution out of individual barrels and conveys them separately to the mixing gun in their correct proportions. The weight of the Monarch pump is 57 kgs and its dimensions are 72 cms x 43 cms x 23 cms.

Three separate lines of hoses are used for connecting the gun with the pumps.

The foam gun, in which the solutions are brought together with compressed air, produces the resin foam. It has three stop valves which are operated by one control handle. The bigger 500 type gun has a capacity of 5 litres/sec., weights 11 kgs. and produces 1m^3 of foam in approximately 3 minutes. The required compressed air pressure range is between 0.3 to 1.24 MPa at 60 - 114 litres/minute quantities.

The resin and foaming solutions are delivered in 200 litres capacity polyethylene drums, and are colour coded yellow (resin) and red (foam) respectively. The costs of one drum of resin and foam are £142 and £40 respectively and the total cost of the equipment is £2,589. One drum of resin and 3/4 drum of foam will produce 10m^3 of Isoschaum at a price of £172.

Application of Isoschaum in Roof Cavities of Longwall Faces.

The sequence of activities is as follows:-

Drill 114mm dia. 1m long holes in the immediate roof at 1m distance.

Fit 76mm x 76mm RSJ's into the holes in such a way that they reach over the top of the cantilevers of the powered supports.

Make a platform over the girders and powered supports by using cover boards, split bars and brattice sheets.

Cut to size tubes made of Brattice cloth or flexible ventilation ducting.

Tie the ends of the ducting in order to form a sausage shaped container and make a few small holes to allow the air out when pumping.

Pump the chemicals from the gate nearest the fall, along the face to the site of application.

Place the tubes on top of the platform and inflate them from the middle with Isoschaum.

If the cavity is not filled by the "sausages", wait until the foam hardens and then push the probe through into the cavity and fill it up with new foam. It is often difficult to judge if the cavity has filled up to the top. One indication is the formation of free water which flows down onto the face once the foam begins to be compressed during pumping.

TRUSS BOLTING

This method of reinforcement is used mainly in retreating roadways with rectangular cross sections and solid non-yielding ribsides but it has been applied successfully also in face opening-out headings. In all cases truss bolts are supplements to the normal standing supports and their best applications are in broken roof conditions, to prevent 'dead weighting' of the immediate roof.

Truss bolts can be very useful in roadways having a soft floor where support legs are likely to penetrate into the floor. Other advantage is that they don't obstruct the roadway which could be very beneficial in the vicinity of longwall face entrance points.

The truss bolts are manufactured by Torque Tension Ltd., and consist of two steel wing dowels, two coupling boxes, two 19 mm steel rods and a threaded turnbuckle. See figure 6.

COMPONENTS OF A TRUSS BOLT

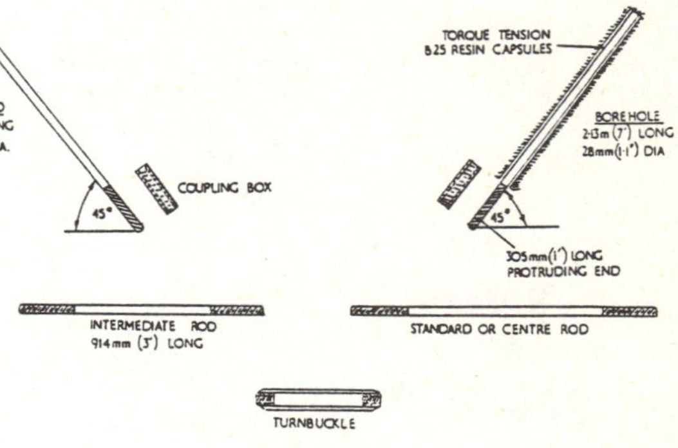

FIGURE 6.

In order to explain the reinforcing effects of truss bolts, the mechanism of roadway deformation especially that of roof lowering is explained below and illustrated in figure 7.

MECHANISM OF ROOF DEFORMATION

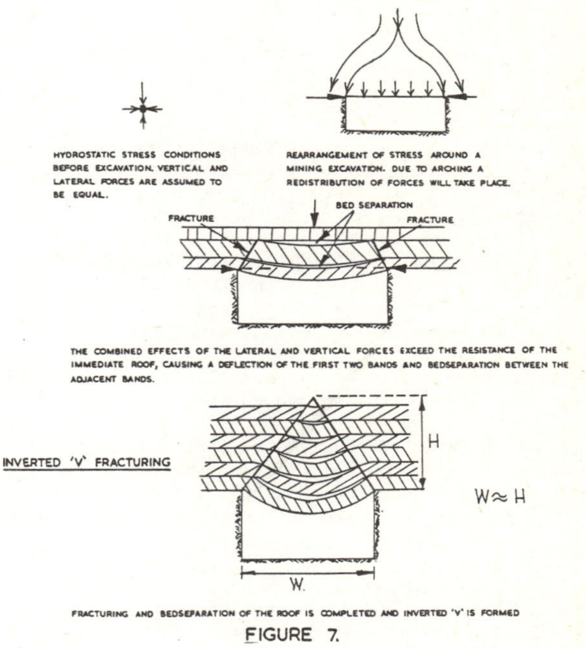

FIGURE 7.

Vertical and horizontal loading of the immediate roof tend to deflect it in the middle of the roadway causing roof lowering. The effect of this movement will be roof expansion and the development of tensile breaks. Roof lowering followed by bed separation will progress upwards to successive layers and it will cause the sagging of the upper roof too. When this movement exceeds the elasticity of the strata an inverted 'V' shaped fracture will develop starting at the sides of the gate and meeting in the middle approximately at the same distance above the roof as the width of the roadway. Silvester (1975). If support legs allow the roof to lower at the sides of the gate the inverted 'V' breaks will open up and this wedge of the roof will tend to slip out, causing 'dead weight' conditions.

The overall effects of truss bolts are, (See Figure 8) to strengthen the immediate roof by putting it under compression, which tend to minimise or neutralise the build up of tensile stresses. Kmetz (1970)

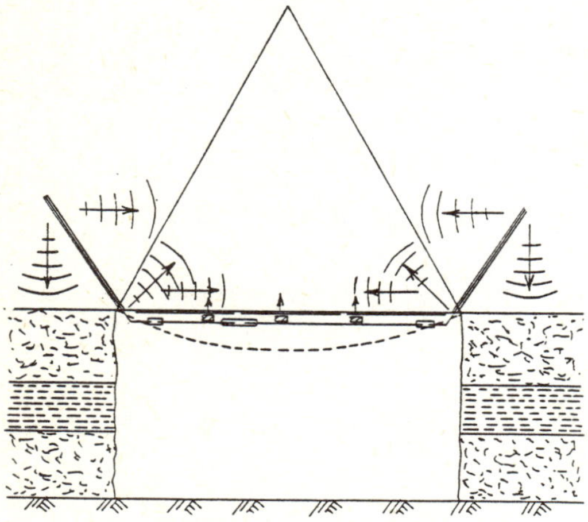

THE PATTERN OF FORCES INDUCED BY TRUSS BOLT AND THE INVERTED 'V' FRACTURE OF THE ROOF

FIGURE 8.

By reducing the inward deflection of the roof, bed separation is minimised and a stronger strata beam is maintained.

If truss bolting is installed after the development of inverted 'V' shaped breaking of the roof or despite such reinforcement the fracturing takes place, the beneficial effects of the truss bolts will be:

(a) The horizontal component of the force will tend to keep the 'roof jaws' closed, therefore, increasing the friction along the break lines.

(b) Keep the wedge shaped broken roof in position by the supporting action of the truss bolts.

The combination of weldmesh lining and truss bolts will prevent the scaling and smaller falls of the roof.

The downward forces in the roof will be transferred via the wing dowels into the ribsides of the roadway. If these forces exceed the strength of the sides of the gate a slackening of the trussbolts will take place and their reinforcing action will be lost.

The following collieries used truss bolting successfully in Barnsley Area: North Gawber, Ferrymoor Riddings, Rockingham, Royston and Park Mill. The latter colliery has applied this technique with good results also in a 6.10m wide face opening-out heading. N.C.B. Broadsheet No. 30 (1978).

A new type of coupling box was developed and tried which allows more flexibility in coupling the bolt and leadsto reduced time for installation.

Average performance by two men installing eight truss bolts per shift which represents up to 4m of advance at an overall cost of £128.

CONCLUSIONS

Bolts with point anchorage should not be used for strata reinforcement, full column resin bolts or dowels are recommended instead. Tandem dowels with their longer reach can be successful in reducing the yield or expansion and the consequent fracturing of strata in the abutment area. After being the first Area in this country to introduce and perfect cement injection for the purpose of strata reinforcement, now the trials of the first cementaceous capsules have been completed. These capsules are used in combination with wood dowels for the reinforcement of a face heading. Being inorganic, these capsules are safer than the resin types, less spinning force is required for their installation and the cost is reduced by approximately 50p in the case of a fully bonded 1.8m long dowel. However, the need to soak these capsules in water, prior to their installation and the time limit in soaking and during installation, cause some inconvenience in their application.

The Cemicron 2000 capsule looks a safe and promising competitor and a potential replacer of the resin based capsules.

Injection dowelling gives the additional benefit to normal resin capsule dowelling that it consolidates the weak strata around the borehole. However, vertical holes cannot be injected by this method.

Strata injection through steel pipes and in combination with mechanical seals and two stop valves were found to be the most successful method of roof reinforcement. It is slower and more costly than dowelling, therefore, injection is only recommended in severely broken roof conditions or in fall areas in excess of 2.0m height. In a number of cases the combination of the two reinforcing techniques is advisable. Dowelling can be used in the fringe areas of the falls and it can also fix the injected and consolidated immediate roof to the stronger bed above.

In Barnsley Area cement grout is used mainly for strata reinforcement because of its cheapness, versatility and adequate qualities. Three types of Pozament cement mixes are used currently the GP1, NCB No. 2 mix and the TGT. Their cost, when compared to the Selfix M 100 type injection grout, is between 13 - 16 times lower.

In the majority of cases the injection was performed through horizontally placed steel pipes which offered additional strength to the roof measures. In most cases, where roof falls occur on a longwall face, there is no high loading in evidence except highly fractured strata. The main objective of strata reinforcement in such situation is to keep the roof in position until the powered supports can get under. Shield supports, used in an IFS system or with high tip load hydraulic forward cantilevers, are the recommended type of supports in strata crippled by interaction.

In the strata infusion technique no inserts are put in the boreholes for extra strength and it is mainly used for consolidating the immediate floor of a roadway, thereby reducing floor lift and the rate of support leg penetration.

Pack reinforcement may reduce roadway deformation, however, it must be applied in the earliest practicable time after the completion of the pack. In order to save the extra work involved in sealing off the face side of the pack, the most ideal place of injection is the last but one finished pack.

Cavity filling is becoming an importantant strata control technique. The cost of the materials and their applications is insignificant when compared with the cost of lost production of a longwall face.

Inflatables are cheap, simple and quick in application but their present construction is vulnerable to sharp objects and subsequent puncturing. It is established that even a small pressure of 0.0069 MPa can provide effective support, providing the whole inside area of the cavity is covered.

Aerostem has been used underground for the first time in Britain for cavity filling and as a sealant around the perimeter of a roadway. Approximately 60% volumetric expansion was achieved and the aerated Hardstem applied successfully in both sprayed and pumped forms.

Surface trials are being carried out for the development of an aerated cement product which expands 4 - 5 times its original volume when treated with a foaming agent. It is intended to be used in a pumped form for cavity filling and for packholes.

The load bearing capacity of arches could be increased economically by filling the attached envelopes with an aerated inorganic product therebye avoiding point contact with the roof.

The use of Isoschaum foam in Barnsley Area has proved to be very quick, safe and effective. The filled bags helped to establish forward lateral support to the roof, thereby preventing the formation and opening of breaks in the strata ahead of the face. The foam in the free space above the bags helped to cushion falls of strata once the supports were moved and prevented ingress of large quantities of broken material onto the face.

Truss bolts could be very effective supplements to normal standing supports in retreating roadways and face headings when conditions are favourable for their applications.

During the last 24 months, 225½ tonnes of Pozament GP1 cement has been used in Barnsley Area at an approximate cost of £11,169. The cost of the same volume of Celtite Selfix M100 grout would have been around £178,704. This represents an approximate saving of £167,535.

In the majority of cases the beneficial effects of strata reinforcement is not doubted but the length of time required for its execution, due to slow drilling rates, and the non-availability of skilled labour at short notice are letting down this increasingly important aid to successful mining operations.

It is believed that the important role of Strata Reinforcement, for maintaining continuous output, will be increased in future as more and more mining operations are likely to be hampered by interactions and geological problems.

The author wishes to thank Mr. J. B. Keirs, Director, Barnsley Area, N.C.B. for permission to publish this paper.

REFERENCES

Carr, F. "Strata reinforcement using resinous material", Not published, April 1976

Carr, F. and Silvester G.V. "Strata reinforcement" Paper No. 10, Fifth international strata control conference, London 1972

Broadsheet No. 30, N.C.B., H.Q. Mining department, production and productivity directorate "Truss Bolting at Park Mill Colliery, Barnsley Area", January 1978.

Broadsheet No. 44, N.C.B., H.Q. Mining department, production and productivity directorate "Method of driving face headings and commissioning faces in difficult conditions in the Silkstone seam at Barrow Colliery, Barnsley Area"., February 1980.

Kmetz, W. J. "Roof Trusses support problem strata". coal age, January 1970.

Silvester, G. V. "Scale model studies of strata movement around mine roadways: Part 9: Roof dowelling" MRDE report No. 57, 1975

ROOF CONTROL WITH POWERED SUPPORTS IN AN INDIAN MINE

T. N. Singh and B. Singh

Central Mining Research Station
Dhanbad-826001, India.

and

A. K. Gulati

Moonidih Project
Bharat Coking Coal Limited
Moonidih, Dhanbad, India.

SUMMARY. The massive sandstone roof of Indian coal measure formation posed a challange to the longwall system of mining as the average production never exceeded 150 tons/day in the past. The powered supports - 280 tons, 4 legged chocks were introduced at Moonidih Colliery of Jharia coalfield to improve the production and productivity. The caving was very systematic with these supports and the face production up to 1550 tons/day and face oms to 45 was achieved. The chocks in panel PS-1 and PS-2 were found to be working under 60% of the yield load and the average face convergence was less than 30 mm/m. The face was free from fissures and the flaking was rare. The support resistance appeared to be more than adequate for the condition.

INTRODUCTION

Longwall system of mining introduced over 25 years ago in Indian coal mines, failed to make impact even to date because of difficult strata condition and resultant low production. The Moonidih Project of Jharia coalfield has worked over 14 caved longwall panels within last 10 years in XVII Top seam. The average face advance in this mine has been 10-15 m/month with output of 100-150 tons/ day, insp te of the ideal working height (1.7-1.8 m), gentle gradient (1.in 7) and experienced miners. The faces were supported by 40 tons friction props, with support density of 0.9 to 1.0 m^2 per prop and additional inclined props and runner bars.

The powered supports- 280 tons, 4 legged Dowty chocks were introduced at this juncture with a view to improve the production, productivity and safety. This was the first trial with powered supports in any Indian mine where the coal measure formation is massive and competent. The experiment has been up to the expectations inspite of a number of bottlenecks with face oms up to 45 tons and the production up to 1550 tons per day. The roof control has been very effective even though, the faces have to cross the faults and advance galleries.

DETAILS OF THE PANELS

The panels PS-1 and PS-2 have been worked so far in XVII Top seam in between 280-400 m horizons while the work in the third panel PS-3 is in progress. The details of the first two panels covered in this paper are summarised in Table 1.

Table 1 Geo-mining details of the panels

	Panels	
	PS-1	PS-2
Seam, Thickness	XVII Top, 1.7-1.8 m	
Average depth	345 m	375 m
Face length	140 m	146 m
Panel length	530 m	560 m
No.of chocks installed	107	110
Setting load	20 Tons	20 Tons
Yield load	70 Tons	70 Tons
Face sequence	Retreating	Advancing

The layout of the panels is shown in Figure 1 which also includes the grabben faults traversing the face PS-1. The panel PS-2 is shifted away to mitigate the faulted zone which created operational difficulty during its crossing.

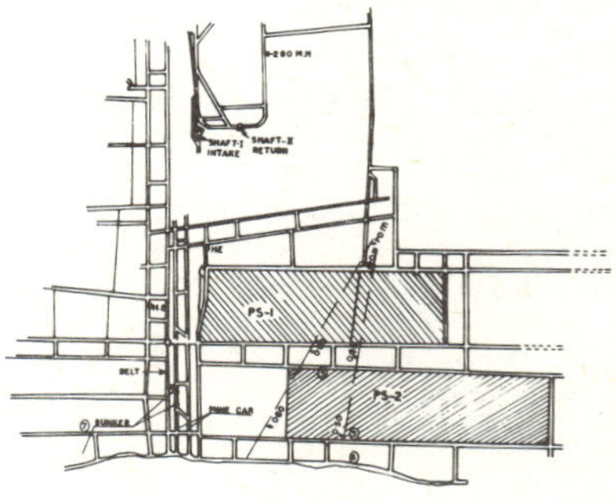

LAYOUT OF THE POWERED SUPPORT PANELS

Figure 1.

Coal measure formation

The strata section above the panels under reference is shown in Figure 2, according to which, the immediate 5.5 m roof is the stratified shale and sandstone, overlain by fine to medium grained sandstone. The immediate formation is weakened by micaceous lamination and a prominent slickenside 0.8-1.0 m above the seam. The physico-mechanical property of the coal measure formation is included in Table 2. A number of major faults of 38 m to 160 m throw formed the western and eastern boundary of the property.

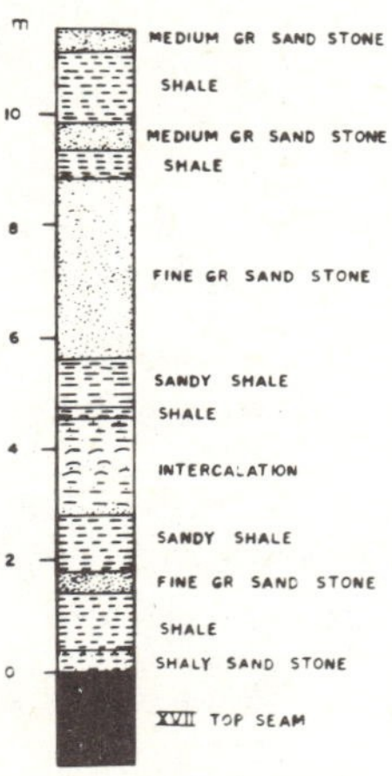

STRATA SECTION ABOVE XVII TOP SEAM

Figure 2.

Table 2 - Physico-mechanical properties of rock mass

Sl. No.	Run m	Formation				
1.	0-0.57	Shaly sandstone	440	43	40	2.29
2.	0.57-1.66	Shale	292	92	45	2.26
3.	1.66-2.13	Sandy shale	544	97	64	2.72
4.	2.13-3.05	Sandy shale	420	52	41	2.64
5.	3.05-4.01	Sandy shale	363	76	68	2.60
6.	4.01-4.84	Shale	460	70	45	2.57
7.	4.84-5.29	Sandy shale	587	70	39	2.58
8.	5.29-6.07	Intercalation	471	92	45	2.69
9.	6.07-6.94	Intercalation	406	103	105	2.53
10.	6.94-7.74	Sandstone(F.G.)	731	131	77	2.76
11.	7.74-8.64	Sandstone(F.G.)	700	151	84	2.68
12.	8.64-9.50	Sandstone(F.G.)	389	73	62	2.34
13.	9.50-10.4	Sandstone(M.G.)	383	88	82	2.25
14.	11.30-11.93	Shale	423	106	74	2.62
15.	11.93-12.64	Intercalation	472	85	69	2.60

Panel performance schedule

The faces were equipped with 280 tons - 4 legged chocks, armoured face conveyor, double ended ranging drum shearer of 150 kw and 610 mm web. The stage loader of 650 mm width discharged the coal to the belt conveyor and ultimately to a bunker.

The coaling was done in two shifts while the third shift was a maintinance shift. Horizontal and vertical transport of the coal was the main bottleneck in achieving the targetted productionfrom the faces. Their performance schedule is summarised in Table 3.

Table 3 - Performance schedule of the panels

	PS-1	PS-2
Average rate of advance	1.6 m/day	1.8 m/day
Maximum advance	4.2 m/day	3.9 m/day
Average output	650 tons/day	725 tons/day
Maximum output in a day	1350 tons	1550 tons
Maximum face oms	40 tons	45 tons

Strata Control Investigation Programme

The effectiveness of the powered supports for the massive strata was investigated for the life of panels PS-1 and PS-2. Under this programme, the pressure over the front and the rear legs in selected chocks was measured. Roof convergence was measured in addition to the qualitative assessment of the strata condition. For the purpose of pressure and convergence measurement, the face was subdivided into three sections - top, middle and bottom and some chocks were selected for the observation. The pressure in each leg of the selected chocks was measured once in a shift, preferably in the morning shift.

PS-1 PANEL OBSERVATIONS

Qualitative assessment

The face advance to 19.3 m along the bottom gate was associated with a minor fall behind 13-16 chocks. The fall extended to chock 48 after 22 m face advance and packed up to the goaf line. During the falls, only 3 m of the immediate shale roof caved in. The goaf up to the tail end was filled after 32 m advance.

The roof started caving regularly thereafter, with casual overhang in patches. The major falls were indicated after 41 m, 63 m, 90 m, 115 m and 141 m face advance. The roof condition during these falls remained satisfactory though active movement was marked in the upper horizon by the cracking and rolling sound of the superincumbent mass.

Quantitative measurement

Pressure variation over front legs

The pressure over the front legs in the bottom section varied from 1000 psi to as high as 7000 psi (Figure 3A). The average pressure over the chocks 3, 13, and 23 was 2700 psi which was 35% over the setting pressure.

The pressure over the chocks 33,43,53 and 63 located in the middle section, varied from 1750 psi to 7250 psi (Figure 3B), with average pressure of 4250 psi.

The pressure variation over the chocks of top section, covering 73,83,93 and 103 is shown in

Figure 3C. The pressure over these legs fluctuated in between 1250 psi to 7250 psi with average value of 2650 psi.

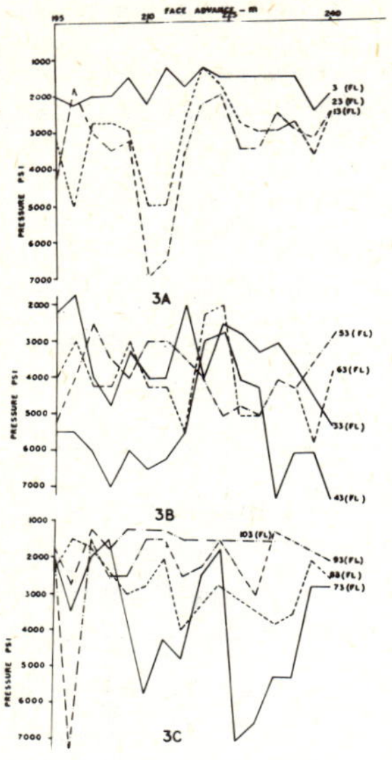

PRESSURE VARIATION OVER THE FRONT LEGS

Figure 3 (A,B,C)

The pressure over the front legs of the chocks located in top and and bottom sections was nearly equal and was 30% over the setting load. As the distance from the barriers increased, the pressure also increased, indicating the influence of the ribs. In the middle section, the average pressure increased to 4250 psi, nearly 110% over the setting pressure. The maximum pressure in all the three zones reached to 7000 psi or more at one or the other stage which appeared to be due to block failure which some times caved up to the rib.

Pressure vatiation over reat legs

The pressure over the rear legs of chocks 3,13,and 23 located in zone I, varied from 1000 psi to 7500 psi (Figure 4A). The average pressure over these chocks was 3500 psi nearly 75% more than the setting pressure.

In the middle section, the pressure over the rear legs varied from 2750 psi to 7500 psi, the average being 5000 psi (Figure 4B). The lower pressure values were noted immediately after the shifting of the chocks. This zone was likely to be free from the rib influence and the pressure was likely to be the representative resistance under the condition of Moonidih strata condition.

In zone III, the pressure varied from 1500 psi to 6750 psi (Figure 4C), the average value being 3500 psi only. The pressure fluctuation in this zone was very erratic, indicating block failure of the immediate roof at periodic intervals.

The pressure variation in different zones over the front and the rear legs is summarised in Table 4. This shows that the chocks on an average worked within 60% of the yield load though, at some stage or the other, the pressure increased up to near yield value.

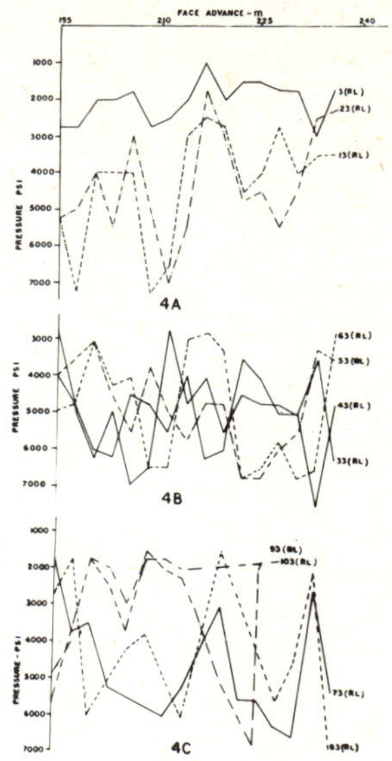

PRESSURE VARIATION OVER THE REAR LEGS

Figure 4 (A,B,C)

Table 4 - Pressure variation in different zones

Zones	Legs	Pressure in psi			% of yield pressure
		Maximum	Minimum	Average	
Bottom-I	Front	7000	1000	2700	35
	Rear	7750	1000	3500	46
Middle-II	Front	7300	1750	4250	56
	Rear	7500	2000	5000	66
Top-III	Front	7250	4250	2675	35
	Rear	7500	1500	3500	46

Roof convergence

The roof convergence in bottom section observation points, corresponding to chocks 3,13, and 23 interspace varied from 3 mm to 129 mm (Figure 5A). Except an extreme condition, the convergence value remained below 60 mm with average convergence value of 29 mm/m face advance. The high convergence value of 129 mm was due to dislocation of the block in which the observation point was fixed.

The convergence values in the middle section covering the observation points near 33,43,53 and 63 chocks varied from 7 mm to 85 mm (Figure 5B). The average convergence in this zone was 15 mm, indicating uniform settlement of the roof without much of overhanging and block dislocation.

The convergence was spairingly available in the top seftion because of the excessive deposition of the gummings. The available convergence values (Figure 5C) varied from 6 mm to 126 mm. The average convergence value in this zone was 28.5 mm.

The observation in this section was limited to 20 settings only and therefore the variation trend is not very conclusive.

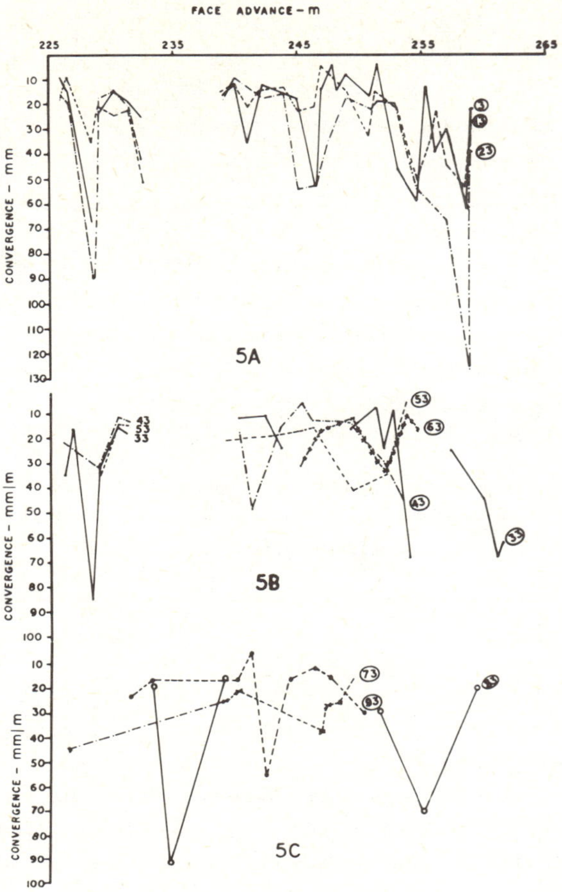

CONVERGENCE VARIATION WITH FACE ADVANCE

Figure 5 (A,B,C)

The convergence observation was not very systematic because of the limited space left over the face, faster rate of face advance and the presence of coal debris behind the conveyor. On the basis of the limited observation, the convergence variation trend is resolved as in Table 5.

Table 5 - Roof convergence in different zones

Zone	Convergence				Roof deflection
	Maximum mm	Minimum mm	Average mm	Average mm/m	
I	129	3	29	29	1°40'
II	85	7	15	15	50'
III	126	6	28.5	28.5	1°38'

The convergence in the middle of face was 50% of the edge portions, indicating severe dislocation of the roof blocks in these zones. The roof sagged without cracks gently by 50' in the middle section while in the edge zones, the deflection was double of the same. The average convergence value for the whole face was 24 mm/m, equivalent to 1.3% of the working height.

PS-2 PANEL OBSERVATIONS

Qualitative assessment

The roof was induced to cave in after 15 m advance in panel PS-2, when a minor fall occurred behind 13-29 chocks. The goaf was tightly packed after 20 m face advance, except 15 m near the ribs. The exposed roof was free of cracks all along with no flaking in advance of the canopy. Overhang of goaf up to 3-5 m was frequent, particularly under the influence of the barriers.

Quantitative measurement

Pressure variation over front legs

The pressure over the front leg of the chocks of zone I (Figure 6A) varied from 1000 psi to 5750 psi. The chock 3 located near the barrier was under the average pressure of only 1875 psi in comparison to the zone average of 2600 psi.

In zone II, chocks 33,43,53 and 63 were selected for the pressure study. These chocks located in the middle of the face were subjected to uniform loading, recorded pressure over which varied from 1750 psi to 7250 psi (Figure 6B). The average pressure in this zone was 4250 psi only.

In zone III, in which the chocks 73,83,93 and 103 were selected for the observation, the pressure varied from 1000 psi to 7000 psi (Figure 6C). The pressure over the extreme chock 103 was again the lowest with average value of 2800 psi in comparison to 3600 psi average pressure in this zone. The pressure over the chocks 73 and 83 was nearly equal to the average pressure observed in zone II. The edge effect thus appeared to be confined up to 93 chock only.

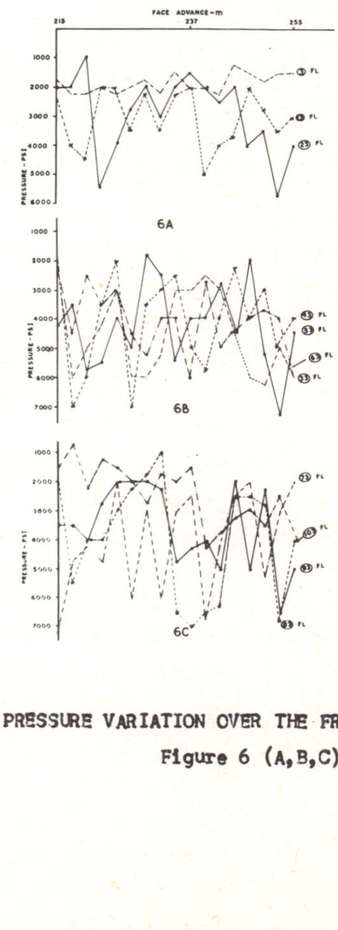

PRESSURE VARIATION OVER THE FRONT LEGS

Figure 6 (A,B,C)

Pressure variation over the rear legs

The pressure variation over the rear legs of the selected chocks in zone I varied from 2000 psi to 7000 psi (Figure 7A) with the average value of 3600 psi. The pressure over the extreme chock 3 was again the lowest - 2500 psi in comparison to 3600 psi average value of the zone. The pressure over the other chocks 13 and 23 was 60% more than that of the third chock.

In zone II, the pressure over the selected chocks varied from 1500 psi to 7500 psi, the lowest being in chock 33 at one stage only (Figure 7B). The average pressure in the rear legs of the chocks of this zone was 4900 psi, which was nearly 66% of the yield pressure. All the chocks of this zone were uniformly under equal load throughout the panel life.

In zone III, the pressure over the rear legs varied from 2000 psi to 7500 psi (Figure 7C) with the average pressure value of 4625 psi. In this case, the pressure over the extreme chock 103 was only marginally low.

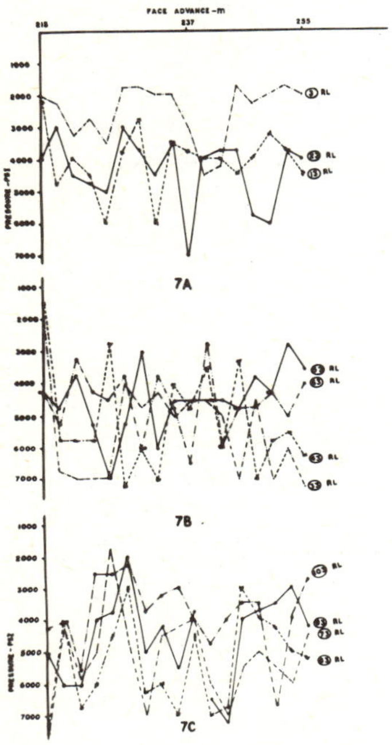

PRESSURE VARIATION OVER THE REAR LEGS

Figure 7 (A,B,C)

The pressure over the front and the rear legs of different zones are summarised in Table 6, which shows that the rear legs were invariably under higher resistance in comparison to that of the front legs.

The highest pressure recorded over the chocks of the different zones was nearer to the yield pressure and its frequency was only 0.5%. At no stage , all the four legs of the chocks were under the pressure over 5000 psi. On an average the front legs for a particular setting were under 4250 psi while the rear legs were under the pressure of 4900 psi only.

Table 6 - Pressure variation in different zones

Zone	Legs	Pressure in psi			% of yield
		Maximum	Minimum	Average	pressure
I	Rear	7000	2000	3600	47
	Front	5750	1000	2600	34
II	Rear	7500	1500	4900	64
	Front	7250	1750	4250	56
III	Rear	7500	2000	4625	60
	Front	7000	1000	3600	47

Roof convergence

The convergence observation in Panel PS-2 was very limited and random because of many constraints. The observation spread over 25 days in different zones is summarised in Table 7. These values were noted after 24 hours, corresponding to 2 m face advance.

Table 7 - Convergence variation in PS-2 panel

Zone	Convergence value				Roof
	Maximum	Minimum	Average		slope
	mm	mm	mm	mm/m	
I	52	3	13	6.5	23'
II	145	6	30	15.0	52'
III	34	6	15	7.5	26'

CONCLUSIONS

The pressure variation over the front and rear legs of zones I-III in PS-1 and 2 panels were taken to visualise the optimum support resistance for the strata condition of Moonidih mine. These values are given in Table 8. It shows that the chocks in zones I and III were working within 50% of the yield load and even the rear legs of the middle section were on an average working within 64% of the yield load. If the average of all the four legs of the middle section were considered, the chocks were working within 60% of the yield load. Thus the optimum requirement of the chock capacity in this case was 180 tons only.

Table 8 - Pressure record in different zones

Zone	PS-1		PS-2	
	Front legs	Rear legs	Front legs	Rear legs
I	2700	3500	2700	3600
II	4350	5000	4200	4800
III	2675	3500	3200	4200

The roof condition with the resistance has been satisfactory with no flaking or crack formation. The caving was regular and no indication of periodic weighting was observed. The supports had nearly 40% reserve capacity to render effective strata control during crossing of the fault and existing gallery in advance of the face.

PRESSURE VARIATION DURING CROSSING OF THE FAULT

The face crossed a grabben fault of about 1.3 m throw in the middle of the face. The strata was very disturbed at this junctions and the floor and roof were highly undulating. The pressure variation during the crossing of the falt was taken to substantiate the optimum support requirement and to visualise necessary reserve capacity of the same.

The variation of pressure while corssing the fault plane was very erratic with wide variation in maximum and minimum values. The maximum pressure in all the zones reached to 7250 psi. The other features of this sector are summarised in Table 9.

Table 9 - Pressure variation in faulted zone

Zone	Legs	Pressure in psi			% of yield pressure
		Maximum	Minimum	Average	
I	Front	7500	1000	3350	43
	Rear	7750	1250	4200	55
II	Front	7750	1500	4000	52
	Rear	7250	1500	4550	60
III	Front	7250	1000	2500	33
	Rear	7250	1500	3650	48

The average pressure during crossing of the fault remained nearly equal to that of the normal strata condition in that mine. Even the middle section rear legs worked under 60% of the yield load. Therefore, the reserve capacity of over 40% does not seem to be justified. The roof of Moonidih XVII top seam as such could be controlled effectively by 180 tons chocks even in adverse conditions.

Acknowledgement - The investigation was financed by the funds made available from the Ministry of Energy, Government of India. Thanks are due to Mr. A. K. Singh of Moonidih Project and Mr. M.A. Rafiqui of CMRS, Dhanbad, for their co-operation. Thanks are also due to the Director, CMRS, Dhanbad and the Managing Director, B.C.C.Ltd. for their kind permission to present the paper.

The views expressed are those of the authors and not necessarily of the organisations to which they belong.

APPLICATION OF THE MRDE DATA LOGGER

TO FACE SUPPORT MONITORING

D N Bigby and D S Cocker

NATIONAL COAL BOARD
Mining Research and Development Establishment
Stanhope Bretby
Burton-on-Trent
Staffs

SUMMARY

The 'Underground Monitoring System' is a valuable tool for sampling and recording data from analogue transducers. The paper introduces the system and the associated data processing techniques which have been developed at MRDE. Results from the instrumentation of shield and chock shield supports on three coalfaces are presented. Accurate measurements of support setting pressures, setting load densities and mean load densities have been possible and these are discussed in terms of theoretical support requirements. Also, the effects of increased setting pressures and the fitting of base extensions in soft floor conditions are examined. Recommendations for future face investigations utilising the monitoring system are made and possible future hardware/software developments are discussed.

INTRODUCTION

The MRDE Strata Properties Group has been closely involved with studying the support of longwall coalfaces over many years and particularly with the development of theoretical models to describe the influence of hydraulic powered supports on strata behaviour. This work has led to the publication of certain recommendations concerning the choice of setting and yield loads depending upon the application planned for such supports; Ashwin (1970), Wilson (1975), Evans (1977), Wilson (1980). The Group recently became involved in the study of shield and chock shield type supports and their application in the British Coalfields. There has been a special interest in the performance of these supports on the soft floors sometimes encountered in this country. Departure from the previous standard chock type design of face supports has also led to investigation of possible alternatives to the concept of a maximum prop free front distance, Wilson (1981).

Experience in the use of hydraulic supports on the coalface as 'load cells', to obtain information on the behaviour of the surrounding strata, led the Group to request a portable electronic means of data acquisition. It had become apparent that if theoretical work was to be verified in the field, it would be necessary to monitor closely a wider sample of supports on a face than was possible with the instrumentation then available. This comprised various types of single channel chart recorders, as described by Shepherd and Ashwin (1968), which were labour intensive in terms of installation, servicing and data analysis. The new equipment, known as the 'MRDE Underground Monitoring System', was developed by Instrument Development Group and became available for underground use towards the end of 1979.

DESCRIPTION OF INSTRUMENTS

The Underground Monitoring System comprises a data logger, up to two interface units, two battery modules, a portable test unit and associated transducers as described by Cocker & Bigby (1981). A schematic diagram of the system is shown in Figure 1.

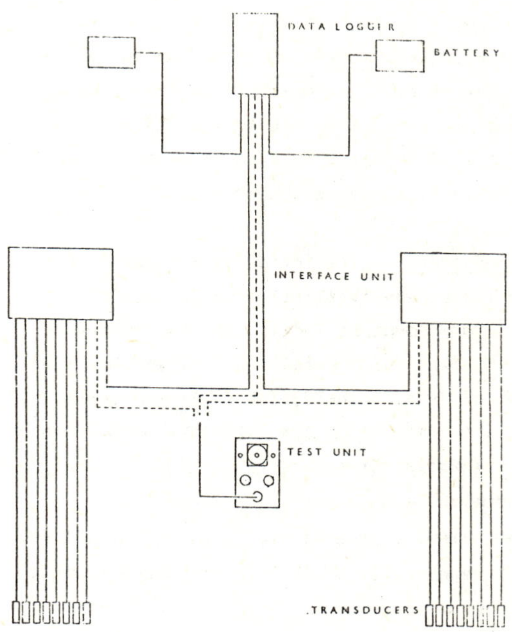

Figure 1 Schematic Diagram of System

Data Logger

The data logger, Figure 2 item D, forms the nucleus of the system. It will accept analogue outputs from up to 16 transducers in the range 0-5 volts. The inputs to the data logger are sequentially multiplexed, converted into digital form, formatted and stored on tape cassette. Approximately 160,000 readings can be stored on one cassette.

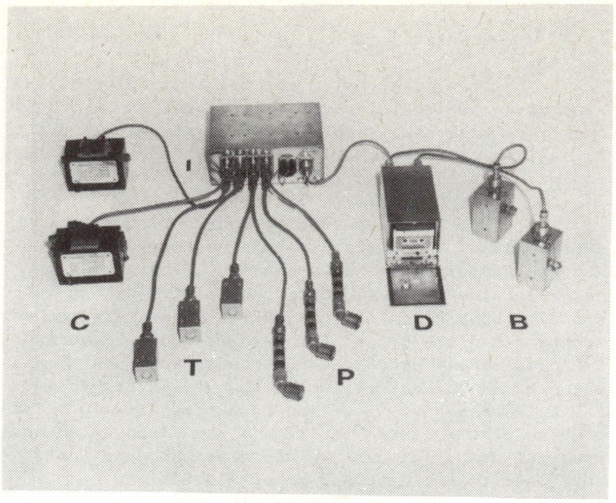

Figure 2 The Underground Monitoring System

A timing circuit, developed at MRDE and mounted internally, enables the transducer outputs to be scanned at regular time intervals. A switch mounted below the tape transport unit enables the user to select time intervals between scans of 4, 30 or 90 seconds, as required. By altering an internal link arrangement, time intervals of up to 5 minutes can be achieved. When all 16 channels are utilized, the logger, scanning every 4 seconds, will operate for eight hours. In the 90 seconds scanning mode the logger will operate for 3 to 4 days, extending to over one week with a scanning interval of 5 minutes. The recording technique employed is such that the tape cassette moves only when the transducers are being scanned. This conserves power and no tape is wasted. The underground data logger is built around a Datel Ltd LPS-16 logger which together with additional circuitry is enclosed in a stainless steel housing for mechanical protection. All input and output connections are made via connectors mounted at the rear of the logger. Access to the tape deck, for changing cassettes underground, is provided by a hinged front cover, which also prevents any excessive ingress of dust.

Interface Unit

The interface unit, Figure 2 item 1, contains 8 signal conditioning circuit cards. Thus, to fully utilise the 16 channels of the data logger, two interface units are required. These cards energise their associated transducers and amplify the transducer outputs to levels suitable for the logger. Four types of conditioning cards are currently available, for use with strain gauges, pressure, closure and tilt transducers. Any combination of the types of available sensors can be used with the interface unit. Because of the requirements of the Intrinsic Safety (IS) Certificate, the unit cannot be opened underground and calibration of the conditioning cards must therefore be carried out on the surface.

Power Supplies

The data logger and the interface unit have separate external power supply modules. This enables them to be employed independently where necessary. When connected as shown in Figure 1 the data logger controls the supply to the interface unit and transducers such that they are only energised when the data logger is scanning the inputs. This ensures that the interface unit power supply will last over long periods. The rechargeable battery module, designated B in Figure 2, has been developed so that it may be carried to the installation and the battery modules changed underground. This allows the remainder of the system to be left undisturbed for long periods. Alternatively the system may be powered from suitable IS power supplies such as the BAT2, DC2, etc.

Transducers

Any compatible certified IS analogue transducer may be used with the system providing the supply to the transducer does not exceed 25 volts. Four transducers are currently in use - pressure, tilt, closure and strain gauges.

Pressure

The pressure transducer, item P in Figure 2, has a range of 0 to 70 MPa. This means that with an 8 bit resolution pressure variations of 0.27 MPa can be detected. It is a rugged solid state strain gauge transducer housed in a robust casing. The transducer was designed such that a standard 'Conflow' connector could be used for attaching the device to the hydraulic test point on the support.

Tilt

The tilt transducer, item T in Figure 2 acts like an electronic spirit level. The core of the transducer is a closed glass tube containing a bubble of gas in an electrically conducting liquid and electrodes to make external contact.

The bubble settles at the highest point in the tube. As the transducer it tilted the bubble stays at the highest point and its position relative to the electrodes changes. The change is detected electronically and is used to derive an electrical signal which is proportional to the angle of the transducer. The linear range of the transducer is typically 10 degrees, and changes of 3 minutes of arc can be detected. The glass tube is housed in a robust case, machined to give a flat surface and cadmium plated.

Generally the transducers are bolted to metal plates, which are attached to the supports by an adhesive compound. This enables the plates to be mounted horizontally prior to attaching the transducer, and ensures an equal range in both directions. It also prevents the need for any drilling of the support and enables the metal plates to be removed and reused elsewhere.

Closure

The closure transducer, item C in Figure 2 provides an electrical signal proportional to the linear extension of a stainless steel cable. Two types of closure transducer have been used to date. One is a coarse movement version and is capable of detecting movements of 2 mm over a 500 mm range. The second version enables movements of 0.4 mm to be detected but, while retaining a range of 500 mm, only relative rather than absolute displacements are measured. It is usually sited in conjunction with the coarse version in order that both an absolute and an accurate recording can be made. Retraction of the cable is effected by means of a coiled spring which maintains a constant tension on the cable. The transducer is housed in a heavy-duty steel case, which is fixed, using a chain attachment, to the stationary section of the support leg. Initial worries that the cable would be 'plucked' causing spurious recordings have proven unfounded in the tests to date.

Strain Gauges

Foil type full-bridge strain gauges are mounted to the relevant section of the support before it is taken underground. Typical strain gauge sensitivity is 0.5 mV/V excitation. These were not used in the investigations described in this paper.

Test Unit

This unit, connected as shown in Figure 1, enables the operator to check that the system is operating correctly. When connected to the system, it can be used to indicate the signal level on any selected channel and also the condition of the batteries. Housed in a stainless steel case it is normally carried in and out of the pit when a check on the system operation is being made.

DATA PROCESSING

When filled with data, cassettes are taken out of the pit and sent to MRDE. Here they are read into a minicomputer and the data analysed.

The philosophy adopted for the software was that it should be flexible yet simple enough for use by non-computer orientated personnel, after some initial instruction. The particular program required is selected from a menu such as that shown in Figure 3. The programs are written in FORTRAN IV PLUS for use on a PDP 11 minicomputer under the RSX 11M Operating System.

```
>;
>; THE FOLLOWING CODES REPRESENT THE DIFFERENT PROGRAMS AND ROUTINES
>; THAT ARE AVAILABLE TO YOU.
>;
>; 1) CASLST4    (TERMINAL LISTING OF CASSETTE FILES ON SYSTEM)
>; 2) FILEX4     (TRANSFER OF SERIAL INPUT FROM CASSETTE READER TO DISC)
>; 3) GINT4      (CONVERSION TO ENGINEERING UNITS & INPUT TO TEST DATA BANK)
>; 4) GRAPH4     (PLOT DATA AGAINST TIME. TERMINAL MUST BE 4014 OR GT2000)
>; 5) SORT4      (TO DETERMINE WHEN DATA MEETS CERTAIN CRITERIA)
>; 6) LOAD4      (CONVERSION OF MICROSTRAIN TO LATERAL LOAD)
>; 7) GFD4       (TO SELECTIVELY DELETE &/OR LIST FILES FROM YOUR WORKSPACE)
>; 8) DTBNK4     (FOR GENERAL I/O WITH DATA BANKS)
>; 9) STATS4     (FOR STATISTICAL ANALYSIS OF DATA & INPUT TO STATS DATA BANK)
>;1?) HELP       ( A '1' FOLLOWED BY A  PROGRAM CODE [1,2,3,4,5,6,7,8 OR 9]
>;              WILL DESCRIBE THAT PROGRAM IN MORE DETAIL)
>;10) MCR        (ALLOWS YOU TO ADDRESS MCR TO TIDY UP FILES ETC.
>;              [SEE RSX11M OPERATORS AND UTILITIES PROCEDURES MANUALS])
>;20) EXIT
>;
>* PLEASE ENTER THE REQUIRED CODE  [D R:1.-20.]:
```

Figure 3 Computer Program Selection Menu

The suite of programs provides the following functions:

1) Convert raw data to engineering units.

2) Plot data from each channel (or mean of several channels) against time, with hardcopy available.

3) Determine times at which data satisfied certain criteria. For example, the times at which the parameter measured dropped below a certain level or at which the rate of increase between subsequent scans exceeded a chosen rate. The latter test is useful for identifying support setting times and pressures.

4) Calculation of nominal setting and yield load densities for a given support type on a particular face.

5) Analysis of individual support cycles for calculation of:

   i) Setting pressures
   ii) Mean setting load density
   iii) Mean load density (time weighted)
   iv) Change of base inclination
   v) Maximum and mean toe and heel loads

6) Miscellaneous file deletion and listing services.

Provided the appropriate parameters have been monitored during a test this analysis is possible for any of the following support types:

4 - leg chock shield

4 - leg shield

4 - leg chock

5 - leg chock

6 - leg chock

It is also possible to calculate and plot pressure maintenance and deficiency profiles when the hydraulic supply has been monitored. These provide a measure of the ability of the hydraulic supply to meet the demands imposed by chocking and related face activities.

The system is supported by four data banks covering faces, supports, tests and statistics. The face and support data banks contain technical data concerning the various underground installations. The test data bank contains information such as transducer layouts and calibrations for each recording made and is automatically updated each time a fresh set of data is converted to engineering units. The stats data bank holds the results of the mean load density related to calculations listed in items 4 and 5 above. Manual input and output of data to and from the data banks is supervised by the databank program DTBNK.

Additional software is planned to provide the following:

i) Graphs of position and magnitude of support reaction against time

ii) Calculated rate of loading over given time periods following setting

iii) Statistical analysis and report generation to cover the full period of an underground test program

UNDERGROUND INSTALLATIONS

The underground monitoring system is now in wide use by MRDE in various underground applications. Over the last two years the system has achieved over 10,000 hours of recording face support behaviour of which nearly 50% has been carried out by MRDE Strata Properties Group. Discounting the proving trials which took place on HO3s at Houghton Main Colliery, Strata Properties Group's investigations centred on three coalfaces, B90s and B46s at Rossington Colliery and W65s at Daw Mill Colliery. These were all thick seam faces extracting between 2.8 and 3.5 metres of coal and were equipped with shield or chock shield supports. We are particularly interested in the application of this type of support to the British mining environment and have a special interest in the performance of these supports on the soft floors often encountered in this country.

One of the primary objectives of the underground monitoring system was to enable the detailed monitoring of a wider sample of supports on a face than had been possible with the instrumentation previously available. There is an inevitable conflict between the size of such a sample and the detail required for each support. Manpower limitations and data processing restraints also play a part in determining the strategy of investigations. The policy adopted in view of these restraints was to monitor in detail two batches of between three and five supports on each face over relatively long periods (2 - 3 months), sometimes moving these stations along the face at intervals within the investigation. Each station was equipped with a 16 channel system, a typical transducer layout being pressure transducers on all leg circuits and one on each hydraulic supply line, tilt transducers on alternate support bases and canopies, and closure transducers on front and rear legs of one or two supports. Stations were usually sited at approximately one third and two thirds the face length, to reduce end effects. Data from the instrument stations was supplemented by Bristol chart recorders and spot pressure gauge readings from the rest of the face.

ROSSINGTON B90s (NCB DONCASTER AREA)
February - May 1980

This was a 2.8 m extraction advancing face in the combined Barnsley/Dunsil seam and was equipped with Gullick 4/300 tonne shield supports. The 220 metre face was cut by a Double Ended Ranging Drum Shearer equipped with 1.98 m x 0.69 m drums, which also eliminated the maingate stable. A Single Ended Ranging Drum Shearer was employed at the tail gate to maintain an inline ranged rip. The maingate rip was a conventional ripping lip and both packs were 3 metres wide and built of 'Aglite' blocks.

Floor tests on this face, using a Dobson dynamometer prop on a 152 mm base, had shown the immediate floor to be extremely soft (c 1.9 MPa). Although its bearing capacity appeared to increase a little with depth, the upper layers down to 0.6 m were soft and badly affected by water. The 4/300 tonne shield had a poor theoretical base load distribution and problems due to penetration and ploughing of the floor had been predicted.

Following a short pilot scheme, carried out by Doncaster Area Strata Control Team involving the monitoring of four adjacent supports (46-49) with pressure and tilt transducers, the instrumentation was expanded in February 1980 to cover two batches of four supports each. Both batches, station A(46-49) and station B (104-107) were instrumented with pressure transducers on all leg circuits, tilt transducers normal to the face line on all bases and tilt transducers on the front legs and canopies of alternate supports.

Theoretical calculations had shown that the fitting of 200 mm extensions to the support bases should give toe loadings low enough to prevent damage to the floor. With nominal setting pressure (13.8 MPa) assumed in all legs and assuming linear base load distribution, the toe pressure generated would be approximately 1.8 MPa normally and 1.2 MPa if a base extension was fitted. The influence of the base extension was much more dramatic if zero pressure was assumed in the rear legs. (General observations indicated that the rear legs often had little or no pressure in them.) In this case the toe load generated would be approximately 22.9 MPa normally and 1.8 MPa with the base extension.

It should be noted that due to the complex geometry of the toe region of this support the theoretical calculations of maximum toe load without base extensions can only be very approximate and only indicate the order of magnitude that might be expected. Calculations with the base extension are more reliable as the geometry is simplified.

Concurrent with installation of the support monitoring system, base extensions were fitted to a sample batch of supports, including those comprising station A. In this way it was hoped that a comparison could be obtained between the performance of supports with and without base extensions.

Throughout the ensuing monitoring period, February to May 1980, conditions on B90's face were extremely poor, with major roof falls occurring regularly throughout the face. Horizon control was particularly poor, with the amount of floor cut throughout the face varying between a few centimetres and 1 metre. A large amount of loose gummings accumulated under the Armoured Face Conveyor (AFC) and the bases of the supports due to poor clean up. This resulted in difficulties in advancing supports and as a consequence the length of time unsupported roof was exposed while the supports were back (supports in Q condition), was very high.

On conclusion of the first batch of tests in May 1980, a report was issued to the colliery and area management outlining the results obtained during the month of March and making certain recommendations. Supplemented by results from April 1980, which were not then available, these may be summarised as follows.

In general the supports were not being set at the available supply pressure of 13.8 MPa. The average setting pressure at station A was 8.3 MPa and at station B it was 9.0 MPa. These averages include periods when bad roof existed above the supports and it would have been difficult to achieve a good setting pressure.

The Mean Load Densities, calculated from the pressure recordings were 0.29 MPa at station A and 0.31 MPa at station B. The nominal Setting Load Density, by comparison, was 0.31 MPa. These poor Mean Load Density figures, generally reflected low setting pressures.

Where the supports were being set at the available setting pressure, (this occurred in general where the floor in front of the supports had been manually cleaned up), the load in the legs was building up towards yield pressure, (39.5 MPa).

The inclination results showed the bases to be tilting down at the front, indicating toe penetration into the floor or floor gummings. The average change in inclination of the support bases during a face cycle was 9 minutes of arc at station A and 17 minutes of arc at station B. Where the support had been set effectively and a good build up of load occurred, the average change in base tilt was 20 minutes of arc at station A and 35 minutes of arc at station B. These results could also indicate the rear of the bases lifting up somewhat.

The tilt in the bases was, to a lesser amount, accompanied by tilt of the canopies and front legs in the same sense, indicating that there was a tendency for the supports to rotate as a whole under load, down at the front and up at the rear. When these supports were tested in the laboratory at Bretby a similar tendency had been observed.

It was concluded that the main strata control problems on this face, apart from those inherent in the proximate geology, were due to poor clean up by the shearer loader, resulting in horizon control and gummings problems, and the length of time unsupported roof was exposed combined with inadequate setting pressures, resulting in excessive coal spalling and subsequent roof falls. The following recommendations were made.

1 Fit base extensions throughout the face

2 Ensure, where possible, that supports were properly set, particularly the rear legs

3 Achieve improved clean up and quicker advance of supports

Following the issue of this report a meeting was called by the Deputy Director (Mining) of Doncaster Area. As a consequence of this meeting it was determined that all supports on B90's would be fitted with 200 mm extensions and that positive set valves would be fitted, triggered by the front legs.

Strata Properties Group were requested to reinstall the instrumentation on B90's face once these alterations had been made and to monitor the results. This was done in July/August 1980 and monitoring continued until October 1980.

August - October 1980

Initially two measuring stations were set up, station A on shield nos 25-29 and station B on nos 30-34. Later, due to hydraulic faults in the supports, station B was moved to support nos 36-40. At the beginning of September 1980 when positive set valves had been fitted to the main gate half of the face, station B was again moved, this time to support numbers

59-63. When the stations were installed in July 1980 about 80% of the face supports, including the measuring stations, were fitted with base extensions.

Both stations were instrumented identically. All supports had pressure transducers fitted to front and rear leg circuits. Tilt transducers were fitted to the base, front leg and canopy of support nos 28 and 39 (62 when station B was moved). These supports were also fitted with closure transducers on both front and rear legs. In addition tilt transducers were fitted to the bases of supports 29 and 40 (63 when station B was moved). At the beginning of September the pressure transducers on shield 25 were transferred to measure the low and high pressure pump supply pressures.

The positive set system installed on the face was designed such that the supports were set utilising a low pressure line (c 15 MPa) to supply the initial large quantity of hydraulic fluid. Positive set was triggered via a 'SEE-TRU' valve monitoring front leg pressure. This was originally set at 10.4 MPa, but was subsequently reduced during August 1980 to 8.3 MPa, as difficulty in triggering positive set had been experienced due to multiple demand and a large pump scan. Once triggered the support pressures were topped up from a high pressure line. When the positive set facility became operational on 5 August the high pressure pump supply pressure was 20 MPa. This was increased to 23.5 MPa on 14 August 1980 when station B became operational on supports 36-40 and was again increased to 27.6 MPa on 19 August 1980. The valve gear was such that a support continued to top up throughout the face cycle until either the positive set system was turned off at the support, the high pressure pump was turned off, or "legs lower" was selected at the rotary control valve.

Throughout this period of measurement, face conditions varied. Roof falls when they occurred, were generally on the tail gate half of the face and were commonly about 2 m high in front of the support canopies. The repair work required slowed down advance which over September and October 1981 was on average 3 m/week. Roof falls did not often occur around station A but did affect station B (59-63).

As during the first investigation, horizon control was generally poor, in this case the amount of floor cut varying from 600-1500 mm. The roof horizon remained reasonably consistent due to the presence of a thin dirt band with a clean parting. An excessive amount of floor gummings continued to accumulate in front of and underneath the AFC and support bases, although conditions around station A were not as severe as at station B.

Before positive set was fitted, the results of base and canopy tilts, mean load densities and setting pressures, with the exception of slightly improved setting pressures, showed little change from those found in the earlier investigation (February-May 1980).

After positive set was fitted a considerable number of hydraulic faults developed due to the generally higher hydraulic pressures imposed on the system. As these were progressively repaired during August and September and as the positive set pressure was increased, so the mean load densities achieved also increased. This is shown in Table 1.

Period	Pump Pressure MPa	Mean Load Density		
		Station A (25-29)	Station B	
		MPa	MPa	Support Nos
Pre Positive Set	15	0.32	0.34	30-34
19 August	20 - 23.5	0.36	0.39	36-40
19 August-31 August	27.6	0.41	0.46	36-40
September	27.6	0.52	0.47	59-63
October	27.6	0.52	0.53	59-63

Table 1 Mean Load Densities - Rossington B90's

When positive set pressure was increased to 27.6 MPa and the majority of the hydraulic faults had been repaired the following features of the system remained reasonably consistent.

a) While the high pressure pump was on, unless a sudden load transfer or increase occurred, the leg pressures did not increase above positive set pressure.

b) When the high pressure pump was switched off, if the leg pressures had not increased above positive set pressure, a pressure drop of between 1 - 7 MPa occurred, generally followed by a steady increase in pressure. The rate of pressure loss varied and could occur in either one or both leg circuits. During October, of the 18 leg circuits monitored, 32% of front leg circuits and 61% of rear leg circuits exhibited such pressure drops.

c) Advancing of adjacent supports had no significant influence on a recently set support. However, if with time the pressure had built up above setting pressure and an adjacent support was advanced and positively set, then a reduction of 4-7 MPa would be experienced in leg pressures. This effect was hardly noticeable one support away.

d) The passing of the machine taking the top cut had little effect on the leg pressures.

The results from the base-mounted tilt transducers indicated that generally the higher leg pressures generated by positive set resulted in greater base tilt downwards towards the face. The higher setting pressures appeared to cause an increase in base 'toeing' of 15-30 mins of arc at setting and generally less 'toeing' during the rest of the cycle. At station B, where more gummings accumulated, total 'toeings' over the whole cycle had been in the order of 36 mins of arc before positive set and was of the order of 75 mins of arc after positive set was fitted. In the majority of cases the canopy did not move much after the support was set provided the roof extended to the rear of the canopy. However, where the roof did not extend this far the supports tended to 'jack knife' during setting until positive set pressure was achieved in the rear legs.

Conclusions from Rossington B90's Investigations

The installation of a high pressure positive set system resulted in much improved setting and mean load densities.

The lack of build-up in leg pressures after positive set and the pressure drops which occurred when the pump was switched off indicated faulty or inefficient hydraulic components, requiring investigation by hydraulics engineers.

Trigger pressure for the positive set system needed to be as low as safety permitted. Also the setting and scan of the low pressure pump needed to be checked regularly.

Increased setting presure led to increased base 'toeing', although the fitting of base extensions probably reduced the severity and consequences.

The modifications to the supports did not substantially alter face conditions. The rate of advance remained poor.

It was quite possible that the high setting pressures had a deleterious effect on the fragile roof strata. A guaranteed set as opposed to high pressure set system may have been more successful, in terms of strata control and reliability.

It remained the opinion of Strata Properties Group that the two main causes of poor face conditions were:

a) Accumulation of gummings due to poor clean up leading to difficult support operation and horizon control problems. The use of cowls on the machine might have improved clean-up.

b) Lack of early forward support to exposed roof. The length of time exposed roof was unsupported was increased due to the unidirectional operation of the main face machine, roof fall repairs and cleaning up.

DAW MILL W65's (SOUTH MIDLANDS AREA) MARCH 1981

Face Operations and Instrumentation

W65's face at Daw Mill Colliery in the Warwickshire thick seam was installed with Dowty 4 x 450 tonne chock shields and a double-ended ranging drum shearer operating a bi-di advancing system. A second machine, a single ended ranging drum shearer cut a buttock for the main machine at the maingate end of the face. The face ends were both range ripped and Thyssen pumped packing systems were in use. At the time of the investigation (March 1981) the extracted section was approximately 3.5 m. This was entirely within the seam giving a coal roof and a coal floor.

The Dowty 4 x 450 tonne Chock Shields were fitted with the following facilities:

Contact Advance - A low pressure (1.24 MPa) was maintained in the legs and side flaps during advance, restricting leg lowering and reducing the ingress of dirt by maintaining the side flap in contact with the adjacent support.

Guaranteed Setting Pressure - This facility ensured that provided the trigger pressure was reached all legs would be set at the pump pressure available at that time. The system was triggered by a pressure of 5.5 MPa in the front maingate leg circuit. Once the legs were set, the positive set was cancelled. It was possible however to override the positive set cancel valve and have the system 'topping up' after setting until maximum pump pressure was reached.

Face Sprag and Powered Forepole - This combined unit incorporated a face sprag which could be pressurised against the coalface to prevent large lumps spalling off the coalface, and a powered forepole which could be extended to contact the exposed roof very close to the coalface.

Station A was installed on support number 33 to 35 towards the maingate end and station B support numbers 103 to 105 towards the tailgate end of the face. A standard transducer layout was used. Initially the supply line pressure was monitored at both stations but latterly at station A, the second transducer was moved to measure the pressure in the powered forepole of support 34.

Two problems associated with coal cutting were (1) obtaining a consistent roof horizon and (2) maintaining a clean track prior to panzer advance. The absence of either a prominent marker band near the roof horizon or a clean parting at roof level made control of the roof horizon more difficult and steps of 15 cm were sometimes left in the roof. Panzer advance was restricted by bottom coal which frequently sheared off the coalface into the clean track immediately after the machine had cut past. This, combined with small amounts of top coal spalling resulted in canopy tip to face distances averaging 1.34 m after the supports had been advanced compared with the nominal distance of 0.74 m.

Generally roof conditions during the investigation were very good, no roof falls occurring and only the occasional isolated roof break being observed. Waste overhangs of the immediate roof did develop but these were generally less than 1 m long and extended for 5 or 6 supports only.

Conditions in and around the supports were generally clean, support advance was unhampered and the contact pressure acting on the side flaps was very effective in preventing the ingress of material from the roof.

Some concern was felt over the lack of use of the powered forepoles. On several occasions the forepoles were either not extended to support exposed roof or had been extended but not set. Although at the time of these investigations roof conditions remained very good it was felt that problems might arise should the roof become disturbed or broken.

Leg Pressures

Figure 4 shows the leg pressures for support 105 over a nine hour period during which the support was lowered advanced and reset three times at the points 1, 2 and 3. The influence of the machine cut, causing a sudden increase in the rate of pressure build-up can be clearly seen at the points A. The points B indicate a further increase in the pressure build-up caused by the advancing of adjacent supports. After reset the pressure built up rapidly at first, followed by a lower rate of build-up which continued until the machine cut past or some other event took place which transferred load to the support Generally the rear legs took more load.

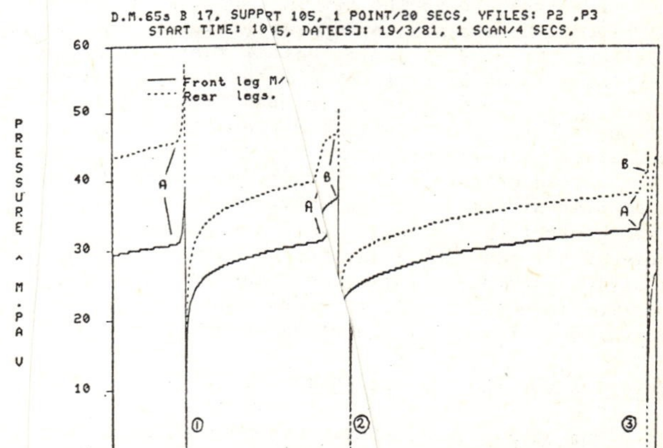

Figure 4 Leg Pressures Against Time
Support 105 - Daw Mill W65's

The influence of machine pass and support advance varied with the speed and continuity of operation. Where the supports were advanced quickly behind the machine (as occurred prior to the first lower and reset on Figure 4) the influence of the cut and adjacent support advance cannot be separated. On the other hand the two events A and B can be clearly seen prior to the second lower and reset when the adjacent supports were advanced about 25 minutes after the machine pass. The rates of build-up after reset varied considerably and rate of support advance was one of the main influencing factors.

During the investigation faults developed in the front maingate legs of supports 33 and 104 and pressure drops related to supply pump switch-off occurred. Figure 5 shows such pressure decays in the front maingate leg of support 104 and superimposed on this graph is the supply line pressure over the same period. The pump pressure was maintained while the pump was on but decayed immediately the pump was switched off. Subsequently a weekend test by MRDE Mining Investigations Branch, Lester M. (1981), indicated approximately 16% of front maingate legs were not maintaining setting pressure. A fault was found in the positive set module causing a very slow leakage and this was later repaired.

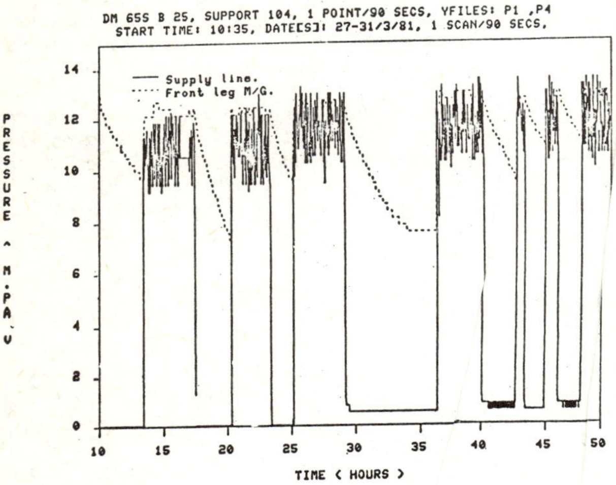

Figure 5 Comparison of Leg Pressure and Supply Pressure - Support 104 Daw Mill W65's

Powered Forepole

The pressure in the setting capsules of the forepole on support 34 was monitored over the latter half of March 1981 and the results indicated that for the majority of the time, load was developing on the forepole tip and quite often reached yield, representing a tip load of 10 tonnes. Because of the limited elevation of the cantilever tip (130 mm above the line of the canopy at full extension) steps in the roof causing the canopy to tilt down at the front could prevent the cantilever tip from contacting the roof. Similarly steps in the roof between the set cantilever and the canopy tip could restrict cantilever retraction. For effective use of the powered forepole it was therefore necessary to maintain a consistent cutting horizon at the roof.

Base, Canopy and Leg Movements

The base tilt transducers recorded almost no tilting of the bases during the working cycle, the only changes occurring at advances where the floor horizon had changed. This confirms the general clean condition of the floor, with only small amounts of dirt around the support bases. Any accumulation of dirt in front of the supports, mainly as a result of changes in the floor cutting horizon was quickly pushed between the bases and left behind in the waste.

The tilting of the canopies during the cycle was related to the magnitude and direction of loading of the support as reflected in the leg pressures. Generally the canopies lowered down towards the waste after reset as the rear leg pressure developed at a faster rate than the front. Material did not accumulate on the canopies and this allied with the clean floor conditions accounted for the rapid build up of support resistance which occurred after setting.

The only major changes in the attitude of the supports occurred at advances and were due to sudden changes in the cutting horizon.

The behaviour of the legs was closely related to the leg pressure, more rapid closure taking place immediately after setting and continuing during the cycle at a slower rate as the leg pressure increased. The contact advance pressure in the legs restricted leg lowering during advance to 60 mm or less.

Hydraulic Supply

A pressure maintenance profile and pump deficiency index were calculated from measurements made at station B on 1 April 1981. Station B was chosen, as this was the measuring position most remote from the power pack. The pressure maintenance profile is shown in Figure 6. This is a method, devised by MRDE Engineering Principles Branch, of showing the proportion of time during a chocking cycle that the available pump pressure exceeded certain levels. These levels are expressed as percentages of nominal unload pressure. The area between the curve and the reference line is a measure of performance of the hydraulic supply system, at the actual demand levels imposed by the support advance procedure. Areas are positive where the curve is displaced above the reference line and negative where it is below. The greater the algebraic sum of these areas, the better is the system.

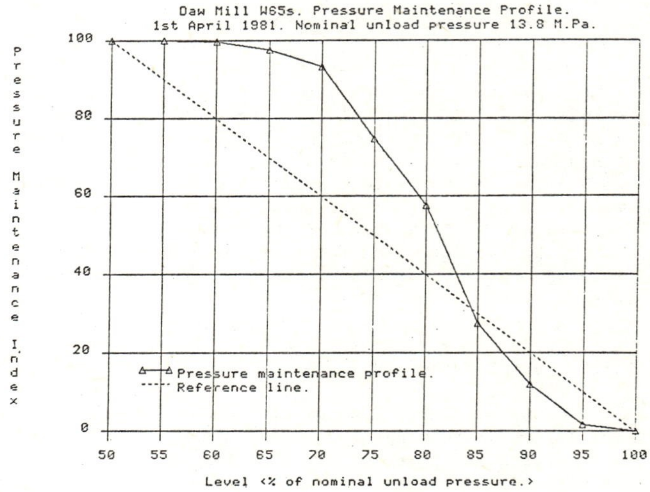

Figure 6 Pressure Maintenance Profile Daw Mill W65's

It can be seen from Figure 6 that the system performance was good at lower pressure levels but less good at higher levels. The transition occurred at 84% of nominal unload pressure.

The pump deficiency index was calculated to be 25%. This represents the proportion of the chocking cycle where the available pressure was below reload pressure. In this case reload pressure was taken to be 75% of nominal unload pressure (ie 10.35 MPa). The lower the index, the better the ability of the hydraulic system to maintain an adequate supply. In comparison with other installations a pump deficiency index of 25% is good, indicating adequate pump capacity to deal with demand.

Setting Pressures

Table 2 shows the front leg, rear leg and mean setting pressures for each support monitored, except number 33. Results from support 33 are omitted as the rear leg pressure transducer was faulty. Also shown in Table 2 are the overall means, weighted according to the number of settings recorded for each support. Only results from 4 second scan interval tests are included as these provide the most reliable values of setting pressure. Cycles during which either leg was not actually set are excluded. In fact only two such cycles were recorded at this scan rate.

Support No	Number of Cycles	Mean Front Leg Setting Pressure MPa	Mean Rear Leg Setting Pressure MPa	Mean Setting Pressure MPa	Mean Setting Load Density MPa
34	12	12.26	12.46	12.36	0.146
35	12	9.86	11.31	10.59	0.125
103	7	12.05	12.21	12.13	0.140
104	7	11.55	11.48	11.52	0.133
105	8	11.73	10.72	11.23	0.131
Station A Weighted Mean	24	11.06	11.89	11.48	0.136
Station B Weighted Mean	22	11.77	11.44	11.61	0.135
Overall Weighted Mean	46	11.40	11.67	11.54	0.136

NB. Only results from 4 sec scan tests included, and excluding cycles where support not set

Table 2 Setting Pressures and Mean Setting Load Densities for Daw Mill W65's Face (4 March - 1 April 1981)

The standard deviation between supports was low, showing a high level of consistency between individual supports monitored. Similarly, there was little deviation between station A and station B and also little deviation between front and rear leg setting pressures. This can be attributed to the effectiveness of the guaranteed set system.

The mean hydraulic supply pressure available at the supports over this period was measured as 11.6 MPa. The overall mean setting pressure achieved was 11.54 MPa. This again highlights the effectiveness of the guaranteed set system.

It is interesting to note that this mean setting pressure was 84% of nominal pump unload pressure, corresponding to the transition point on the pressure maintenance profile (Figure 6).

Mean Setting Load Density (MSLD)

The overall MSLD achieved was 0.136 MPa. This was 74% of nominal setting load density (NSLD) which was 0.183 MPa. Again there was little deviation between supports.

Mean setting load density is not only directly dependent upon mean setting pressure, but also inversely dependent upon supported roof area. The reason for a lower ratio of achieved to nominal setting load density (74%), than that of achieved to nominal setting pressure (84%) is that the measured supported roof area at setting was 13% greater than nominal. This was due to insufficient advance caused mainly by face spalling after machine pass reducing AFC advance. This 13% greater supported roof area implies that the exposed roof at setting (canopy tip to face) was on average 82% greater than nominal.

Figure 7 is a pie-chart showing the contribution of these various factors to the 26% difference between actual and nominal setting load densities. It shows 12% to be due to extra supported roof and 14% to be due to lower than nominal pump pressures. The contribution of setting below available pump pressure is too small to be included. As discussed earlier it is unlikely that much improvement could have been achieved in setting pressures as the pump profile and guaranteed set systems were good. However a reduction in supported roof area at setting would have resulted in a definite improvement.

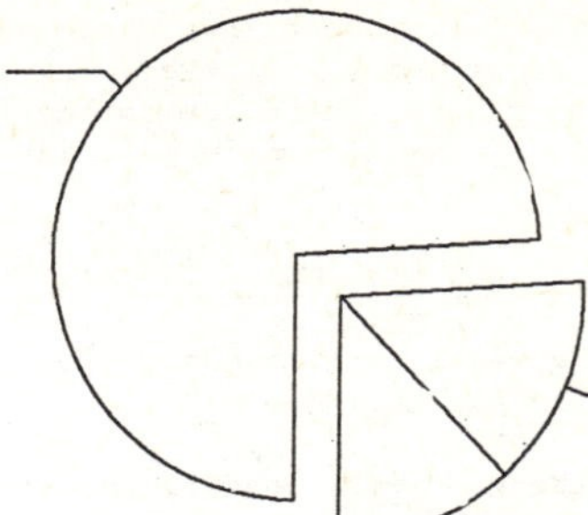

Figure 7 Analysis of Nominal Setting Load Density - Daw Mill W65's 4/3/81 - 1/4/81
Values in MPa. NB APP = Available Pump Pressure

Mean Load Densities (MLDs)

The measured MLDs over the full period of the tests are shown in Table 3. When more accurate information was not available, it was assumed that the time from machine pass to support advance was 10% of the cycle (ie P:Q = 9:1).

Support	Number of Cycles	Mean Load Density (Weekdays) MPa	Mean Load Density (Weekend) MPa	Number of Cycles
34	90	0.3278 (51.58%)	0.4858 (76.44%)	5
35	90	0.3243 (51.03%)	0.4636 (72.95%)	5
103	46	0.3355 (52.79%)	0.4379 (68.91%)	3
104	46	0.3158 (49.69%)	0.2590 (40.76%)	3
105	43	0.3589 (56.48%)	0.4772 (75.09%)	3
Station A Weighted Mean	180	0.3261 (51.31%)	0.4717 (74.70%)	10
Station B Weighted Mean	135	0.3362	0.3914	9
Overall Weighted Mean	315	0.3304 (51.99%)	0.4337 (68.24%)	19

Table 3 Comparison of Weekday and Weekend Mean Load Densities

The MLDs shown in Table 3 are broken down into weekdays and weekends. It can be seen that the overall weighted mean load density for weekdays was 0.330 MPa, approximately 52% of nominal yield load density (Q condition). There was very little deviation between individual supports or between stations, the standard deviation being 0.012 MPa.

The overall weighted MLD for weekends was 0.433 MPa, 68% of nominal yield load density (Q condition). The larger MLD for weekends reflects the longer period during which the supports were stationary and the consequent larger proportion of cycle time near yield. For weekends, the standard deviation between supports was 0.078 MPa, much higher than for weekdays. This is partly due to the very low weekend MLDs experienced by support 104, caused by its leaking front leg, which became a more predominant factor at the slower load build-up experienced at weekends. If this support is ignored the standard deviation reduces by half to 0.039 MPa and the weekend MLD increases to 0.468 MPa

(74% nominal yield load density).

The trend during a typical week was that MLDs on Mondays were regularly lower than those on Fridays, the mean daily increase being 0.01 MPa. This indicates that redistribution of load took place during the weekend stand, probably by settling of upper beds and compaction of the waste.

Conclusions from Daw Mill W65's Investigation

The following general conclusions can be drawn from this investigation on Daw Mill W65's.

1) Despite the good roof conditions existing during the investigation, more systematic use of the powered forepole was necessary particularly where exposed roof was excessive. Measurements indicated that the average area of exposed roof was 82% greater than the nominal. Also in order to maintain forepoles in contact with the roof more consistent horizon control was required.

2) The measured mean load densities reflected a healthy build-up of load on the supports.

3) It is likely that a more even distribution of support load and a reduction in face spall would have occurred had there been less stoppages on the face and had the supports been advanced in a more regular sequence.

4) During this investigation the performance and capacity of the hydraulic supply were both good and the guaranteed set system operated efficiently setting the supports very close to the available pump pressure.

5) The contact advance system maintained a clean travelway and prevented build-up of debris on the canopies.

6) No appreciable base penetration occurred.

ROSSINGTON B46's (DONCASTER AREA) JUNE-AUGUST 1981

Face Operations and Instrumentation

B46's Face at Rossington Colliery in the Barnsley/ Dunsil Seam was equipeed with 130 Gullick Dobson 4/300 tonne Shield Supports similar to those on B90's Face, but without base extensions. The face was cut by a double ended ranging drum shearer operating a bi-di advancing system and fitted with powered cowls (unlike that on B90's). The maingate end of the face was a conventional ripping lip and at the time of this investigation the tailgate end was being converted from an in-line ranged rip to a conventional ripping lip. The height of extraction was generally 3.0 - 3.2 m during the investigation with approximately 0.5 m of coal left in the roof and varying amounts of floor cut. The floor was generally stronger and more competent than was encountered on B90's face in the same seam, due to thinning of the weak immediate floor in that part of the mine.

Station A was installed on supports 62-66 near the middle of the face with station B on supports 104-108 towards the tailgate end. Pressure transducers were fitted to the front and rear leg circuits of all supports. Tilt transducers were fitted normal to the faceline on the bases and canopies of supports 63, 64, 105 and 106, and closure transducers were fitted to the front and rear legs of supports 64 and 106.

The supply line pressure was also monitored at both stations. Measurements were made over the period 19 June to 20 August 1981.

Horizon control at floor level was good and very little material accumulated under the AFC. Control of the roof horizon was not as consistent and problems were encountered where the dirt parting did not always fall away, producing uneven roof conditions. This resulted in loose material accumulating on the support canopies, whereas very little material accumulated under the bases. Measurements were regularly made of the amount of material under the bases and on the canopies at the front leg of the supports at both measuring stations. The average canopy dirt was 106 mm and the average base dirt was 12 mm.

When the face machine cut past it was not uncommon for top coal to shear off the face into the track restricting panzer advance. This was the main reason for the large canopy-to-face distances measured, averaging 1143 mm at station A and 1191 mm at station B, compared with the nominal distance of 395 mm.

Generally roof conditions during the investigation were good, roof falls on the face occurring only twice. These falls were small and were restricted mainly to loss of roof coal of which there was normally 300-500 mm. The average face advance at the beginning and end of the investigation was 5 m/week but fell below this figure in the middle of the investigation due mainly to problems of consolidating the tailgate end as it was changed from an in-line ranged rip.

Conditions in and around the support bases were dirty, with considerable material falling from between the supports at advance. Despite this material the advance of the supports was generally not hampered, the main problem being restriction of the travelway particularly towards the maingate end. The supports were not fitted with linear pressure gauges or positive set. The nominal supply pressure was 16.55 MPa at the beginning of the test but this was increased to 18.62 MPa on the 21/7/81.

One aspect of face operations was the ineffectiveness of the powered forepoles to support exposed roof. The powered forepole had an extension of 600 mm and at this extension could only be raised 65 mm above the line of the canopy. As a result of this limited vertical movement there were many occasions where, due to the attitude of the canopy or the unevenness of the roof, the powered tip could not contact the exposed roof.

It was also apparent that the supports' rear legs were often not set. A pressure survey of forty supports between numbers 62 and 108 indicated 20 with zero rear leg pressure and 4 more at less than setting pressure at a time when roof conditions were good.

Leg Pressures

Figure 8 gives a general picture of leg pressures at support 105. The points A represent the influence of the machine pass causing an increase in load on the support. A further increase, shown by the points B, occurred when adjacent supports were advanced. After reset the pressure can be seen to build up quite rapidly at first followed by a lower rate of build-up which continued until the machine passed.

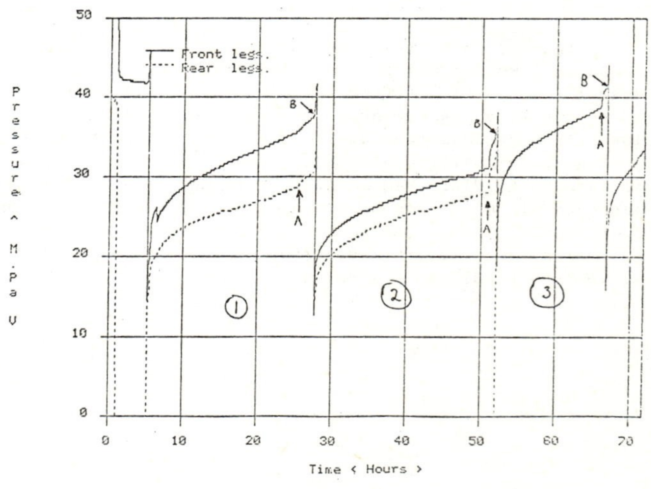

Figure 8 Leg Pressures Against Time - Support 105
Rossington B46's (4-7/8/81)
Start time: 10: 18, 1 scan/min

The influence of the machine pass was not always as pronounced as that shown in Figure 8 but varied with the speed and continuity of operation and the amount of coal cut by the shearer. Where top coal had sheared off the machine pass might not expose fresh roof. It then had very little, if any, effect on the support loading. In general there was much less material under the bases than was encountered on B90's face and this probably accounted for the better build-up of resistance after setting. Where the load build-up was poor it was felt that accumulation of excessive material on the canopies was one of the main factors causing this.

The front and rear leg pressures were generally similar and over a weekend stand quite often reached yield. Although some faulty leg circuits were found and some of the yield valve settings appeared low (eg 34 MPa on 63 front leg circuit compared with a nominal level of 40.5 MPa), the hydraulic condition of the supports was reasonable.

Base, Canopy and Leg Movements

The base tilt transducers showed virtually no tilting of the bases during the working cycle under normal circumstances.

Figure 9 shows the base and canopy movements for support 105 over the same time period as Figure 8. It is noticeable that there was no base tilting on cycles (1) and (2) but on cycle (3) when the rear legs were not set, the base of the support 'toed in' by approximately 15 minutes of arc during the cycle.

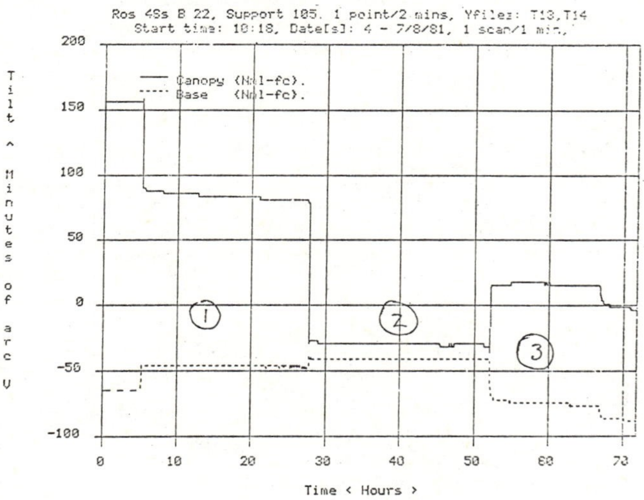

Figure 9 Base and Canopy Tilt Against Time Support 105 Rossington B46's

Changes in the attitude of the canopy throughout the cycle were generally small. However the changes in overall canopy attitude at advance tended to be much greater than those of the base, reflecting the less consistent roof horizon.

Leg closures and pressures were closely related, rapid closure taking place immediately after setting and continuing more slowly during the cycle. When a support was advanced, leg lowering was manually controlled but was generally less than 125 mm. However, as the side flaps of the supports were not maintained in contact during advance this amount of leg lowering permitted material from the roof and canopies to fall between the supports.

Hydraulic Supply

As stated previously the nominal unload pressure was increased from 16.55 MPa to 18.62 MPa on 21 July 1981. The following analysis of the hydraulic supply deals with the period after 21 July. Only measurements made of supply pressure at station A are included as an intermittent fault occurred on the transducer used at station B.

The mean available pump pressure at station A was 14.19 MPa (76% nominal unload). The measured unload pressure was 16.96 MPa (91% nominal) and the measured reload pressure was 13.55 MPa (73% nominal unload).

Figure 10 shows a Pressure Maintenance Profile (PMP) calculated from pressure data recorded on 23 July 1981. It is interesting to note that, as for Daw Mill W65's the transition point on the PMP (77% nominal unload pressure) was very similar to the mean available pump pressure.

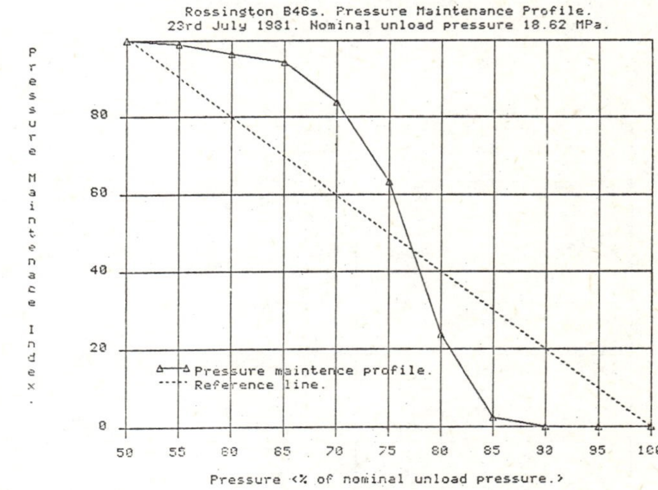

Figure 10 Pressure Maintenance Profile Rossington B46's

The Pressure Deficiency Index (PDI), calculated from the same pressure data was 37%. Both the PDI and PMP show reasonable pump performance under the demands actually imposed by face operations.

Setting Pressures

The measured setting pressures are shown in Table 4. This table is complicated due to the change in nominal unload pressure during the period of measurement and because two sets of figures are shown, one including and one excluding cycles where rear legs were not set. The rear legs were not set on 19% of the cycles. The mean setting pressure including these cycles was 12.11 MPa (65.5% nominal unload pressure), excluding them it was 13.33 MPa (72.08% nominal unload pressure).

It is reasonable to conclude that the mean setting pressure would have been 10% higher had rear legs been regularly set.

(20.6.81 - 20.8.81) Only results from 10 sec scan tests included.
Figures in brackets represent percentage of Nominal Unload Pressure.

Support	Including Cycles where Rear Legs not Set				Excluding Cycles where Rear Legs not Set			
	No of Cycles	Setting Pressure (MPa)			No of Cycles	Setting Pressure (MPa)		
		Front legs	Rear legs	Mean		Front legs	Rear legs	Mean
Before 21.7.81 Nominal Unload Pressure 16.55 MPa								
Weighted Mean Stns A and B	2	12.58 (73.98)	12.58 (73.98)	12.58 (73.98)	2	12.58 (73.98)	12.58 (73.98)	12.58 (73.98)
After 21.7.81 Nominal Unload Pressure 18.62 MPa								
Weighted Mean Stn A	27	13.48 (72.40)	10.27 (55.14)	11.87 (63.77)	21	13.28 (71.32)	13.20 (70.32)	13.24 (71.11)
Weighted Mean Stn B	2	14.47 (77.71)	15.42 (82.81)	14.95 (80.29)	2	14.47 (77.71)	15.42 (82.81)	14.95 (80.29)
Overall Weighted Mean	29	13.55 (72.77)	10.63 (57.05)	12.08 (64.91)	23	13.38 (71.88)	13.39 (71.93)	13.39 (71.91)
Overall Mean of before 21/7 & after 21/7	31	13.49 (72.84)	10.76 (58.14)	12.11 (65.50)	25	13.31 (72.06)	13.33 (72.09)	13.33 (72.08)

Table 4 Analysis of Setting Pressures for Rossington B46's

The standard deviation in setting pressures between supports was approximately 1 MPa. This was not excessive for a face without guaranteed set. However the mean setting pressure achieved, particularly after the increase in pump pressure (12.08 MPa), was significantly less than the mean available pump pressure (14.19 MPa) a direct consequence of the absence of guaranteed set.

Mean Setting Load Densities (MSLD's)

Table 5 shows MSLD's calculated from measured canopy tip to face distances and measured setting pressures.

(20.6.81 - 20.8.81) Only results from 10 sec scan tests, where P condition face to canopy tip distances were measured, are included. Figures in brackets represents percentages of Nominal Setting Load Density.

Period	Support	Including cycles where rear legs not set		Excluding cycles where rear legs not set	
		No of Cycles	Mean Setting Load Density (MPa)	No of Cycles	Mean Setting Load Density (MPa)
Before 21.7.81	Weighted Mean	2	0.198 (62.06)	2	0.198 (62.05)
After 21.7.81	Weighted Mean Stn A	8	0.199 (54.92)	4	0.250 (69.01)
	Weighted Mean Stn B	2	0.236 (66.38)	2	0.236 (66.38)
	Overall Weighted Mean	10	0.206 (57.21)	6	0.245 (68.13)
Overall mean of before 21.7.81 and after 21.7.81		12	0.205 (58.02)	8	0.233 (66.61)

Table 5 Analysis of Mean Setting Load Densities for Rossington B46's

The nominal setting load densities were as follows:

NSLD = 0.319 MPa prior to 21/7/81
= 0.360 MPa after 21/7/81
= 0.353 MPa (weighted mean allowing for number of cycles used for calculations before and after 21/7/81).

The overall mean setting load density achieved was 0.205 MPa (58.02% nominal). Excluding cycles where rear legs were not set this becomes 0.233 MPa (66.61% nominal). The measured canopy tip to face distance was on average 148% greater than nominal when the supports were forward (P condition). Assuming caving at the rear of the canopy this figure implies a supported roof area at setting 24.6% greater than nominal.

The standard deviation for MSLD between supports was 0.029 MPa.

Figure 11 is a pie chart showing a breakdown of the various factors which contributed to the difference between achieved MSLD and nominal. It can be seen that almost half of this difference, 19.8% NSLD, was due to the rear legs not being set and from setting below available pump pressure. This loss could have been removed by fitting a guaranteed set system to the supports.

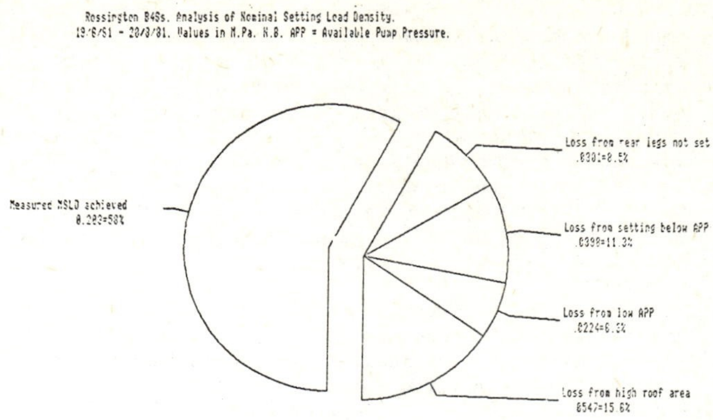

Figure 11 Analysis of Nominal Setting Load Density Rossington B46's

Mean Load Densities

The MLD's, achieved over the full period of the tests, are shown in Table 6. When more accurate data was not available, it was assumed that the time from machine pass to support advance was 10% of the cycle (ie P : Q = 9:1). The MLD's shown in Table 6 are broken down into weekdays and weekends. It can be seen that the mean load density for weekdays was 0.282 MPa (45% of nominal yield load density). The overall weighted MLD for weekends was 0.358 MPa (57% of nominal yield load density).

(20.6.81 - 20.8.81) Figures in brackets represent percentages of Nominal Yield Load Density (Support standing back)

Support	Number of cycles	Mean Load Density MPa (Weekdays)	Mean Load Density MPa (Weekends)	Number of cycles
62	25	0.275 (43.90)	0.369 (58.91)	3
63	25	0.298 (47.57)	0.428 (68.33)	3
64	25	0.279 (44.54)	0.320 (51.09)	3
65	25	0.287 (45.82)	0.380 (60.66)	3
104	9	0.281 (44.86)	0.308 (49.17)	2
105	7	0.362 (57.79)	0.365 (58.27)	1
107	9	0.187 (29.85)	0.303 (48.37)	2
Station A Weighted Mean	100	0.285 (45.50)	0.374 (59.71)	12
Station B Weighted Mean	25	0.270 (43.10)	0.318 (50.76)	5
Overall Weighed Mean	125	0.282 (45.02)	0.358 (57.15)	17

Table 6 Comparison of Weekday and Weekend Mean Load Densities - Rossington B46's

The standard deviations in MLD between supports for weekdays and weekends for 0.033 and 0.043 MPa respectively. These indicate that except where the rear legs were not set, load was developed fairly evenly across the supports.

In Figure 12 average MLD's at station A are plotted with respect to their time of occurrence relative to the beginning of the week. The gradients of the linear regressions are 0.022 MPa/day including weekends and 0.016 MPa/day excluding weekends. The correlation coefficients of these regressions are 0.475 and 0.389 respectively. This indicates that support loading follows a weekly cycle with a gradual rise during the week and a reduction following the weekend stand.

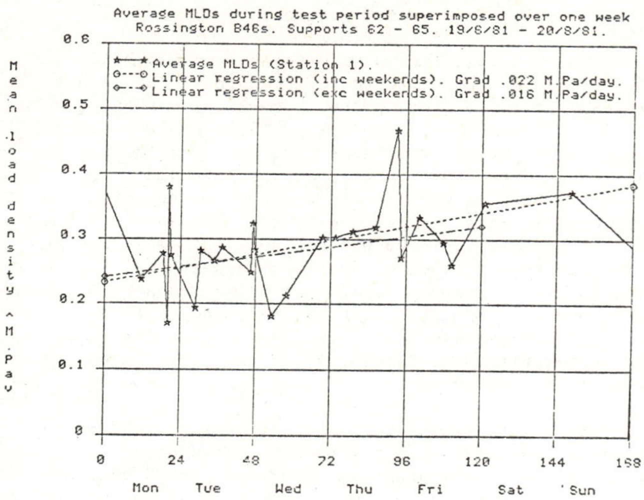

Figure 12 Average MLD's during Test Period Superimposed over one week. Rossington B46's Supports 62-65 (19/6/81-20/8/81)

Conclusions from Rossington B46's Investigation

The following general conclusions can be drawn from the investigation on B46's Face.

1) The mean setting pressures were significantly less than the mean available pump pressure due mainly to poor manual setting and rear legs not being set. The loss from rear legs not set and from setting below available pump pressure (representing 19.8% of the nominal setting load density) could have been removed by a guaranteed set system. One reason contributing to the rear legs often not being set both on B46's and B90's was that the nearest position on the valve block to 'Support advance' was 'Front leg set' and not 'All legs set'.

2) Despite reasonable roof conditions during the investigation, the powered forepoles could have been used more effectively to support exposed roof. To ensure effective operation of the forepoles, considering their limited extension and vertical travel particular attention must be paid to roof horizon control and to keeping the canopies clear of loose debris.

3) The average face to canopy tip distance was 148% greater than the nominal and this was mainly due to AFC advance being inhibited by face spalling after the machine pass.

4) The measured mean load densities indicated a build-up of load on the supports conducive with controlled roof-to-floor convergence. There was a tendency for the mean load density to increase over the week from Monday to Friday. More effective support setting could possibly have been achieved if the support canopies had been regularly cleaned.

5) During this investigation the performance and capacity of the hydraulic supply were adequate to meet the demands put upon the system.

6) Base penetration occurred only when the rear legs were not set, and even then this did not present a problem in advancing the supports.

THEORETICAL SETTING AND YIELD LOAD DENSITIES

Several different approaches have been suggested to calculate the setting and yield load densities required on a longwall face. The most common are briefly described below and then applied to Daw Mill W65's and Rossington B46's.

Required Setting Load Density (RSLD)

1) Bates (1978) has suggested extrapolation of recorded leg pressure curves, in order to determine the setting pressure required to eliminate initial rapid pressure build up due to convergence after setting. The ideal setting pressure thus found can then be translated into a required setting load density.

2) Wilson (1980) has suggested an equation developed from the 'roof block concept' which takes into account friction of canopy on roof and of rock on rock, position of support thrust, height of extraction, face width and angle of rise.

$$\text{RSLD}=0.05\ \text{MF} \times \frac{(\ell/2\ \cos\delta+\text{M}\sin\delta)(1-\mu^2)+\mu\ell(\mu\cos\delta+\sin\delta)}{(\text{p}\ \cos\beta+2\text{M}\sin\beta)(1-\mu^2)+\mu\ell(\mu\cos\beta+\sin\beta)}\ \text{MPa} \quad (1)$$

where P represents		support thrust	(M.N.)
ℓ	"	rear of canopy to face distance	(m)
s	"	support spacing	(m)
F	"	factor of safety	
μ	"	coefficient of friction rock on rock	
β	"	angle of support thrust to the normal to the roof	($^\circ$)
δ	"	angle of rise in direction of face advance	($^\circ$)
p	"	distance of resultant support thrust from rear of canopy	(m)
M	"	height of extraction	(m)

Wilson suggested that values of 0 for β, 0.4 for μ, and 1 for F might be appropriate.

3) A draft NCB Mining Department Production Instruction, presently under discussion, calls for a setting load density in MPa of 0.074 times the extracted height in metres.

4) Ashwin et al (1970) suggested a nominal setting load density of 0.27 MPa in thick seams (up to 2.8 metres).

Required Yield Load Density (RYLD)

1) Wilson (1980) suggested that the same equation put forward for setting load density, equation (1) above, was appropriate, if a factor of safety of 1.5 was introduced and μ was taken as zero.

2) Wilson (1981) later suggested a much simplified equation.

$$\text{RYLD} = 0.05\ \frac{\text{M}\ell}{\text{P}}\ (\text{MPa}) \quad \text{....(2)}$$

where the variables are as defined for equation (1).

This method would give results very similar to method 1 for level seams if the factor of safety were increased to 2.

3) The draft NCB PI suggests a yield load density in MPa of 0.15 times the extracted height in metres.

4) Ashwin et al (1970) called for a yield pressure/set pressure ratio of 1.25:1, although they only considered seams of up to 2.8 metres.

Application of RSLD and RYLD Calculations to Daw Mill W65's and Rossington B46's

	Required Setting Load Density (MPa)				Required Yield Load Density (MPa)			
Method of Calculation	1 BATES	2 WILSON	3 NCB PI	4 ASHWIN	1 WILSON(1)	2 WILSON(2)	3 NCB PI	4 ASHWIN
Daw Mill Colliery W65s	Insufficient Data	0.26	0.26	0.26	0.55	0.72	0.53	0.41
Rossington Colliery B46s	0.28	0.20	0.22	0.27	0.43	0.58	0.45	0.28

Table 7 RSLD and RYLD for Daw Mill W65's and Rossington B46's

A comparison of the required, nominal and measured setting and yield load densities for these two faces is provided in Table 8 below.

Colliery and Face	Daw Mill W65's (MPa)	Rossington B46's (MPa)
Mean RSLD	0.26	0.24
Nominal SLD	0.18	0.35
Mean Measured SLD	0.14	0.21
Mean RYLD	0.55	0.43
Actual YLD (Q)	0.64	0.63

Table 8 Required, Nominal and Measured Setting and Yield Load Densities for Daw Mill W65's and Rossington B46's

Table 8 shows that the yield load densities for both faces were well above the mean theoretical requirements. Wilson's second method gives the highest requirement for yield load density as it incorporates the largest factor of safety.

The mean measured setting load density for Daw Mill W65's was well below the theoretical requirement, as was the nominal SLD. This was due to the relatively narrow bore of the support legs. The nominal SLD for Rossington B46's, however, was well above the mean theoretical requirement and although this nominal SLD was not achieved, due to high supported roof area and no guaranteed set, the actual SLD measured was only just below the mean theoretical requirement. It is noteworthy that both face conditions and performance were better on Daw Mill W65's than on Rossington B46's although the mean theoretical requirements for setting load density were much closer to being satisfied on the latter.

Of the different methods presented for calculating theoretical SLD, Bates' tends to give the higher value and Wilson's the lower, although selection of the input variables to Wilson's equation critically affects the value obtained. Bates' method is a practical one, provided sufficient measurements are made and their variance is acceptable. However it is not of direct use in designing new faces and experience has shown that in most cases it will produce a theoretical requirement greater than the measured SLD however high this may be. This is because time dependent compaction of roof and floor post setting almost invariably produces a convex pressure curve.

CONCLUSIONS AND FUTURE DEVELOPMENTS

The MRDE Underground Monitoring System is proving of great value in Strata Properties Group's investigations into support and strata behaviour on longwall coal faces. It is now possible to record, analyse and store a large volume of data on powered support behaviour, enabling accurate calculation of mean setting load densities, mean load densities and support movements. It must be stressed however, that such information is of value only when combined with observational data from the face. Also, use of the support as a medium for measuring strata behaviour has limitations, although there is still a large scope for new transducers such as for measuring face-to-support distances and canopy and floor debris compaction. It is hoped to combine use of the Support Monitoring System with borehole extensometer work and measurement of total convergence on future face investigations, though this presents practical problems in terms of manpower and logistics.

The investigations at Rossington and Daw Mill Collieries have highlighted the fact that face supports are only a single item, although an important and costly one, in the complex system comprising a modern longwall face installation. The major problems identified on these faces were horizon control and inadequate AFC advance, which in turn led to problems of support often reflected in large canopy tip-to-face distances. It has not yet been established whether coal spalling on high faces can be regulated by increased support resistance. Certainly this was not apparent on Rossington B90's. The powered forepole systems on the supports were generally inadequate due to insufficient vertical travel and vulnerability to damage, given the variable nature of roof horizon control.

It was also found that the forepoling and face sprag systems that were fitted were often under-utilised. The test programme confirmed the advantages of guaranteed set and contact advance although further work is still necessary to identify optimum setting loads. Of the three faces covered by these investigations floor penetration by the support bases was a cause for concern only on Rossington B90's. There was some penetration on Rossington B46's, but only when rear legs were not set. The problem on B90's was alleviated to some extent by fitting base extensions. Floor penetration by shield supports is only likely to cause concern on very soft floors or where large amounts of 'gummings' build up and it is felt that in most circumstances this could be reduced to an acceptable level by care in the design of support bases and by good machine clean-up.

High pressure positive set systems have been installed on both Daw Mill W65's and Rossington B46's since this work was completed, and it is intended to return to these sites to monitor their effects.

Even though over 10,000 hours of data has been recorded using this equipment, data logging underground is still in its infancy. However, the experience obtained by the use of this system will provide the expertise necessary for further developments. Experience has shown, for example, that although tape cassettes can store large quantities of data on a non-volatile medium, they can be unreliable when used in a harsh environment. Humidity and the ingress of dust can affect both the performance of the tape cassette and the moving parts of the cassette drive. Isolating the tape and tape transport from the environment by sealing prevents the user from changing cassettes underground. It is hoped that any future developments will contain no moving parts and will incorporate low power electronic memory devices to store data.

It has been found that in many circumstances a large percentage (up to 90%) of the data recorded is redundant. Development of a microprocessor-based data logger will enable the user to program the device underground so that only selected data will be recorded. This facility will enable a much smaller memory to be used and will provide a far more versatile system.

ACKNOWLEDGEMENTS

We wish to thank our colleagues at MRDE and the management teams at Rossington and Daw Mill Collieries for their invaluable assistance, without which this work would not have been possible. We also wish to thank the Director of the MRDE for his permission to present this paper. The views expressed are those of the authors and not necessarily those of the NCB.

LIST OF REFERENCES

1. ASHWIN et al., 1970, 'Some Fundamental Aspects of Face Powered Supports Design', Mining Engineer, 659-675.

2. BATES, J J, 1978, 'An Analysis of Powered Support Behaviour', Mining Engineer, 137/203, 681-694.

3. COCKER, D S and BIGBY, D N, 1981, 'A Portable Underground Face Support Monitoring System Developed at MRDE', Colliery Guardian, 229/3, 83-86.

4. EVANS, I, 1977, 'Face Support Requirements - A Problem in 'Arching'', Int J Rock Mech Min Sci and Geomech Abstr. 14, 1-5.

5. LESTER, M, 1981, 'Daw Mill Colliery - Powered Supports Condition Test, 65's Face, 1-3 May 1981'. MRDE Tech Memo No IB/TM478.

6. SHEPHERD and ASHWIN, 1968, 'Measurement and Interpretation of Strata Behaviour on Mechanised Faces', Colliery Guardian, 795-804.

7. WILSON, A H, 1975, 'Support Load Requirements on Longwall Faces', Mining Engineer, 479-491.

8. WILSON, A H, 1980, 'The Stability of Underground Workings in the Soft Rocks of the Coal Measures', Ph.D. Thesis, Nottingham University.

9. WILSON, A H, 1981, 'An Alternative to the 2 m Prop Free Front', Colliery Guardian, 148-150.

STRATA CONTROL PROBLEMS ON LONGWALL FACES - AN OVERVIEW

J S Oram

National Coal Board
Mining Research and Development Establishment
Ashby Road, Stanhope Bretby, Burton-on-Trent, Staffs

SUMMARY. Investigations carried out by the Strata Properties Group of the Rock and Coal Mechanics Branch, MRDE, have indicated in many cases that problems on longwall faces are arising because the basic requirements for strata control are not being maintained. The paper outlines these basics and their relationships with day to day operations. Some of the problems encountered are described and their sources analysed. Some recent developments in equipment and facilities which help to improve the effectiveness and reliability of face equipment are described and areas where improvements could be made discussed. In conclusion the areas where manufacturers, management, and research workers should be directing their attention in order to bring about improvements are outlined.

INTRODUCTION

Since the early 1960's there has been a steady development of face supports and the hydraulic systems to operate them. We have progressed from the early simple frame supports offering a minimum of facilities to the present day heavy duty shield supports which are highly sophisticated and offer a tremendous variety of additional facilities. These facilities include increased support load density, guaranteed set, contact advance, powered forepoles and face sprags to name but a few. Similar developments have taken place with regard to hydraulic pumps and hydraulic distribution systems. However despite all these improvements strata control problems on longwall faces still persist. These problems which are costly to the industry in terms of lost production and damage to equipment also present a safety hazard as recent serious accident figures show; out of a total of 166 serious accidents which occurred underground over a 13 week period of 1981, 35 (21%) were caused by falls of ground at the face. It is important therefore to identify the cause or causes of these problems and effect a cure if one exists.

Over the past 4-5 years the Strata Properties Group of the Rock and Coal Mechanics Branch at MRDE have been carrying out investigations into strata behaviour around coalfaces, and as a result of these investigations have found in an increasing number of instances for a variety of reasons, that the basic requirements for effective strata control were not being achieved. The object of this paper is to highlight these reasons and to put forward recommendations on how to handle some of the problems.

In order to appreciate how the problems arise it is essential to describe the basic requirements for strata control and how they interrelate to certain face operations.

BASIC STRATA CONTROL REQUIREMENTS

Figure 1 is a simple representation of the strata around a fully caved longwall face. Ahead of the coalface, due to the redistribution of stresses an area of increased stress exists, commonly known as the zone of the front abutment. Within this zone fracturing of the immediate roof beds occurs. Thus in the area above the face supports the immediate roof consists of blocks which if unsupported would collapse. The upper beds, which may also be fractured are assumed to bridge over into the waste bending down over the coalface as the waste compacts. The first major requirement for strata control is therefore that the face supports should develop sufficient resistance to support the immediate roof beds, restricting roof lowering and bed separation.

The second major requirement for effective strata control is the early forward support of exposed roof. When the face machine cuts past an area of roof is exposed. If this exposed roof is not supported quickly, small sections of the immediate roof may fall out. Although such a small roof fall in itself may not constitute a strata control problem, the majority of major roof falls initiate from small falls of this kind. Figure 2 shows a number of break formations and the typical falls that could generate from them if exposed roof was left unsupported. It is important therefore to secure effective forward support by keeping the support as close to the coalface as possible and making effective use of forepoling.

These two main requirements for effective roof control are not independent of face operations but can be influenced considerably by the quality of clean up in the machine track and horizon control. The interdependence of these four parameters is shown diagrammatically in Figure 3, which indicates to neglect any one or more of them could lead to strata control problems. Some of these problems and examples of underground installations where they arose are discussed in the following section.

PROBLEMS AND THEIR SOURCES

Let us consider first problems associated with horizon control and clean up. Figure 4(a) shows the influence of a 76 mm step in roof at the point (A) on the attitude of the support canopy of a Dowty 4 x 450 tonne shield support. In this situation at full extension the powered cantilever tip would be unable to contact the roof leaving an extensive area of roof from the point A to the coalface unsupported.

Problems of this type were observed on W65's face at Daw Mill Colliery. On this face because of a heavy spall of coal into the machine track restricting panzer advance, and shearing of top coal, the average face to canopy tip distance after advance was 1.35 m. The use of the powered forepole to support this exposed roof was important, particularly where roof breaks were present. Areas were observed on this face where the attitude of the canopies prevented forward support by the forepoles. There was a much greater risk of roof falls initiating in these areas. Another problem which can arise from changes in roof horizon is the accumulation of debris on the canopy from spalling and crushing in the vicinity of the steps. This material on the canopy can reduce the build up of support resistance and allow additional roof lowering to occur as the debris is compressed.

Figure 4(b) shows problems associated with changes in floor horizon. Steps in the floor at B and C when encountered by the support base or the AFC will be ploughed up if the floor is soft or highly laminated. This material accumulating in front can restrict AFC and support advance, resulting in greater face to canopy tip distances. If the steps are negotiated they will alter the attitude of the support base and AFC and allow material to accumulate underneath as shown in Figure 4(b). Such changes in the attitude of the AFC not only influence the subsequent cutting horizon of the face machine but also reduce its ability to leave a clean track. If this problem is allowed to go unchecked the situation represented in Figure 4(c) can result where substantial material has accumulated under the AFC, and the support base. The floor material is increasingly compressed under the influence of the AFC, the front legs of the support and finally the rear legs of the support. This difference in compressibility results in the support base 'toeing' in, particularly if the support design generates high toe loads. Apart from the slow build up in support resistance which generally results, problems are also encountered with ploughing of material in front of the support base and in advancing the support itself. An example of this situation was encountered on B90's face in the Barnsley/Dunsil Seam at Rossington Colliery. Face investigations were carried out using the MRDE Underground Monitoring System over the period February-April 1980 during which face conditions were poor with considerable material accumulating above and below the supports. This resulted in poor support resistance and penetration of the support bases into the floor dirt. Figure 5(a) shows the general poor build up in support leg pressures, while Figure 5(b) shows the base of the support tilting down at the front at setting and during the cycle. Face advance over the period of the investigation was very poor with roof falls occurring regularly. Despite the subsequent fitting of base extensions to resist toeing in and the installation of high pressure positive set to improve support setting resistance, conditions in general did not improve as the clean up by the face machine remained ineffective and problems of horizon control continued. This lends support to the statement made earlier that to secure effective strata control it is necessary to maintain all the parameters indicated on Figure 3.

This is an appropriate point at which to make some comments concerning base penetration and the bearing capacity of mine floors. It is possible if certain assumptions are made to calculate the loads imposed on the floor by the base of a support for specific leg loading. Such calculations show that for any support properly set on a clean level floor the loads imposed on the floor are insufficient to rupture the average mine floor. The exceptions to this rule are:

a) where the floor is a clay seatearth with an 'in-situ' strength below 3.5 MPa,

b) where excessive toe loads are generated by a badly designed base or rear legs not being set,

c) where the strength of the floor has been considerably reduced by the presence of water.

Although the average mine floor can have sufficient strength to resist base penetration the strength parallel to the bedding in laminated material is very low and as mentioned earlier where steps are formed in the floor ploughing of the upper floor layers by the support base can occur. In many instances where it was claimed that supports were penetrating the floor it was found that the supports were merely compressing loose material which had accumulated under the base of the support. Another problem allied to material accumulating under the base is toppling or tilting parallel to the face line. An example of this occurred at Bold Colliery on T15's face in the Trencherbone Seam. This retreat face had a dip of 1 in 10 along the face and material was accumulating under the high side of the support bases. At advance any material which accumulated in front of the bases would be ploughed on the low side which was taking the weight of the support, while on the high side material would accumulate under the base. On this particular face the situation had been permitted to progress unchecked such that the stabiliser rams fitted between each pair of supports were having no effect. The loading under the low side of the support base would be quite intense as the bearing area had been considerably reduced and no improvement in stability could be envisaged until the general cleanliness of operation was improved.

Finally, problems related directly to maintaining adequate setting pressures fall into three main categories;

i) legs not being properly set by the operator,

ii) hydraulic faults in the leg circuits,

iii) insufficient supply pressure due to either multiple demand problems or inefficient supply pump operation.

These faults have the general effect of reducing the initial setting load density of the support and the poor resistance offered to roof lowering often results in a worsening of roof conditions.

An example of (i) above occurred on B90's face at Rossington Colliery. As mentioned earlier considerable material had accumulated above and below the supports thus on resetting a support it was necessary to hold the manual control handle for a considerable time to allow the setting pressure to build up as this material was compressed. In the absence of linear gauges to indicate when setting pressure was achieved supports were frequently left well below setting pressure. Later this problem was largely overcome by the installation of linear gauges and positive set modules which guaranteed the setting pressure.

CURES

Some of the problems discussed above are related to inefficiencies in face operation and others to design and maintenance of equipment and the areas where improvements have been and can be made are discussed below.

Support setting load density

With regard to securing effective support setting pressures three major advances have been made in recent years:

a) Guaranteed set

As mentioned in the previous section by fitting positive set modules to the support valve gear the setting pressure can be guaranteed in all leg circuits provided a relatively low trigger pressure is achieved.

b) Linear pressure gauges

These gauges, fitted into the leg circuits have proved to be extremely useful in indicating to the operator that setting pressure has been achieved, and where a guaranteed set facility is fitted that the positive set valve gear is operating efficiently. These gauges are also useful when carrying out weekend hydraulic condition tests on the leg circuits.

c) Computer design of hydraulic systems

The design of hydraulic distribution systems for powered supports involves many parameters (pump capacity, system demand, hose layout and diameters etc), and in order to design these systems for optimum efficiency, MRDE Bretby have developed a computer program which can be used to give the best layout parameters for a particular installation. The use of this computer design and ring main circuits has been effective in reducing the problems of multiple demand - a major source of low setting pressures.

In relation to the capital cost of a modern powered support installation the cost of the use and installation of the above facilities, particularly linear gauges, is very low compared with the obvious benefits that can be achieved with respect to ensuring the effiency of the support system. The importance of management making maximum use of these facilities therefore cannot be over emphasised.

Securing the efficient operation of these and other support facilities however will put an added emphasis on the maintenance of hydraulic systems and equipment and consequently management will have to set up regular maintenance and servicing schemes.

Horizon control

At the moment the steering of most machines is a matter of judgement by the machine operator and in the case of a double ended ranging drum shearer in a thick seam in the absence of a reliable marker band or a clean roof parting, it is extremely difficult to maintain a consistent cutting horizon. The manufacturers can fit a simple device which will indicate the height of the drum relative to the machine body or the panzer but many of these devices are either not fitted or are extremely difficult for the operator to observe. If such devices are to be of any use they must be redesigned so that the output from them is easily seen and interpreted by the operator from his normal operating position.

However, probably of equal importance in relation to horizon control is the urgent need for some simple signal device which could relate to the operator the position of the drum within the seam in its entirety. This is particularly important where the seam thickness varies and difficulty is experienced in maintaining a consistent thickness of coal at either or both top and bottom horizons.

Clean-up

Although many of the problems of machine clean up are associated with horizon control, some are due directly to incorrect drum speeds or haulage speeds. The Coal Cutting Group at MRDE have as a result of extensive model work, produced a computer simulation program which can be used to obtain the relative drum and haulage speeds to give optimum loading characteristics for a particular shearer installation and use should be made of this service wherever applicable.

Concerning the general cleanliness around supports the development of hydraulic side flaps giving skin to skin contact of supports is effective in reducing the ingress of dust and dirt at lower and advance provided the flaps remain in contact during advance. Such a system of contact advance operated very effectively on 65's face at Daw Mill Colliery. The low level contact pressure of 1.25 MPa also restricted leg lowering during advance.

Early forward support

Improvements in horizon control and machine clean up will allow more effective use of forepoling systems but many of the forepoling systems are themselves badly designed. Blade type cantilevers generally have insufficient extension or vertical travel to meet most circumstances, and when roof contact is made the area of contact is very small. A much improved design is the type fitted to the Dowty 4 x 450 tonne chock shield supports. This forepole has an extension of 1 metre, a pad area of 390 mm x 1310 mm, and a tip load at setting of 6.4 tonne. Despite these improved features, at full extension the tip can only be extended 130 mm above the line of the canopy. The support of exposed roof is fundamental to good strata control and the manufacturers should be looking to forepole designs which will cater for the uneven roof conditions which will inevitably be encountered, irrespective of the quality of horizon control.

CONCLUSIONS

This paper has presented the argument that many of our present day problems of strata control are a result of not maintaining the basic requirements as illustrated in Figure 3. In the future as mining goes to greater depths as it inevitably will, it is not foreseen that our support system and face operations will change dramatically, and the basic requirements for strata control will remain unchanged. However, with increase in depth a proportional increase in the front abutment pressure will occur and the increased fracturing of the immediate roof and floor beds which is likely to occur will place even greater importance on the basic strata control requirements, particularly early forward support.

It is important therefore that manufacturers, management and research workers consider how improvements can be brought about now. Some of the improvements that can be made with respect to support load density were outlined in the previous section and a major recommendation must be, considering their cost effectiveness, that those facilities should be used with every future face installation.

The manufacturer for his part has to look at the design of more efficient forepoling systems and the design of support bases for soft floor applications. With the increased facilities available on modern supports, they must avoid the support becoming a confusing jumble of hoses and valve gear, but develop simple systems which are reliable and easy to maintain.

The management must ensure that face operations are carried out as efficiently as possible with attention to early forward support, horizon control and machine clean up. The efficient operation of the supports and their facilities will be dependent on a high level of hydraulic maintenance. Maximum use should be made of special facilities where available and it will be important to train face workers in the proper use of these facilities and their relevance to strata control. With regard specifically to early forward support it must be stressed that the exposed roof must be secured as quickly and as effectively as possible at all times. To neglect such precautions often leads to major roof falls with a corresponding loss of output while repairs are effected.

With regard to research into strata behaviour around coal faces there is a need to understand more about the formation of induced fractures in the immediate roof beds and the seam and how these are influenced by the properties of the different strata and the magnitude of the front abutment. Although it is accepted that there is a certain minimum support resistance for effective support of the roof block (there are guide lines available for determining minimum setting load densities), very little is known concerning the influence of the magnitude and position of the support thrust and of powered forepoling on the behaviour of the immediate roof strata. A better understanding of this behaviour can only be gained by carrying out measurements of the movements occurring in the immediate roof and how these movements are related to events within the mining cycle. The onus on research workers is therefore to develop the techniques necessary to carry out those measurements.

ACKNOWLEDGEMENTS

The author would like to thank the Director of MRDE of the NCB for his permission to present this paper. The views expressed are those of the author and not necessarily those of the NCB.

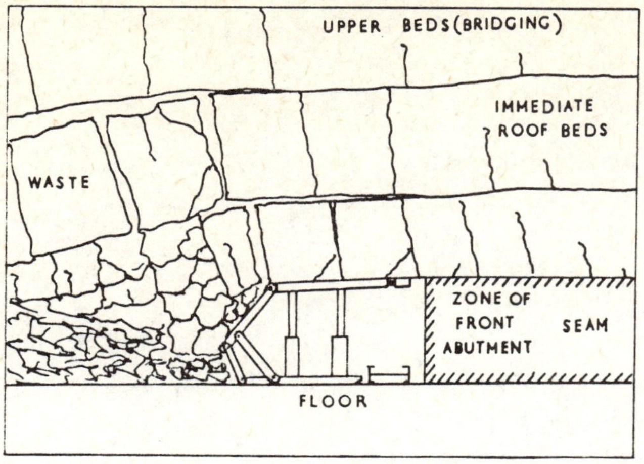

Figure 1 - Strata around a fully caved longwall face

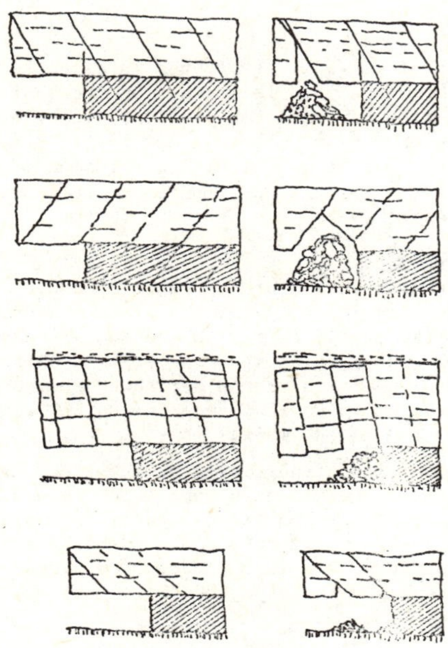

Figure 2 - Falls resulting from various break formations

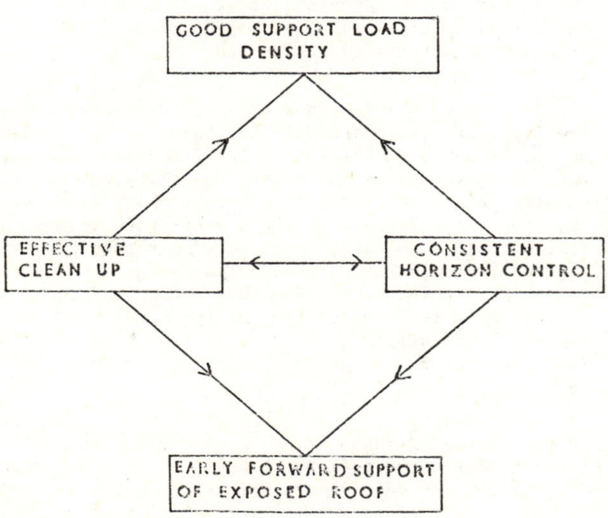

Figure 3 - Basic requirements for effective strata control

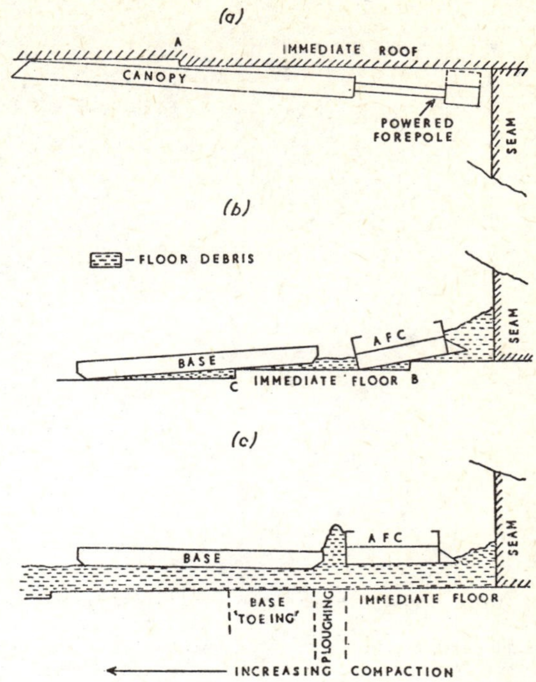

Figure 4 - Problems associated with horizon control and clean up

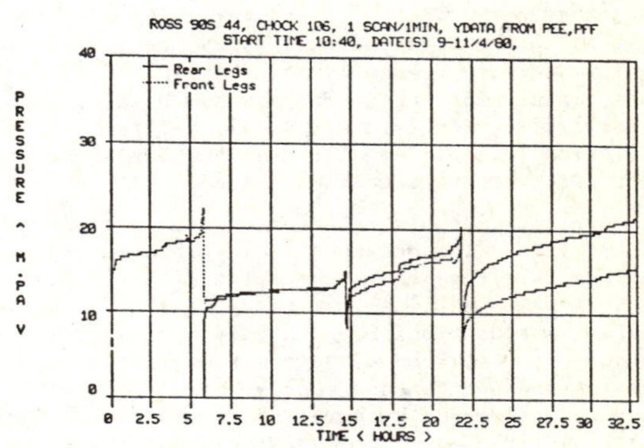

Figure 5(a) - Graph of leg pressures against time

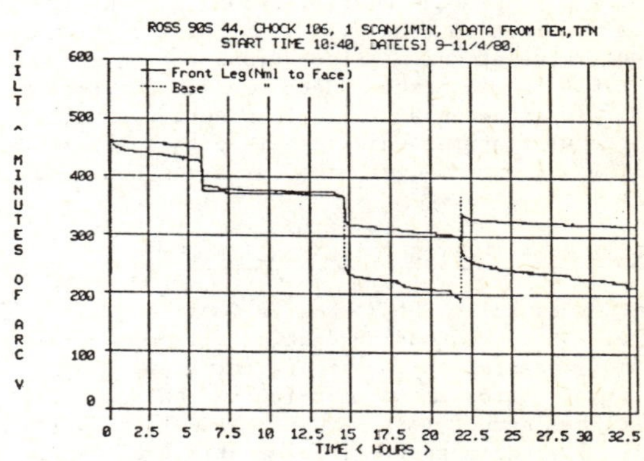

Figure 5(b) - Graph of front leg and base inclination against time

RELATIONS BETWEEN GEOLOGY AND THE STABILITY OF FACES AND ROADWAYS IN THE BARNSLEY SEAM

PAUL HOLMES

DEPARTMENT OF MINING ENGINEERING
UNIVERSITY OF NEWCASTLE UPON TYNE
NEWCASTLE UPON TYNE NE1 7RU

SUMMARY

The relation between the geological environment and the stability of longwall structures is discussed. Particular emphasis is given to the influence of cleat orientation on such mine openings. It is shown - on the basis of underground observation in the Doncaster Area - that the major factor affecting face and roadway stability is the relative orientation of the face and roadway direction with the strike direction of the major cleat. Stability problems are likely to occur when the angle between these directions is less than 30°. It is recommended that the direction of all faces in the Barnsley Seam should be at angle between 30° and 60° with the major cleat direction.

INTRODUCTION

The Barnsley Seam has been mined in Yorkshire since Roman times and is still the most important seam in the East Pennine Coalfield. Although it has been largely exhausted to the West of Doncaster there are extensive reserves to the north, around Selby, and to the East. These latter are presently being exploited by several mines, mainly Rossington and Markham Main, and to a lesser extent, Hatfield and Bently collieries. Other collieries in the area also mine the Barnsley Seam but their reserves are almost exhausted.

Exploitation of the Barnsley Seam has not been trouble free. The occurrence of strata control problems is often seen to be related to the geological environment as well as to mining practice. The major problems include roof falls, face spalling, waste flushing, severe roadway deformation, high gas emission and spontaneous combustion; the occurrence of which is influenced by both the seam structure and near seam lithology. However, the severity of many of these problems is related to the orientation of the coal face with respect to major structural features, the most important being the major cleat planes. Where the face line is parallel to the major cleat plane conditions in the gate roadways tend to be good, but face conditions poor. The reverse situation is also true. It is, therefore, important to determine at which orientations the influence of the cleat is minimised and to explain the reasons for this. It is necessary also to identify those other geological features which may lead to instability and thus allow movement to occur along the cleat planes.

STRUCTURE OF THE YORKSHIRE COALFIELD

Structurally the Yorkshire Coalfield is a shallow basin, extending Eastwards from the Pennines. Around Doncaster this basin is interrupted by a series of narrow and highly faulted NW-SE striking anticlinal structures, such as the Askern-Spital belt. A younger group of folds strike NE-SW. These include the Don Monocline and Don faults. Both fold belts result in extremely high dips of up to 1:3 and are associated with severe fracturing. Almost all of the faults are normal and have an average dip of 60°. The only reversed fault is to the South of Askern Colliery shafts and is part of the NW striking Morley-Campsall structure.

The orientation of the major tectonic structures and discontinuities, joint and cleat planes, are similar. They strike mainly to the NE and NW, with some minor faulting and jointing striking North. These orientations are inherited from older Lower Palaeozoic structures and have resulted from an E-W stress field (Moseley and Ahmed, 1969).

These discontinuities give the rock mass a distinctive strength anistropy. The fold and fault belts dissect the coalfield, describing large blocks of relatively undisturbed coal. On a smaller scale, the cleat forms distinct planes of weakness which significantly affect the rock mass response to mining activities.

Seam Structure

The structure of the Barnsley Seam changes throughout the area around Doncaster. There are many split lines in the North, East and South of the Area. Over the bulk of the region the seam consists of five major units, the lowest being the Dunsil Seam, followed by the Bottom Softs, Top Hards, Top Soft and the Barnsley Rider. A generalised section and the split lines are shown in Figure 1:

The major changes in seam structure occur where the Top Softs split from the Low Barnsley and the latter splits from the Dunsil Seam. The Top Softs part along an East-West Line to the North of Bentley and Hatfield shafts. Along a similar line the Barnsley Rider parts from above the Top Softs. This separation increases to the South. The splits are also coincidental with the deterioration of the Low Barnsley section and result in the thickening of the Top Softs. In this area the latter becomes the major part of the Barnsley Seam worked, especially at Askern Colliery, where the Top Softs and the Rider coal combined are known as the Warren House Seam.

The other major change in the seam structure occurs to the East of Markham Main and Rossington shafts. The Low Barnsley parts from the Dunsil Seam along a North-South line, the separation increasing to the West. It is to the East of this split line where the major working Barnsley Seam faces are situated, exploting the Barnsley/ Dunsil combined seam.

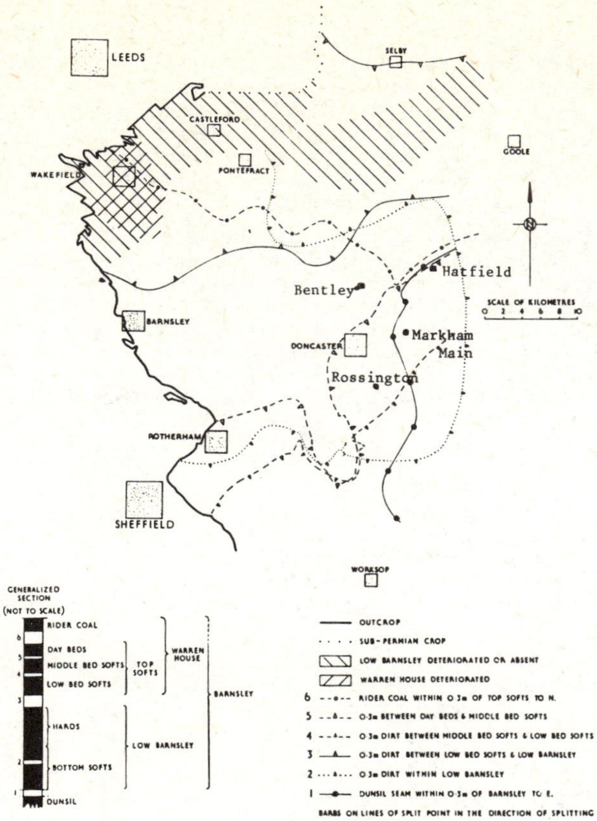

Figure 1 The structure, variation and lines of splitting in the Barnsley Seam (After Pearson, 1974).

Near Seam Geology

In the area around Markham Main and Rossington collieries, the structure of the roof remains predictable. It consists of 6m $\pm$ 1m of mudstones, overlain by 3m+ of silt and sandstones. However, at Hatfield Colliery, where massive sandstones encroach within 1m of the Top Softs, the mudstones have become highly sheared, especially along the base of the sandstones. An intermediate situation occurs around Bentley Colliery, where the massive sandstones are situated 2-3m above the Top Softs and the intervening strata is extremely weak with numerous coal partings.

The character of the floor measures is somewhat less variable than that of the roof. It is dependent upon the position of the Dunsil Seam with respect to the base of the Low Barnsley. Where the Dunsil Seam is remote, as around Bentley Colliery, there is only a thin horizon of weak seatearth between the coal and the underlying siltstones. However, where the two seams are combined the floor can exceed 3m in thickness.

The various near seam geologies described above are shown in Figure 2. The sections are idealised, each being constructed using information from several boreholes.

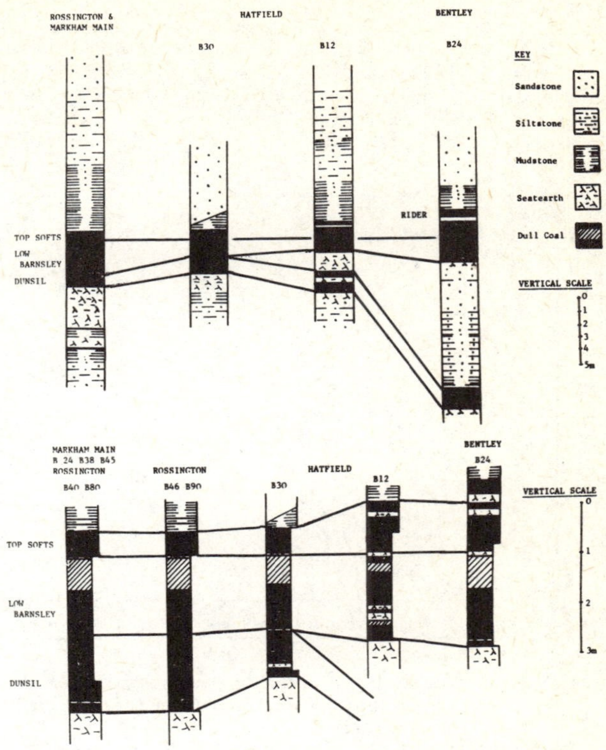

Figure 2 The near seam geology (Generalised)

STABILITY OF WORKING EXCAVATIONS

Gate Roadways

One of the major problems encountered when mining the Barnsley Seam is the severity of gate roadway closure, the major component of this being floor lift. The amount of closure experienced is dependent upon both mining and geological parameters. These include:

- the shape, size and position of the roadway,
- the location and method of formation of the roadway,
- the type of support and packing, both face and ribside,
- the degree of interaction, both vertical and lateral,
- the near seam geology, especially the thickness of seatearth,
- the nature of the floor material, mineralogy and moisture content,
- the orientation of the roadway with respect to tectonic structures, especially cleat,
- the presence of free water in the roadway,
- the depth of the roadway and the stress regime.

At both Markham Main and Rossington collieries the stability of both the face and gate roadways is affected particularly by the direction of face advance. Panels which are on-end tend to have poor roadways, whilst those panels which are on-bord have good roadways. The reverse is true in the case of face conditions.

Two faces at Markham Main Colliery, $B24^s$ and $B45^s$, and one face at Rossington Colliery, $B40^s$, are presently being worked within 3km of each other. They employ the same system of mining and have similar geologies and depths (775m). Their directions of advance are all different and their gateroads demonstrate the influence of cleat upon roadway stability.

It can be seen from Figures 3 and 4 that as the angle between the strike of the major cleat planes and the roadway increases, the severity of gate roadway deformation decreases. This is especially true in the case of width loss (Figure 3). Where the angle between the major cleat planes and the gate roadway is less than

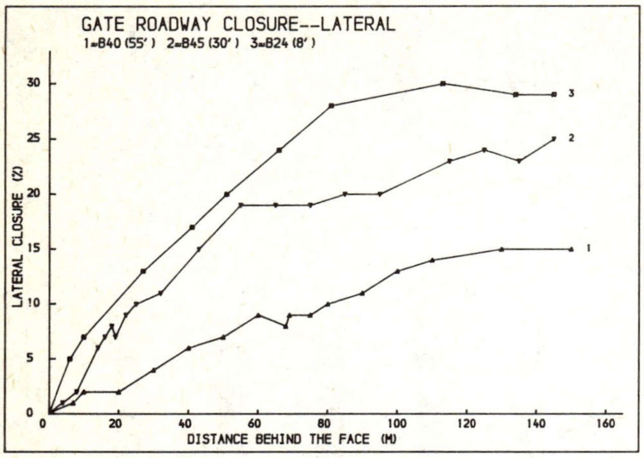

Figure 3 Lateral gate roadway closure expressed as a percentage of the original width.

25° the strength of the coal ribside is reduced. The redistribution of stresses within the ribside result in fracturing which is preferentially developed along the cleat planes. Such fractures exacerbate the extrusion of the coal ribside into the roadway. Such extrusion results in the bulking of the ribside arch leg. Arch deformation of this type is more evident in $B24^s$ tailgate, Markham Main than in $B45^s$ and $B40^s$ tailgates at Markham Main and Rossington respectively.

The influence of cleat orientation upon height loss is only evident before the dint (Figure 4). After this point there is no significant influence.

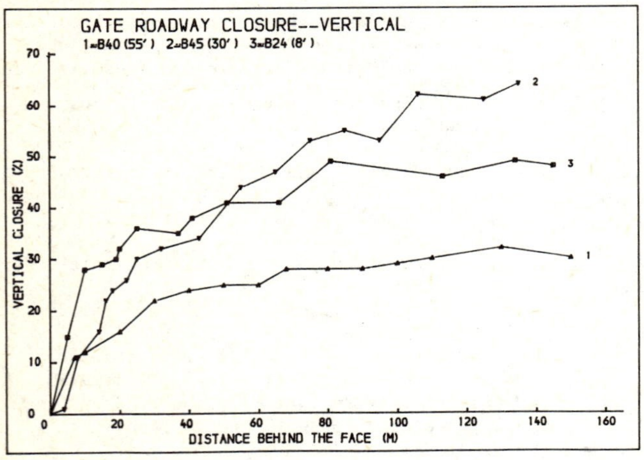

Figure 4 Vertical gate roadway closure expressed as a percentage of the original height.

The majority of the height loss experienced in these three tailgates can be attributed to floor heave. The mechanism of failure of the floor strata is a combination of psuedo-plastic flow of the weak mudstone seatearth units and buckling of the more competent horizons. At the face line the immediate floor measures are comprised of coal, the Dunsil Seam. Outbye of the face this horizon buckles and fractures; these fractures strike sub-parallel to the roadway. Where cleat strikes within 20° of the roadway (as in $B24^s$ tailgate) movement is accommodated along the cleat planes and floor heave appears to occur more readily. However, at the dint the Dunsil Seam is extracted exposing a mudstone seatearth floor. The strength of this floor is not influenced by the cleat plane developed within the coal, hence the orientation of the gate roadway with the master cleat planes does not influence the amount of height loss experienced.

It may be concluded that the deformation of gate roadways is affected by the direciton of the cleat. The condition rapidly worsens where the angle between the roadway and the cleat strike is less than 30°. This is particularly true in the case of sideways deformation. To a lesser extent the severity of floor heave is also affected if the immediate floor measures are composed of coal. Where mudstone seatearths form the roadway floor, the relative orientation of the cleat and the gate roadway is not influential.

Face Stability

The incidence and severity of face spalling is related to three main parameters. These are the extraction height, the angle of the face line with respect to cleat, and the disposition of hard and soft coals and seatearth horizons within the extracted section.

The influence of cleat orientation upon the degree of face spalling is well known. Faces having a direction close to the strike of the master cleat planes tend to experience an excessive amount of face spalling. This is the case along $B24^s$ face at Bentley Colliery and $B12^s$ and $B30^s$ faces at Hatfield Colliery. The angle between the face line and cleat in the other panels in the area is greater than 30°. Along these faces the cleat planes have little influence upon the incidence of face spalling. Indeed, spalling is only a problem on two of these faces, $B46^s$ and $B90^s$ at Rossington Colliery.

The severity of spalling on these faces is the result, in part of their height of extraction (3.3m) but mainly because of the anatomy of the extracted section. The coal seam is fractured in the front abutment zone; this results in the formation of two major fracture systems. One set is near vertical and strikes parallel to the face line. The second is horizontal; these form shear planes between different coal horizons, especially where these horizons are separated by a thin seatearth. It can be argued that the shear strength of both coals and seatearths are dependent upon the shear resistance of the bedding planes. This is strongest in the case of durain, followed by vitrain, durain and coaly shale combinations, and mudstone seatearths (Jeremic, 1980). In the Barnsley Seam, shear zones tend to have developed within the mudstone seatearth horizons which are located beneath the Dunsil Seam, between the Dunsil and Low Barnsley Seams and between the Low Barnsley and Top Softs.

Face spalling tends to result from lateral displacement along these shear zones and from partings along the vertical fractures. As can be seen from Figure 2, the whole of the Low Barnsley and Dunsil seams are extracted on $B90^s$ and $B46^s$ at Rossington Colliery, whereas, on the other zones (except Hatfield, B12) only part of the Dunsil Seam is extracted. On these faces, the Low Barnsley Seam is able to extrude and spall, the Dunsil Seam is restricted by the floor coal. Failure in this case is usually on a small scale, only extending for two or three supports. The individual slabs of coal rarely exceed 50mm in thickness. Usually, only the Bottom softs fail, the Hards falling where they overhang more than 500mm.

The situation is somewhat different on Rossington $B46^s$ and $B90^s$ faces where both the Low Barnsley and Dunsil seams are extruded and excessive spalling results. However, the movement of the Dunsil Seam is restricted by the debris accumulated between the A.F.C. and the face. Despite this, large blocks, often in excess of 500mm in thickness, are detached from the face. These blocks often include the whole of the Dunsil and Low Barnsley Seams and may extend for several supports.

It may be concluded, therefore, that serious face spalling occurs when the strike of the major cleat planes is within 30° of the face line, or when stronger sheared mudstone seatearths occur in the extracted section.

Face roof stability

In all except one face, Hatfield B30, the strata immediately above the Top Softs includes several metres of mudstones. These often contain several coal bands and many lystric surfaces, especially above the Top Softs. In general, they are unable to sustain the minimal loading of their own weight. In order to prevent these mudstones from falling the Top Softs are left as a roof coal.

The practise of leaving roof coal is not employed at Hatfield B30^{s} face. Here, the immediate roof measures comprise of less than 1m of mudstones which part readily from the overlying massive sandstones. Minor roof falls frequently occur, as the supports are advanced. The face was initially designed with 250mm of roof coal; however, this proved ineffective and also fell along with the mudstones.

In addition to the weak roof, B30^{o} face is on-bord. This has resulted in face spalling also being a major problem. Excessive spalling increases the prop-free front distance and thus may aggravate problems of roof control. In an attempt to alleviate this, the face has been turned through 40°. This has improved face conditions and helped to reduce roof stability problems.

Where a coal roof is left it is crucial that it is maintained as a competent plate between the face and the support canopy. If this plate fails, the resultant roof falls are progressive, tending to migrate both upwards and forwards. At Bentley, B24^{s} face, this is a particular problem; roof falls migrating upwards to a height of 4m where they are curtailed by massive sandstones.

The failure of this roof beam usually results where it is too thin; inadequate support; or more commonly, because the face line is within 30° of the strike of the master cleat planes. In order to prevent failure, the minimum thickness of roof coal has been found to be 450mm. The 250mm thick coal roof used at Hatfield B30^{s} proved to be totally inadequate.

The most common type of face support used at the four collieries are Gullick Dobson 6 x 240 chocks, with wrap-around shields. These supports replaced the Dowty 6 x 180 chocks used on some of the older faces. The other types of supports presently being used are Gullick Dobson 4 x 360 chock shields at Hatfield B12^{s}, and Gullick Dobson 4 x 300 shields at Hatfield B30^{s} and Rossington B46^{s} and B90^{s}. The latter have also a positive set capability. It is on this face that major support problems have occurred. The support reaction of a shield is towards the face; on B90^{s} face this has resulted in the front of the support base 'toeing' into the soft floor. In addition to this (prior to the introduction of positive set) very low setting pressures were attained, as low as 40% of the design pressure (Oram, 1981). These two factors contributed towards the poor support performance which may have caused failure of the coal roof.

One of the major factors inflencing the strength of the coal roof is the orientation of the cleat planes with respect to the mining induced stressfields. This stressfield results in the formation of induced cleavage planes (Faulkner and Phillips, 1935). Over the bulk of the face, these strike parallel with the face line, but towards the sides of the panel tend to curve towards the ribsides. It has been reported that where the face line lies between 20° and parallel with the major cleat planes induced cleavage is not formed within the coal seam and this movement then occurs along the cleat planes (Midland Counties Institute of Engineers, 1936). Hence, the roof coal will be weakended in those faces which are within 20° of being on-bord, and these faces will tend to be more susceptible to roof falls.

In order to quantify the incidence of roof falls a Roof Stability Index has been used. The face is surveyed and any roof falls which are greater than 300mm high and occur between the front legs and the face are noted. The Roof Stability Index is the length of broken roof expressed as a percentage of the total survey length.

As the angle between the face line and the cleat planes decreases the incidence of roof falls and, therefore, the value of the Roof Stability Index increases. Information collected from faces in the Doncaster Area shows that this relation can be expressed as a hyperbolic function (Figure 5). The figure indicates that the influence of cleat orientation becomes significant when this angle is less than 30°.

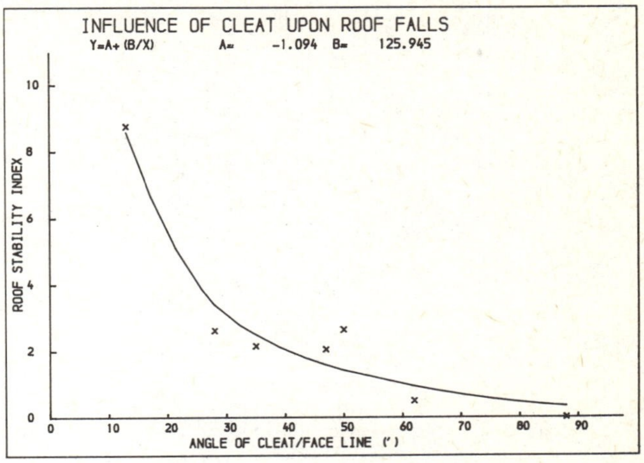

Figure 5 The relation between face orientation, with respect to the master cleat, and the incidence of roof falls.

It should be noted that only the 6 x 180 and 6 x 240 chock supports are included in this analysis, the shield supports being omitted.

The orientation of the cleat not only influences the general occurrent of roof falls but also the most probable location along the face line. In general, the majority of roof falls occur in the central section of the face; there are no significant concentrations of falls towards the gate ends. This is the case with Bently B24^{s} (Figure 6a). However, where the angle of cleat is between 15° and 35°, as at Rossington B06^{s}, significant concentrations of roof falls tend to occur towards one of the gate ends (Figure 6b). Such concentrations occur within the zones where the planes of induced cleavage can be postulated to bend towards the ribsides. The falls result from the interaction of these cleavage planes and the coal cleat.

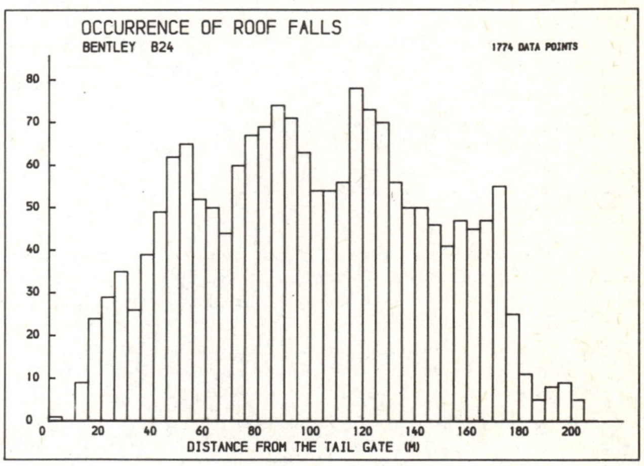

Figure 6a The distribution of roof falls across Bentley B24^{s} face (cleat 14° from the face line).

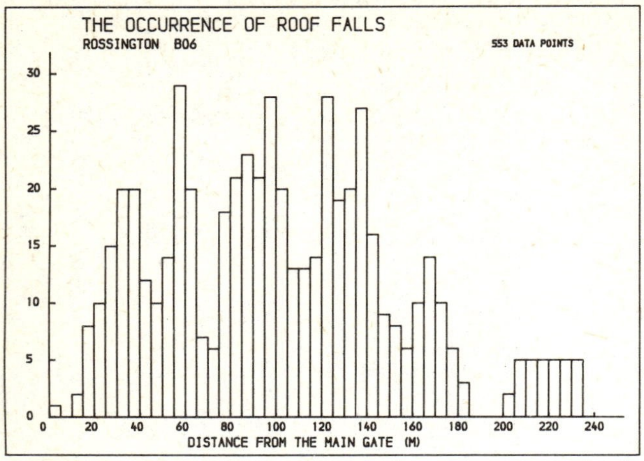

Figure 6b The distribution of roof falls across Rossington B06 face (cleat 30° from the face line).

Longwall faces that are developed on-end, with the minor cleat planes parallel with the face line, are much more stable than those on-bord. Rossington B80s is such a face and to date it has experienced no major roof falls. This face has also broken the colliery production records. However, its gate roadways are not in good conditions, the main gate requiring 2 dints after only 80m advance.

It may be concluded, therefore, that roof stability problems are exacerbated where the major cleat direction lies at an angle between 0°-30° with the face line. In this case, imposed abutment fractures ahead of the face interact with the cleat planes, causing major planes of weakness in the roof.

CONCLUSIONS

The conditions of the gate roadways is affected by the coal cleat; their condition deteriorates when the angle between the strike of the major cleat planes and the gate roadway direction is less than 30°. The influence of the cleat is more evident in the case of the latter; the influence of the cleat/roadway orientation is apparent only where the roadway floor is coal.

The severity of face spalling is also influences by the orientation of the cleat. It becomes a serious problem when the angle between the cleat and the face line is less than 30°. Face conditions are also exacerbated when strongly sheared mudstone seatearth horizons occur within the extracted section. This is a particular problem when these horizons are located at the base and top of the face.

The incidence of roof falls at the face is greatly reduced when a coal roof of at least 450mm thickness is used. However, where the angle between the cleat and the coal face is less than 30°, roof stability problems may occur.

The occurrence of roof falls along the face line is also influenced by the orientation of the cleat planes. Those faces which are on-bord experience the majority of roof falls in the central section of the face line. However, as the angle between the face line and the cleat approaches 20°, concentrations of roof falls may occur towards the gate ends.

It may, therefore, be concluded that all faces in the Barnsley Seam which suffer from severe gate roadway closure should be orientated at an angle of between 30° and 60° to the strike of the major cleat planes. However, those faces which do not experience excessive roadway closure achieve greater roof stability if they are developed on-end.

ACKNOWLEDGEMENTS

The author gratefully acknowledges Dr I.W. Farmer for his advice and encouragement given during this study. The author would also like to thank the managers and surveyors at the collieries visited and the Doncaster Area Geologists for their cooperation during the collection of the data.

REFERENCES

1. Faulkner R. and Phillips D.W., 1935, "Cleavage induced by mining". Trans. Inst. Min. Eng., V89, pp. 264-88.
2. Jeremic M.L., 1980, "Influence of shear deformation structure in coal on selecting methods of mining". Rock Mechanics, V13, 23-38.
3. Midland Counties Institute of Engineers, 1936, Third Progress Report: "The influence of the nature of the nether roof on falls of side at the face". Trans. Inst. Min. Eng., V91, pp 93-103.
4. Moseley F. and Ahmed S.M., 1967, "Carboniferous joints in the North of England and their relation to earlier and later structures". Proc. Yorks. Geol. Soc., V36, pp 61-90.
5. Pearon H.W., 1974, In the Geology of Minerals Resources of Yorkshire (ed.) D.H. Raynor and H.E. Hemmingway, York Geological Society, pp 309-27.
6. Oram, J.S., 1981, "Report of investigations into strata and support behaviour around shield supports on B90's face at Rossington Colliery". M.R.D.E. Technical Memorandum, No. RCM/B(81)4.

GROUND CONTROL ASPECTS OF LONGWALL DESIGN

S. P. Singh

Department of Mining Engineering
University of Utah, Salt Lake City,
Utah 84108, U.S.A.

SUMMARY. Recognizing the important role to be played by longwall mining in achieving the projected doubling of U.S. coal production, it has been endeavoured to outline the basic considerations for longwall design with a special focus on ground control measures. The influence of the geological features, geotechnical considerations, seam characteristics and support system on the longwall design has been discussed. The importance of design and stability of gate roadways has been given due consideration. Special emphasis has been given to the significance of ground control measures in the success of a longwall design.

INTRODUCTION

Due to the energy crisis and large in place reserves, coal is emerging as an important alternative to supplement the liquid and gaseous fuel needs of U.S.A. The circumstances dictating the higher demand of coal has focused the attention of engineers and scientists to develop a suitable coal mining technology ensuring productivity, economy and minimum impact upon the environment. The increased demand of coal and more difficult mining conditions at greater depths indicate that longwall mining is the obvious choice and will continue to be in the foreseeable future.

The aim of high productivity can be achieved by laying a sound foundation in the form of a good longwall design. The variations in the combination of geological and mining conditions make the process of longwall design more significant and difficult. The purpose of this paper is to outline the basic considerations for longwall design with a special focus on ground control measures.

Though longwall mining has numerous advantages with a wide range of application, yet it is not a panacea for all mining conditions. Several problems are associated with the application of longwall mining; the most important and serious are commonly related to strata conditions. Therefore, any longwall design within the guidelines of safety and economy can be achieved only if the ground control problems are effectively taken care of.

APPROACH TO LONGWALL DESIGN

Best mine design is that which gives best results in terms of economy, recovery and safety. The main objective of the design process is to achieve high order of production consistency and reliability supplemented by effective strata control.

Longwall design is a multi-dimensional process and is a function of numerous geological, mechanical and operational variables, some of which are interdependent. Knowledge and understanding of all these variables facilitates economical planing and optimum mine design. Figure 1 depicts a general approach to longwall design.

Geological considerations

Geological factors cannot easily be resolved into quantitative data and no one factor alone is diagnostic of the problems associated with longwall design. Geological features of particular interest are:

- Orientation of joints
- Occurrence of clay veins
- Sandstone or claystone filled channels
- Occurrence of faults
- Abnormally high concentrations of gas and water

Joints are geological planes of weakness in a rock mass. The coal winning process, strata behaviour and caving behind the

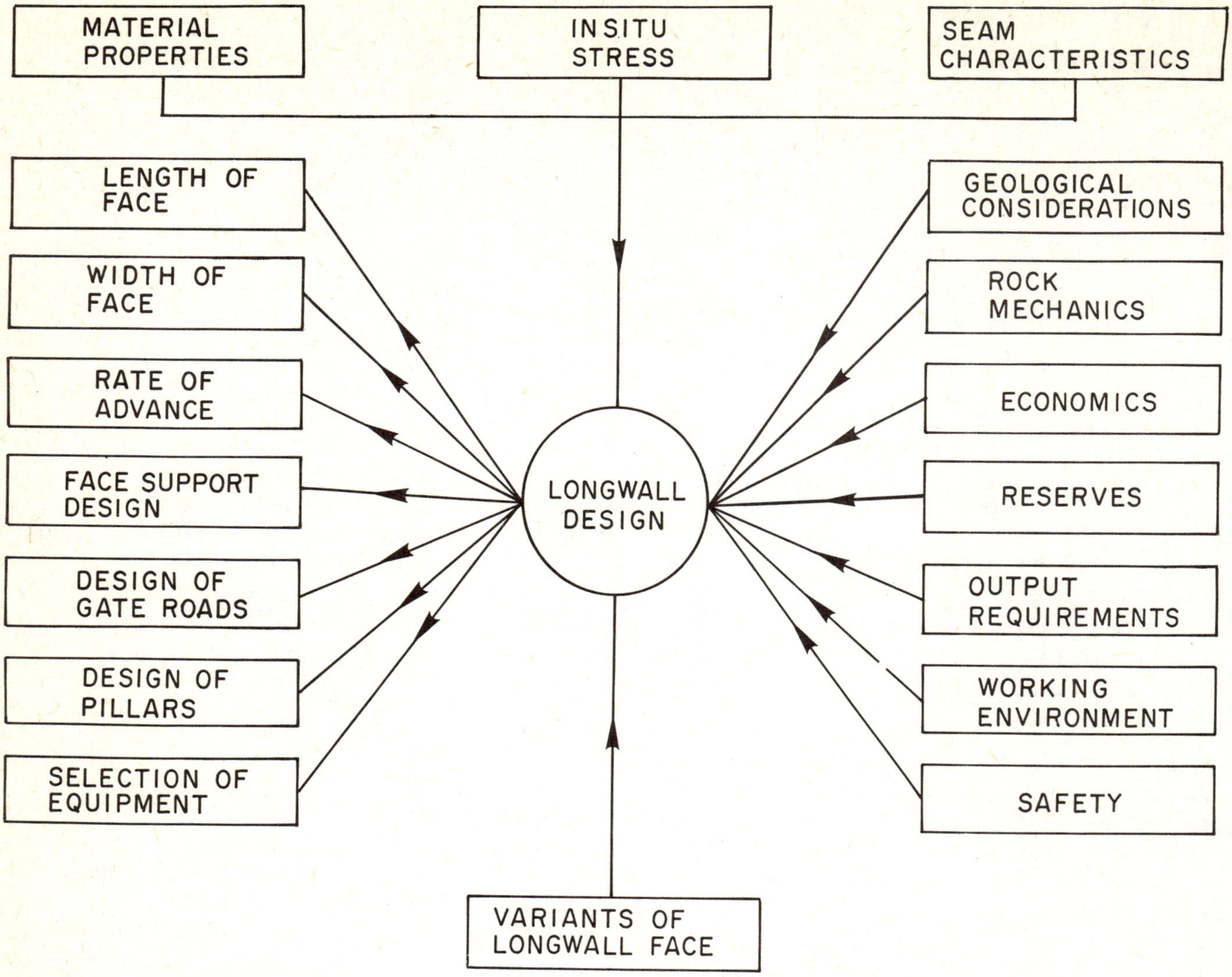

Figure 1. A general approach to longwall design.

powered supports is greatly influenced by the presence of natural jointing in the roof and ribs of the mine. Generally, a longwall face oriented at an angle to the preferred joint direction (Figure 2) results in successful mining from the point of view of ease of winning, ground stability and proper caving.

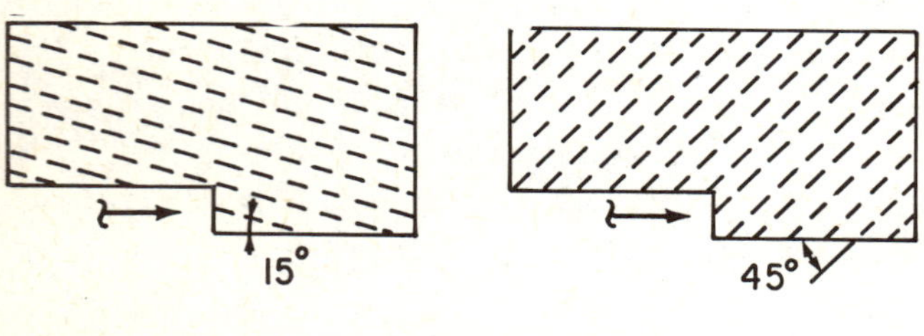

Figure 2. Face orientation with reference to the preferred joint direction

Clay veins or sedimentary dykes are a particular problem in mining because they may intersect so as to form isolated blocks of coal that may act as gas cells. Clay veins consist of calcareous claystone breccia conglomerates and are highly slickensided (Peng, 1978). For this reason they tend to fall and form cavities in the immediate roof as soon as the underlying coal is mined out. To ensure roof stability in such areas, suitable reinforcement technique should be used.

Faults may displace a coal bed and also create areas of abnormally high gas and water accumulations. The presence of sandstone channels interrupts coal production. There is strong evidence that a combination of stress concentrations and joints contributes to the high incidence of roof failure under some stream valleys (McCulloch et al, 1975).

High concentration of methane poses a risk of explosion and diverts attention from productive mining operations. Methane drainage should be carried out prior to mining of such areas, rendering it safe for mining operations. Excessive water deteriorates roof and floor conditions resulting in poor ground control. Effective pumping and adequate drainage systems will diminish the seriousness of the problem. Hydrology of the mine strongly influences the stability of gate roads and the support requirements for a longwall mine.

Seam characteristics

It is a well known fact that the seam characteristics play a significant role in the design and performance of a longwall mine. The depth, gradient, geological abnormalities, gas emission, surrounding strata of a coal seam are all fixed by nature and are beyond the control of a design engineer. But timely recognition of these inherent characteristics facilitates the judicious selection of variable parameters like length of the face, length of the panel, thickness of extraction, pillar and roadway dimensions, system of support, orientation of layout and rate of extraction resulting in an optimum longwall design.

Geo-mechanics considerations

The most common design concept has been to duplicate mine design under similar conditions (Dahl et al, 1976). This concept makes use of the past operating experience but is reluctant to introduce changes. This reluctance is partially based on the lack of understanding of the mechanics of failure and stability of underground structures. The inability to react to changing conditions often leads to severe ground control problems, loss of reserves and unreasonable design parameters.

Strengths and elastic constants are commonly used for roof stability evaluation. Tensile and shear strengths are important because typical roof failures in rectangular openings are tensile failure at the center of the opening and shear failure at the intersection of the roof line and rib line on both sides of the opening (Peng, 1978). For effective roof stability evaluation, isopach maps of tensile and shear strengths must be made separately.

Without the controlled movement of the rocks surrounding a mining excavation, any mine design is impossible within the limits of economics and safety. Geomechanics is a strong scientific tool which facilitates more rational mine design by equipping us with the knowledge of the mechanics of failure and stability of underground structures.

SUPPORT SYSTEM

A good powered support should be able to control the untimely failure of roof, ensure high stability against lateral forces, induce minimum floor pressure, move freely and offer a large operating range at a reasonable cost (Peng, 1978). In the U.S.A. support capacity on longwall faces varies from 200 to 700 tons at yield.

Powered supports in a longwall face hold together a beam of immediate roof extending from the frontabutment above the coal seam to the waste edge. The waste edge is ideally at the rear of the powered supports but its position can vary both towards the face in the case of inadequate supports, to several feet behind the supports in the case of some stronger roofs and in thinner sections (Lewis and Stace, 1981). The thickness of the immediate beam of the roof to be supported by the powered supports depends upon the thickness of extraction, nature of roof beds and natural strata partings. The roof layers above this beam normally remain competent and are supported by the coal seam and in between strata in the front and the broken waste at the rear of the longwall face. The irresistible downward movement of these competent beds results in surface subsidence and is an important factor in the evaluation of a support system.

Roof-support interaction

The understanding of the mechanism of interaction of roof and supports in the longwall faces will undoubtedly facilitate the optimum design of supports for these faces (Bordia and Ghose, 1974).

- The resistance of a support system basically depends upon the capacity and direction of application of the loading point and the distance from the face. An adequate support system should offer maximum resistance near the face and minimum resistance towards the gob.
- The state of stress in roof strata plays a vital role in the roof support interaction. A powered support for a longwall face should be so designed that it does not impede the flow of roof in the gob due to ground stresses.
- The magnitude of frictional resistance generated between the roof and support at the gob edge should be low to increase the efficiency of the caving process.

Capacity of a support system

Commonly, the capacity of a support system is expressed in terms of setting and yield load density. Setting load density is a measure of the force applied to per unit area of exposed roof and floor by the supports when they are initially set. Yield load density is a measure of the force applied per unit area of exposed roof and floor when sufficient convergence has occurred. In addition to the capacity of the support, its dimensions should be such that during the entire operation it may at all times be lowered, advanced and reset between roof and floor.

Two approaches are generally used to assess the setting and yield loads for powered supports. One method is to make an estimate of the greatest volume of rock which is liable to need support at any one time (Ashwin et al, 1970). The second method is to measure the resistance offered by those supports which in practice have been found to be satisfactory.

LENGTH OF LONGWALL FACE

Experience indicates that it is better to have longer face in undisturbed measures than to have higher rate of advance. The longer the face, the better will be the ratio of tons to various costs. But the trend towards longer faces has its limits imposed by strata loading and face machinery particularly the face conveyor.

There are numerous factors which need due consideration while selecting the optimum length of a longwall face. But this paper will concentrate on those factors which influence the length of the longwall face from the strata control point of view.

- In general longer the face, slower the rate of advance and more serious the strata control problems. Shorter faces ensure better supervision, thereby allowing more attention for effective ground control measures.
- In longer faces there is a greater probability of encountering geological abnormalities which remain in the longwall face for a longer period of time.
- Difficulties in working steep and/or thick seams limit the length of the face.
- Roof which is strong and breaks regularly with the advance of the face allows longer faces.
- Shorter face lengths should be selected in weak floors liable to heaving. Strong massive roof and weak floors pose serious problems and limit the face length.
- Difficult hydro-geological conditions also limit the length of the face.
- The longwall length will be restricted in the case of contiguous seams.

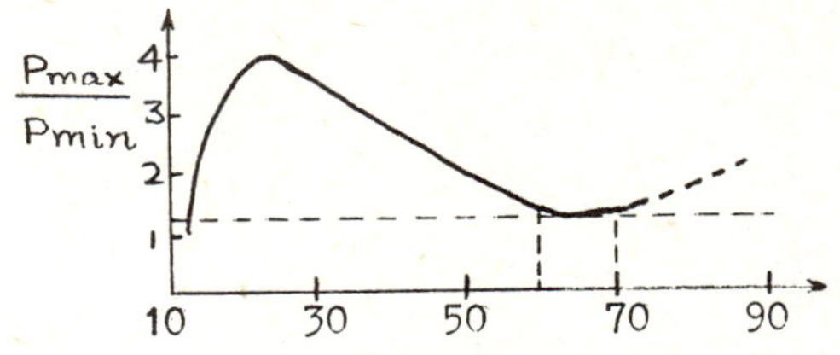

Figure 3. Length of longwall face vs. loads on support (after Kovrishin et al, 1974)

In addition to other considerations, there is probably an optimum longwall length at which the loads on the support units are distributed relatively uniformly (Kovrizhin et al, 1974) as shown in Figure 3. The ratio of the maximum support resistance to minimum support resistance should not be high to ensure effective ground control. It is difficult to estimate this ratio at the mine design stage but the observations on other working faces under similar conditions can be of considerable help.

RATE OF ADVANCE

Rate of advance plays a predominant role in the safety and productivity of a longwall face. Higher rate of advance is synonymous to the successful longwall operation and offers following favourable conditions:

- Higher production
- Low cost per ton of coal produced
- Less number of working faces for the same output
- Less chances of spontaneous combustion
- Allows the selection of shorter faces with several advantages
- Concentration of output ensures better utilization of equipment and effective ground control.

From strata control point of view, as the rate of advance increases, the convergence decreases and so also the stress on the face but at the same time the peak pressure is brought closer to the face (Lama, 1977). Actually rate of advance of the face, rigidity of roof beds, viscosity of roof, floor and coal seam and support resistance influence the pressure developed ahead of a longwall face. To ensure strata control all these factors must be given due consideration.

DESIGN AND STABILITY OF GATE ROADS

Gate roadways play a vital role in the performance of longwall faces. In many cases low productivity on longwall faces has been caused by ground control problems in gate roadways, which are more severe in tailgates. The commonly encountered problems are:

- Excessive floor heave
- Shearing of roof strata at the rib line
- Rib sloughing
- General deterioration of roof and floor conditions

Strata movement in a gate road is two-fold (Lewis and Stace, 1981). First is due to the downward pressure of the block of strata above the seam which is almost equal in thickness to the block of strata held intact by the powered supports. Secondly it is due to the major strata movement above the face excavation. The side and front abutments intersect at the T-functions of the head and tail entries and superimpose to become peak abutment pressures (Peng, 1978).

As mentioned earlier, the most serious problem in the longwall design is the protection of tailgates. These problems are induced by the pressure of adjacent worked out panels, so that more the number of worked out panels, severe will be the tailgate problems. No longwall design is adequate unless it ensures the safety of tailgates. Some of the practical measures being taken in the coal mines are mentioned below:

- Yield pillar concept
- Making each panel a first panel
- Maintaining the longwall face at an angle to the gate roads
- Multi entry system
- Reinforcement of the strata
- Using additional supports of yielding design
- Injection of resin into the coal surrounding the entries
- Locating the gate roadways under old gobs in case of multi seams
- Achieving higher rate of advance
- Increased application of retreat mining
- Maintaining gate roads height slightly less than the height of extraction

In the yield pillar concept, a small "yield pillar" is placed adjacent to the tail entry and a large pillar next to the previously worked out panel. When the abutment load exceeds certain limits, small pillar yields transferring the load to block of unmined coal and large pillar. Though large

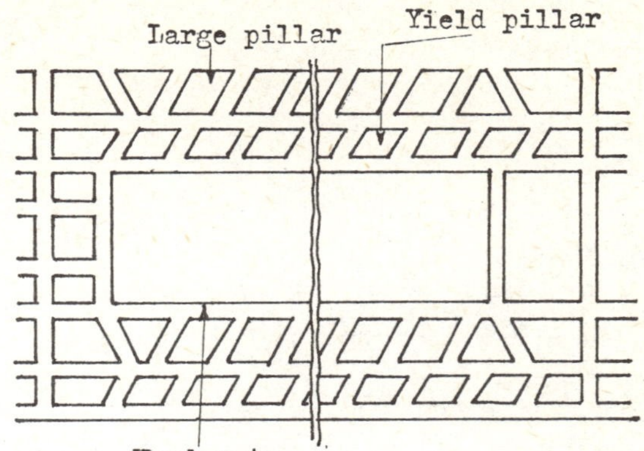

Figure 4. Longwall panel design using yield pillar concept

differential displacements take place on either side of the entry to be protected, yet it has been successfully used in many mines.

Leaving sufficient coal between panels ensures no interference between panels and makes each panel a new panel. This method ensures stability of gate roadways at the cost of blocking large amounts of coal underground and increased development per panel.

U.S. mining laws do not permit the use of single entry system. Therefore multi entry system eases compliance with the law and allows the utilization of yield pillar concept. At the same time, it lowers resource recovery, increases development, creates potential conditions for spontaneous combustion and causes severe ground control problems for upper and lower seams to be mined later.

Keeping the longwall face at an angle to the gate roads. According to the experience at a mine in the Pocahontas number 3 coal bed, such adverse conditions as strong convergence and rock falls in the gate roads can be avoided when the tail gate is kept about 55 feet ahead of the head gate (Curth, 1974).

Reinforcement of strata means to enhance the inherent strength of rock and is different from providing external support. Generally rock bolts and injected resin grouts are used for the purpose. Rock is relatively strong in compression but weak in tensile, shear and flexural strength (Murphy, 1981). Strata reinforcement improves these characteristics but the degree of reinforcement depends on local conditions.

Depending upon the requirements, additional supports of yielding design should be used to keep an entry open and in safe condition during the mining operation. Cribs used in many U.S. longwall entries have exhibited unfavourable support characteristics (Euler, 1979). They bear the load only after excessive convergence. Wooden posts bear load early but break at little convergence. Single hydraulic props have been successfully used in many mines.

By locating the gate roadways under old gobs in case of multiseams, avoids the ground control problems due to high stress zones caused by remnant pillars in the worked out seams.

Many times the seriousness of ground control problems is time dependent and the higher rate of advance will help to diminish the occurrence of roof control difficulties. With the higher rate of advance convergence decreases as peak pressure is kept at one point for a shorter span of time.

By maintaining gate roads height slightly less than the height of extraction, the area of the unsupported opening is reduced. In Deer Creek mine in Utah, the height of development headings is held to 9 ft. to maintain rib stability, whereas the average height of face is 10 ft. 1 inch.

In retreat mining an assessment of the ground conditions is made during development. Based on that assessment effective ground control measures are planned in advance to ensure the stability of gate roadways during retreat.

CONCLUSIONS

Longwall technology promises lower labour and operating costs, higher coal bed recovery, greater output and improved safety. This paper reviews some of the important principles applicable to longwall design with particular reference to coal mining in the U.S.A. Success of a longwall design is a function of numerous geological, geo-technical and operational variables, some of which are interdependent. No longwall design is adequate without effective ground control measures which facilitate economical planing and optimum system output.

REFERENCES

Ashwin, D.P. et al, 1970, "Some fundamental aspects of face powered support design", Mining Engineer, 129, 659-675.

Bordia, S.K. and Ghose, A.K., 1974, "Mechanics of roof-support interaction in longwall faces", Colliery Guardian, 3, 83-84.

Curth, E.A., 1974, "Roof support problems in longwall mining: a study in U.S.A. and Germany in 1971", USBM, IC 8630, 101-114.

Dahl, H.D. and von Schonfeldt, H.A., 1976, "Rock mechnics elements of coal mine design", 17th U.S. Rock Mech. Symp.

Euler, W.J., 1979, "Improving longwall entry development and operation", Mining Congress Journal, 5, 21-27.

Kovrizhin, A.K. et al, 1974, "Efficient lengths of mechanized longwalls for inclined and steep seams in the Kuxbass", Soviet Mining Science, 669-672.

Lama, R.D., 1977, "Principles of underground coal mine design--an approach", Colliery Guardian, 6, 333-344.

McCulloch, C.M. et al, 1975, "Geologic investigations of underground coal mining problems", USBM, RI 8022.

Murphy, J.M., 1981, "Roof control with resin anchored rock bolts, dowels and injection systems", Colliery Guardian, 229, 15-19.

Peng, S.S., 1978, "Coal mine ground control", New York, Wiley.

CONSIDERATIONS FOR THE STABILITY OF LONGWALL CHAIN PILLARS AND ADJACENT ROADWAY

P. R. Sheorey, Tara Nand Singh and B. Singh

Central Mining Research Station
Dhanbad-826001, India.

SUMMARY. The old subject of stability of chain pillars and a roadway protected by them is dealt with using a new approach. The simple method of the theory of beams on elastic support is used to estimate the one-sided pressure distribution over the chain pillars the goaf being only on one side. It is shown how the mechanical properties involved in the method can be determined measuring roof-seam contact displacements in situ. It is found that even if the pillars per se are stable, the roadway which they protect may still show signs of distress, as given in a case study. A modern rock mass classification has been applied to assess roadway stability.

INTRODUCTION

Chain pillars are often left on the side of a working longwall panel for protecting a roadway. These pillars thus undergo one-sided abutment pressures, as the other side is solid coal. Since the roadway is likely to come under the influence of the abutment pressure, it is necessary to consider the stability, or otherwise, of the roadway, although the pillars may be designed to be safe. The cost of supporting the roadway and the cost of coal lost in the pillars, depending on roadway location, will sometimes decide the pillar width provided the width so obtained is stable. The best approach would be to first consider the stability of the protecting pillar and then estimate whether the roadway would be stable and whether any support measures would be required.

CHAIN PILLAR STABILITY ASSESSMENT

Wilson's formulae (1972) are available for the strength of flat pillars, e.g. chain pillars, which are susceptible to progressive, rather than sudden, failure. He has also given an approximate method for estimating the average pressure over chain pillars, but only when goaves exist on both sides. One way of obtaining the pressure distribution with the goaf only on one side, as in the present instance, is to use some numerical or analytical method, as is done in S. Africa, Wagner (1980). The main problem with such methods is the difficulty in estimating realistic strata and coal properties, especially behaviour of the goaf material in situ, which are required for the input. Use of laboratory properties or unrealistic in situ estimates can often lead to absurd answers. Probably the best way to overcome these difficulties is to correlate the properties with underground observations of strata behaviour.

In this paper is shown the applicability of the theory of beams on elastic support, Timoshenko (1956) and Sheorey (1975), for estimating the pressure distribution over the chain pillars, using in situ observations for determining the mechanical properties involved.

Two situations, Figure 1(a) and (b), can exist, depending on the total extraction width Z in the panel, requiring two different locations of the coordinate system (x,y). L' is the distance from the face or solid coal side at which goaf pressure stabilises. Wilson (1972) has given a value of L'= 0.3H, where H is depth from the surface.

Then the beam theory gives the following expressions for strata displacements v and v' along y for the two situations:

(a) $Z \geqslant 2L'$

Over the seam:

$$v = \frac{q}{k} + e^{-\alpha x}(A_1 \cos\alpha x + A_2 \sin \alpha x) \qquad (1)$$

Over the goaf:

$$v' = \frac{q}{k'} + e^{\alpha' x}(A_3 \cos \alpha' x + A_4 \sin \alpha' x) \qquad (2)$$

(b) $Z < 2L'$

Over the seam:

$$v = \frac{q}{k} + e^{-\alpha x} (A_1 \cos \alpha x + A_2 \sin \alpha x) \qquad (3)$$

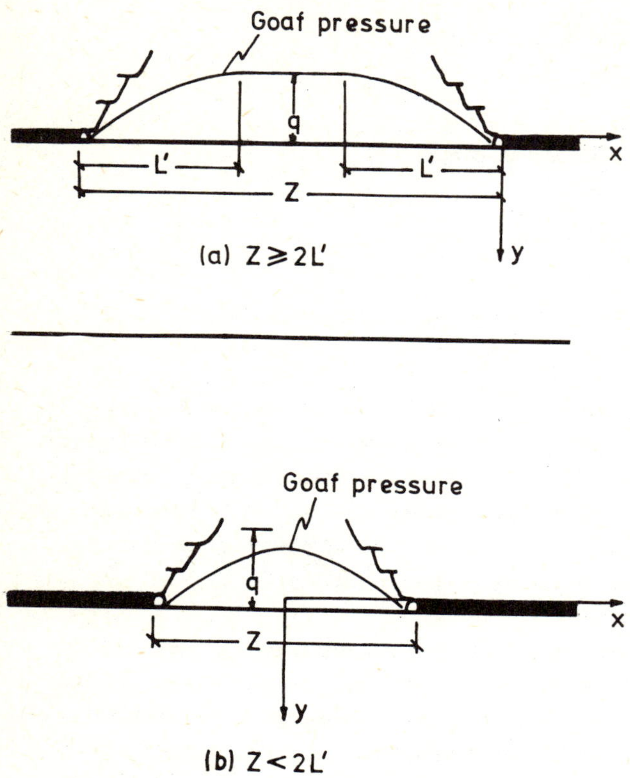

SCHEMATIC SECTION OF A LONGWALL PANEL PARALELL TO THE FACE

Figure 1

Over the goaf:

$$v' = \frac{q}{k'} + A_3 \cosh \alpha' x \cos \alpha' x + A_4 \sinh \alpha' x \sin \alpha' x \qquad (4)$$

In these equations

q = cover pressure,

$k = \dfrac{2E_c}{h(1-\mu_c^2)}$

E_c, μ_c = in situ elastic constants of seam,

h = seam thickness or working height,

$\alpha = \left(\dfrac{k}{4D}\right)^{\frac{1}{4}}$,

D = flexural rigidity of strata,

k', α' = parameters similar to k, α for the goaf

A_1-A_4 = arbitrary constants.

Eq. (1) and (2) for situation (a) satisfy the conditions

at $x = \infty$, $v = \dfrac{q}{k}$

at $x = -\infty$, $v' = \dfrac{q}{k'}$

Eq. (3) for situation (b) satisfies the first condition above and Eq.(4) the condition of symmetry.

The four constants A_1-A_4 in situation (a) can be determined from the four continuity conditions

$$\text{at } x=0; \quad \left| \begin{array}{l} v = v', \quad \dfrac{dv}{dx} = \dfrac{dv'}{dx}, \\[2ex] \dfrac{d^2v}{dx^2} = \dfrac{d^2v'}{dx^2}, \quad \dfrac{d^3v}{dx^3} = \dfrac{d^3v'}{dx^3}. \end{array} \right| \qquad (5)$$

These conditions, when applied to Eq.(1) and (2), give 4 simultaneous linear algebraic equations, their solution giving $A_1 - A_4$. In the case of situation (b), the same conditions are applied at $x = Z/2$.

The pressure distribution over the pillar of width w is then given by

$$p(x) = kv \qquad (6)$$

and the average pressure by

$$P_{ave} = \frac{qB}{2w} + \frac{B+w_1}{w_1} \cdot \frac{1}{w} \int_0^w p(x)\, dx$$

where B = width of cross cuts in the chain and

w_1 = length of pillar.

In the case of long pillars, $w_1 >> w$

$$P_{ave} = \frac{qB}{2w} + \frac{1}{w} \int_0^w p(x)\, dx \qquad (7)$$

Using Wilson's pillar strength formulae (1972), the safety factor of the pillar can be obtained, once the pillar loading is obtained.

Estimation of properties involved

According to the beam theory, the pressure distribution over an elastic continuous support or "foundation" is of the form shown in Figure 2, e.g. if $p(x)$ is plotted it will have this general form. This wave-like distribution damps very quickly to the constant ultimate value q, the distance being one wave length i.e. $2\pi/\alpha$ or $2\pi/\alpha'$ as the case may be. However, at half this distance, the wave crosses the constant value and gives values slightly less than q further. So for all practical purposes the pressure stabilises at a distance $L = \pi/\alpha$ over the seam and $L' = \pi/\alpha'$ over the goaf.

The distances L and L' can be readily measured under similar underground conditions elsewhere or in the same mine, giving α and α'.

Now, an effective value of the flexural rigidity of roof strata D, which is involved in the definition of α and α', is difficult to determine. Its value is, however, not required directly since we only require k

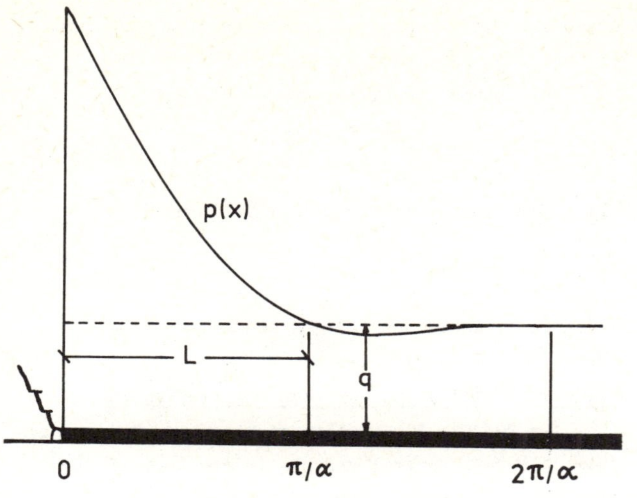

RIBSIDE VERTICAL PRESSURE AS PER THE THEORY OF BEAMS

Figure 2

and k' besides α and α'. These can be determined from

$$k = \frac{2E_c}{h(1-\mu_c^2)}$$
$$k' = \left(\frac{\alpha'}{\alpha}\right)^4 k \qquad (8)$$

where E_c and μ_c are elastic constants of the seam, either determined experimentally in situ or estimated using a rock mass classification approach from laboratory properties, Deere et al. (1966) and Bieniawski (1978). The second of Eq.(8) has the advantage in caving panels that the "height" of the goaf material, like the seam height h in the expression for k, need not be known.

It is felt that, using these methods to obtain values of the property parameters, pillar loading in the situation under discussion can be obtained realistically.

ROADWAY STABILITY

The stability of the roadway, which is to be protected by the chain pillar, can perhaps be best assessed by a rock mass classification approadh, Barton et al.(1974). A suitable support system can also be designed using this method, if required. Roadways under normal cover pressure elsewhere in the mine may become unstable if they are located within mine may become unstable if they are located within the abutment pressure zone of the working panel, because of a reduction in the rock mass quality Q as defined by Barton et al. (1974).

Using Barton's equation for the safe unsupported span B_1 of an opening,

$$B_1 = 2.\text{ESR}.\ Q^{0.4} \text{ metres} \qquad (9)$$

the stability of the roadway in question and the corresponding supports required, if any, can be assessed. In Eq.(9)

ESR = excavation support ratio,

$$Q = \frac{RQD}{J_n} \cdot \frac{J_r}{J_a} \cdot \frac{J_w}{SRF} \qquad (10)$$

RQD = rock quality designation,

J_n = joint set number,

J_r = joint roughness number,

J_a = joint alteration number,

J_w = water reduction number and

SRF = stress reduction factor.

As compared to roadways under normal cover pressure elsewhere in the mine, the only factor which normally changes for the roadway under the influence of the abutment pressure is SRF, reducing the rock mass quality Q in inverse proportion. Once this value of a decreased Q is obtained, a decision can be taken regarding the type and spacing of supports.

A worked out practical example is given below to explain the approach outlined.

EXAMPLE OF MOONIDIH COLLIERY

Figure 3 shows the plan of a powered support longwall panel at this mine with the chain pillars and the roadway to be maintained adjacent to the pillars.

The following data is available:

Seam thickness h = 1.7 m

Pillar width w = 18 m

Roadway width B_1 = 4.8 m

Extraction width Z = 155 m

Depth from surface H = 280 m

Cover pressure q = 7085 kN/m^2

In addition, the experimental data below was obtained:

The distance from face at which abutment pressure stabilises L = 35–40 m (mean 37.5 m)*

The distance from face at which goaf pressure stabilises L' = 80–90 (mean 85 m)*

Estimated in situ elastic modulus of seam E_c = 7.68 x 10^5 kN/m^2

Poisson's ratio of seam (assumed) μ_c = 0.25

Then

α = 0.083 m^{-1}

α' = 0.036 m^{-1}

k = 9.21 x 10^5 kN/m^3

k' = 3.01 x 10^4 kN/m^3

Average pressure on the pillar

Since $Z > 2L'$, Eq. (1) and (2) of situation (a)

* Courtesy N. M. Raju and K.P. Mukherjee, CMRS.

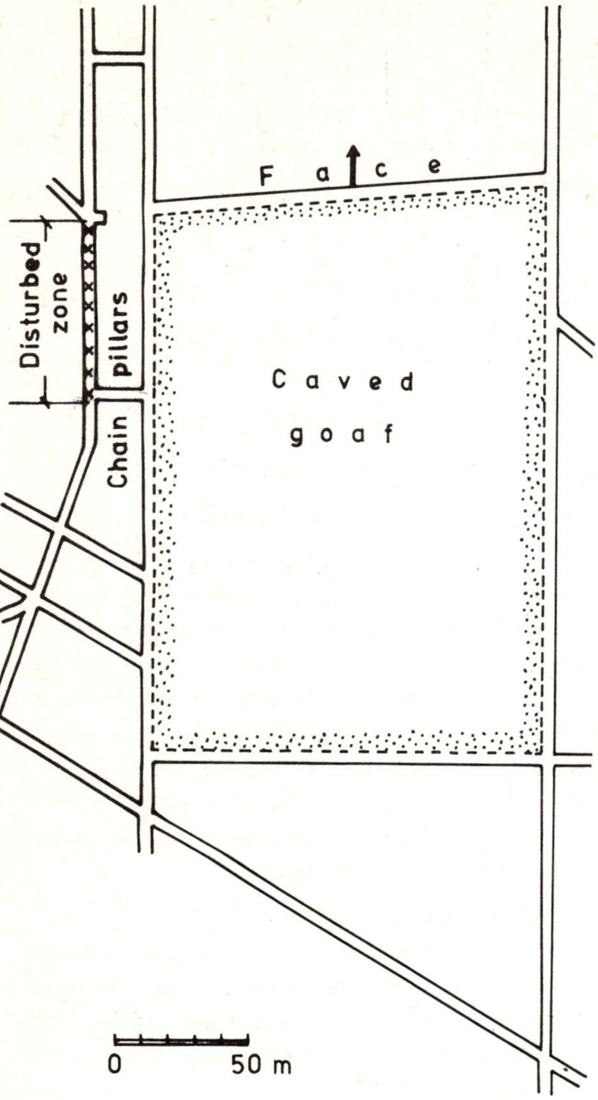

PLAN OF POWERED SUPPORT PANEL AT MOONIDIH

Figure 3

are applicable here. Applying conditions (5) to these equations and solving the resulting 4 simultaneous equations, we get

$A_1 = 0.035$

$A_2 = -0.014$

The chain pillars in this panel can be treated as long pillars, so Eq.(7) is used here for calculating the average pressure on the pillar. Then from Eq.(1),(6) and (7)

$$P_{ave} = \frac{qB}{2w} + q + \frac{k}{w}\int_0^w e^{-\alpha x} (A_1 \cos\alpha x + A_2 \sin\alpha x)\, dx$$
$$= 17,835 \text{ kN/m}^2.$$

Pillar strength and safety factor

The strength of long chain pillars is defined by Wilson (1972) as

$$S = \frac{4\gamma H}{w} (w-1.5 \text{ h H} \times 10^{-3}) \text{ tons/ft}^2$$

where $\gamma = 0.0707$ tons/ft^3 is the average rock density, all other symbols being in ft.

This gives the pillar strength

$S = 229$ tons/ft^2

or $S = 24,600$ kN/m^2

The safety factor of the pillar is then

$$F = \frac{S}{P_{ave}} = 1.38$$

which is greater than 1.0 and hence is sufficient when using Wilson's formula.

Roadway stability and wall support requirement

The roadway in question at this mine had started showing strata control trouble in the form of side spalling and opening up of roof joints as the longwall face advanced. The roof was immediately supported adequately using timber cross beams. Since the zone of influence of the side abutment pressure is about 35–40 m at this mine, this roadway would come under high induced stresses. Under normal cover load, the same roadway was reasonably stable. These facts immediately give us the rock mass quality using Eq. (9):

$$B_1 = 2 \,.\text{ESR}.\ Q^{0.4} \text{m} \qquad (9)$$

in which

$B_1 = 4.8$ m is the stable unsupported span and

ESR = 3 is the excavation support ratio for temporary mine openings, since the roadway was to be maintained for 1–2 years.

Thus Q = 0.57.

This value of Q, which corresponds to the "medium stress" condition (normal cover pressure at 280 m depth), will change due to the "high stress" induced by the abutment pressure.

From Eq. (10)

$$Q \propto \frac{1}{SRF}$$

and SRF = 1.0 for medium stress, SRF = 2.0 for high stress, Barton et al. (1974).

Thus due to the influence of the side abutment pressure, Q will have a value

$$Q = \frac{0.57}{2.0} = 0.285$$

Using the equation for wall support pressure of Barton

$$P_{wall} = \frac{2}{J_r} (2.5\ Q)^{-0.33} \text{ kg/cm}^2 \qquad (11)$$

and substituting $J_r = 1.5$ corresponding to "rough planar joints" in coal, we get

$P_{wall} = 1.5$ kg/cm^2 (147 kN/m^2)

The bolt spacing a in metres is then obtained from

$$P_{wall} = \frac{1}{a^2} \text{ kg/cm}^2$$

as a = 0.81 m, say 0.8 m

The length of the bolts was determined as

$$= 2.0 + 0.15 \frac{h}{ESR}$$

= 2.085 m, say 2.5.

It was, therefore, decided that two rows of 2.5-m bolts 0.8 m apart, with the same spacing between bolts, were required on both walls of the roadway. Pre-tensioned fully grouted bolts were recommended considering slightly spalling coal conditions.

DISCUSSION

The theory of beams on elastic support affords a simple method for estimating the pressure on chain pillars with a panel being extracted only on one side. As is the case with numerical and analytical methods in general, only a realistic estimate of the mechanical properties can lead to a reliable solution employing this approach. A method is given to estimate the properties, which define in situ seam and goaf behaviour, from underground observations.

The stability of the chain pillars must be considered together with that of the roadway which is to be protected by the pillar. Obviously, a roadway which is to be located outside the influence of the side abutment pressure will normally be stable, but this will mean leaving an excessively wide pillar. It is better to locate the roadway within the abutment pressure zone, provided a reasonably economical support system can be designed for the roadway and provided the pillar has a sufficient safety factor.

The rock mass classification approach is a simple and reliable method to estimate roadway stability vis-a-vis support requirement. The stress reduction factor SRF in Eq. (10), however, needs to be defined more accurately for mining situations to increase the reliability of the method. Also Eq. (11) for wall support pressure depends on the rock mass quality Q of the roof and is independent of the wall quality. Although Barton et al. recommand different multiplying factors for Q (2.5 in Eq.(11) in the case of Moonidih) which must be used in Eq.(11), depending on the value of Q for the roof, the equation is still approximate. A possible modification will be to determine Q for the wall independently.

The sample problem considered in the paper offered the case of a roadway which was just stable prior to the influence of the side abutment pressure. This fact could give the value of the rock mass quality directly using Eq.(9) for the safe unsupported span. In general, however, it will have to be evaluated using Eq.(10), measuring all the parameters RQD, J_n etc. in situ.

Acknowledgement - Thanks are due to the Director Central Mining Research Station for permission to publish this paper. The opinions expressed in the paper are those of the authors and not necessarily of the CMRS.

REFERENCES

Barton, N., Lien, R. and Lunde, J., 1974, "Engineering classification of rock masses for the design of tunnel support", Rock Mechanics, 6, 189-236.

Bieniawski, Z.T., 1978, "Determining rock mass deformability -experience from case studies", International Journal of Rock Mechanics and Mining Sciences and Geomechanics Abstracts, 15, 237-247.

Deere, D.U., et al, 1966, "Design of Surface and near-surface construction in rock", Proc.8th Symp. Rock Mech., Minnesota.

Sheorey, P.R., 1975, "An approach to thick beam analysis for roof strata", International Journal of Rock Mechanics and Mining Sciences and Geomechanics Abstracts, 12, 373-377.

Timoshenko, S., 1956, "Strength of materials", part II, New York, Van Nostrand.

Wagner, H., 1980, "Pillar design in coal mines", Journal of South African Institute of Mining and Metallurgy, 80, 37-45.

Wilson, A.H., 1972, "Research into the determination of pillar size", "Mining Engineer, 131, 409-417.

ULTIMATE STRENGTH OF CONCRETE
SHAFT LININGS AND ITS INFLUENCE
ON DESIGN

F. Alan Auld

Cementation Mining Limited,
P.O. Box 22,
Bentley House, Doncaster,
South Yorkshire, DN5 OBT,
England.

SUMMARY. The paper sets out formulae for determining the elastic, pure plastic and Coulomb plastic stresses in a thick cylinder subjected to external pressure.

It is then shown how the three types of stress derivation can each be applied in the design of concrete shaft linings. Consideration is given to both the long term permanent loading and the short term, backwall grouting situation.

An appraisal is made of the different methods leading to recommendations for improved design.

INTRODUCTION

Fundamental principles for the design of concrete shaft linings are set out in a previous paper, Auld (1979). The earlier paper concentrates initially on the determination of stresses, deformations and strains in a thick concrete cylinder based on an elastic analysis. Design of concrete shaft linings can be carried out using the elastic design concept but, at depth, the method produces extremely conservative lining thicknesses which are wasteful in terms of excavation required and concrete needed in addition to being impracticable from a construction aspect.

When pressure is applied from the outside, the tangential stress in a thick cylinder based on an elastic analysis is maximum at the internal face, decreasing through the wall to a minimum at the external face (Figure 1a). The radial stress is zero on the inside increasing to a maximum value, equal to the applied pressure, at the outside. Failure in a shaft lining is initiated when the tangential stress at the internal face exceeds the strength of the concrete. Dividing the concrete strength by an appropriate factor of safety and equating it to the calculated maximum elastic tangential stress allows the relationship between wall thickness and depth to be determined for any shaft.

However, bearing in mind that in this method of design only the internal face of the shaft lining is assumed to reach its ultimate load, most of the wall is stressed to an extent which is below the failure limit. There remains an additional strength within the lining which is only reached when the concrete is

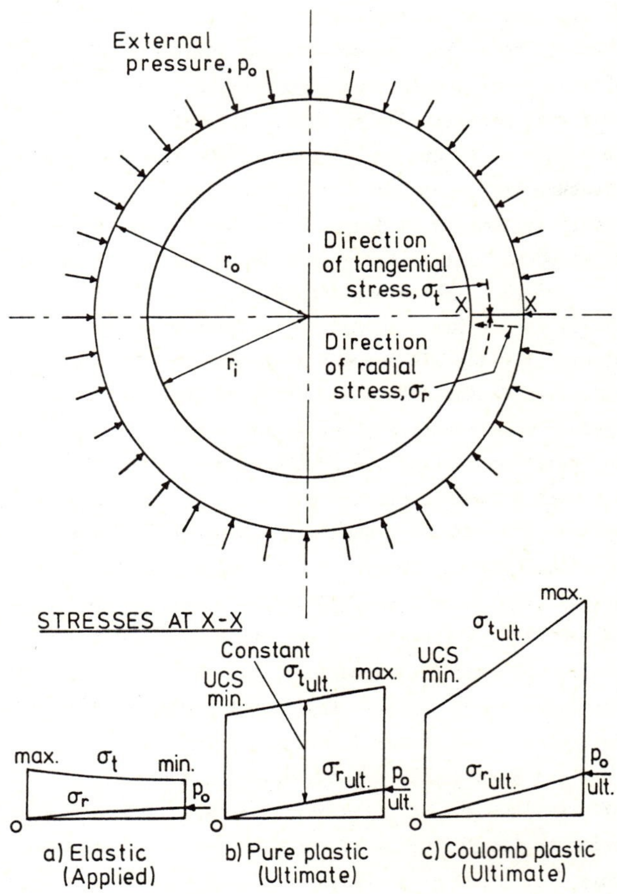

Figure 1. Stresses in a thick cylinder.

stressed to failure across the full width of the wall. The overall ultimate load for the structure, therefore, remains undefined and the applied factor of safety is unrealistic when based on the elastic design, maximum tangential stress method.

In the second part of the previous paper, Auld (1979), an ultimate load failure mechanism was defined for a thick concrete cylinder wall. It was based on principles of pure plasticity, as outlined by Timoshenko (1976), whereby at ultimate failure a constant shearing stress is assumed to exist throughout the wall which is equal to the difference between the tangential and radial stresses divided by 2 (Figure 1b). Expressions for the tangential and radial plastic stresses were derived and qualified by using the unconfined compression strength, UCS, divided by 2 as the constant shearing stress.

Having defined an ultimate limit state failure mechanism, it was then possible to provide shaft lining designs in accordance with the current British Standard Code of Practice for concrete design, CP.110:Part 1 (1972), which is based on ultimate limit states. The partial safety factors for load, γ_f, and strength, γ_m, could be applied directly to provide a more economical design than with the elastic approach.

However, even this method does not fully introduce the true strength of the wall into the design. A further approach, based on Coulomb plasticity which is referred to by Jaeger and Cook (1969), allows the biaxial concrete strength to be utilized whereby the failure stress in one direction increases with increase of confinement pressure in the direction at right angles. The objective of this paper, therefore, is to set out the stress equations for the three methods and to evaluate their usefulness and validity in design.

STRESSES IN A THICK CYLINDER

Elastic stresses

Derived from a two-dimensional analysis approach, Timoshenko (1976), the elastic stresses in a thick cylinder can be represented by:

$$\left.\begin{aligned}\sigma_r &= p_o \frac{r_o^2}{r_o^2 - r_i^2}\left(1 - \frac{r_i^2}{r^2}\right)\\ \sigma_t &= p_o \frac{r_o^2}{r_o^2 - r_i^2}\left(1 + \frac{r_i^2}{r^2}\right)\end{aligned}\right\} \quad \ldots\ldots\ldots(1)$$

Where: σ_r = radial stress;
σ_t = tangential stress;
p_o = applied external pressure;
r_i = internal radius;
r_o = external radius;
r = radius to the point in the wall at which stresses are determined.

Lamé is accredited with the original solution of these equations in which the stresses are always positive and hence compressive. The notation is illustrated in Figure 1.

When $r = r_i$, Equations (1) become:

$$\left.\begin{aligned}\sigma_{r_{min}} &= 0\\ \sigma_{t_{max}} &= \frac{2p_o r_o^2}{r_o^2 - r_i^2}\end{aligned}\right\} \quad \ldots\ldots\ldots\ldots\ldots\ldots\ldots(1a)$$

and when $r = r_o$, Equations (1) give:

$$\left.\begin{aligned}\sigma_{r_{max}} &= p_o\\ \sigma_{t_{min}} &= p_o \frac{(r_o^2 + r_i^2)}{r_o^2 - r_i^2}\end{aligned}\right\} \quad \ldots\ldots\ldots\ldots(1b)$$

Pure plastic stresses

Again using a two-dimensional method of analysis, Timoshenko (1976), the ultimate pure plastic stresses in a thick cylinder subjected to external pressure are given by:

$$\left.\begin{aligned}\sigma_r &= 2\tau_{Y.P.} \log_e \frac{r}{r_i}\\ \sigma_t &= 2\tau_{Y.P.}\left(1 + \log_e \frac{r}{r_i}\right)\end{aligned}\right\} \quad \ldots\ldots\ldots\ldots\ldots(2)$$

where: $\tau_{Y.P.}$ = constant shearing stress at plastic yield $\left(= \frac{\sigma_t - \sigma_r}{2}\right)$

Stresses from Equations (2) are always positive and therefore compressive.

When $r = r_i$, Equations (2) become:

$$\left.\begin{aligned}\sigma_{r_{min}} &= 0\\ \sigma_{t_{min}} &= 2\tau_{Y.P.}\end{aligned}\right\} \quad \ldots\ldots\ldots\ldots\ldots\ldots\ldots(2a)$$

and when $r = r_o$, Equations (2) give:

$$\left.\begin{aligned}\sigma_{r_{max}} &= 2\tau_{Y.P.} \log_e \frac{r_o}{r_i}\\ \sigma_{t_{max}} &= 2\tau_{Y.P.}\left(1 + \log_e \frac{r_o}{r_i}\right)\end{aligned}\right\} \quad \ldots\ldots\ldots\ldots(2b)$$

In addition, the pressure which is required to bring the entire wall of the cylinder into a state of plastic flow is:

$$p_{o_{ult}} = \sigma_{r_{max}} = 2\tau_{Y.P.} \log_e \frac{r_o}{r_i} \quad \ldots\ldots\ldots\ldots(2c)$$

Coulomb plastic stresses

The pure plastic method of analysis does not fully represent the true shear failure mechanism of brittle materials such as rocks or concrete. According to Jaeger and Cook (1969), Coulomb in 1773 proposed that, for the shear failure of rocks, the shear stress tending to cause failure across a plane is resisted by the cohesion of the material and by a constant times the normal stress across the plane. According to this concept, the criterion for shear failure in a thick cylinder can be shown to be represented by:

$$\sigma_t = UCS + A\sigma_r \qquad \text{.....................(3)}$$

where:

UCS = uniaxial compressive strength = $2\tau_{Y.P.}\tan\alpha$;

A = a constant = $\tan^2\alpha$;

$\alpha = \frac{\pi}{4} + \frac{\phi}{2}$;

ϕ = angle of internal friction for the material.

Based on the Coulomb criterion, the ultimate plastic stresses in a thick cylinder subjected to external pressure are:

$$\left.\begin{aligned} \sigma_r &= \frac{UCS}{(A-1)}\left[\left(\frac{r}{r_i}\right)^{(A-1)} - 1\right] \\ \sigma_t &= \frac{UCS}{(A-1)}\left[A\left(\frac{r}{r_i}\right)^{(A-1)} - 1\right] \end{aligned}\right\} \text{.................(4)}$$

Stresses from Equations (4) are always positive as they are compressive.

When $r = r_i$, Equations (4) become:

$$\left.\begin{aligned} \sigma_{r_{min}} &= 0 \\ \sigma_{t_{min}} &= UCS \end{aligned}\right\} \text{.....................(4a)}$$

and when $r = r_o$, Equations (4) give:

$$\left.\begin{aligned} \sigma_{r_{max}} &= \frac{UCS}{(A-1)}\left[\left(\frac{r_o}{r_i}\right)^{(A-1)} - 1\right] \\ \sigma_{t_{max}} &= \frac{UCS}{(A-1)}\left[A\left(\frac{r_o}{r_i}\right)^{(A-1)} - 1\right] \end{aligned}\right\} \text{............(4b)}$$

The pressure which is required to bring the entire wall of the cylinder into a state of plastic flow is:

$$p_{o\,ult} = \sigma_{r_{max}} = \frac{UCS}{(A-1)}\left[\left(\frac{r_o}{r_i}\right)^{(A-1)} - 1\right] \text{...........(4c)}$$

DESIGN OF CONCRETE SHAFT LININGS

Elastic design

When designing concrete shaft linings on an elastic basis, the wall thickness t, can be determined from the $\sigma_{t_{max}}$ value of Equations (1a) by substituting t for $r_o - r_i$ to give:

$$\frac{\sigma_{t_{max}}}{SF} = \frac{2p_o\,(t + r_i)^2}{t\,(t + 2r_i)} \qquad \text{...................(5)}$$

where SF is a factor of safety over the lining strength. Equation (5) can be transposed to provide a direct solution for the thickness:

$$t = r_i\left\{\sqrt{\left[\frac{\sigma_{t_{max}}/SF}{(\sigma_{t_{max}}/SF) - 2p_o}\right]} - 1\right\} \text{.............(5a)}$$

It can be seen from Equations (5) and (5a), therefore, that if the external pressure, p_o, is known, then a wall thickness may be determined on an elastic design basis to resist this pressure. The stress distribution within the wall can subsequently be evaluated using Equations (1).

In using Equations (5) and (5a), three fundamental parameters must be evaluated. The first priority is to determine the design load, p_o, correctly. Deduction of the concrete lining true strength in situ is the second aspect and, thirdly, a realistic factor of safety needs to be applied.

Design load. Two conditions of loading enter into the design of concrete shaft linings. One is the long term loading which the structure is subjected to throughout its life. The other is the short term situation whereby backwall grouting is carried out to fill voids behind the lining, to seal off water and eventually to tighten up the lining into the surrounding rock under pressure.

In competent (self standing when excavated) strata it is not normally necessary to design for rock pressures on shaft linings. However, where aquifer zones occur within the strata, hydrostatic pressures prevail which are usually related to the head from the ground surface. Therefore, unless the ground is incompetent, concrete shaft linings need only be designed for hydrostatic pressure down to a level at which water no longer occurs in the strata. Below this level only a nominal lining will then be required.

The hydrostatic pressure, once it occurs on the lining, will remain for the life of the structure and is considered to be equivalent to a long term dead load. Grouting pressures are also related to the hydrostatic pressure in so far as their magnitude must be marginally greater to force out water from the voids behind the lining. As an upper limit, grout pressures are normally restricted to a specified maximum value of 1.25 x hydrostatic pressure. This is in accordance with generally accepted principles for structural design whereby a 25% increase in permissible stress is allowed for short term loading.

Lining strength. The actual compressive strength of concrete in a structure is equivalent to 0.85 x characteristic cylinder strength, CEB-FIP (1970), the characteristic value being the lower limit below which not more than 5% of the results would fall based on a statistical analysis of samples tested. In the case of a shaft lining, reducing the characteristic strength by 0.85 takes account of the difference between instantaneous loads on cylinders at an age of 28 days (the normal test criteria) and loads applied for a longer duration on specimens of the same age.

British Standard practice uses the 28 day cube test, BS 1881 (1970), as a means of strength control. A correction factor of 0.8 is needed to convert the cube strength to the equivalent cylinder strength. Hence, in relation to the characteristic cube strength, f_{cu}, the actual in situ strength of a shaft lining is represented by 0.68×characteristic cube strength, the value of 0.68 being equal to 0.85 x 0.8. A figure of 0.67 is employed in the current British Standard Code of Practice for concrete design, CP 110:Part 1 (1972).

Factor of safety. Partial safety factors for load, γ_f, and strength, γ_m, are used in the present ultimate limit state approach to the design of concrete structures, CP 110:Part 1 (1972). On the basis of its long term, permanent nature, the value of γ_f for hydrostatic loading is 1.4. For γ_m, which is introduced to account for possible strength differences between test specimens and the actual structure caused by such aspects as insufficient compaction and differences in curing, the specified value is 1.5.

Returning now to Equation (5), replacing p_o by the hydrostatic pressure of 0.01 N/mm^2 per metre of depth multiplied by the depth, d, and incorporating the lining strength and factor of safety parameters gives:

$$\frac{0.67\ f_{cu}}{1.4 \times 1.5} = \frac{2 \times 0.01d\left(t + r_i\right)^2}{t\left(t + 2r_i\right)} \quad \ldots\ldots\ldots\ldots(6)$$

where 1.4 x 1.5 = 2.1 is the factor of safety for the lining against reaching the crushing strength of 0.67 f_{cu} at the inside face. Equation (5a) becomes:

$$t = r_i\left\{\sqrt{\left[\frac{0.67f_{cu}/(1.4 \times 1.5)}{(0.67f_{cu}/(1.4 \times 1.5))-2 \times 0.01d}\right]}-1\right\} \ldots(6a)$$

If cylinders are used as a means of determining the characteristic strength, f_{cu}, instead of cubes, then the value of 0.67 should be replaced by 0.85.

Shaft lining thicknesses generally range from a minimum of 300mm to a maximum of 1200mm, the upper limit being determined by what is practicable to construct within the confined shaft sinking environment. Between the extremes, increments of wall thickness of 150mm are convenient.

Concrete strengths should always be greater than Grade 25, CP 110:Part 1 (1972), in order to provide a dense, impermeable, durable material. Grade 45 is the recommended upper limit for specifying concrete strength bearing in mind the restricted placing conditions requiring high workability.

Values of d for the various wall thicknesses in a 7.315m diameter concrete shaft lining in Grade 45 concrete subjected to hydrostatic pressure, as derived from Equation (6), are given in Figure 2.

Pure plastic design

In the case of concrete shaft lining design using the pure plastic method, it is necessary to determine the parameter $\tau_{Y.P.}$ for use with Equation (2c). As

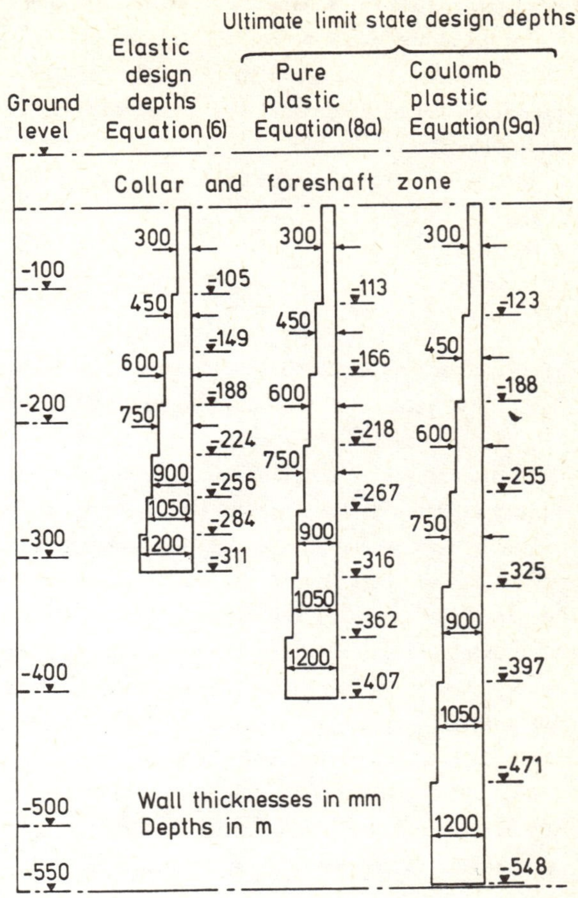

Figure 2. Hydrostatic lining design depths and thicknesses for a 7.315m diameter shaft in Grade 45 concrete.

defined previously, in relation to Equation (2), the maximum shear stress, $\tau_{Y.P.}$, is equal to $(\sigma_t - \sigma_r)/2$. This represents the difference between the maximum and minimum principal stresses divided by 2. For an unconfined compression strength test, the minimum principal stress is zero and therefore the maximum shear stress is given by:

$$\tau_{Y.P.} = \frac{UCS}{2} \quad \ldots\ldots\ldots\ldots\ldots(7)$$

In design, the UCS is replaced by the characteristic concrete strength, f_{cu}, and if this is then related to the lining itself, Equation (7) becomes:

$$\tau_{Y.P.} = \frac{0.67\ f_{cu}}{2} \quad \ldots\ldots\ldots\ldots(7a)$$

Substituting the value of $\tau_{Y.P.}$ from Equation (7a) into Equation (2c) and introducing the partial safety factors for load, γ_f=1.4, and strength, γ_m=1.5, results in the following expression:

$$p_{o\,ult} = 1.4\,p_o = \frac{0.67f_{cu}}{1.5}\log_e\frac{r_o}{r_i} \quad \ldots\ldots\ldots(8)$$

where 1.4 x 1.5 = 2.1 is the factor of safety against reaching the ultimate strength of the lining based on a pure plastic method of design. As $\log_e(r_o/r_i)$ = 2.3026 $\log_{10}(r_o/r_i)$, the above expression defined as

Equation (8) can be simplified further. On replacing p_o by 0.01d and r_o by $t + r_i$ the relationship between lining thickness and depth to resist hydrostatic pressures can be determined for a concrete shaft lining on a pure plastic basis:

$$0.01d = \frac{2.3026\times0.67f_{cu}}{1.4\times1.5}\log_{10}\left(1+\frac{t}{r_i}\right) \quad ..(8a)$$

Alternatively:

$$t = r_i\left[e^{\left(\frac{0.01d\times1.4\times1.5}{0.67\ f_{cu}}\right)} - 1\right] \quad ......(8b)$$

If cylinders are employed to determine f_{cu}, instead of cubes, the value of 0.67 should be replaced by 0.85.

Using Equation (8a), values of d for the various wall thicknesses in a 7.315m diameter concrete shaft lining in Grade 45 concrete subjected to hydrostatic pressure have been evaluated and are shown in Figure 2.

Having first designed the shaft wall thickness to be capable of carrying the ultimate load, the pure plastic stress distribution throughout the lining can be reviewed using Equations (2), (2a) and (2b) with the relationship $2\tau_{Y.P.}$ replaced by 0.67 f_{cu} (alternatively 0.85 f_{cu} where cylinders are used as the control test). With the 2.1 factor of safety provided, the actual stress distribution under the applied load will be elastic, values being given by Equations (1).

Coulomb plastic design

The ultimate load for the design of concrete shaft linings by the Coulomb plastic method is given by Equation (4c). For concrete, using a value of the angle of internal friction, ϕ, equal to 30° allows the constant A to be determined. In conjunction with Equation (3), A was defined as being $\tan^2\alpha$ where $\alpha = \pi/4 + \phi/2$. This gives a value of A approximately equal to 3. Substituting 0.67 f_{cu} for UCS and introducing the value of A together with the partial safety factors for load and strength into Equation (4c) produces:

$$p_{ult} = 1.4p_o = \frac{0.67f_{cu}}{1.5\times2}\left[\left(\frac{r_o}{r_i}\right)^2 - 1\right] \quad ....(9)$$

where 1.4 x 1.5 = 2.1 is the factor of safety against reaching the ultimate strength of the lining based on the Coulomb plastic method of design. It should be noted that for the case of no internal friction, where $\phi = 0$, Equation (9) would reduce to Equation (8).

Introducing p_o = 0.01d and $r_o = r_i+t$ into (9) gives:

$$0.01d = \frac{0.67\ f_{cu}}{0.4\times1.5\times2}\ \frac{t\ (t+2r_i)}{r_i^2} \quad ....(9a)$$

Transposing:

$$t = r_i\left\{\sqrt{\left[\left(\frac{0.01d\times1.4\times1.5\times2}{0.67\ f_{cu}}\right)-1\right]}-1\right\} \quad ....(9b)$$

If cylinders are used to determine f_{cu}, instead of cubes, the value of 0.67 should be replaced by 0.85.

Equation (9a) has been employed to determine values of d for the various wall thicknesses in a 7.315m diameter concrete shaft lining in Grade 45 concrete subjected to hydrostatic pressure. The results are included in Figure 2.

From Equations (4), (4a) and (4b) the Coulomb plastic stress distribution throughout the shaft wall can be obtained by substituting 0.67 f_{cu} for UCS (0.85 f_{cu} for control tests using cylinders) and replacing A with its value of 3. Again with the 2.1 factor of safety provided, the actual stress distribution under the applied load will be elastic, values being given by Equations (1).

APPRAISAL OF DESIGN METHODS

Figure 2 indicates the depths to which it is proposed the various shaft lining thicknesses could be taken for a 7.315m diameter shaft in Grade 45 concrete if hydrostatic pressures were to occur. It can be seen from the comparison of the three different design approaches that the elastic method gives the most conservative results but the Coulomb plastic scheme would be the most competitive from a cost point of view. However, to study further the validity of each approach, it is necessary to observe the inter-relationship between the design load stresses and the predicted ultimate load stresses through the lining thicknesses. As a particular case study, the 1200mm wall has been chosen and the results are illustrated in Figures 3, 4 and 5 for the three different design methods.

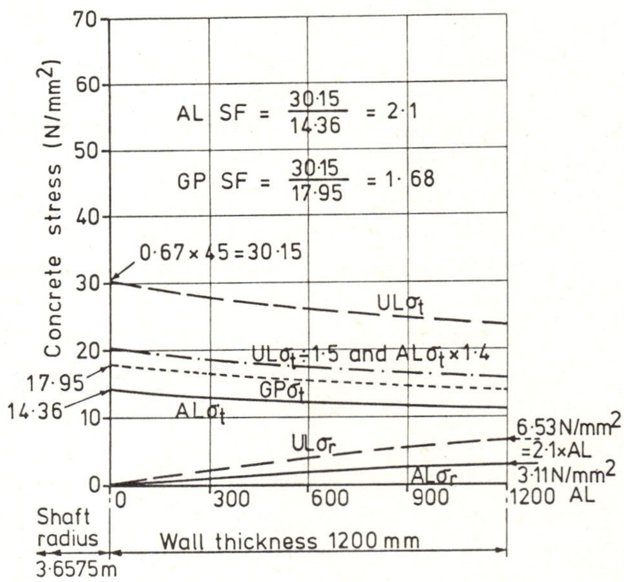

Figure 3. Stresses in a 1200mm thick shaft lining of 7.315m diameter in Grade 45 concrete, the wall thickness having been determined on an elastic design basis to resist hydrostatic pressure.

In each case, the particular design load is the hydrostatic pressure at the bottom of the 1200mm wall

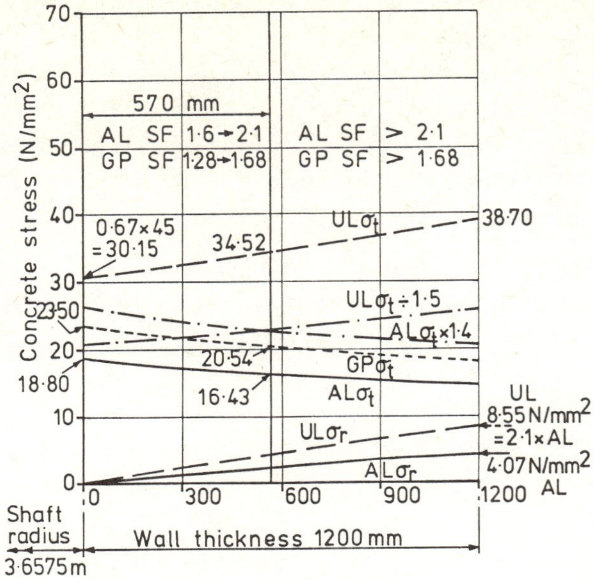

Figure 4. Stress in a 1200mm thick shaft lining of 7.315m diameter in Grade 45 concrete, the wall thickness having been determined on a pure plastic design basis to resist hydrostatic pressure.

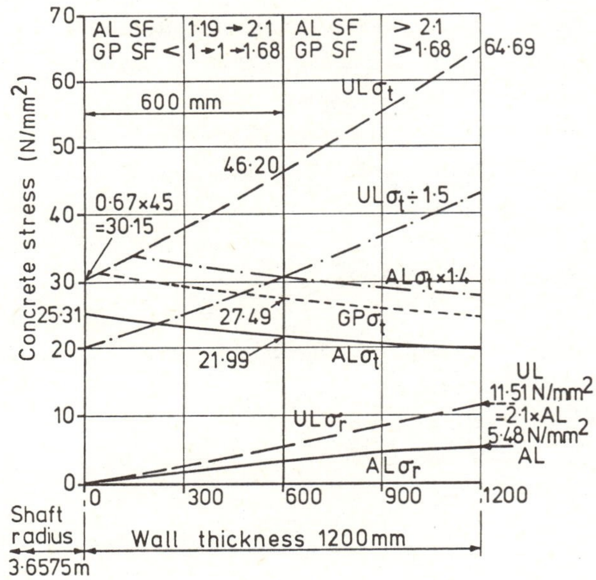

Figure 5. Stresses in a 1200mm thick shaft lining of 7.315m diameter in Grade 45 concrete, the wall thickness having been determined on a Coulomb plastic design basis to resist hydrostatic pressure.

section.

In Figure 3, the design load for the 1200mm thick wall is shown to be 3.11 N/mm^2. This is the value for the hydrostatic pressure at a depth of 311m for the elastic method of design (Figure 2). Plotted in Figure 3 are the design (applied) load radial and tangential stresses, $AL\sigma_r$ and $AL\sigma_t$ respectively, together with the assumed ultimate load equivalent values, $UL\sigma_r$ and $UL\sigma_t$. The ultimate stresses, however, are not true values as the failure strength of the concrete, equal to 0.67 x 45, is only reached at the inside face of the lining. Throughout the rest of the wall, the stresses remain elastic. Having provided a 2.1 factor of safety, in relation to the inside face tangential stress, directly in the design, the $UL\sigma_t \div 1.5$ and $AL\sigma_t \times 1.4$ lines coincide.

In the case of backwall grouting, the tangential stress, $GP\sigma_t$, is equal to 1.25 $AL\sigma_t$. This is equivalent to increasing the permissible stress by 25% which is in line with normally accepted principles for structural design when considering short term loading. The factor of safety is reduced from 2.1 to 1.68 at the inside face.

For the 1200mm wall designed by the pure plastic method, the design pressure is 4.07 N/mm^2 (Figure 4). From Figure 2 this can be seen to be the hydrostatic pressure at the maximum depth of 407m. The design and ultimate load radial and tangential stresses are shown in Figure 4 together with the grouting tangential stress profile.

From Figure 4, it can be seen that, in the case of the lining designed by the pure plastic method, the $UL\sigma_t \div 1.5$ and $AL\sigma_t \times 1.4$ lines intersect and cross over at a point 570mm from the inside face of the wall. To the left of the 570mm line the design (applied) load factor of safety varies from 1.6 at the inside face of the wall to 2.1 at the 570mm position. Beyond the 570mm line, towards the outside face of the wall, the factor of safety is everywhere greater than 2.1. In the case of the grouting pressure, the factor of safety changes from 1.28 at the inside face of the wall to 1.68 at the 570mm line and increases beyond 1.68 towards the rear of the lining.

The Coulomb plastic design method hydrostatic pressure value is 5.48 N/mm^2 from the depth of 548m in Figure 2. Figure 5 shows the stresses in the 1200mm thick wall for this case. The $UL\sigma_t \div 1.5$ and $AL\sigma_t \times 1.4$ lines again cross over, this time at a point 600mm from the inside face. It can be seen that for the design (applied) load, the factor of safety is now reduced to 1.19 at the inside face of the wall. Furthermore, the grouting pressure would cause failure at the inside face of the wall as the lining strength is exceeded at this position. This can be seen by the fact that the $GP\sigma_t$ line intersects the $UL\sigma_t$ line adjacent to the inside face.

CONCLUSIONS AND DESIGN RECOMMENDATIONS

1. The elastic method of design is conservative as only the inside face of the shaft wall is assumed to reach the ultimate strength of the concrete, the rest of the

wall remaining in an elastic state of stress.

2. Using the elastic method of design, a minimum factor of safety equal to 2.1 is provided for the wall. The minimum value occurs at the inside face of the lining, everywhere else within the wall having a factor of safety greater than 2.1 (it should be noted that this minimum value also occurs at the lowest extremity of the lining thickness being considered, any point above possessing a greater factor of safety because the hydrostatic pressure is less).

3. From a consideration of the pure plastic method of design, it can be seen that there is an inner zone of the shaft wall in which the factor of safety is less than 2.1 under the design load. Having commented earlier, under Item 1 of this section, that the elastic design method is conservative, in recommending an improved method of design, whereby the increase in strength through the wall is utilized, it is necessary to accept this fact.

4. With regard to specifying the minimum value for the factor of safety which would be acceptable at the inside face of the wall for the pure plastic method of design, the author would recommend 1.5. This is based on the fact that a hydrostatic load has a clearly defined magnitude and is not likely to be exceeded once it is determined. On this basis a γ_f value equal to 1 would be acceptable, the $\gamma_m = 1.5$ providing the factor of safety over the material strength. With the pure plastic method of design, the minimum factor of safety is in fact 1.6, this value increasing rapidly throughout the wall from front to back.

5. For the short term grouting situation, in the case of a wall designed by the pure plastic method, the factor of safety would be reduced to an unacceptable level at the inside face of the lining if pressure grouting was to be carried out at 28 days. To maintain the previously mentioned 1.5 factor of safety at the inside face for the grouting situation, it is necessary to include for a modification factor within the design formula, Equation (8a). Instead of replacing p_o in Equation 8 by 0.01d, it should be replaced by 0.86 x 0.01d, where 0.86 is the modification factor. The value of 0.01d should subsequently be replaced by 0.86 x 0.01d in Equations (8a) and (8b).

6. No account of increase in strength with age has been assumed in the calculations. Although in a normal shaft sinking situation it may be possible to leave the backwall grouting until longer periods than 28 days have elapsed, in very wet conditions this may not be possible.

7. The Coulomb plastic method of design cannot be used as it will result in failure at the inside face of the wall during the backwall grouting stage.

8. The recommended method of design, therefore, is the pure plastic one incorporating a modification factor for pressure grouting. However, the Coulomb plastic stress analysis can subsequently be used to substantiate the design as it indicates the additional strength which is inherent in the shaft wall provided the inside face does not fail.

ACKNOWLEDGEMENTS

The author wishes to thank Mr. J.C. Black, Managing Director of Cementation Mining Limited, for permission to publish this Paper. Mr. Black's substantial experience and knowledge of sinking procedures and shaft lining design were responsible for the Author's initial interest in the subject of shaft lining design and have acted as a guide throughout the study.

REFERENCES

1. AULD, F.A. Design of concrete shaft linings. Proceedings of The Institution of Civil Engineers, Part 2. Vol.67. September, 1979. pp. 817-832.

2. BRITISH STANDARDS INSTITUTION. The structural use of concrete. British Standards Institution, London, 1972, CP 110, Part 1.

3. BRITISH STANDARDS INSTITUTION. Methods of testing concrete. Part 4. Methods of testing concrete for strength. British Standards Institution, London, 1970, BS 1881.

4. COMITE EUROPEEN DU BETON - FEDERATION INTERNATIONALE DE LA PRECONTRAINTE (CEB-FIP). International recommendations for the design and construction of concrete structures. Principles and Recommendations. June 1970: FIP Sixth Congress, Prague. p.41.

5. JAEGER, J.C. and COOK, N.G.W. Fundamentals of rock mechanics. Methuen & Co. Ltd., London, 1969.

6. TIMOSHENKO, S. Strength of materials. Part II. Advanced theory and problems, 3rd edn. Van Nostrand Reinhold, New York, 1976.

STABILITY OF A THICK RING OF GROUTED OR FROZEN GROUND

A H Wilson

Head, Rock and Coal Mechanics Branch
Mining Research and Development Establishment
National Coal Board, Ashby Road, Stanhope Bretby
Burton-on-Trent, Staffs, DE15 0QD

SUMMARY. The stability of ground which has been grouted or frozen to hold back water from excavations is considered. The application of the Lamé equations for thick-walled cylinders is shown to be unrealistic for very thick cylinders of ground, but the concept of an outer cylinder in an elastic condition, stabilised by the formation of an inner yield zone conforming to the laws of soil mechanics, is shown to be feasible. The stresses in the yield and the elastic zones are analysed, and diagrams as a guide to the design of grouted or frozen layers are given.

NOMENCLATURE

a	radius of excavation (inner radius)
b	radius of impermeable layer (outer radius)
k	triaxial stress factor = $(1 + \sin \phi) \div (1 - \sin \phi)$
p	resistance offered by supports on radius a
p'	factor based on apparent cohesion of granular material = $\frac{\sigma_o'}{k-1}$ = approximately 0.1 MPa
q	water pressure acting on outer radius b
r	radius in general
t	thickness of grouted or frozen zone = b - a
y	radius of boundary between yield zone and elastic zone
ϕ	angle of internal friction
σ_o	unconfined compressive strength in-situ
σ_o'	intercept on stress axis of the graph of failure strength against confining pressure for a granular material
σ_r	radial stress in unfailed region
σ'_r	radial stress in failed region
σ_θ	tangential stress in unfailed region
σ'_θ	tangential stress in failed region

INTRODUCTION

While drifting or sinking through water-bearing strata, recourse is often made to the establishment of a zone of frozen or grouted ground to hold back the water during excavation and subsequent lining of the opening. If the water pressures are low, the required thickness of impermeable ground can be calculated using the equations developed by Lamé (1852) for stresses in a thick cylinder, details of which can be found in most engineering text books, for example Salmon (1931).

As coal mining in the central part of England is extended towards the east, the weak water-bearing sandstones of the Permian System are encountered at ever increasing depths, and the resultant water pressures are now approaching a magnitude at which traditional thick-wall theory obviously cannot be applied in assessing the required thickness of the impermeable zone.

It is the objective of this paper to consider the equations for stress in thick-walled cylinders, to modify them to take into account the formation of a zone of yielded ground in the highly stressed inner portion, and to deduce in what circumstances such a yield zone will augment the capabilities of the impermeable zone to resist water pressure.

ELASTIC THICK CYLINDER

If an elastic thick-walled cylinder of inner radius a and outer radius b is subjected to an internal pressure p and an external pressure q, then the stresses at any point in the cylinder according to the Lamé equations are:

$$\sigma_r = \frac{(qb^2 - pa^2) - (q - p)\, a^2\, b^2/r^2}{b^2 - a^2} \quad (1)$$

$$\sigma_\theta = \frac{(qb^2 - pa^2) + (q - p)\, a^2\, b^2/r^2}{b^2 - a^2} \quad (2)$$

where σ_r is the radial stress, σ_θ the tangential stress, and r is the radius in question.

If the external water pressure q is greater than the internal support resistance p, then the maximum stress occurs tangentially on the inner radius a, and has a magnitude:

$$\hat{\sigma}_\theta = \frac{2q - p - \frac{a^2}{b^2} p}{1 - \frac{a^2}{b^2}} \qquad (3)$$

As the thickness of the impermeable zone is increased, the maximum stress at the inner boundary reduces, but it has a lower limiting value of $\hat{\sigma}_\theta = 2q - p$ as b approaches infinity. The resistance p given by the supports is usually small compared with the hydrostatic head q, hence a grouted sand of strength, say, 7 MPa, cannot hold back a water pressure in excess of 3.5 MPa without failure of the inner wall, regardless of the extent of the grouting.

PARTIALLY YIELDED CYLINDER

The following analysis considers the stresses in a thick cylinder of ground in which a zone of yielded ground has developed adjacent to the inner boundary. The yielded ground is assumed to conform with the laws of soil mechanics, ie it acts as a granular material, while the unyielded ground beyond continues to remain in the elastic state. The nomenclature adopted is as given at the beginning of the paper.

Stress in the yield zone

In a failed zone immediately surrounding a circular opening, Airey (1977) and others have shown that the stress at any point can be found from the equations:

$$\sigma'_r = (p + p') \left(\frac{r}{a}\right)^{k-1} - p' \qquad (4)$$

$$\sigma'_\theta = k\,(p + p') \left(\frac{r}{a}\right)^{k-1} - p' \qquad (5)$$

This build up of stress with increasing distance from the side of the excavation is a consequence of the internal friction between the grains or fragments of broken material, and is independent of the conditions in the elastic zone beyond.

Stresses in the elastic zone

In an unfailed cylindrical elastic medium, the stresses at any point are given by Salmon (1931) and others as:

$$\sigma_r = A - \frac{B}{r^2} \qquad (6)$$

$$\sigma_\theta = A + \frac{B}{r^2} \qquad (7)$$

where A and B are constants depending on the boundary conditions.

The elastic zone will extend from the yield-elastic boundary ($r = y$) to the outer limit of the impermeable zone ($r = b$). At the yield-elastic boundary, the ground will be at the point of failure, and assuming the ground behaves as a Coulomb type material, then:

$$\sigma_\theta = k\,\sigma_r + \sigma_o \quad \text{at } r = y$$

At the outer boundary, the radial stress will equal the water pressure q, therefore:

$$\sigma_r = q \text{ at } r = b$$

These boundary conditions allow A and B to be determined, and by substitution in equations (6) and (7) give:

$$\sigma_r = q - \frac{q\,(k-1) + \sigma_o}{\frac{1}{y^2}(k+1) - \frac{1}{b^2}(k-1)} \cdot \left(\frac{1}{r^2} - \frac{1}{b^2}\right) \qquad (8)$$

$$\sigma_\theta = q + \frac{q\,(k-1) + \sigma_o}{\frac{1}{y^2}(k+1) - \frac{1}{b^2}(k-1)} \cdot \left(\frac{1}{r^2} + \frac{1}{b^2}\right) \qquad (9)$$

Determination of q in terms of y

At the yield-elastic boundary, the radial stresses must balance. Hence by putting $r = y$ and equating equations (4) and (8), it can be shown that:

$$q = \tfrac{1}{2} \left[\left\{ (p+p') \left(\frac{y}{a}\right)^{k-1} - p' \right\} \cdot \left\{ (k+1) - \left(\frac{y}{b}\right)^2 (k-1) \right\} + \sigma_o' \left\{ 1 - \left(\frac{y}{b}\right)^2 \right\} \right] \qquad (10)$$

This expresses the sustainable water pressure q in terms of the known quantities and the position of the yield-elastic boundary y. Unfortunately, except for special values of k such as $k = 3$, it is not possible to express y in terms of q.

IMPLICATION OF DERIVED EQUATION

To study the implications of equation (10), typical values of the parameters were assumed and inserted. In the time between the excavation and the installation of the lining, only a limited amount of support is given to the sides. Compared with the other pressures involved, the support resistance p can be considered negligible. Even broken material, however, still tends to hold together, and gives rise to an apparent cohesion equivalent to a resistance $p' = \frac{\sigma'_o}{k-1}$ (see Airey 1977), where σ'_o is the intercept on the stress axis of the graph of failure strength of a granular material against confining pressure. Examination of a number of coal measure rocks gave p' equal approximately to 0.1 MPa. The inner radius a has been taken as 5 m, the unconfined compressive strength in-situ σ_o of grouted ground as 5 MPa, and a value of 3.5 has been assumed for the triaxial stress factor k, equivalent to an angle of internal friction of 34°. The external radius b and the radius y of the yield-elastic boundary are considered variable.

Figure 1 shows the theoretical stress distributions across cylinders of impermeable ground of thickness 5 m and 15 m. At 5 m thick, failure on the inner surface of the elastic cylinder occurs at an external water pressure q equal to 1.9 MPa. The failure spreads immediately right across the cylinder as shown by the condition for q equal to 2.0 MPa. The failed cylinder is capable of generating a radial pressure of only 0.5 MPa on its outer surface, less than the water pressure it must resist.

The 15 m thick cylinder fails on its inner surface at an external water pressure of 2.3 MPa, but on this occasion failure does not spread immediately across the cylinder, but spreads in a controlled manner as the external water pressure is further increased. At q = 3.0 MPa, the yield-elastic boundary has a radius y = 11.8 m, and a maximum value of q = 3.4 MPa occurs when y = 16.9 MPa. Thereafter instability occurs, and the yield zone spreads right across as shown by Figure 1.

This is further illustrated in Figure 2, where equation (10) has been used to show the relationship of q with y for a 15 m thick cylinder. As the yield zone spreads from y = a = 5 m, q increases until a maximum value is achieved, then falls away. If the equivalent graph for a 5 m thick cylinder is drawn, the maximum occurs in the hypothetical area to the left of the 5 m radius, and q continously decreases from y = 5 m to y = 10 m. Hence the instability of the 5 m thick cyliner.

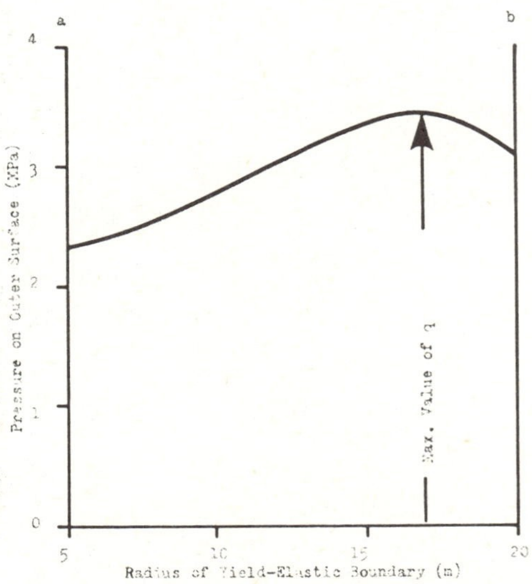

FIGURE 2. VARIATION OF SUSTAINABLE PRESSURE WITH RADIUS OF YIELD-ELASTIC BOUNDARY

(a = 5 m, p = 0, p' = 0.1 MPa, σ_o = 5 MPa, k = 3.5)

MAXIMUM VALUE OF WATER PRESSURE RESISTED

The maximum value of q can be obtained by differentiating equation (10) with respect to y, equating to zero, and substituting the condition so obtained back into equation (10).

It can be shown that $\frac{dq}{dy} = 0$ when:

$$b^2 (k^2 - 1) (p+p')y^{k-2}$$
$$= (k^2 - 1) (p+p')y^k + 2a^{k-1} \quad \sigma_o - p' (k-1) \quad y \qquad (11)$$

Unfortunately once again a solution for y cannot be found except by numerical methods, but a solution for b is possible, namely:

$$b = y \left\{ (1 + 2 \cdot \frac{\sigma_o - p' (k-1)}{(k^2-1) (p+p')} \cdot \left(\frac{a}{y}\right)^{k-1} \right\}^{\frac{1}{2}} \qquad (12)$$

Equation (10) will give the maximum value of q when equation (12) is satisfied for values of y lying between a and b. In executing this procedure, however, certain values of q may have to be rejected as equation (12) also indicates minima and points of inflexion. These can occur with certain combinations of the parameters, but are readily detectable.

Design guide

As a guide to the design of grouting and freezing patterns, Figure 3 has been produced for a 5 m radius excavation, indicating the maximum water pressure which can be resisted by a given thickness of grouted or frozen ground. It covers a range of values of triaxial stress factor and in-situ unconfined compressive strength. The dotted line shows the minimum value of thickness for the formation of a stable yield zone. To the left of the dotted line the cylinder is elastic throughout; to the right a partial yield zone exists adjacent to the excavation. The curves for the latter condition were obtained by selecting a series of values of y between y = a and y = b, calculating the corresponding values of b from equation (12), substituting these values of y and b in equation (10), then plotting the relevant values of q against t = b-a.

Figure 3 illustrates the importance of the internal friction present in the material. If the triaxial stress factor is low, as some measurements made by the NCB suggest for parts of the Bunter Sandstones (k = 3), then if the formation is encountered in this condition in any great thickness at depth, very wide grouted or frozen zones will be required. In contrast, some sandstones have a triaxial stress factor of k = 4.5 or more. On this occasion, relatively thin grouted or frozen zones will suffice, regardless of the unconfined compressive strength or hydraulic head encountered.

The analysis put forward in this paper applies to a long cylinder of ground. Frequently the aquifers occur in strata of limited thickness, in which case impermeable stronger strata above and below can give substantial restraint. This condition has been looked at separately by Airey (1981) and will be the subject of a further publication.

CONCLUSION

It has been apparent for some time that the basic Lamé equations for thick cylinders are not applicable on their own to zones of ground rendered impermeable in order to hold back high hydrostatic heads of water. The concept of a yield zone forming adjacent to the excavation which stabilises an elastic zone beyond, does give thicknesses in line with current practice. Caution is indicated, however, should substantial thicknesses of weak sandstone with low internal friction properties be encountered at great depths.

ACKNOWLEDGEMENT

The author thanks the Director of Mining Research and Development, National Coal Board, Mr P G Tregelles, for permission to present this paper. The opinions expressed by the author are his own, and do not necessarily represent the views of the National Coal Board.

REFERENCES

Airey, E M, 1977. "A study of yield zones around roadways", Appendix 1, NCB Final Report on ECSC Research Project 6220-AB/8/802, CEC, Brussels.

Airey, E M, 1981. Private Communication.

Lamé, 1852, "Leçons sur la théorie mathématique de l'élasticité des corps solides", Paris, p 188.

Salmon, E H, 1931. "Materials and structures", London, Longmans Green, Vol 1, pp 38-40.

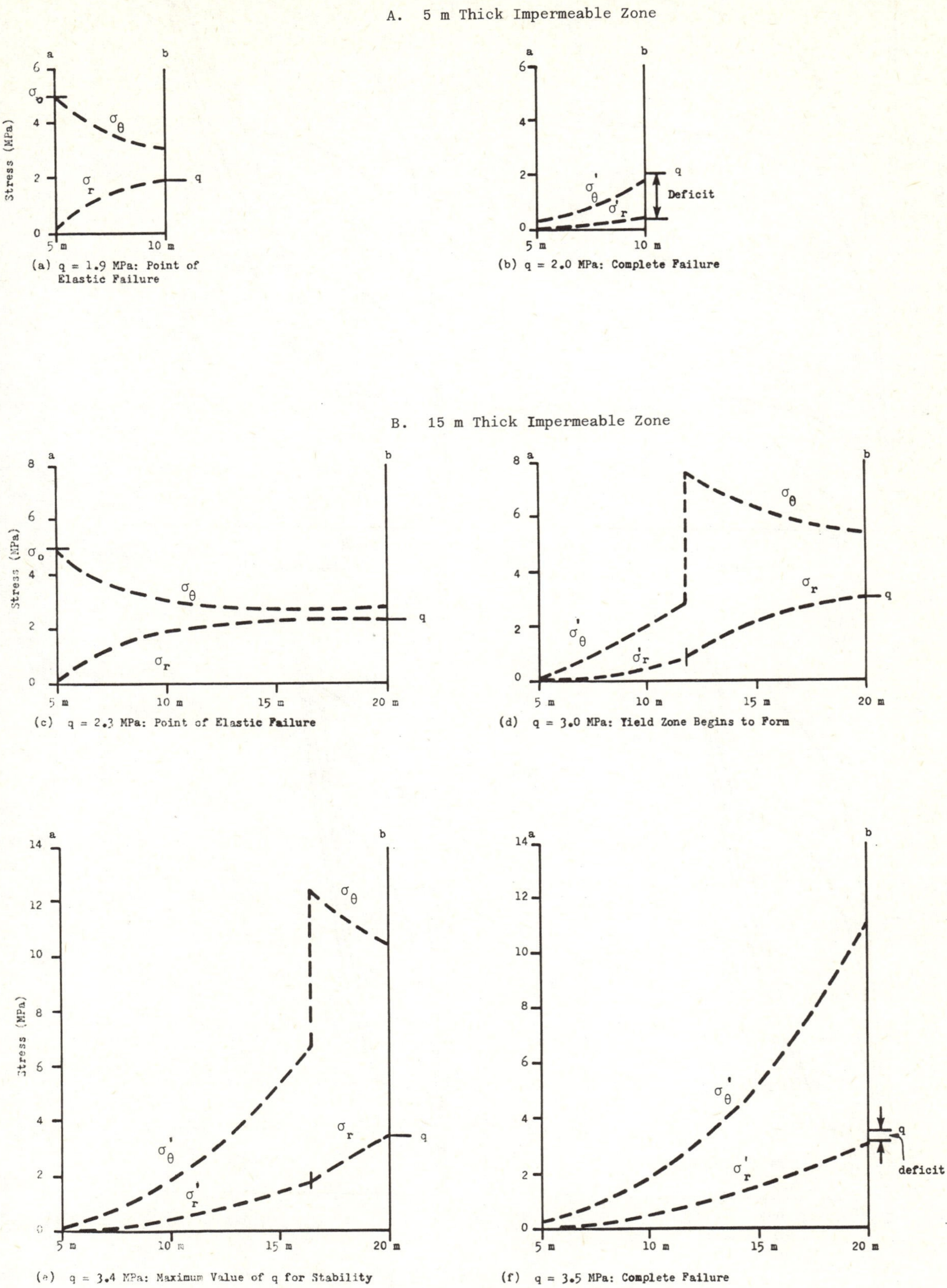

FIGURE 1. STRESS ACROSS IMPERMEABLE ZONE WITH INCREASING WATER PRESSURE

($p = 0$, $p' = 0.1$ MPa, $\sigma_o = 5$ MPa, $k = 3.5$)

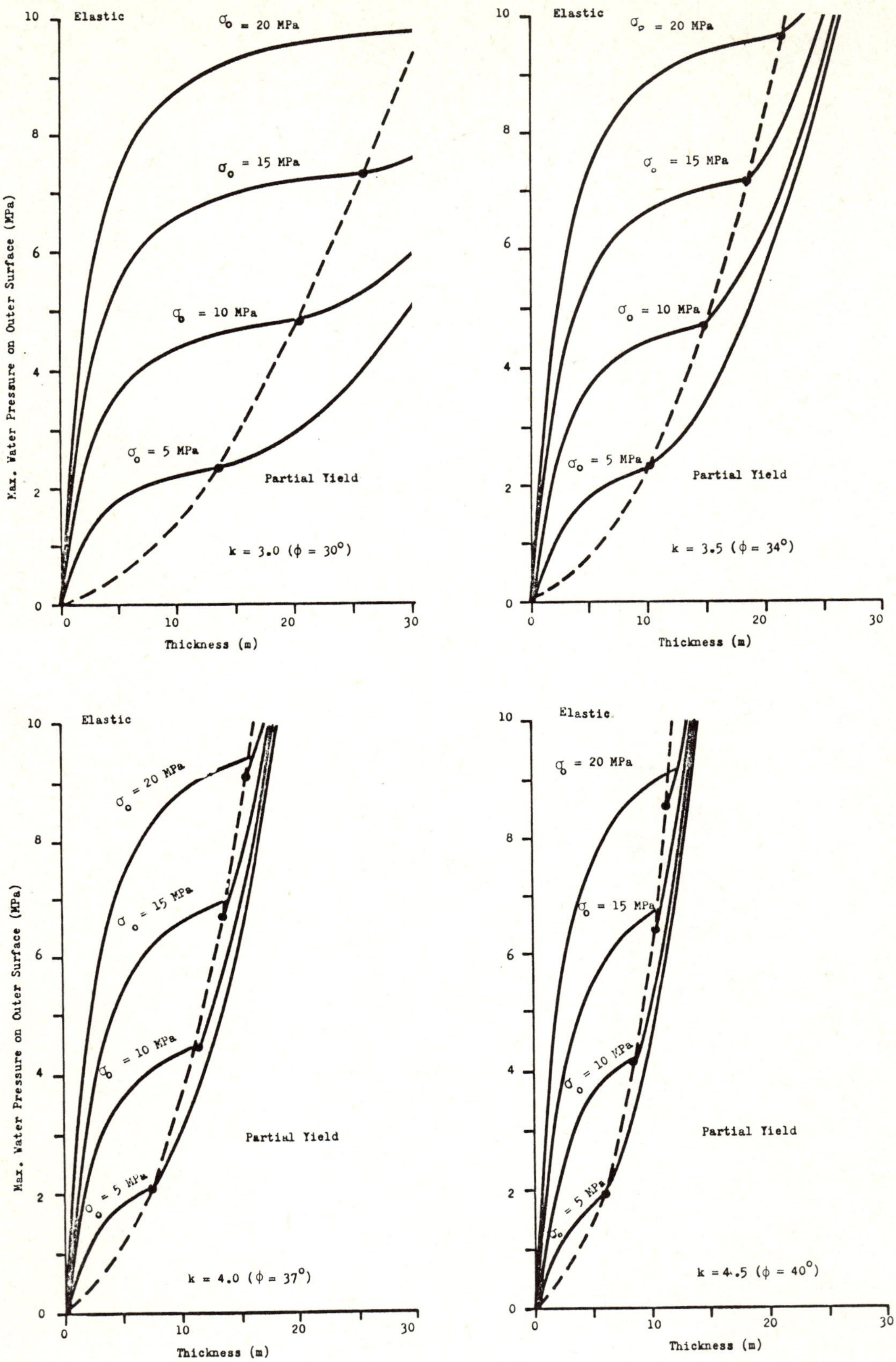

FIGURE 3. MAXIMUM WATER PRESSURE WHICH CAN BE HELD BY A GIVEN THICKNESS OF IMPERMEABLE ZONE

(a = 5 m, p = 0, p' = 0.1 MPa)

PRESENT STATE OF FREEZE SHAFT DESIGN IN MINING

Dr.-Ing. Joachim Klein

Bergbau-Forschung GmbH, Essen
D 4300 Essen 13

SUMMARY. The ground freezing technique is the best way for shaft sinking in non-stable ground and water-bearing layers. Qualified methods of design and security under economical aspects dictate the use of such a method in practice. One of the most interesting questions is the necessary frozen wall thickness in dependence of stratigraphical geology. Besides heat conduction problems involving latent heat the description of the mechanical strength of the frozen soil itself is of priority. The approach to the design of retaining frozen earth structures like shafts is to assume a thick hollow circular cylinder and base the dimensions on strength criteria. This paper deals with the influence on design using different mechanical idealizations for the frozen soil concerning the failure conditions. In addition recent R & D results for deep freeze shafts in German coal mining are demonstrated from theoretical and practical point of view. Especially the measurements at Voerde shaft - 600 m freezing depth - and in particular the lining system are represented.

INTRODUCTION

Present shaft-sinking techniques in West-German hard-coal mining, in particular in the Ruhr district, is logically matched to the zone of mining activities which is moving northwards. These conditions imply difficulties which are not only a function of the steadily increasing depth but also of the increasing thickness of the overburden. While the average final depth of newly sunk shafts was of approximately 800 m in the sixties, the 1000 m limit is largely exceeded meanwhile (Fig. 1). Simultaneously, the useful cross-section of the shafts was increased by approximately 60 % to 50 m², and the shaft diameters are normally of 8.0 m.

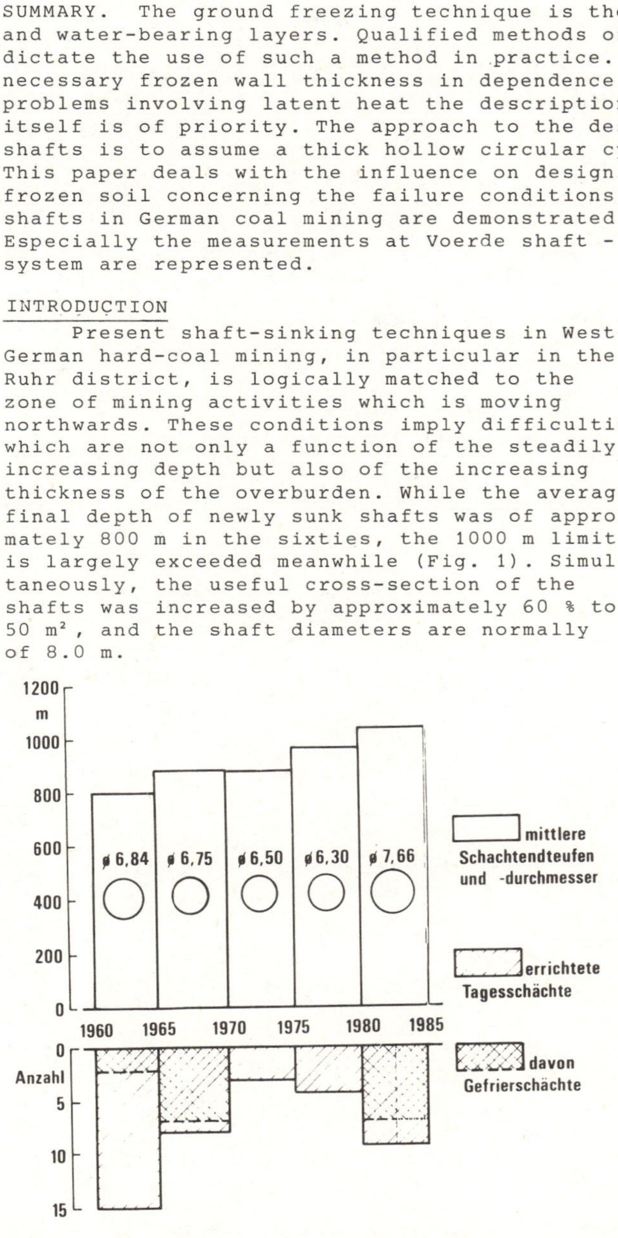

Figure 1. Development of shafts in coal mining

Supposing that, to the end of the century, still approximately 30 shafts (1) need to be sunk in order to assure the planned hard-coal mining capacity, the statistics predict that 20 to 30 % of these shafts will require freezing. The necessity to apply this technology is of course primarily dependent on the choice of the shaft site and of the geotechnical conditions prevailing. Accordingly, freezing will become necessary above all in non-stable water-bearing overburden strata. In particular when working through layers of quicksand or loose earth with high water content, a frozen cylinder around the shaft zone enables safe sinking.

The shafts the sinking of which implies freezing are now to be found near the river Lippe:the shafts Haltern 1 and Haltern 2 (sunk by Messrs. Deilmann-Haniel and Gebhardt & Koenig), Polsum 2 (sunk by Thyssen Schachtbau) and Voerde (sunk by Gewerkschaft Walter). The Voerde shaft with 6,0 m diameter was sunk over a depth of 600 m with ground freezing technology. This is the record for modern freeze shaft performance and the Voerde shaft accordingly will be mentioned repeatedly.
In German coal mining over the past 20 years 16 freeze shafts were sunk (Fig. 2).

SCHACHT	BAUZEIT	LICHTER DRM.	ENDTEUFE	GEFRIERTEUFE
NIEDERBERG 4	1960-1962	5.50	535	297
SOPHIA-JACOBA 6	1961-1962	6.75	620	268
AUGUSTE VICTORIA 8	1963-1966	6.75	1056	218
LOHBERG 3	1963-1967	6.50	910	0-204 UND 383-450
SCHLÄGEL & EISEN 8	1964-1965	7.50	800	100
NIEDERBERG 5	1964-1966	6.75	823	230
GENERAL BLUMENTHAL 8	1964-1967	7.50	980	115
NORDSCHACHT	1964-1967	7.30	1020	320
ALTENDORF	1967-1969	5.00	837	115
PROSPER 10	1977-1980	8.00	1070	140
AN DER HAARD	1978-1980	8.00	1115	153
LAUTERBACH	1978-1980	7.00	950	220
POLSUM 2	1979-	8.00	656	99
HALTERN 1	1979-	8.00	1135	217
HALTERN 2	1980-	8.00	1077	217
VOERDE	1980-	6.00	1060	600

Figure 2. Completed freeze shafts (1960-1980)

(1) Weber, H.: "Künftige bergtechnische Aufgaben des deutschen Steinkohlenbergbaus und Ansätze zu deren Lösung." Glückauf 117 (1981) S. 1001/19.

This corresponds to almost 4000 m of sinking performance in frozen strata. For comparison: over the same period a total of 30 shafts was sunk. Since the early sixties freeze shafts in Germany are no longer lined with cast-iron segments but by non rock connected technology which implies a bitumen mass between inner and outer shaft lining. Initially, a rigid preliminary concrete lining was used for safety of men and material and from the construction period for the Auguste Victoria 8 shaft onwards prefabricated concrete blocks were preferred. This new technology was called "Gleitschacht" (literally translated "sliding shaft") because the bitumen around the shaft assures a flexibility of the total shaft relative to the surrounding strata and in this way realized a planning of mining operations near the shaft. The bottom end of the freeze shaft needs not necessarily to be situated in the overburden/coal measures interface. It is tried in any case to keep the freezing length as short as possible, i.e. to the extent enabled by stable and watertight strata. Furthermore the strata into which the freeze shaft foundation is set, needs to exhibit a certain minimal compressive strength for withstanding the foundation pressures.

The slide shaft sunk within a frozen protective zone thus rests on a foundation and is separated from the surrounding strata by the bitumen mass. Consequently, it is logical to categorize such a shaft design, from the engineers point of view, as civil engineering construction. The dimensioning of the shaft lining corresponds exactly to statical calculation procedures as used in civil engineering. The respective directives are closely related to the regulations for reinforced concrete and to the standards for steel structures. The loads involved are allowed for firstly in terms of the hydrostatic load imposed by the viscous sliding mass secondly in terms of an additional horizontal rock-pressure component, and lateron in terms of unsymmetric loads conditioned by mining activity. For the construction period the frozen body itself is supposed to withstand loads resulting from water and strata pressure. Since the sliding shaft, in its present day configuration, is built up from the bottom upwards, the construction period for a deep freeze shaft may be longer than one year. This is one of the characteristic features which makes German shaft sinking technology different from many methods practised abroad, e.g. in Great Britain where the freeze shaft is given the final monolithic concrete lining in the course of shaft sinking already. Conditioned mainly by the water loads increasing with depth the frozen wall and the freeze shaft lining needs to be increasingly thick. Fig. 3 shows schematically the necessary wall thickness of a 500 m shaft. Due to the varying acceptable compressive strength values of concrete and the rock strata, the wall thickness does not increase proportionally.

FREEZING AND LINING TECHNOLOGY

Before discussing the mechanical aspects of the dimensioning of the frozen body with consideration of strength criteria, some features of the freezing technology as well as some details relative to lining techniques should be pointed out. The freezing techniques in mining imply the use of calcium chloride brine as cooling liquid down to - 40 °C. Towards of the end of the last century already the ground freezing process was invented by Hermann Poetsch and, in 1883 a patent application for this process was issued by the German Imperial Patent Office. In two years therefore this process will be 100 years old and, inspite of this, we are to state that the basic concept did not change meanwhile.

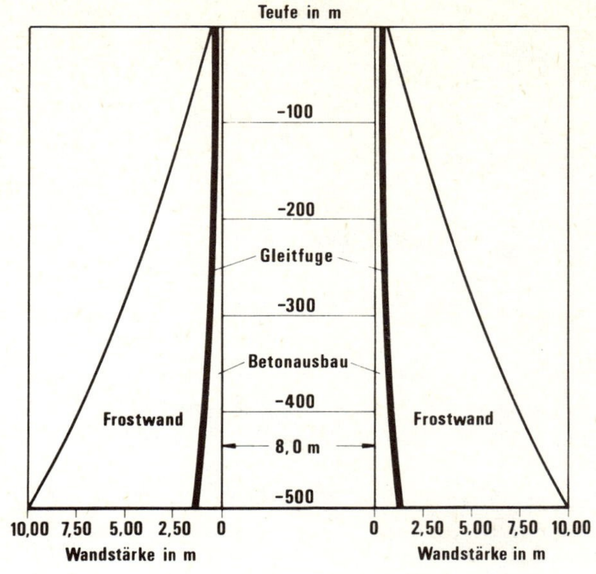

Figure 3. Necessary thickness of shaft lining and frozen wall

The safety by the frozen wall as well as the clean technology of the process itself will assure that also for the next century no other method will replace ground freezing.

The making-up of a frozen body in the required place within the strata starts with drilling of the freezing boreholes along a freezing-main circle. Normally, the boreholes are drilled vertically and equidistantly. Fig. 4 shows the freeze tube circle of the Voerde (2) shaft and the additionally required boreholes for temperature measuring, in this case even four.

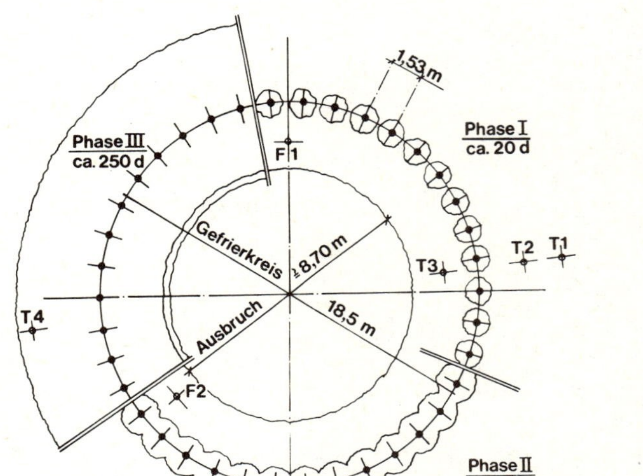

Figure 4. Freezing circle (Voerde)

All drillings had to be brought down for the Voerde shaft within a tolerance of 0.25 % of the depth. The vertical tube diameter implies a freezing circle of 18.5 m corresponding to a required thickness of the frozen wall of 10 m. The cooling brine reaches the bottom of the borehole by the inner tube and then mounts through the concentric outer tube back to surface thus evacuating heat from the strata. In this way a frozen wall is created which has not only a sealing effect against water inflow but also staticly carrying functions. The frozen

(2) Gewerkschaft Walter: "Informationsmaterial zum Gefrierschacht Voerde." (1981) S. 1/14.

strata may exhibit compressive strength values which correspond for example approximately to 1/10 of the compressive strength of concrete.

In the cooling plants normally piston compressors or screw compressors are used. According to requirements the cooling performance is varied in the range between 200.000 and 600.000 kcal/h per unit. For efficiency reasons several units are installed on each site. While in the active freezing phase initial full capacity is run, the performance is reduced when the frozen body is created and only needs to be maintained. The installed cooling performance for the Voerde shaft site is rated to 2.9 Mio. kcal/h. This rating corresponds to approximately 10.000 household refrigerators. The temperature structure according to various depths, and in particular the progress of the frozen zone, can be determined relatively exactly by means of thermo-couples inserted in the measuring boreholes. Monitoring the build-up of a frozen body of such dimensions requires a considerable volume of measuring equipments simply for the sake of safety. More than 200 thermocouples are installed in selected depths in order to give an exact information on the temperature profile in the body to be frozen, and of the heat withdrawn from the strata. Synchronously with temperature measurements additional ultrasonic measurements and inclinometer deformation measurements are carried out.

This paper does not deal with the problems involved in dimensioning of the inner shaft lining. Since, however, knowledge on the sinking techniques and shaft lining design are of some importance for the understanding of the calculation methods for freeze shafts, some explanations may be given in the following. The prefabricated concrete blocks are of predominant importance for the shaft lining system. These concrete blocks are made of high concrete qualities - e.g. B45 and B55 - and are supplied ready for use to the site. By means of these blocks a well-fitting but also flexible lining is obtained which is to protect personnel during sinking and lining work. According to the shaft diameter these preshaped concrete blocks are conical. The flexibility of this preliminary lining is assured by wooden plates made of hot-pressed flax and glue. These buffer plates exhibit a high dilatability even under water, and assure high radial deformability of the preliminary or outer lining system. The concrete block lining is built up from support rings upwards, i.e. against the direction of sinking and the working cycles must be matched accordingly.

Up to present the inner shaft lining was made of 8 mm welded steel shells and inner reinforced concrete rings. A bituminous film painted on the inside of the steel shells enables movement relative to the reinforced concrete lining. In this way shaft bending radii of 3000 m do not lead to destruction. The bitumen mass between inner lining and outer lining is of 13 kN/m³ density and, by its pressure, counteracts the pressue exerted by water and horizontal rock pressure.

Normally, the preliminary concrete block lining is not to assume static functions, but a modified version of this concept is applied to the deep Voerde shaft. According to a new patent (3) the preliminary support of the concrete blocks is used for counteracting rock-pressure loads. According to the stratigraphic conditions prevailing several rows of preshaped concrete blocks are fitted in several sections, according to the soil-physical parameters investigated by soil mechanics and foundation engineering institute of Bochum university. The concrete-block lining here is no preliminary lining but should be regarded as supporting outer lining which by its resistance, implies less stress for the frozen body. Furthermore, the density of the bitumen liquid between inner and outer lining can be reduced to the one of water. This in turn affects positively the dimensioning of the inner lining. The additional inner steel shell fitted to the Voerde shaft (Fig. 5) became necessary due to the considerable hydrostatic pressure. It is not intended to fit reinforced concrete linings of more than 1 m of thickness and therefore a steel/reinforced concrete compound structure will be adopted.

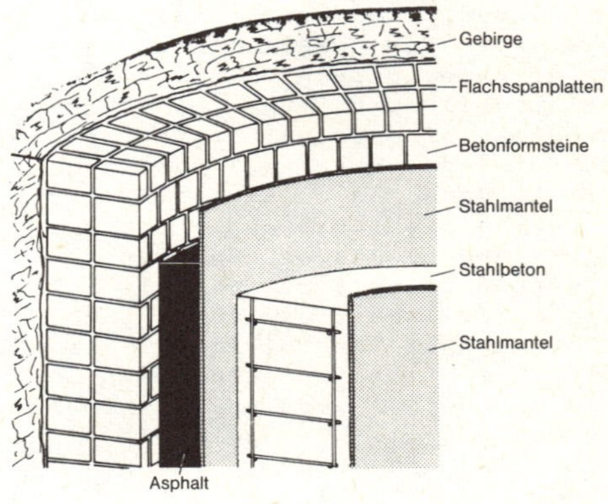

Figure 5. Lining system (Voerde)

THERMAL AND STRESS ANALYSIS

General Survey

Two global problem areas are of decisive importance when practising the freezing method in shaft sinking. On the one hand there are strength criteria based on soil mechanics, where data like compressive strength or creep behaviour and the various material characteristics are of predominant importance. On the other hand the thermophysical interdependences control the necessary capacity rating. Both fields are to be seen in one context in the end, however, are considered separately. Thus by stress calculations isothermal conditions are assumed, based on a thick-walled tube idealization with constant freezing temperature over the total wall thickness. The circumferential spacing of the freezing tubes is determined by thermophysical calculations and the important distance of the freezing tubes from the shaft side is determined by strength criteria. For determination of wall thickness in freezing technology for shaft sinking the mechanical component, is by far the more important one. When applying this method the priortiy of approaches will always read:

1. Which thickness of the frozen wall is required at all?
2. How can a frozen wall of that thickness be realized?

Cost-effectiveness of an optimized freezing shaft will always depend on reducing the thickness of the frozen wall. This trend will also go on for even deeper shafts and consequently the sequential-lining technology will be assigned priority over temperature calculations.

For strength calculations relatively little soil-specific knowledge is available. This shortage of information prevails even more with respect to the thermo-technical questions. Conductivity a frozen strata for instance is not yet sufficiently investigated up to present. The correlations of freezing-tube spacing and other geological data are influenced by so many feedback effects that very

(3) Lütgendorf, H.O.: "Bemessung des Gefrierschachtausbaus - System Auguste Victoria mit tragendem Außenausbau -." Glückauf-Forschungsh. 42 (1981) S. 259/61.

often a wide value range results from measurements. Furthermore the calculation methods, once specific data on convection influences, transient phenomena, and control of cooling systems are also considered, become so complicated that they do not correspond anymore to requirements of practice.

One essential reason for this is also the complexity of the mathematic correlations, i.e. the heat conduction calculations of the frozen zone which enable analytical solutions only for the individual tube or the straight temperature front. Furthermore a statement on temperature zones is not possible and the assumptions remain limited to stationary states. New possibilities were found by means of numerical calculation methods as the finite-element method. In this way a larger variety of marginal conditions can be allowed for and, above all, even non-stationary correlations can be simulated. It is of course that thinkable to apply the finite-element method on stress analysis. In particular with difficult multiaxial stress situations for investigation of the stresses over the free shaft length (Fig. 6) in the frozen zone this method turned out to be practicable. No prophet is required to predict that this method will be made use of quite frequently, in particular in shaft lining design where deformations are to be minimized. Within this paper only analytical methods will be discussed for better reproducibility, and the interested reader is referred to specialized literature (4).

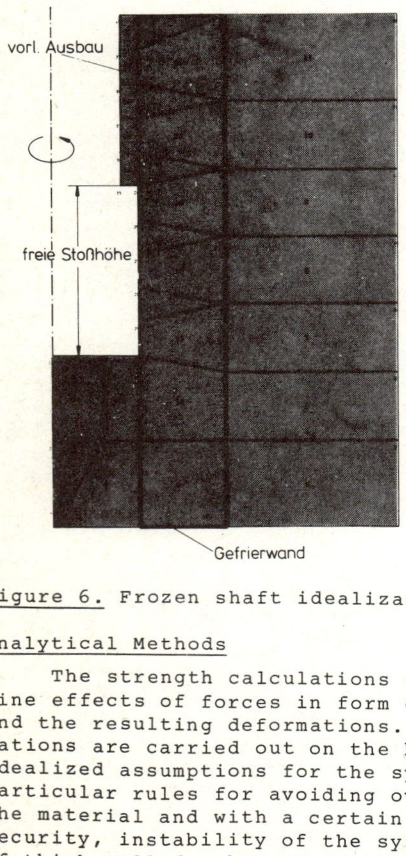

Figure 6. Frozen shaft idealization with FEM

Analytical Methods

The strength calculations are to determine effects of forces in form of stresses and the resulting deformations. These calculations are carried out on the basis of idealized assumptions for the system. In particular rules for avoiding overstress of the material and with a certain degree of security, instability of the system. In case of thick-walled tubes - the simplified model of the frozen retaining structure stability problems, i.e. physical nonlinearity - are neglected, however, particular significance is assigned to a correct description of materials. Without considering any specific problem three conditions must be met for each static calculation: equilibrium, kinematic and constitutive relations. The conditions 1 and 2 above are identical for all calculations involved in freeze shaft calculations; only the different interpretation of the soil behaviour is the reason for varying results in terms of dimensioning, and consequently this comparison requires special consideration.

It is understandable that with different description of the soil behaviour of frozen strata also different dimensioning formulae are obtained. The answer to the question why there are so different dimensioning formulae may be found in the behaviour of the frozen strata which do not only exhibit linear elastic properties but also undergo plastic deformations. Furthermore the stress/strain diagram is not linear and, in addition, a function of time and temperature. According to the shaft sinking/construction method applied (e.g. whether the final lining is applied in a sooner or a later phase) considerations must be made whether, besides the not time-dependent deformation, also time-dependent parameters need to be allowed for. Over a longer freezing period the creeping deformations may reach an order of magnitude that they account for a considerable share in the overall deformation. This phenomenon is important in particular for deep freeze shafts the definite lining of which is fitted from bottom to top thus involving longer periods during which the shaft remains with concrete block lining only.

Since all matter contained in the subsoil exhibit different behaviour the specific properties need to be investigated a priori in the laboratory. This of course cannot be done, for each meter of depth but does make sense for certain stratigraphical layers. Very often it is neglected that for a comprehensive statement with respect to dimensioning a correspondingly large number of adequate soil parameters needs to be known. The knowledge of the uniaxial compressive strength K ist not sufficient in normal case. Therefore for the Voerde shaft covered by the above-mentioned research project, an experimental program synchronized with Bochum university and Karlsruhe university was started. For better knowledge of the in-situ conditions core samples were taken directly from the frozen walls of the Voerde shaft for being soil-mechanical lab-tested afterwards. From these tests time-dependent material properties can be defined for being used lateron as a basis for dimensioning formulae.

Before discussing the various results of different dimensioning formulae and methods for determination of the frozen-wall thickness reference should be made to the idealized system. The freeze shaft is idealized as a thick-walled tube (Fig. 7).

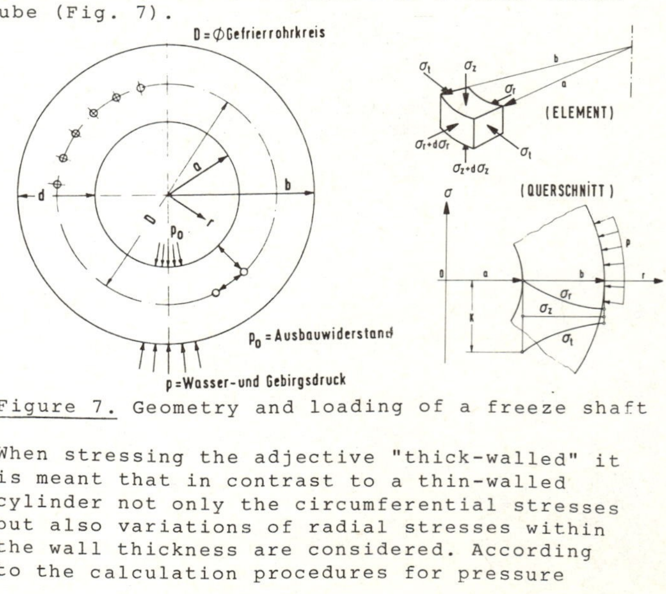

Figure 7. Geometry and loading of a freeze shaft

When stressing the adjective "thick-walled" it is meant that in contrast to a thin-walled cylinder not only the circumferential stresses but also variations of radial stresses within the wall thickness are considered. According to the calculation procedures for pressure

(4) Klein, J.: "Finite Element Method For Time-Dependent Problems of Frozen Soils." International Journal For Numerical And Analytical Methods in Geomechanics, Vol. 5 (1981) S. 263/83.

vessels a cylindrical boiler can be regarded as thick-walled if the ratio outside radius/inside radius $\frac{b}{a} > 1{,}2$. This is normally the case for freeze shafts. The load p comprises water and rock pressure loads, and the supporting inside pressure p_o is caused e.g. by the concrete block lining.

In case of axisymmetric loading on a thick-walled freeze shaft the circumferential stress σ_t, the longitudinal stress σ_z, and the radial stress σ_r act within the system. Without a supporting lining the radial stress reaches zero inside and corresponds to the actual pressure, outside. Supposing a linearly elastic material behaviour over the cross-section the circumferential stress σ_t reaches its maximum inside. The highest stress deviator therefore is found inside and, when using the dimensioning formula developed by Lamé, must there not be greater than the uniaxial compressive strength K.

Time-Dependent

A description of the material behaviour of frozen soils by rheological models as e.g. bodies according to Maxwell or Kelvin, seems possible. By combining spring/dashpot systems, complicated model structures may be set up for approximation of the strain characteristics in the tests. The combination of various types of model led to differential equations which could be solved by various methods and which are found back within the framework of the technical mechanics described in the chapter on linear visco-elasticity. Since, however, for frozen soil a linear-viscous stress-strain relation exists only very rarely, the rheological model body (just to remain with the Maxwell-model) needs to be converted in a way that the viscosity model itself is time-dependent. This, however, can no longer be solved by the known mathematical methods e.g. by means of the Laplace-transformation. Therefore, frequently the functional description of the time-dependent creeping properties by a potential law in form of a creep equation ($\dot{\varepsilon} = A\sigma^B$) which then enables analytical solutions for axially symmetrical systems is recommended.

For B = 1, i.e. linear visco-elasticity, the stress distribution corresponds to the elastic one, while $B = \infty$ corresponds to the ideal plastic conditions. A diagram for the dimensioning of freeze shafts with (Fig.8) viscous frictionless material properties can be established. It became standard to indicate the ratio d/a, i.e. the required frozen-wall thickness relative to the inner radius of the frozen wall. In this case the time-dependent uniaxial strength K (t) is considered besides the differential pressure $p-p_o$. It is obvious that with small p/K ratios in a linearly viscous case with no regard to hydrostatic stress level considerable wall thicknesses become necessary.

Besides the uniaxial compressive strength K(t) the influence of friction for a correct structure-mechanical description of the frozen soil is decisive. For this reason recently a dimensioning model (5) was presented in which the viscous properties of a non-Newtonian fluid are complemented by additional friction parameters of the material. In this case the wall-thickness ratio d/a is besides the time-dependent cohesion also a function of friction angles. Fig. 9 shows a design chart for a stress/strain nonlinearity of B = 2 including the time-dependent design equations.

(5) Klein, J.: "Die Bemessung von Gefrierschächten in Sandformationen vom Typ B=2 mit Berücksichtigung der Zeit." Glückauf Forschungsh. 42 (1981) S. 112/20.

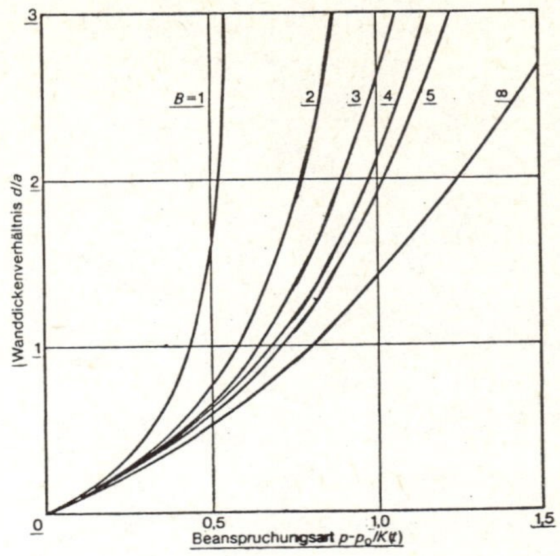

Figure 8. Design chart for viscous frozen soils

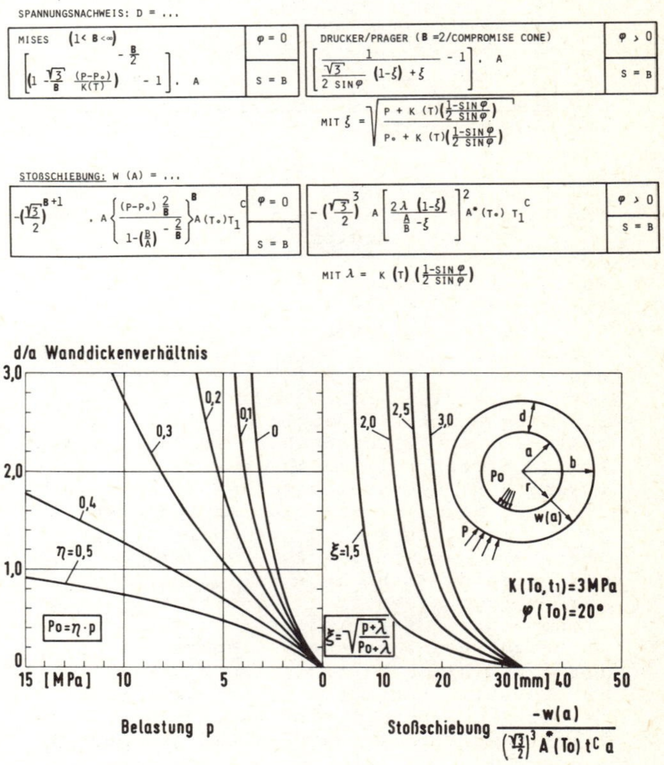

Figure 9. Design chart for viscous frozen soils with friction and the time-dependent formulae

Time-Independent

For time-independent calculation methods it is assumed that the frozen body from a certain stress level does fail and that so-called plastification occurs. In this method the time factor is not allowed for, and statements as to the wall deformation in time are not possible either. Nevertheless this stress calculation is the most widespread method for dimensioning of frozen wall cylinders. Triaxial pressure tests for determination of the characteristic values of friction angle and cohesion are prerequisites for a comprehensive material description. The increase of the maximum deviator stress $\sigma_1 - \sigma_3$ with increasing hydrostatic lateral for friction material becomes evident.

Since in this context also the terms "cohesion" and "friction angles" are mentioned, the essential features of two failure criteria by means of the Mohr-diagram are explained. On top of Fig. 10 you see the

Tresca-criteron within the so-called three-dimensional stress space. The hexagon, as cross-section, does not change over the length, however, the Mohr-Coulomb criterion - three-dimensionally - forms a pyramid.

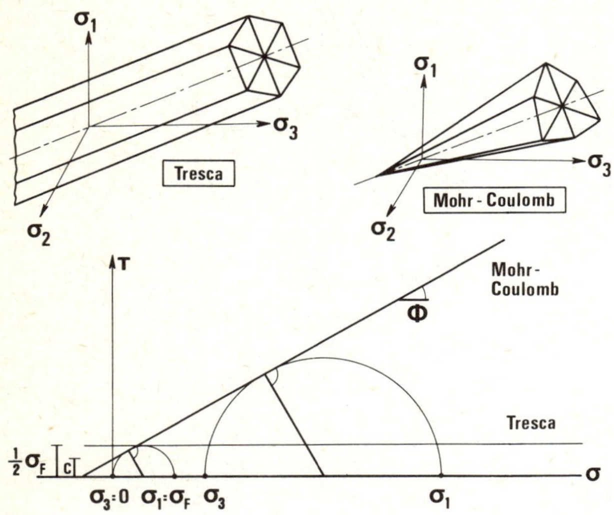

Figure 10. Failure criteria of Tresca and Mohr-Coulomb

The cross-sections or - more specially - the deviator planes increase in this case with increasing aperture or hydrostatic stress level. As shown on the bottom sketch, also the admissible stress deviator increases with increasing lateral stress, when considering a friction angle. If now a failure criterion is selected which is inappropriate to the frozen soil matter the resulting strength values are consequently wrongly defined. Normally all rock strata even if frozen, exhibit a friction angle greater than zero, except for water-saturated clay may be zero. With the existing variety of stress hypotheses it is important to select those constitutive models for calculations which represent in the best-possible way the material characteristics of frozen soils.

Usually the thick frozen wall (Fig. 11) is with regard to the stresses subdivided into an inner plastic zone and an outer elastic one. On this basis also Domke (6) developed his well-known approximation formula in 1915 supposing arbitrarily that the limit zone is just situated at the mean geometric distance $s = \sqrt{a \cdot b}$.

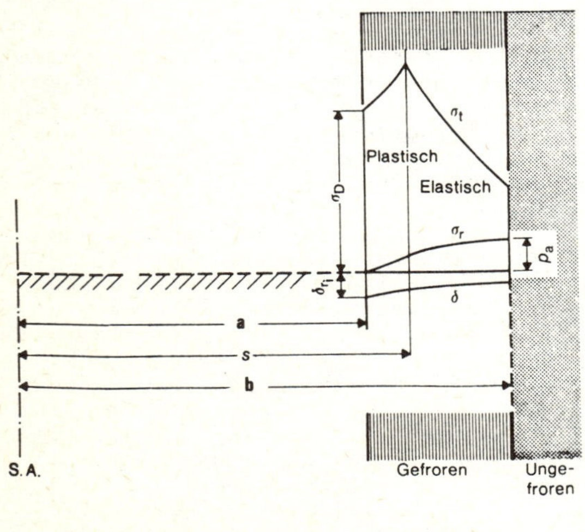

Figure 11. Plastic/elastic stress distribution in a cross section of a frozen wall

(6) Domke, O.: "Über die Beanspruchung der Frostmauer beim Schachtabteufen nach dem Gefrierverfahren." Glückauf 51 (1915) S.1129/35.

The uniaxial strength K as characteristic material value of the frozen strata is determined as failure condition according to Tresca, i.e. difference between greatest and smallest main stress without consideration of the friction angle and of the average longitudinal stress. In this way for the plastic part of the frozen cylinder a hypothesis is used considering only the cohesive character of the soil sufficiently. The frictional properties of frozen soils are contained implicitely in K, however, do not result in increased strength in case of higher hydrostatic stress conditions.

This disadvantage of the Domke assumption which, by the way corresponds, to reality in the very minority of cases, becomes apparent in particular if the strata contain predominantly non-cohesive soils. In the following two examples illustrated by Fig. 12 the yield point is not defined before, and instead of the Tresca-criterion the Mohr-Coulomb-criterion is used.

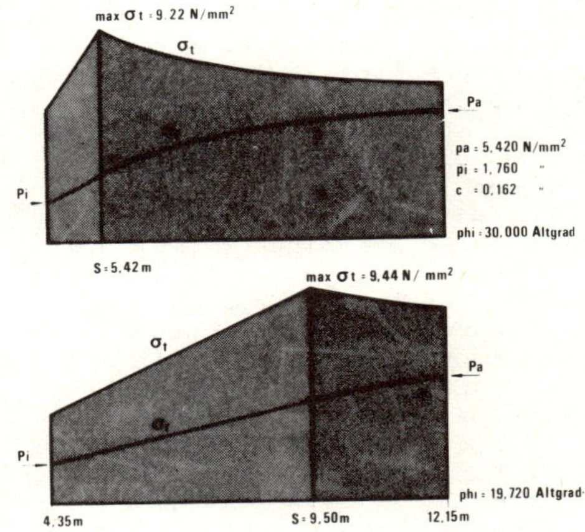

Figure 12. Examples with Mohr-Coulomb criterion for the plastic zone

An outside pressure p_a of 5.42 N/mm² prevailed and a lining resistance p_i = 1.76 N/mm² is allowed for. In this case for frozen sandstone with a friction angle of 30° a relatively small plastic zone of 4.35 m to 5.42 m results. Even with a smaller friction angle a stable equilibrium state is reached due to the inside pressure. Obviously, the plastic zone increases to a radius s = 9.50 m. The fact that the friction angle accounts for a considerable share in the supporting behaviour of the system as well as the advantages of an additional inner support become obvious.

If, as Domke did, the limit radius $s = \sqrt{a \cdot b}$ is chosen, for the plastic zone, however, the Mohr-Coulomb-criterion, considerably more favorable dimensioning data are obtained. As shown on Fig. 13 the required wall-thickness ratio d/a decreases as a function of the friction angle. In this way, even without a lining with supporting capacity, acceptable frozen-wall thicknesses are obtained even for depths where, according to the Domke-formula, only unpracticable results were calculated. The simulation could be obtained after numerical iterations of the transcendental equation in Fig. 14. In order to avoid this volume of effort in practice the technically relevant value range was approximated by a square function (7) dependent on P/K in analogy to Domke. Even though a higher-grade polynom would approximate a larger spectrum even better, this idea was rejected to simplify matters. The approximation approach - from the engineering point of view with acceptable deviations -,

(7) Klein, J.: "Zur Berechnung der erforderlichen Dicke der Frostwand im Gefrierschacht."Vortrag am 17.11.1981 im Rahmen des "Bergmännischen Kolloquiums" am Institut für Bergbaukunde I Aachen.

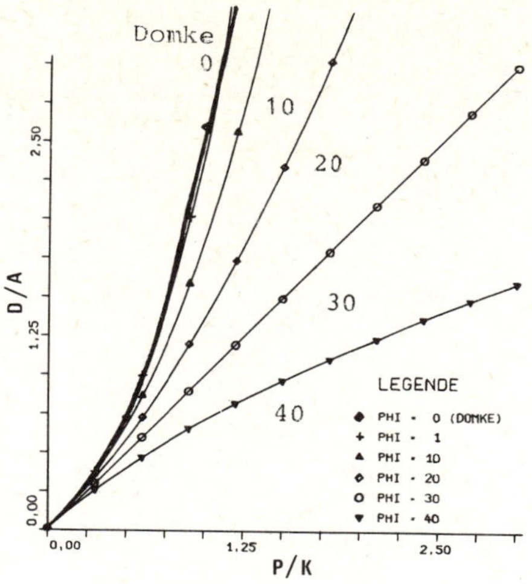

Figure 13. Design chart with Mohr-Coulomb criterion for $s=\sqrt{a.b}$

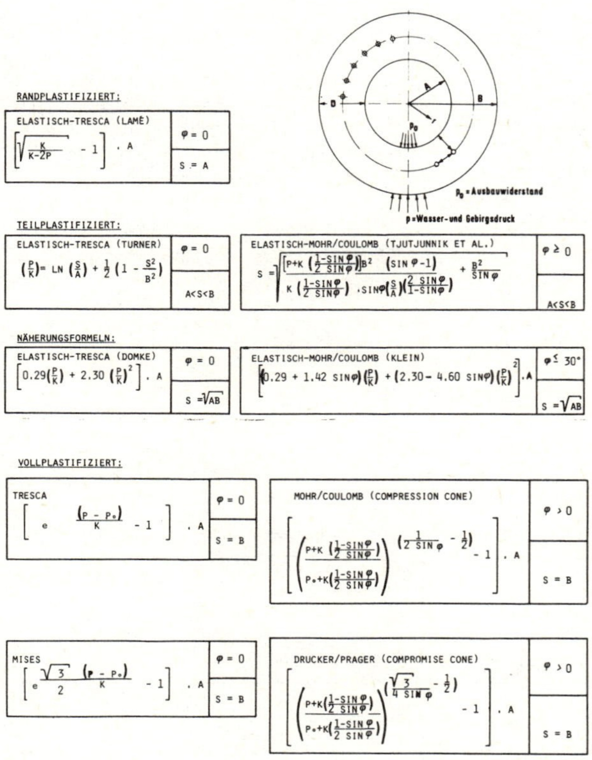

Figure 14. Survey of formulae for freeze shaft design

contains additional terms for the friction angle of the frozen soils and which enable a better dimensioning of the frozen cylinders. Finally looking at the large variety of design possibilities the applicants may direct the attention to the safety aspects, please, when dimensioning the frozen wall, because the freeze shaft "doesn't care a bit" about where the engineers assume the most-favorable limit radius s of plasticity to be.

CONCLUSIONS

For shaft sinking very often stable-frozen strata structures in form of thick-walled cylinders need to be created. The layout of temporarily frozen soil-support systems is a complex task which implies questions requiring in turn fundamental knowledge of artificial ground freezing technology. The basic concept in ground freezing is the removal of heat from the strata in question causing the pore water to freeze and to form a cylinder-shaped frozen body in the earth. Excavation and other work can then be done safely inside a barrier of a strong and watertight frozen wall.

Some of the major considerations involved in the design of ground freezing structures include comprehensive thermal and rheological calculations. In order to simplify the description of the complex thermodynamic relations the calculations are split up in temperature calculations and those of stress/strain relations. By using assumptions along with heat conduction theory, design calculation for the volume of energy transfer for freezing purposes, the time required for freezing and the refrigeration plant capacity can be determined with sufficient accuracy for many cases.

For the structural analysis, design principles for thick-walled cylinders are of foremost importance. Formulae based on the theory of ideal linearly elastic solids are not sufficient, because frozen strata behave visco/plastically. Therefore a general procedure for designing a retaining structure of frozen soil is to determine, in a first phase, the wall thickness, based on the appropriate material idealization.

Today mine shafts in West Germany sunk by the freezing method are lined by a support system which is allowed to slide vertically, i.e. coaxially to the natural shaft wall in order to realize planning of mining operations near the shaft. This method implies the installation of the shaft lining being realized in two phases. During shaft sinking a preliminary/outer and relatively flexible lining is set. When the final depth of the shaft is reached the final support is brought in from bottom to top. This lining is composed of reinforced-concrete, steel-sheet segments, and an asphalt mass coaxially between the preliminary and the final inner lining.

In freeze shaft design different yield criteria known from the plasticity theory may be used in order to determine the strength of the frozen soil. In great depths and if longer periods of construction are involved, a staticly bearing support is significant not only from the viewpoint of cost-effectiveness. Recently, time-dependent constitutive models including the load-carrying capacity of the preliminary support, were developed. In German freeze shaft design Domke's formula is shelved or modfied even for time-independent strain calculations. The influence of internal friction gains access to modern design procedures for frozen retaining walls, and strength criteria like Mohr-Coulomb or other hydrostatic influenced principles are favored.

The biggest shaft ever built in accordance with these new considerations is the 1060 m deep Voerde shaft of Ruhrkohle AG near the Rhine. This shaft is still being sunk. Freezing is practised over 600 m of depth through water-bearing strata. First phase freezing took ten months before sinking could start in April 1981. The preliminary lining consisting of concrete blocks varies over the depth according to the soil characterics. A separate research program including temperature, inclinometer, deformation, and ultrasonic-wave measurements is synchronized to the project till the foundation of the freeze shaft will be set in late 1982.

ACKNOWLEDGEMENT

The measuring and soil-physical investigation program for the Voerde shaft is covered by a research project of Ruhrkohle AG, Bergbau AG Niederrhein, and Bergbau-Forschung GmbH, sponsored by the federal ministry for economics within the project "Tiefe Gefrierschächte".

MEASUREMENT OF STRESSES, STRAIN AND TEMPERATURE IN CONCRETE IN SHAFTS AND INSETS

P.F.R. Altounyan

Department of Mining Engineering
University of Newcastle upon Tyne
Newcastle upon Tyne, NE1 7RU, England.

SUMMARY. An instrumentation system has been developed to measure stress, strain and temperatures in concrete shaft and inset linings, utilising vibrating wire strain gauges. Details of the system are described and specimen results are presented.

INTRODUCTION

Design of concrete linings for underground structures such as shafts and insets requires information on radial geostatic and hydrostatic stresses and on temperature distributions in the hydrating concrete. Temperature measurements are particularly important where linings are in contact with frozen ground where groundwater freezing is used to control shaft water during construction.

Construction of ten shafts in the Selby Coalfield, all passing through frozen Bunter Sandstone, provided an opportunity to obtain design data for future shaft sinking operations. Instrumentation was developed to monitor the most important parameters.

INSTRUMENTATION

Instrumentation in shafts poses several problems the most important of which is inaccessibility. To overcome this and to keep down costs, two important decisions were made in the design of a shaft lining system. The first was to use the same transducer throughout, and the second was to multiplex signals at source to enable continuous monitoring through a single cable.

For reasons of stability, reliability and intrinsic safety the vibrating wire gauge was chosen as the basic unit. This was invented by Potoki (1958) and has since been developed by Tyler (1968) and others and is widely used in civil engineering.

Table 1. Specifications for vibrating-wire transducers

	PIEZOMETER	STRESS CELL	TEMPERATURE GAUGE	STRAIN GAUGE	
Supplier	Soil Instruments Ltd	Soil Instruments Ltd	Gage Technique Ltd	Gage Technique Ltd	Strainstall Ltd
Cost Jan '82	£260	£340	£26	£26	£32
Dimensions	38mm dia x 240mm	300mm diameter	65mm length	140mm length	140mm
Vibrating Wire length	70mm	70mm	63mm	140mm	140mm
Wire Diameter	0.2286mm	0.2286mm	0.254mm	0.254mm	0.254mm
Gauge Factor	7×10^{-5}	7×10^{-5}	6.9×10^{-5}	3×10^{-3}	3×10^{-3}
Temperature Coefficient					
- Free	Insensitive	2.14×10^{-4}	Not Applicable	1×10^{-3}	4.5×10^{-4}
- In-situ	Insensitive	Not Known	Not Applicable	11×10^{-6}	11×10^{-6}
Range	0-8 MN/m^2	0-8 MN/m^2	100°C	3000 microstrain	3000 microstrain
Resolution	.015 MN/m^2	.015 MN/m^2	0.15°C	0.5 microstrain	0.5 microstrain
Type	High entry acoustic	Mercury filled circular	To order	5½" steel embedment Type T/E/S	Type PC 657

Variations of the basic vibrating wire gauge to measure strain and temperature, and transducers for attachment to pressure cells are commercially available. Specifications as for various systems are given in Table 1.

The basic vibrating wire strain gauge comprises a flanged tube containing the vibrating wire. The gauges are installed free or precast into briquettes in the shaft lining. A typical installation is illustrated in Figure 1.

practically unlimited number of transducers to be monitored via a single cable.

The cable is four core screened and armoured for protection. Trials indicate no significant signal loss over 2000m cable length.

The readout is a modified version of a conventional unit providing additional signals for the multiplexer and a count output to control lines for the computer.

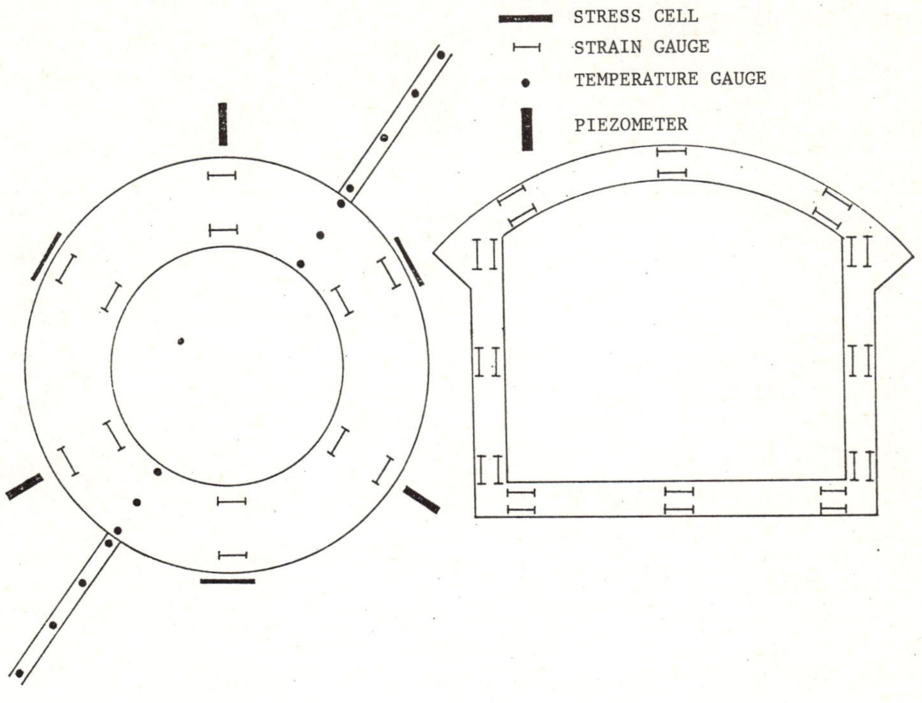

Figure 1. Schematic layout of typical shaft and inset instrumentation

The vibrating wire piezometer consists of a cylinder with porous ceramic surfaces which allows water to pass through and act on a diaphragm which has a vibrating wire transducer attached. The piezometers are installed in boreholes surrounded with bentonite and sealed with grout.

The vibrating wire pressure cell consists of a hollow steel disc containing a small amount of mercury. Pressure on the faces of the cell forces mercury against a diaphragm which has a much higher stiffness (>200 GN/m^2) than rock and may therefore be said to record stress directly. The cells are installed at the concrete rock interface to measure radial stresses.

The vibrating wire temperature gauge consists of a brass tube containing the vibrating wire. Differences in thermal expansion between brass and the steel wire make the gauge temperature sensitive. The gauge is contained in a tube filled with powder to insulate the transducer from external forces, yet not interfere with heat flow. The gauges are installed in rock along boreholes filled with grease and in concrete attached to glass fibre rods (Figure 3).

MONITORING EQUIPMENT

The system for monitoring is shown schematically in Figure 2. This had to be designed from scratch and required considerable R. and D. effort. This has now been completed and the system is currently undergoing I.S. certification.

The monitoring system comprises four main elements: multiplexer, cable, readout and data logger.

The multiplexer behaves as a remote controlled rotary switch selecting up to 20 channels in turn. Multiplexers can be cascaded together allowing a

In addition two interfaces are required between readout and shaft cable and readout and computer.

The data-logger is a microcomputer (HP 85) with V.D.U., printer and tape storage. As well as being cheaper than conventional data-loggers a microcomputer is more flexible and capable of providing useful output on site. (Figure 4c)

INSTALLATIONS

A summary of instrumentation installations either planned or in progress in the Selby coalfield is given in Table 2. The Selby project has been described in detail by Forrest (1978). The geology in the region of the shafts comprises 200-300m of Bunter Sandstone below 20m of glacial drift. This overlies the typical Permian succession of Upper Permian Marls, Upper Magnesian Limestone, Middle Permian Marls, Lower magnesian Limestone, Lower Permian Marls and Basal Permian Sandstones to a depth of about 500m where, the Permian overlies unconformably the Middle Coal Measures. The strata dips at about 5° and thins to the East, so that at the Gascoigne Wood drift, the Bunter Sandstone is virtually absent and the Coal Measures are 150m below the surface. The area is flat and about 10-20m A.O.D., with groundwater levels close to the surface.

Lining design problems, are caused largely by the presence of Sandstone aquifers, namely the Bunter Sandstone, the Basal Permian Sands and several Coal Measures sandstones. Subsidiary design problems have been related to evaporite layers in the Middle Permian Marl. The Bunter Sandstone has an average mass permeability coefficient measured by borehole inflow packer premeability tests of about 5×10^{-6} m/s and a virtually identical horizontal laboratory

Table 2. Summary of site details.

SITE	DEPTH (M)	GEOLOGY	LINING SPECIFICATIONS	INSTRUMENTATION	INSTALLATION DATE	COMMENTS
Riccall No 2 Shaft	325	Middle Permian Marl - Halite	Concrete 1.2m thick σ_c = 45 MN/m^2	4 temperature, 12 strain, 6 hydraulic pressure cells anchor extensometers	March 1980	Proving section described by Gilbert (1982)
Riccall No 2 Shaft	540	Coal Measures - Woolley Edge Sandstone	Concrete 1.2m σ_c = 60 MN/m^2	10 temperature, 12 strain 6 piezometers, 3 stress	To be Installed	Full hydrostatic head, but low permeability relieved by pumping until lining fully tested.
Whitemoor No 2 Shaft	240	Bunter Sandstone	Concrete 0.6m σ_c = 45 MN/m^2	12 strain, 3 stress, 3 piezometers, 28 temperature	March'81	First site trail for multiplexer which failed. Temperature results lost. Ground frozen. Full hydrostatic level. Results show close relationship between stress, strain,and water pressure
Whitemoor No 2 Shaft	265	Bunter Sandstone	Concrete 0.6m σ_c = 45 MN/m^2	28 temperature	April'81	Successful repeat of above failure. Ground frozen. Results show complete hydration of concrete lining.
Whitemoor No 2 Shaft	440	Lower Magnesian Limestone	Concrete 1.050m σ_c = 50 MN/m^2	10 strain, 10 temperature	To be Installed	These two experiments are designed to tie in with laboratory & computer modelling work to build up a detailed understanding of the behaviour of concrete shaft linings in the first 28 days.
Whitemoor No 2 Shaft	470	Lower Magnesian Limestone	Concrete 1.05m σ_c = 60 MN/m^2	10 strain, 10 temperature	To be Installed	
North Selby No 2 Shaft	234	Bunter Sandstone	Concrete 0.6m σ_c = 45 MN/m^2	12 strain, 3 piezometer, 3 stress, 4 temperature	November'80	Identical to Whitemoor Bunter trial; results compatable
North Selby No 2 Shaft	350	Middle Permian Marl - Red Marl	Concrete 1.050m σ_c = 45 MN/m^2	12 strain, 3 stress, 4 temperature	June'81	Possibility of squeezing ground; results show zero strains.
North Selby No 2 Shaft	550	Coal Measures - Ackworth Sandstone	Concrete 1.2m σ_c = 60 MN/m^2	12 strain, 6 piezometer, 10 temperature, 3 stress	To be Installed	Similar to Riccall-Woolley Edge.
North Selby No 2 Shaft	420	Lower Magnesian Limestone	Concrete 1.2m σ_c = 45 MN/m^2	12 temperature	July '81	Heat of hydration trial on new concrete mix "Cemsave". Results show close comparison with numerical models & predicted temperature rise.
Stillingfleet No 1 Inset	662	Coal Measures - Barnsley Seam	Reinforced Concrete 1m σ_c = 35 MN/m^2	90 strain, 18 stress 6 temperature	March-July '81	Results described by Borsah et al, 1982, strains lower than predicted.
Whitemoor No 2 Inset	990	Coal Measures - Barnsley Seam	Reinforced Concrete 1m σ_c = 35 MN/m^2	120 strain, 6 temperature, 12 stress	To be Installed	Level of inset not fixed.

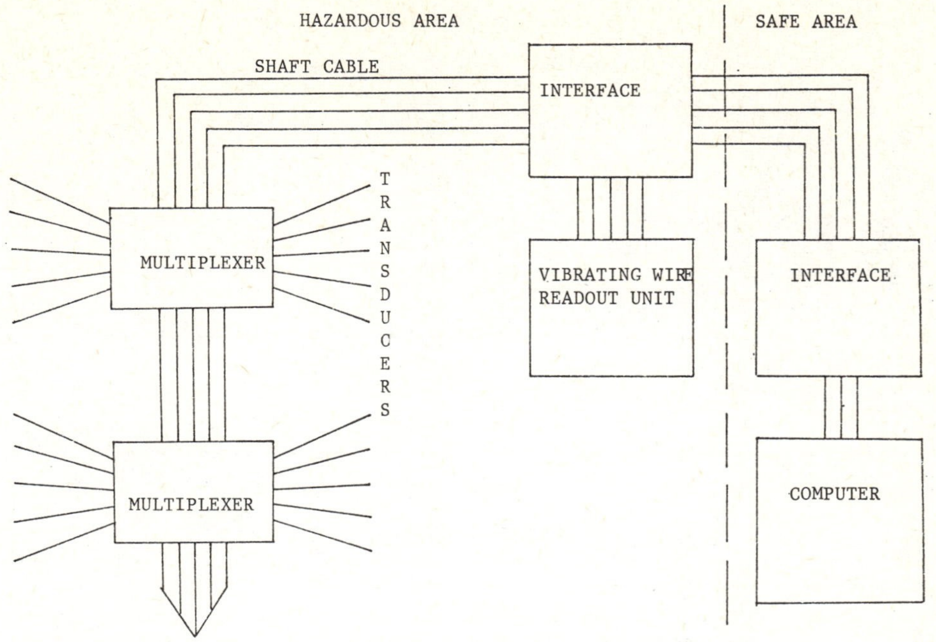

Figure 2. Schematic diagram of monitoring system

Figure 3. Temperature gauges installed in frozen shaft wall inside the shuttering.

permeability, together with an average porosity of 34%. Groundwater freezing (Wild & Forrest, 1981, Forrest & Black 1979) has been used to control groundwater during construction in the Bunter Sandstone in the shafts and during construction in the Basal Permian Sands in the Gascoigne Wood drifts. Grouting and well pointing has been used elsewhere.

The first instrumentation installations were in the Gascoigne Wood drifts and these have been described by Altounyan and Farmer (1981). These and a subsequent installation at Riccall No. 2 shaft (Gilbert, 1982) in the halite used direct reading equipment, and illustrated the need for remote monitoring in shafts. Subsequent installations which have been completed at Whitemoor No. 2 shaft (Altounyan et al 1982), and North Selby No. 2 shaft (Bell 1982) in the Bunter Sandstone, and at Riccall No. 2 shaft have utilised the remote monitoring system. The instrumentation installed in Stillingfleet No. 1 inset (Borsah et al, 1982) was read manually.

RESULTS

Typical site printouts are illustrated in Figures 4 and 5. Figure 4 shows a typical printout of temperature with time during hydration of a concrete lining against an icewall and also illustrates how this can be presented in isometric projection using computer graphics. These results illustrate the long term interaction between the hydrating concrete and the frozen shaft wall. Note that even at the rock concrete boundary temperatures stay above 4°c for at least 20 days.

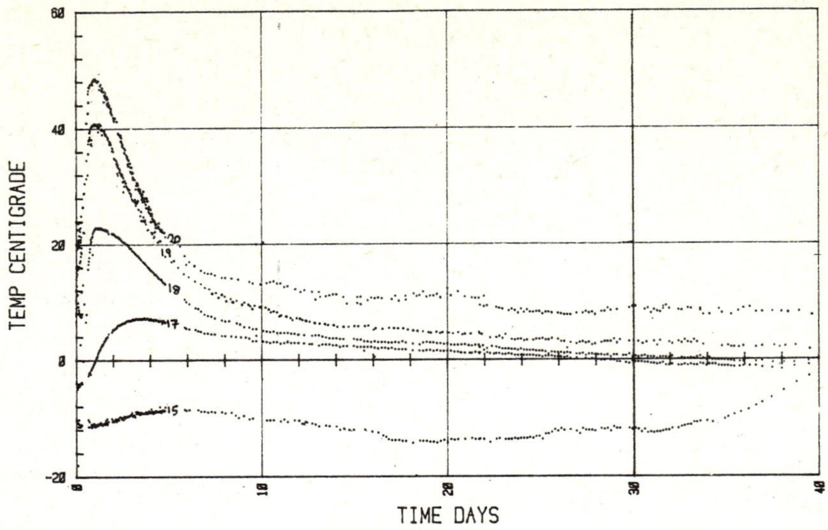

Figure 4(a). Printout of temperature against time for an array of 6 temperature gauges at Whitemoor No. 2 Shaft (see Figure 1).

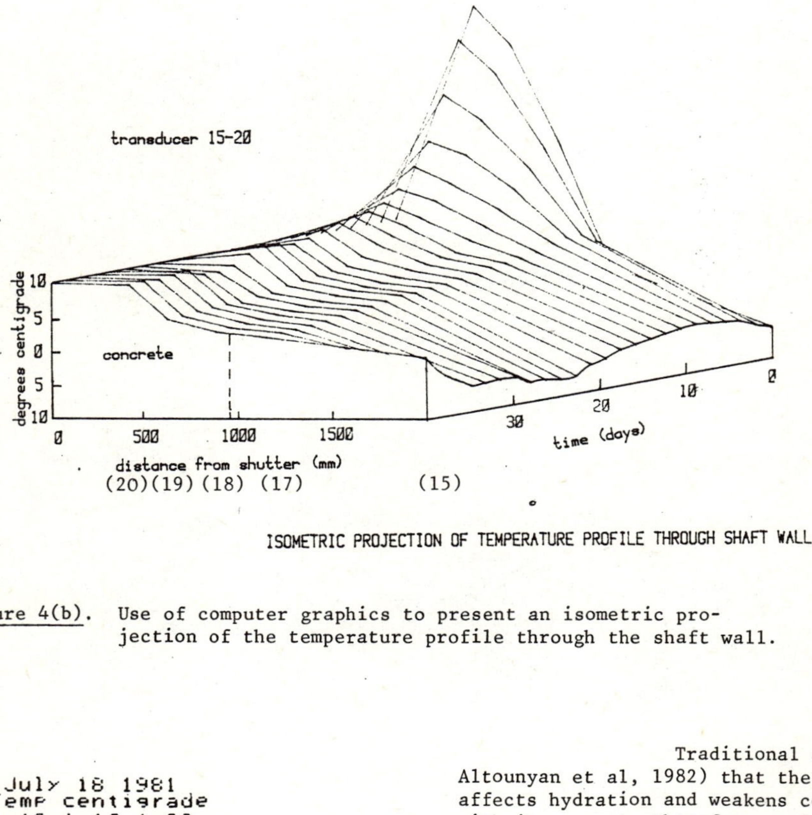

Figure 4(b). Use of computer graphics to present an isometric projection of the temperature profile through the shaft wall.

Whitemoor No2 July 18 1981
Temp centigrade

Gauge	08	12	16	20
1	10.8	10.6	10.6	10.5
2	12.3	12.3	12.1	11.9
3	14.9	14.8	14.5	14.6
4	14.6	14.8	14.9	14.9
5	15.8	15.7	16.1	16.3

Figure 4(c). Typical printout of data at 4 hourly intervals during temperature monitoring.

Traditional assumptions (see Altounyan et al, 1982) that the presence of an icewall affects hydration and weakens concrete in contact with it are not, therefore, valid.

Figure 5 illustrates a typical printout of strain and piezometric pressure during thawing of the Bunter Sandstone icewall at North Selby No. 2 shaft. An important conclusion here (see Bell, 1982) is that strains and lining pressures result largely from groundwater pressures, and that the rock - even in weak strata - has a minimal effect on lining pressure.

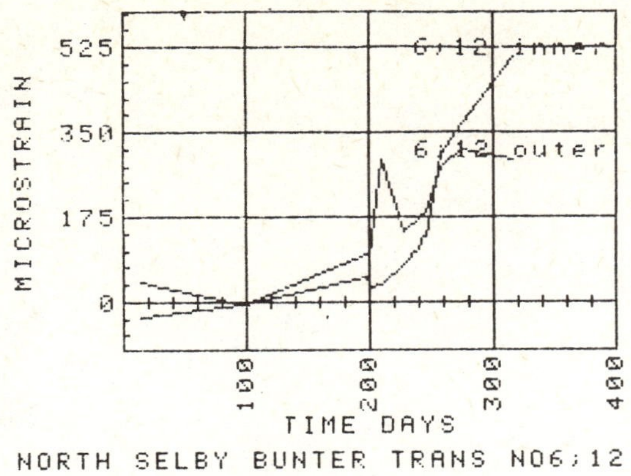

Figure 5(a). Printout of inner and outer surface strain with time during thawing of Bunter Sandstone icewall.

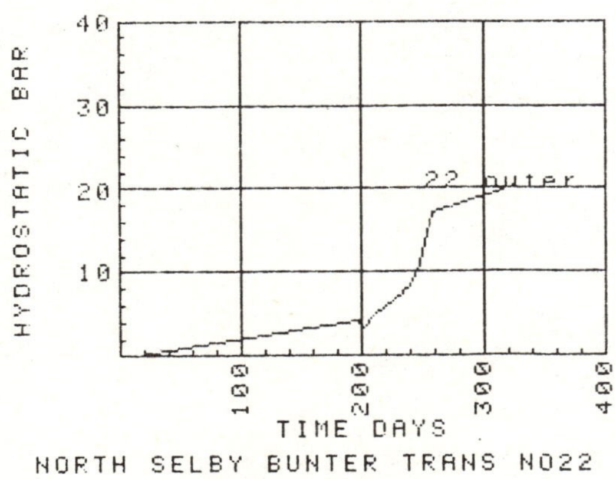

Figure 5(b). Printout of hydrostatic pressure with time during thawing of Bunter Sandstone icewall.

ACKNOWLEDGEMENT

This work is financed by the National Coal Board: Selby Project, and by the European Coal and Steel Commission (YCE 67/13307). Acknowledgements are also due to Rock and Coal Mechanics Branch, M.R.D.E. and all those associated with the Selby project.

REFERENCES

ALTOUNYAN P.F.R., BELL M.J., FARMER I.W., & HAPPER C.J. Temperature, stress and strain measurements during and after construction in concrete shaft linings. Paper to 3rd Int. Symp. Ground Freezing, Hanover, New Hampshire, June 1982.

ALTOUNYAN, P.F.R. & FARMER I.W. (1981), Tunnel lining pressures during groundwater freezing and thawing. Proc. 5th R.E.T.C., San Francisco, 784-800.

BELL M.J. (1982), Design of shaft linings in Coal Measures rock. Proc. Symp. Strata Mechanics, Newcastle upon Tyne, April 1982.

BORSAH J., SHELTON P.D., TOMLIN N. (1982). Deformation of deep shaft insets in Coal Measures rocks during construction, ibid.

FORREST W. (1978), The Selby project, Mining Engineer, 138, 237-245.

FORREST W. & BLACK J.C. (1981) Hydrogeological analysis, ground treatment and special construction techniques at Selby: Gascoigne Wood surface drift mine; Tunnelling '79,Inst. Min. Metall., 256-263.

GILBERT M.J. (1981), pressures and displacements around a lined shaft in weak rock. Proc. Symp. Strata Mechs., Newcastle upon Tyne, April 1982.

POTOKI F.P. (1958), Vibrating wire strain gauges for long term internal measurements in concrete. The Engineer, 206.

TYLER R.G. (1968), Developments in the measurement of concrete bridge structures. Report LR 189. Transport & Road Research Laboratory, Crowthorne.

WILD W.M. & FORREST W. (1981), Application of the freezing process to ten shafts and two drifts at the Selby Project, Mining Engineer, 140, 895-904.

THE DESIGN OF SHAFT LININGS

IN COAL MEASURE ROCKS

M. J. Bell

National Coal Board,
H.Q. Mining Department,
Shaft Sinking Branch,
The Lodge, Doncaster.

SUMMARY. This paper sets out to explain the design methods adopted for modern concrete shaft linings. It identifies that whilst structural strength is of paramount importance other factors such as methods of ground treatment, Chemical Resistance and Watertightness are design parameters which need careful consideration at the design stage.

Figure 1. Diagram of Selby Shafts

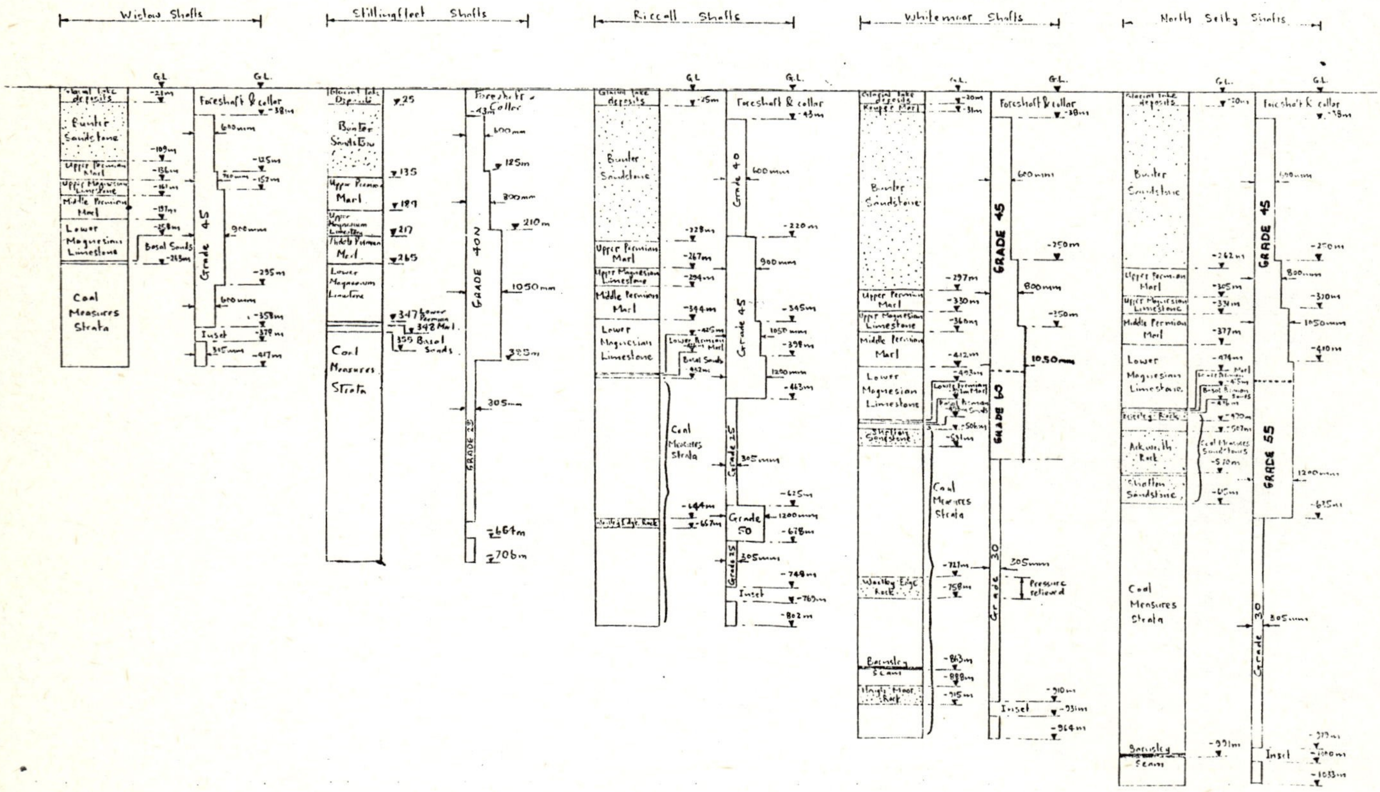

SELBY NEW MINE COMPLEX - SHAFT LINING THICKNESSES AND DEPTHS, CONCRETE STRENGTHS AND STRATA SECTIONS

FIGURE 1

INTRODUCTION

From the late sixties to the mid-seventies the N.C.B. did not commission the sinking of any new shafts, when therefore the Selby Coalfield was given the go ahead in 1976 with 10 no. new shafts to be sunk a review of the previous shaft designs and the problems that had been encountered was undertaken. It was clear that the new coalfield was significantly deeper than most of its contemporaries, being a concealed coalfield, and that over most of the area major aquifers in both the Permian and Coal Measures existed. (See Fig 1).

Initial calculations using only water pressures and Lamé thick cylinder formula indicated that a concrete lining was possible provided a characteristic concrete strength of 40N or $45N/mm^2$ was achievable, coupled with a water cement ratio of 0.45 or lower to stay with the Code of Practice for Sulphate Resistance (4). The design for the deeper part of the later shafts was left in abeyance until the experience on the initial shafts had shown whether the concrete quality required could be achieved. The main design parameter apart from Strength and Chemical Resistance was water tightness, the winding equipment in the shafts being susceptible to water and the long term costs associated with pumping water being high and also difficult to dispose of since most of the water from the deeper aquifers would contain high concentrations of chloride.

Looking at the history of some of the concrete lined shafts sunk in the sixties the problem of concrete attack by highly chlorinated waters seemed to be the most significant, apart from isolated ones where structural failure had occurred due to poor quality control of the construction.

It was decided that a comprehensive concrete specification would be issued and to ensure that this was adhered to, close supervision on site would be exercised. This it was thought, would allow concrete linings to be used throughout the whole development.

DESIGN LOADINGS

The rule in British coalmining is to leave a pillar of coal to protect the shaft from ground movements induced by subsidence, the main design loads are therefore the hydrostatic load, and the rock load.

Hydrostatic Load

Generally this is taken as the measured pressure in the aquifer under consideration from tests carried out in boreholes drilled to determine strata properties. However, in certain instances the pressure recorded has been low and an estimate of the pressure from the static head of water to the surface has been made. In any strata containing water the hydrostatic load is by far the most predominant.

Calculations are then made to determine the external pressure on the lining and the consequential stresses in the concrete

i.e. $P = H \times .0098061\ N/mm^2$

P = external pressure N/mm^2

H = hydrostatic head (metres)

Lamé $$f_c = 2p\,\frac{(t+R)^2}{t(t+2R)} \quad \text{or} \quad t = R\left[\sqrt{\frac{f_c}{f_c - 2p}} - 1\right]$$

f_c = Concrete stress on inner face

t = Concrete thickness

R = Internal radius of shaft

The concrete stress thus calculated is multiplied by a Factor of Safety of 2.73 and divided by an age factor of 1.2 (Ref. 2 &3) thus giving a theoretical safety margin of 2.275 at 28 days against full hydrostatic load and 2.73 at 6 months.

Rock Loads

Where no hydrostatic load exists only the rock load needs to be calculated. Where a hydrostatic load exists there is some doubt as to whether the rock load should be added to this load or ignored. In general terms in this case the rock loads calculated are small in comparison with the hydrostatic and they have been added in total. Experiments now being carried out will determine whether this was a reasonable judgement or not.

The calculation of the pressure applied by the rock to the shaft lining has been the subject of much discussion and many papers have been written on the subject. I do not propose to discuss the merits or otherwise of these at this point since our instrumentation has not yet proved what, if any, loads exist.

The theory used to calculate the stresses on the lining is the one proposed by Coates (5) which is based on a zone of ground around the shaft being in a state of plastic equilibrium with a stress relationship governed by the Mohr Strength Theory. This approach may seem conservative but in the absence of any firmer data seemed to be sensible. The stiffness and hence deflection of the lining is equated to the load applied by the rock. The stiffer the lining becomes the greater the load applied which seems to fit very well with the practice and experience in coal measures. The basic equations are as follows:

$$P_c = \frac{dE_c\left[1 - \left[\frac{r_i}{r_o}\right]^2\right]}{2r_i}$$

$$P_1 = \frac{2\gamma a}{10^3\,A\,(f-2)} \times \left[\frac{f-2}{f}\right]^{f/2}$$

where:-

$\varnothing$ = angle of internal friction of the rock (degrees) a = radius of excavation (= r_o) (mm)

d_f = radial deformation of the surface of excavation(mm)

d = radial deformation of shaft lining (mm)

ν = Poissons ratio for the strata

E_f = mean failure strain for the strata (%)

E_c = modulus of elasticity for the lining (N/mm^2)

r_i = inner radius) of the shaft lining (mm)

r_o = outer radius)

γ = density of the strata (N/mm^3)

P_c = lining reaction (N/mm^2)

P_1 = rock pressure (N/mm^2)

A = the limit of proportionality = Kd (max. value tan ϕ)

K = Constant of proportionality

$d_f = \nu E_v \times \Delta r$ where $\Delta r = r_1 - a$

f = the (Rankine) ratio of vertical to lateral stress

$$r_1 = a\left[\frac{f}{(f-2)}\right]^{\frac{1}{2}} \text{ where } f = \tan^2\left(45 + \frac{\phi}{2}\right) \text{ OR } f = \frac{1 + \sin\phi}{1 - \sin\phi}$$

Out of Balance Loading

The possibility of any out of balance loading was considered remote since every part of the shaft lining is backwall grouted, however, calculations as recommended by Link Lutgendort and Stoff (7) have been carried out as a check, using the following formula:

$$\text{Tangential stress} = F_t = -\frac{P_z r_o}{2 \times F}\left[\left[1+\frac{t}{2r_s}\right] \pm \frac{1}{3}\left[1+\frac{6r_s}{t}\right]\left[\frac{V}{V-1}\right]\right] \quad (1)$$

$$\text{where } V = \frac{3EI}{P_1 r_o r_s^2}$$

and F_t = tangential stress (N/mm^2)

P_1 = uniform load (N/mm^2)

P_z = non-uniform load as a % of P_1 (N/mm^2)

E_c = Youngs Modulus of lining (N/mm^2)

r_o = radius to outside of lining (mm)

r_s = radius to centroid of lining (mm)

t = wall thickness (mm)

F = lining cross section/unit length(mm^2/mm)

(= t x $\frac{\text{unit length}}{\text{unit cross section}}$)

$$I = \frac{bd^3}{12} = \frac{t^3}{12} \quad mm^4$$

N.B. In formula 1 above a negative sign indicates compression and a positive sign tension.

OTHER DESIGN CONSIDERATIONS

Ground Treatment

Two types of ground treatment generally are used to reduce or eliminate any water ingress into the excavated shaft, they are freezing and grouting.

Freezing

This technique increases the strength of the ground significantly whilst ice is present but can reduce the long term strength of the ground on thawing. It is important in any shaft design to know if freezing is to be employed since particularly with concrete lined shafts the cold conditions can affect the hydration process of the concrete. In modern shafts this is particularly significant where the concrete is poured directly onto the frozen strata. The general rule is that provided the concrete can reach a strength of some $5N/mm^2$ before the ice returns to the concrete then the concrete will remain intact and even though its temperature subsequently drops below 0^oC its gain of strength will be slow or even stop, until such time as the temperature rises again.

In the early stages of the Selby Project we were very concerned about this aspect and specified minimum concrete thicknesses of 600mm in the frozen ground to (a) provide a substantial quantity of concrete and hence heat to overcome the effects of the low temperature and (b) provide a layer of sacrificial concrete about 75mm thick. Recent experiments have proved that it takes some 2 weeks for sub zero temperatures to return to the concrete and that the heat of hydration from the concrete affects the ice wall some 1.2 metres back from the interface.

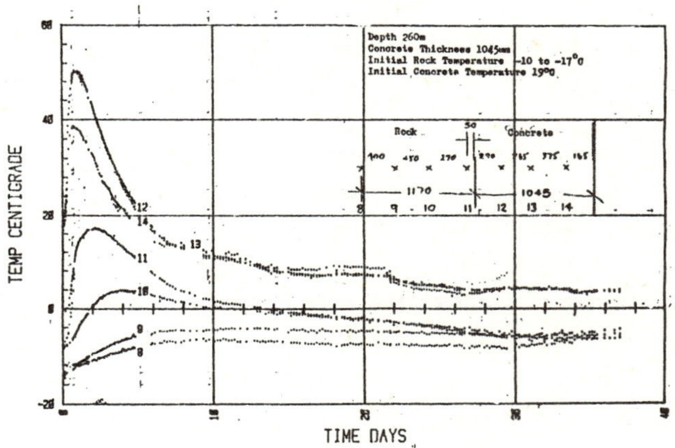

WHITEMOOR NO 2 SHAFT: TRANS NO 8-14

Figure 2 Temperature Profiles in Frozen Ground.

Grouting

Various types of grout are used to stem the flow of water into unlined shafts. These have to be matched to the type of strata involved and are either particulate grouts (cement or clay) or non-particulate (chemical i.e. organic polymers, resorcinal or silicate-metal salt combinations).

Shafts sunk through grouted ground tend not to need any special design criteria, although in certain cases, backsheeting of the excavated faces to lead away excessive water does reduce the bond between the rock and the concrete, and cribs may be needed to support the self-weight of the lining.

Chemical Resistance

In recent years deterioration of existing concrete shaft linings in the Midlands and Yorkshire have been seen. This is due to attack by the chemicals in the ground water on the concrete itself. The exact nature of the attack is not understood fully but certain conditions seem to be needed to initiate the chemical reactions with the concrete.

1. Strata water at high pressure.
2. Strata water at temperatures of over 10°C.
3. High concentrations of chlorides.
4. Large volumes of air being swept past the concrete surfaces.
5. Total dissolved solids within the strata water approaching saturation.

Coal measure waters have all the above, a typical analyses would be:

	mg/litre
Calcium	19,200
Magnesium	5,000
Sodium	48,000
Chloride	123,000
Sulphate	0
Barium	150
Strontium	960
Free CO_2	25
pH	5.6

Figure 3.

The possible causes of deterioration as a result of these could be:

1. Evaporation of water causes the salts within the water to be concentrated on or just below the surface of the concrete, inducing high internal pressures and hence spalling of the concrete. In practice where water can be seen as weepers little damage is caused, possibly because the salts are kept in solution. This affect would only be expected where the concrete quality is low.

2. Concentrated Magnesium or Calcium and to a lesser extent, Sodium Chloride solution may react with Tricalcium Aluminate in Portland Cement to form Chloroaluminates. This reaction product has a larger volume than the compounds it replaces leading to expansion and disruption of the concrete. Since the Tricalcium Aluminate content is small the likelihood of this being a major contribution is also small. Free Lime and Gypsum are more soluble in sodium chloride than water and there may be a tendency for free lime and gypsum to be leached out depositing Calcium Carbonate.

3. Certain aggregates may be susceptible to attack by alkalis from either the cement or the strata water. This expansive reaction in the aggregates can cause cracking of the concrete over long periods.

4. Quantities of dissolved Carbon Dioxide in the strata water can react with the free lime to form Calcium Carbonate or the more soluble Calcium Bicarbonate which may leach out leading to cavities within the concrete.

The possible ways to design a lining against chemical attack would be:-

1. To install a waterproof lining behind the concrete and prevent the strata water coming in contact with the concrete. This solution in general terms, would be very expensive and somewhat impractical since it would be virtually impossible to backwall grout without puncturing the waterproof membrane.

2. Remove the pressure from the lining by constructing a relieved lining. This method is in fact used although the quantity of relieved water can become an embarrassment.

3. Use a low permeability concrete made with cements which have low C_3A contents and on completion of hydration produce small amounts of lime and aggregates which are not susceptible to alkali-silica or alkali-carbonate reactions. This is the solution being followed currently.

Watertightness

The current specification for watertightness calls for a residual water make of 1 gall/min per 100' of shaft. This, from recent experience is readily achievable, however. Bearing in mind the high costs of water disposal obviously the more watertight the lining the better.

In modern times the mass of the concrete is relatively impermeable ($K = 1 \times 10^{-9}$m). It is the joints and cracks which create the overall permeability, it is however probably the consequences of making the concrete of the required strength and impermeability that create much of the cracking problems, a balance between the two must therefore be established.

The joints in a shaft lining are usually every 6 metres vertically, this being a practical length from both a construction and support point of view.

A water bar is used to try and make a reasonable seal at the joint but the fact that the water bar is located in the kerb ring (supporting shutter) and has to have concrete cast up to it from below makes it a very difficult joint to make absolutely watertight. Backwall grouting with cement grout seals most of any remaining seepage paths, and particularly difficult joints can be sealed with chemical grouts.

Cracks in the lining generally are either vertical or horizontal. They occur in random patterns, generally however if they occur, there is a predominant direction.

The causes of the cracking are difficult to determine. The most obvious explanation is the thermal expansion and subsequent contraction of the concrete. These movements are restrained by the rock face and to a lesser extent by the shuttering and set up internal stresses which sometimes exceed early age tensile strength of the concrete which induces cracks. In the circumstances under consideration the constituents of the concrete are of great importance in determining whether or not the concrete will crack. To date investigations have centred around the type of cement used, this being the only source of heat in the mix, although attention is now being paid to the aggregate used.

Cements, it was found, have differing proportions of C_3A and C_2A and a combined percentage of the two constituents of greater than 60% seemed to give consistently more cracking than cements with a combined percentage of say 55% and below. In an attempt to reduce the peak temperatures and hence cracking within the pours (see Fig. 4) even further, a blend of blastfurnace slag cement and OPC have been used but the effect on reducing cracking has to date been disappointing although the long term strength gain of these cements has been much better than with Sulphate Resisting Portland Cements.

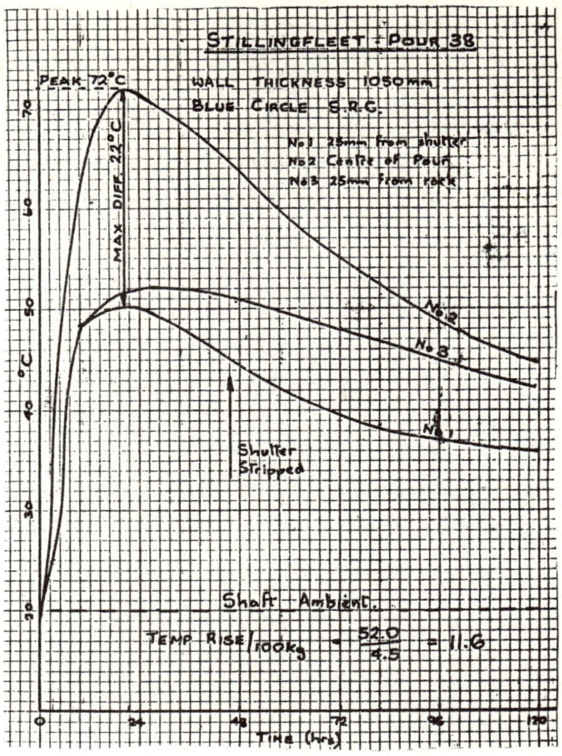

Figure 4(b)

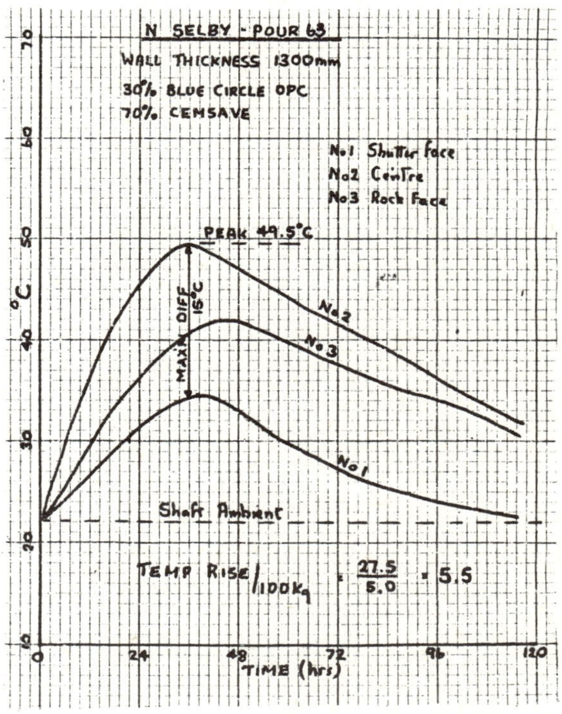

Figure 4(c)

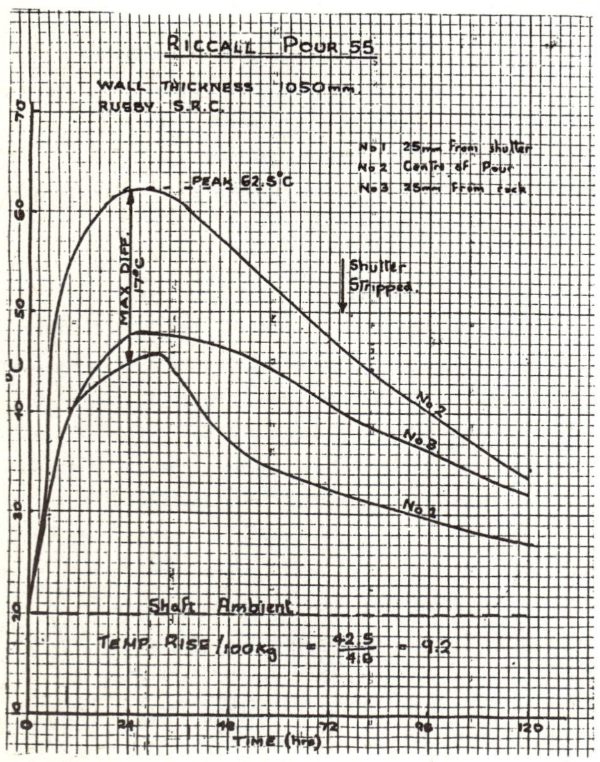

Figure 4(a)

Attention has now been turned to the aggregates within the mix. Until now Limestone aggregate has not been looked upon as a suitable aggregate for use in shaft linings because of the problems of chemical attack. However, the attack that takes place destroys the matrix leaving the aggregate intact, the concrete being useless at this stage. The theory in allowing limestone to be used is that if the attack takes place it will destroy the whole of the concrete rather than just the matrix, and if it is assumed that there is the same amount of material available to attack the concrete, the rate of deterioration should be less. The advantage of using limestone is that concrete made with this aggregate has a significantly less coefficient of expansion than ones made with quartzitic aggregate (cf 12.75 μ strain/°C Quartzitic 9 μ strain/°C Limestone.

Blasting

The shaft sinking programme depends on lining and sinking as consecutive operations, the lining being not more than 20m above the sump level. Significant vibrations are therefore caused in the young concrete by the blasting in the rock below sump level. An attempt to measure the level of vibration (measured as peak particle velocity) has been made and it would appear that it is high enough to cause damage. A Finnish paper (Ref. 8) was the yardstick used for comparison of results. The results obtained by ourselves indicated the p.p.v. in excess of 120mm/sec were occurring whereas the limits specified in the above paper for 5 day old 40N/mm^2 concrete is 110mm/sec and as low as 25mm/sec at 1 day old. The practical situation within the shafts does not suggest that blasting is causing any cracking problems, comparisons having been made between proximity of the blast to the concrete, and the age of concrete and the incidence of cracking and no correlations found. However, some further work on this subject needs to be done in order to define limits.

Instrumentation

In an attempt to determine whether the design loadings used and the resultant stresses in the lining are as calculated by the above methods, a programme of instrumentation has been undertaken to determine hydrostatic loads, total loads and strains in the concrete. A description of this equipment and its installation is given in a paper by Altounyan (Ref. 1). The results obtained from these instruments to date indicate a good correlation between applied hydrostatic loads and concrete strains.

They also indicate that a rock load exists although this is significantly smaller than calculated and could possibly be due to the lining straining under the hydrostatic load, thus reducing the effect of the rock load. It is difficult to ascertain from the limited number of results available whether any out of balance loads exist or not.

The programme of instrumentation is by no means complete and until all the results are known it would be wrong to draw premature conclusions. However, the results are encouraging and at least indicate that the design methods used are giving reasonably realistic results.

CONCLUSIONS

The shafts within the Selby Complex have been constructed using a good quality concrete and should therefore not suffer the deterioration which is occurring within some of the shafts constructed in the late 60's.

It would appear that when concrete is poured against an ice wall there is sufficient heat generated to overcome the effects of the ice for sufficiently long to dispense with any sacrificial layer of concrete.

Early indications suggest that the stresses induced in the lining due to hydrostatic load are close to the calculated stresses. The rock loads imposed appear to be smaller than calculated, no out of balance loading has yet occurred.

Concrete strengths of 45N/mm^2 are readily achievable and strengths of 60 to 65N/mm^2 are possible, and with wall thicknesses of up to 1200mm, hydrostatic heads of up to 650m can be resisted safely.

With regard to Chemical Resistance and Watertightness the paper has outlined some of the possible causes of deterioration, which in general terms can be reduced by improving concrete quality. This has been done to a significant extent but as a consequence cracking of the lining is occurring. These cracks have and can be repaired satisfactorily but the gain achieved by the improved quality is partially lost due to cracking. The aim therefore is to identify the effect of each of the factors involved and ultimately define stricter limitations on the materials and methods used to eliminate the problem.

ACKNOWLEDGEMENTS

The author wishes to thank the Director General of Mining N.C.B. Mr. R. B. Dunn for permission to publish this paper and also the help and assistance received from his colleagues in Shaft Sinking Branch of the N.C.B. Any comments and opinions expressed are those of the author and not necessarily those of the N.C.B.

REFERENCES

(1) Altounyan P.F.R., Measurements of Stresses, Strains and Temperatures in Concrete Shaft Linings (1982).

(2) B.S.I., The Structural Use of Concrete CP 114.

(3) B.S.I., The Structural Use of Concrete CP 110 (1972).

(4) Building Research Establishment, Digest 250, Concrete in Sulphate Bearing Soils & Groundwater.

(5) Coates D.F., Rock Mechanics Principles, Dept. of Energy Mines & Resources, Canada, Mines Monograph No. 874 (1970).

(6) Gutt W.H. & Harrison W.H., Chemical Resistance of Concrete - C & CA Current Practice Sheet - The Properties of Concrete.

(7) Link H, Lutgendorf H.O. & Stoff K., Guidelines for

Calculating Shaft Linings in Incompetent Strata, Verlag Glauhauf.

(8)Oy Finnrock Ab., The Damage Threshold of Concrete which is at The Setting Stage.

DEFORMATION OF A DEEP SHAFT INSET IN COAL MEASURES ROCKS DURING AND AFTER CONSTRUCTION

John K. Borsah, Philip D. Shelton, Nevil Tomlin

Department of Mining Engineering
University of Newcastle upon Tyne

SUMMARY

Results obtained from vibrating wire strain gauges installed within the concrete lining of a 660m deep shaft inset are discussed with reference to measurements of carcass deformations made during the initial excavation. Prior to the placement of the permanent lining, roof lowering and floor heave occurred. The rate of movement,which has been shown to be localised in both roof and floor, decreased significantly prior to the installation of the roof lining. There are some indications that the floor heave was regenerated, slightly, through dinting before the completion of the lining. The vibrating wire strain gauge results support the view that the concrete lining is behaving in a manner similar to a series of fixed-ended, uniformly loaded beams, which are also being subjected to significant hoop strains. The stability of the lining is enhanced by the compressive nature of the hoop strains which neutralise the tensile bending strains where they occur.

INTRODUCTION

Temporary and permanent instrumentation has been installed during excavation and within the concrete lining of the downcast inset at Stillingfleet Mine in the Selby Coalfield.

Work began on the excavation of the inset in December 1980 and was completed in October 1981. During excavation and prior to the placement of permanent support, vertical movement of the periphery of the inset and of points 3m into the rock were measured. As the concrete lining was cast, pairs of vibrating wire strain gauges and fluid pressure cells were incorporated at three sites in one of the inset headings and also in the 12m high wall of the shaft/inset junction.

The objective of this investigation was to determine the behaviour of the rock and the concrete lining around a relatively large opening of complex geometry which has been excavated in the weak rocks of a typical Coal Measures cyclothem in order to aid the final design of other insets within the Selby Coalfield.

GEOLOGY

Stillingfleet downcast inset has been excavated at a depth of 663 metres from ground surface to inset crown. Figure 1 shows the geological sequence in the immediate vicinity of the opening along with results of rock testing (Davies, 1978). It can be seen that the sequence forms a typical, complete Coal Measures cyclothem . The inset crown has been constructed against a relatively strong sandstone/siltstone, with a uniaxial compressive strength of between 70 and 100 MN/m^2, grading into weaker mudstones, while the rocks adjacent to the sidewalls and floor are the much weaker Barnsley coal seam, underlain by seatearths and mudstones, which have compressive strengths of the order of 30 MN/m^2. This sequence, however, is underlain by stronger interbedded sandstones, siltstones and mudstones. The beds dip slightly to the north.

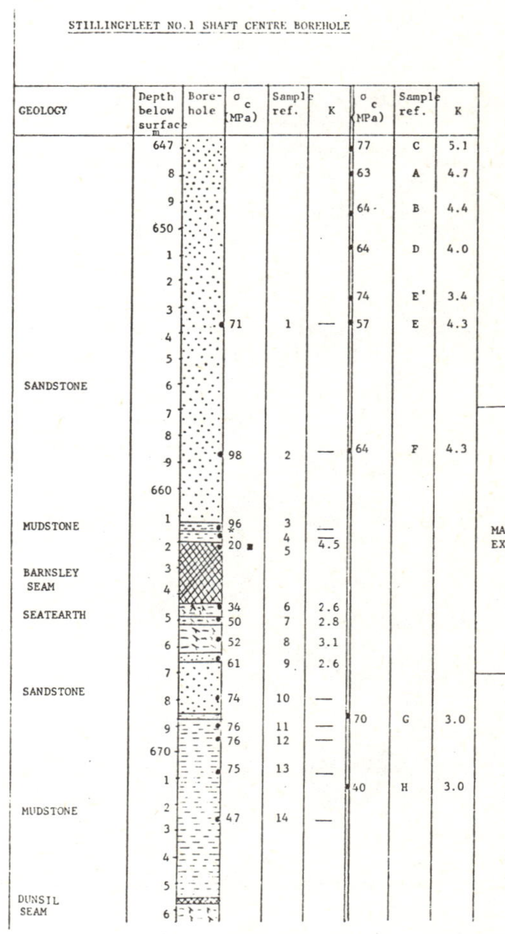

Figure 1 Geological sequence and results of rock testing in the vicinity of No.1 shaft inset.

CONSTRUCTION

In the section and plan of the shaft inset shown in figure 2, the order in which the in-situ lining was installed is indicated by the sequential numbering system

Following the initial widening of the shaft and the placement of the concrete which forms the shaft collar (concrete pours 1a and 1b), two 8 metre wide by 4 metre high headings were excavated (2c to 2a and 2f to 2d) to the full inset length, north and south. These headings were supported temporarily with 30 x 24mm RSJ arches with centre props at 0.9 metre intervals. The permanent concrete lining was then placed in sections working back towards the shaft and linked at the inset "box section" by concrete pours 2g and 2h. The floor and sidewalls were subsequently excavated by first driving a 4 metre wide dint through the temporary floor and opening out in retreat fashion towards the shaft. Subsequent concrete pours for the permanent lining are shown as stages 3 and 4.

The concrete batches were mixed on surface and pumped via the shaft staging into position behind temporarily erected steel shuttering. The nominal thickness of the concrete lining throughout the inset headings is 0.915 metres. Reinforcement within the concrete consists of two mats adjacent to the inner and outer faces of the lining. Additional heavy reinforcement has been installed at the shaft/inset "box section" in the form of two vertical columns illustrated on Figure 2. These were designed to retain the 12 metre high east and west box section walls against horizontal loading.

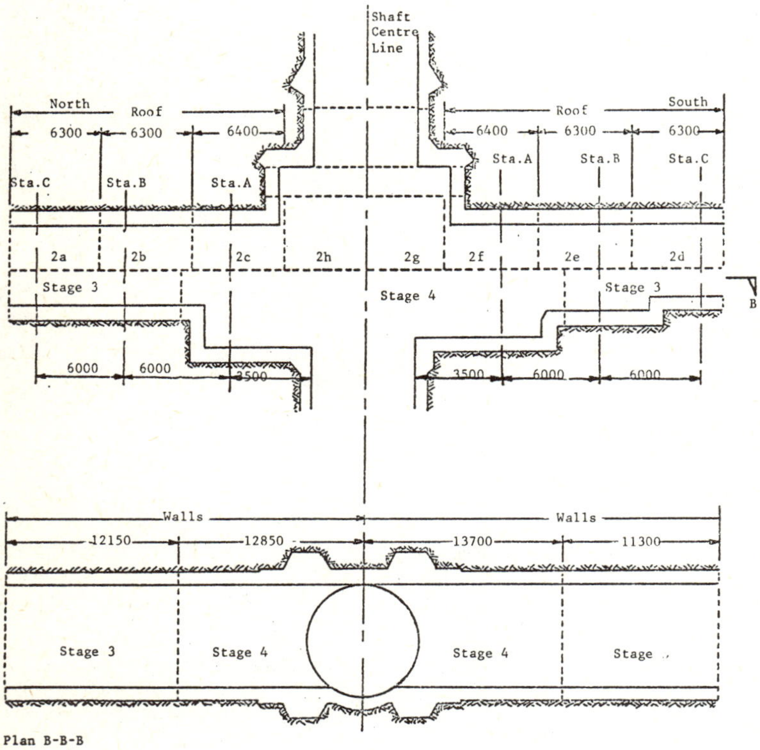

Figure 2 Section and plan of inset showing the sequence of concrete pours.

ROCK-LINING INTERACTION

From the point of view of establishing a design approach for use with respect to deep inset construction, in which closure and instability must be tightly controlled, a model is required with which the behaviour of closely related structures can be compared. Probably the most important consideration to be made in calculating the required concrete thickness and reinforcement quantities, is the prediction of the maximum radial loads which a specific lining must sustain. It must be remembered that it can take many months and even years for full stress redistribution to take place. The magnitude of this load will be dependent upon a number of factors but, in the absence of water pressures, the following will be most significant: initial stress conditions, strength of the surrounding intact rock, the degree of jointing in the rock, the stiffness of the support and the method of excavation and time to lining erection.

In strong rocks the natural geostatic stresses will redistribute themselves in such a way that the circumferential stresses are high, while radial stresses are very low. However, in wall jointed hard rocks, radial pressures can be mobilised from the dead weight of loosened jointed material overlying the roof. In the case of weak rocks, high radial pressures can be developed by the volumetric expansion of failed or failing material within a yield zone, particularly if the failure is of a strain hardening type (Farmer, Price and Youdan, 1980).

A number of solutions, including those of Hobbs (1966), Daemon and Fairhurst (1972), Ladanyi (1974) and Wilson (1980), exist which will either directly or indirectly estimate the magnitudes of loads acting on a stiff support lining which result from the non-elastic behaviour of the surrounding rock. These solutions are based on the equilibrium equation for a thick-walled cylinder and criteria for peak and residual yield for rocks. Care must be taken when employing these types of approach for any situation other than an opening of circular cross-section in homogeneous ground. However, they may give an indication of the possibilities of yield, and estimate their order of magnitude.

In the case of strong jointed rocks, the radial pressures will depend upon the volume of material bearing onto the tunnel soffit. Terzaghi (1946) has suggested that an indication of the radial pressure can be obtained by considering the arching action of a cohesionless soil. Terzaghi developed a well known classification system from which the heights of these arches and hence the rock loads can be estimated. Recently, more sophisticated rock mass classification systems have been developed by Barton, Lein and Lunde (1974) and Bieniawski (1973) among others for the purpose of predicting support requirements, based on case history data. However, as pointed out by Ward (1978), such approaches can neither differentiate between good and bad practice nor can they consider rock lining interaction or the effects produced by various excavation methods.

OBSERVATION AND INSTRUMENTATION PROGRAMME

The observation and instrumentation programme has been undertaken in two stages:

(a) precise levelling and tape measurements to determine carcass deformation of the stage 2 excavation as the headings were advanced and while the opening was supported by arches and centre props.

(b) instrumentation of the permanent concrete lining to monitor lining strains and rock-lining interfacial stresses, using vibrating wire strain gauges and hydraulic, mercury filled pressure cells. The use of these types of instrumentation has been described by Altounyan (1982).

Temporary Instrumentation

The layout of a typical observation station is shown in Figure 3. This complete layout was only achieved at Station A in the south heading. Four other stations were established in the inset to measure absolute roof and floor movements. The locations of these are shown in Figure 2.

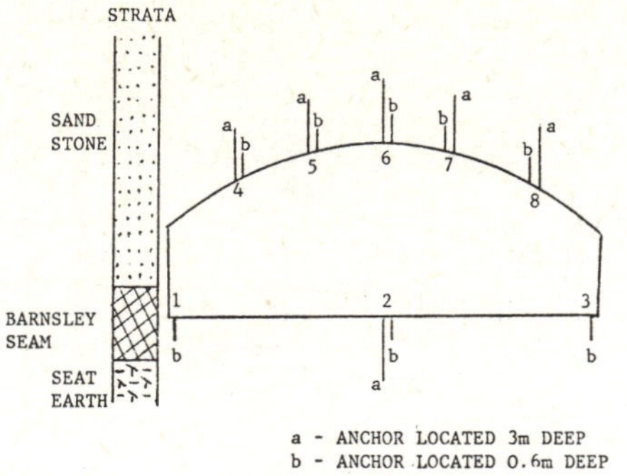

Figure 3 Typical measurement station in the unlined heading.

Permanent Instrumentation

Permanent instrumentation was installed in the concrete lining as it was placed at 3 stations in the south inset heading at approximately the same positions as the temporary observation stations, with the addition of Station C which was 15.5m from the shaft wall. The strain gauges were installed in pairs, adjacent to the inner and outer reinforcement mats and were aligned such that circumferential strains are measured. Figure 4 shows a typical instrumentation array. At each station a pair of temperature gauges were included alongside strain gauges to allow a temperature correction to be applied to the other instruments.

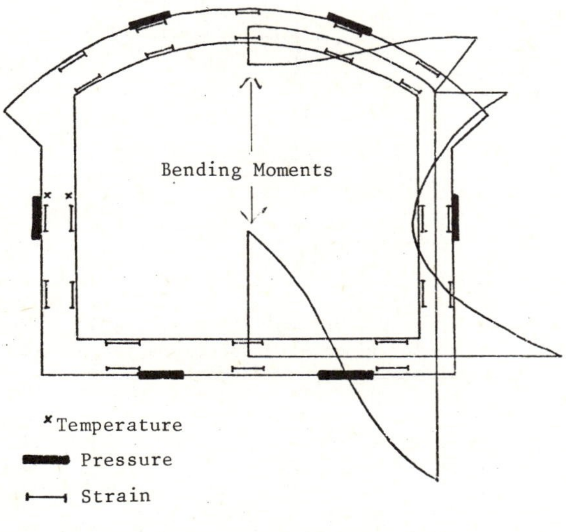

Figure 4 Section of the lined inset showing the location of the instrumentation and the predicted bending moment pattern.

In most cases, the vibrating wire strain gauges were positioned at locations where maximum bending moments have been predicted. These predictions were based on treating the roof, each sidewall, and floor as separate fixed end beams under uniformly distributed load. Figure 4 also shows the predicted bending moment pattern superimposed onto the instrument layout.

The east wall of the inset/shaft intersection was chosen as a site for further instrumentation where a pair of heavily reinforced columns has been constructed in each wall to restrain any horizontal loading. 12 pairs of strain gauges were installed in and between the columns to measure vertical tangential strains.

DISCUSSION OF RESULTS FROM OBSERVATIONS AND INSTRUMENTATION

Temporary Instrumentation

As a reasonably constant rate of face advance, about 1.5 metres per day, was achieved during the excavation of the stage 2 headings, and measurements of strata movements were continued after the heading was completed, strata displacements have been plotted against time. In addition, so that the curves obtained from each station can be compared, the first measurements recorded at each station have been taken as zero displacement and zero time.

Figures 5 and 6 show roof lowering and floor lift respectively, measured in the centre of the inset, for each station in the north and south headings. The characteristic 'S' curve is well established in most cases. The main points which can be raised from these measurements are:

(a) as expected, convergence at Station A is greater than at Station B, which is greater than at C, due to the shielding effect of the face.

(b) floor heave is generally greater than roof lowering. From an examination of the geology it can be seen that, while the relatively strong siltstones and sandstones in the roof are quite capable of accepting the redistributed stresses, while allowing some bed separation to occur, the weaker floor coal is failing and being forced upwards. From measurements of re bars anchored at a depth of 2m into the floor, it was found that floor heave was restricted to the upper layers of floor coal only.

(c) the rate of roof and floor movement reduced rapidly with time (and face advance) being substantially completed within 10-15 days (15-25 metres).

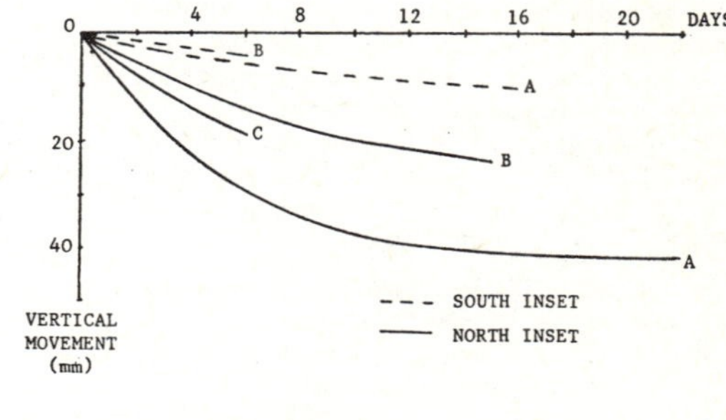

Figure 5 Floor uplift at centre of inset.

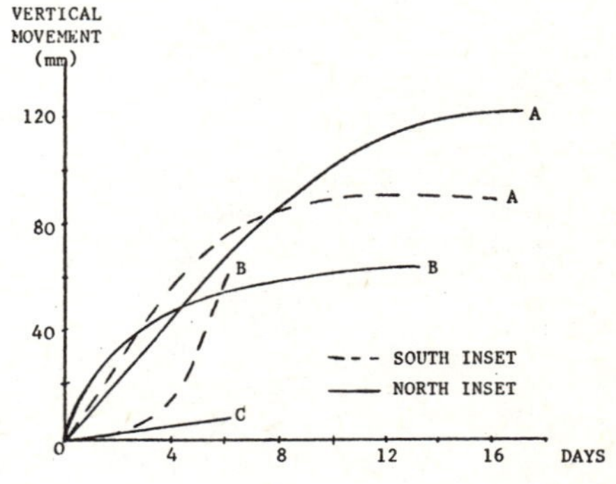

Figure 6 Roof lowering at centre of inset.

Figure 7 illustrates the initial and final heading profiles; the scale of deformation has been exaggerated 10 times. Lateral closures have also been measured and have demonstrated deformation in the order of 40mm.

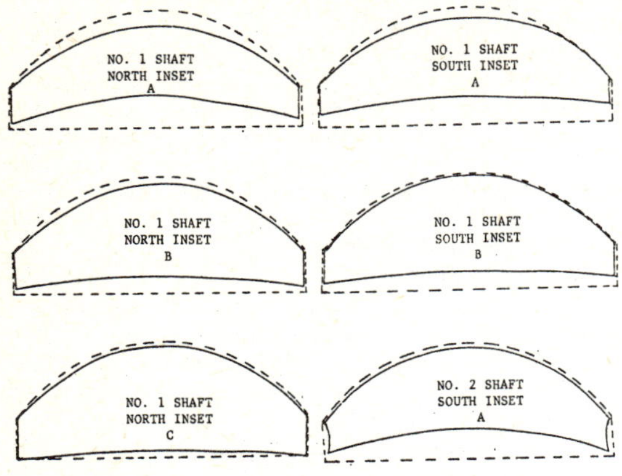

Figure 7 Distortion of cross-sectioned shape for each measurement station.

Permanent Instrumentation

The progressive nature of excavation and concrete placement makes detailed quantitative analysis of the completed structure and indeed the results from any station very difficult. One approach might have been to consider the inset lining as an approximation to a series of thickwalled cylinders. However, analysis of the results from strain measurement and the time delay between the formation of such a cylinder preclude such an approach. Instead, each instrumentation station has been considered in two stages. Firstly, the behaviour of the roof section up to and including the time of the removal of floor material and, secondly, the subsequent and additional strains imposed on the whole unit following completion.

The results from measurements taken on the hydraulic pressure cells have been disappointing. The sensitivity of the read-out unit is particularly poor at low pressures between 0 and 500 kN/m^2. The only basic conclusion then is that radial pressures are of this order of magnitude. Satisfactory measurements have been obtained, however, from the vibrating wire strain gauges installed in the concrete lining. Inevitably, there has been the loss of some gauges, principally from Station B, which is probably due to cable damage during concrete placement.

The concrete lining was cast in three sections, from the end of each inset heading towards the shaft; therefore, Station C was the first to be installed, followed by B and A respectively.

A period of between 7 and 20 days was allowed to elapse before datum readings were taken for the vibrating wire strain guages to allow time for thermally induced strains to dissipate.

Results of strain measurement at points in the instrumentation arrays have been plotted as the mean and difference between inner and outer gauges,where the 'inner'gauge is that which is adjacent to the opening and the 'outer' gauge is that which is adjacent to the rock. A linear variation of strain has been assumed between the strains recorded at the inner and outer positions and plotted results are extrapolations to the inner and outer concrete surfaces.

The mean strains represent the average hoop strains through the concrete section, while differential strains between the inner and outer surfaces represent the bending component.

The convention of compression positive and tension negative has been employed throughout.

Roof Sections

Figures 8(a), 9(a) and 10(a) illustrate the mean and differential strains for Stations A, B and C respectively up to the time when the temporary floor was removed. It can be seen that in all three cases the mean or hoop strains remain fairly constant and compressive across the inset roof at magnitudes of between 50 and 100 microstrain. Differential strains, however, vary cross the inset roof. The effects of these strain differences can be clarified if the mean strains are subtracted from the inner and outer surface strains. These have been plotted in Figures 8(b), 9(b) and 10(b) for Stations A, B and C respectively as inner and outer surface bending strains.

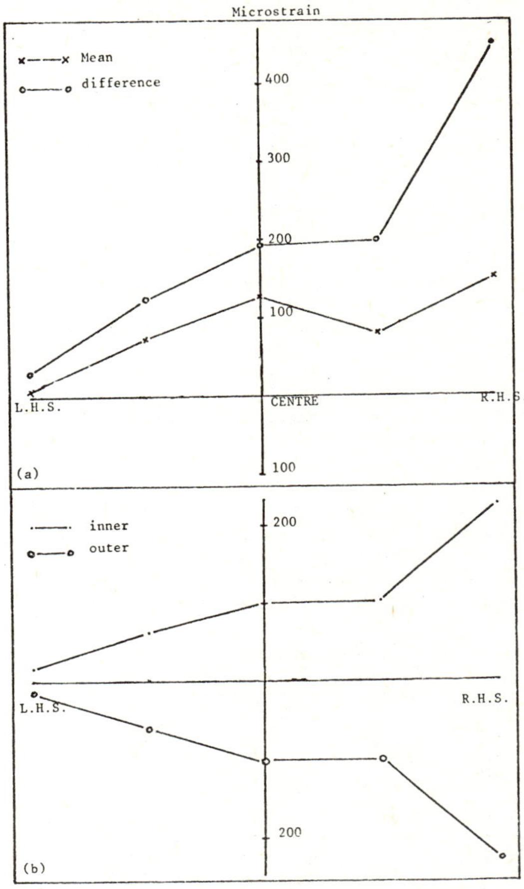

Figure 8. Station A. (a) Mean and differential strains across the inset roof recorded after 103 days, just prior to the removal of the temporary floor. (b) Inner and outer surface bending strains for the same period.

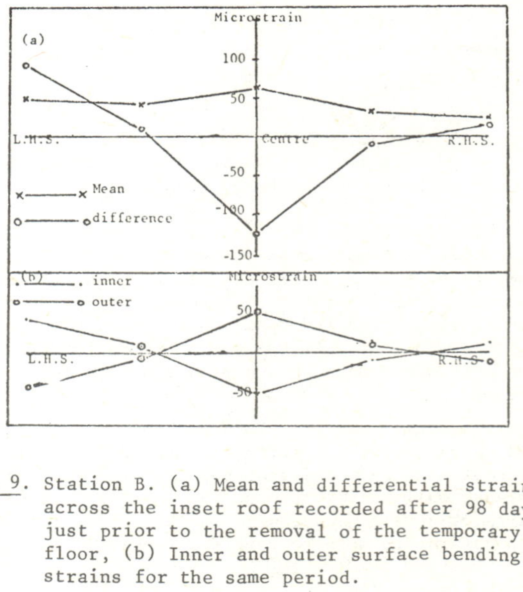

Figure 9. Station B. (a) Mean and differential strains across the inset roof recorded after 98 days, just prior to the removal of the temporary floor, (b) Inner and outer surface bending strains for the same period.

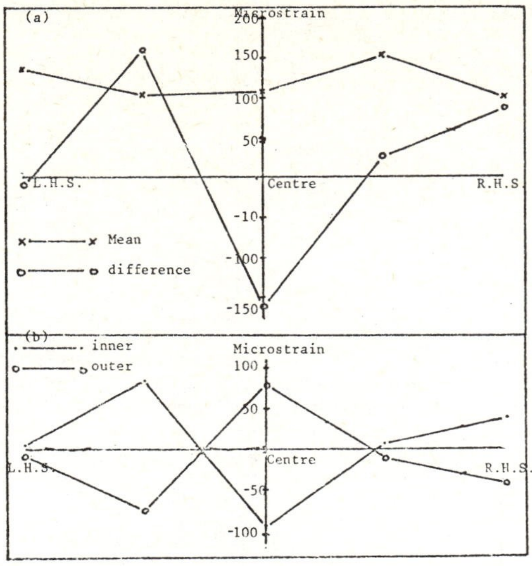

Figure 10 Station C: (a) Mean and differential strains across the inset roof recorded after 85 days, just prior to the removal of the temporary floor. (b) Inner and outer surface bending strains for the same period.

A beam with fixed ends under uniform load will deflect in such a way as to produce tensile strains on the lower surface at midspan and on the upper surface at its ends, and compressive strains at its opposing surfaces correspondingly. In addition, for a normal flat beam, the strains at the fixed ends will be twice those which occur at midspan. These phenomena have been observed in the roof beams. Complex geometry, reinforcement and beam approximation have produced some departures from the theoretical model. However, the pattern is clear.

From simple beam theory it can be shown that the bending moment 'M' for a beam of length 'L' at midspan is given by:

$$M = \frac{qL^2}{24} \qquad \text{.... (1)}$$

where q is the uniformly distributed load per unit length. Similarly, it can be shown that the strain e in the beam can be related to the bending moment by:

$$e = \frac{MZ}{EI} \qquad \text{.... (2)}$$

where E is the elastic modulus, which for concrete is approximately 35 GN/m^2, I is the moment of inertia, and Z equals the distance from the neutral axis. Considering just the bending element of roof behaviour an estimate of the uniformly distributed loads per square metre can be made by combining equations (1) and (2) and calculating:

$$q = \frac{24EeI}{L^2Z} \qquad \text{.... (3)}$$

load per unit length per metre of roof length.

For surface bending strains of magnitude 50 to 75 microstrain which were observed at Stations B and C the uniformly distributed loads would be in the order of 100 to 175 kN/m^2 or 4 to 8 metres of rock. This simple approach, however, does not consider the presence of the reinforcement within the concrete lining. The effect of this is that the distributed loads are over-estimated.

At Station A (Figure 8) the observed roof behaviour has been quite different and is characterised by very high compressive strains on the inner surface of the right side (R.H.S.) of the roof and tensile strains at its outer surface along with high compressive hoop strains (150 microstrain). The bending effects, however, decrease across the roof and zero outer and inner fibre strains are recorded at the L.H.S. roof edge. This situation is similar to the strain pattern which would occur in a cantilever beam fixed at the R.H.S. and allowed to deflect at the L.H.S. Speculating on a possible cause for this change in behaviour, other than instrument malfunction, then it might appear that the keying of the L.H.S. roof edge was not effective and has allowed the weight of the roof to compress into the floor material beneath the sidewall while the concrete was still young.

It might have been expected that the removal of the floor material from beneath the small sidewall sections would have produced a reduction in the subsequent strain measurement through relaxation, particularly at the roof edges. At Stations B and C this has not been so, and only slight reductions have been measured at Station A, where strains have reduced by between 15% and 20%. This indicates that the reactions to roof loading are being taken through the key sections, as shown in Figure 11. Furthermore, these keys assist in the generation of compressive hoop stresses.

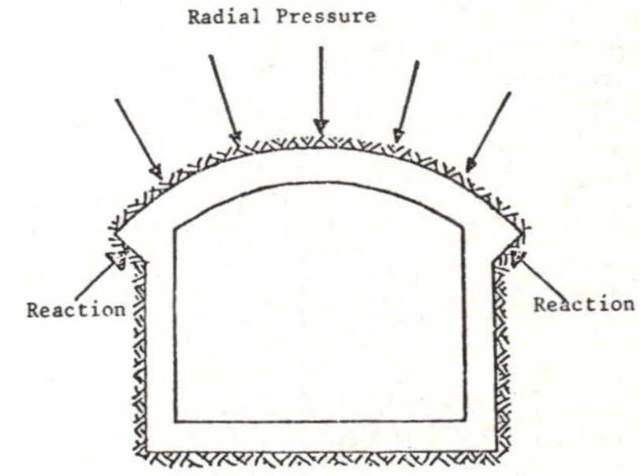

Figure 11 Section of inset showing reaction to roof loading.

Figures 12, 13 and 14 show the total mean and differential and inner and outer surface bending strains as measured to date for stations A, B and C respectively. In general the increases in strain since the completion of the inset have been small and these are mainly attributable to increases in the mean or hoop strains.

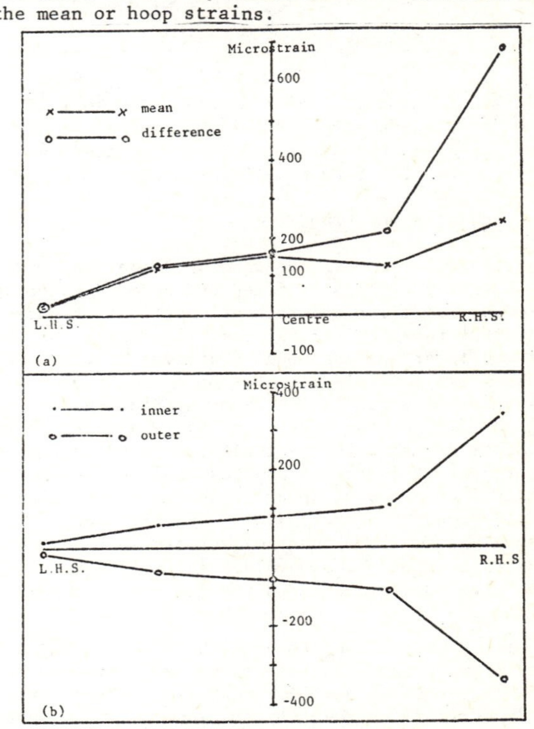

Figure 12 Station A: (a) Mean and differential strains across the inset roof after 259 days. (b) Inner and outer surface bending strains after 259 days.

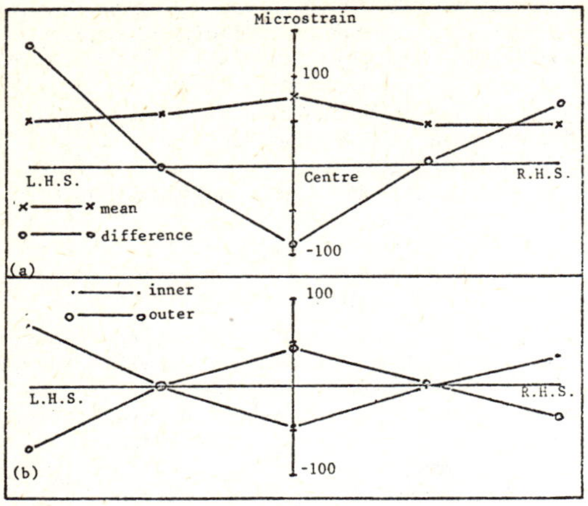

Figure 13 Station B: (a) Mean and differential strains across the inset roof after 266 days. (b) Inner and outer surface bending strains after 266 days.

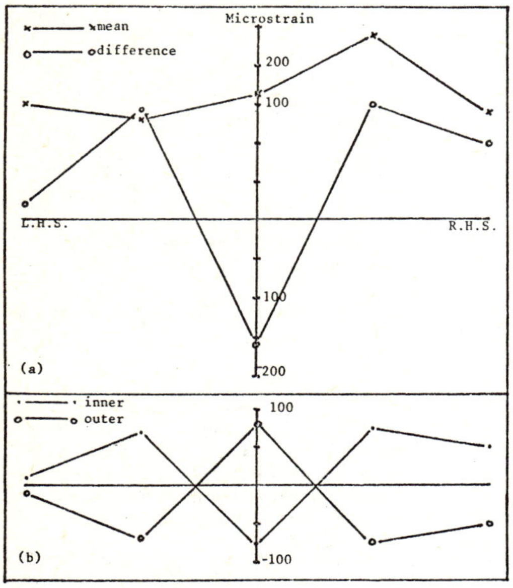

Figure 14 Station C: (a) Mean and differential strains across the inner roof after 273 days. (b) Inner and outer surface bending strains after 273 days.

Sidewalls and Floor

The comparison of the observed behaviour of the roof section to that of a fixed end beam can also be applied to the floor and sidewall sections. Figures 15 and 16 show the mean and differential strain and the inner and outer surface bending strains in the sidewall and floor respectively for Station A and Figure 17 shows the strains measured in the floor at Station C. The magnitude of the hoop strains at Station C are of the same order of magnitude as those generated in the roof, indicating that axial loading has again been generated. At Station A, however, mean strains are fairly low, between 20 and 30 microstrain . Surface bending strains indicate that uniform loads of between 30 and 70 kN/m^2 are being applied. These values are again over-estimates of the true loading based on the absence of any reinforcement.

It is interesting to note that the strains are higher at Station C than at Station A. This may be due to regeneration of floor heave in the seatearth which underlies the concrete at C but which has been removed from underneath Station A.

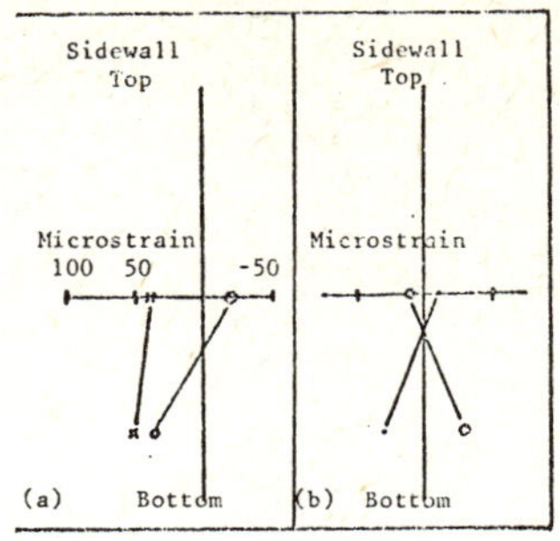

Figure 15 Station A: (a) Mean and differential strains measured in the R.H.S. sidewall after 148 days. (b) Inner and outer surface bending strains after 148 days.

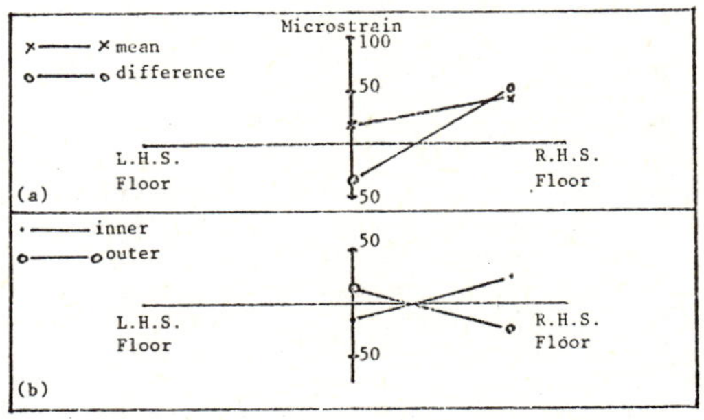

Figure 16 Station A: (a) Mean and differential strains measured in the inset floor after 148 days. (b) Inner and outer surface bending strains after 148 days.

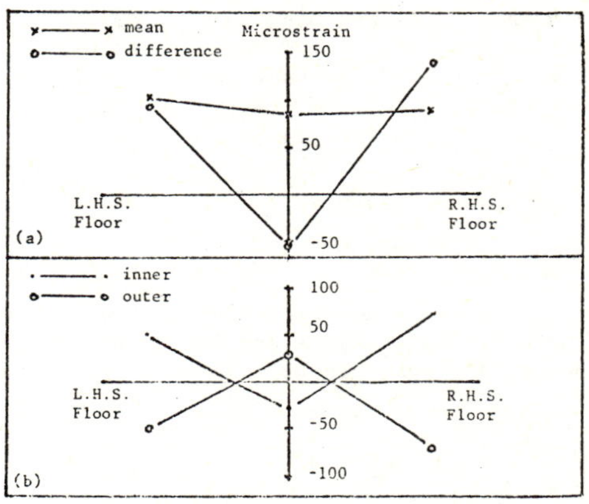

Figure 17 Station C: (a) Mean and differential strains measured in the inset floor after 180 days. (b) Inner and outer surface bending strains after 180 days.

Inset/shaft intersection

Results from the 12 pairs of vibrating wire strain gauges installed in the 12m high wall and reinforced columns in the inset section have yet to show any significant vertical tangential strains in either of the columns or the intersecting wall.

CONCLUSONS

1. From the carcass deformation measurements taken during the excavation of the upper level of the inset, it is clear that differences in strata have produced differing effects in the roof and temporary floor. The relatively strong roof has lowered for a relatively short period of time (8-12 days), by an average of 20-25mm. This has resulted from rock loosening and bed separating which ceased as soon as sufficient restraint was applied to the strata from

the arch support. The relatively weak floor material, however, has been unable to accommodate the stress redistribution resulting in localised floor heave.

2. Results from hydraulic pressure cells indicate that radial pressures on the concrete lining are less than 500 kN/m^2. However, the readout unit is not sufficiently sensitive for accurate measurements to be made in this range.

3. Strain gauge results have been analysed and the pattern of bending strains indicate that the analogy of the roof, floor and sidewall behaviour being comparable to fixed end, uniformly distributed beams is valid. Estimates of the radial loading based on a simplified model indicate floor and sidewall pressures of between 20 and 30 kN/m^2 and roof loads up to 175 kN/m^2. The larger strains recorded in the floor at Station C, nearest the end of the inset, than at Station A, indicate that some regeneration of floor heave has occurred where the floor is particularly weak.

   The rate of strain increase is becoming less with time with most of the strain occurring before 100 days, and in the case of the roof sections prior to the completion of the lining.

4. Results so far indicate that both bending and mean strains are in the area of the inset/shaft intersection.

5. In considering the results of these measurements with respect to the design and construction of other insets the following points can be made:

   (a) The concrete lining can be considered to act as four reinforced beams, consisting of roof, floor and sidewalls. The strength of such beams in bending can be shown to be very low.

   (b) The load bearing capacity of such beams is increased considerably by the presence of compressive mean or hoop strains as in pre-tensioned reinforced concrete. These compressive strains have been shown to be present in the inset linings.

   Provision for the development of compressive hoop strains can be made by arching the roof section and the construction of 'keys' at the abutments to take axial reactions (Figure 9). Similarly, where high floor pressures are anticiapted an inverted arch will, similarly, generate the axial pressures.

   (c) Magnitudes of radial pressures will be less than predicted by yield analysis or by Terzaghi's rock load criterion if the inelastic deformations are allowed to take place prior to support placement.

ACKNOWLEDGEMENTS

This research was supported by the National Coal Board and the European Coal and Steel Commission under contract no. YCE/67/13307. The authors wish to thank the staff of both the National Coal Board and Thyssens (G.B.) Ltd. at Stillingfleet for their assistance.

REFERENCES

Altounyan P.F.R. (1982): Measurement of stresses, strains and temperature in concrete shaft linings. Symposium on Strata Mechanics. University of Newcastle upon Tyne.

Barton N. Lein R. and Lunde J. (1974): Engineering classification of rock masses for the design of tunnel supports. Rock Mech., 6, 189-236.

Bieniawski Z.T. (1973): Engineering classification of jointed rock masses. Trans. S. African Inst.Civ.Engrs., 15,335-44.

Daemen J.J.K. and Fairhurst C (1972): Rock failure and tunnel support loading Proc. Int. Symp. Underground Openings, Lucerne 356-69.

Davies R. (1978): M.R.D.E. Test Report. RCM/B(78)22

Farmer I.W., Price A.M. and Youdan D.G. (1980): Design of tunnels in Coal Measures rock Eurotunnel 80, 13-27.

Hobbs D.W. (1966): A study of the behaviour of broken rock under triaxial compression and its application to mine roadways. Int. Jour. Rock Mech. and Min.Sci., 3, 11-43.

Ladanyi B. (1974): Use of the long term strength concept in the determination of ground pressure on tunnel linings. Advances in Rock Mechanics, Proc. 3rd Int. Conf. I.S.R.M., Denver II, Part B, 1150-156.

Terzaghi K. (1946): Rock defects and loads on tunnel supports. Rock tunnelling with steel supports. Proctor and White (Youngstown, Ohio. The commercial Shearing and Stamping Co.).

Ward W.H. (1978): Ground support for tunnels in weak rocks. Geotechnique, 28, 133-171.

Wilson R.H. (1980): A method of estimating the closure and strength of lining required in drivages surrounded by a yield zone. Int.Jour. Rock Mech. Min. Sci. and Geo. 17, 349-55.

PRESSURES AND DISPLACEMENTS AROUND A LINED SHAFT IN WEAK ROCK

Michael J. Gilbert

OVE ARUP AND PARTNERS,
13, Fitzroy Street,
London, W.C.1.

SUMMARY. Shaft models are commonly based upon the twin assumptions of an elastic lining and an isotropic stress distribution. These assumptions are tested against data from a shaft into which a number of different instruments were placed. It is demonstrated that the behaviour of an actual shaft is significantly different to that predicted by such models. In particular concrete shrinkage, shaft-lining bond and stress anisotropy are shown to be of crucial importance.

INTRODUCTION

For many years now it has been customary to model the closure of lined shafts and tunnels in soft rocks as three phase problems. At some distance from the opening an elastic continuum is assumed. Enclosed within this zone is a distressed region where the elastic stresses violate the chosen failure criteria. The third, inner, region is a second elastic zone representing the lining. Such models have been summarised by Muir-Wood (1979).

Lining pressures and displacements predicted by models of this type are dependent on three factors, the failure criteria, the stress distribution within the distressed zone and the moduli chosen for the elastic regions. Often it is necessary to assume the lining to be installed at the instant of excavation if any lining pressure is to result. This assumption may be partially justified by reference to 'stress shield' phenomena associated with the face (see Abel et.al., 1979). The many elastic models are also dependent on assumptions of this type (see Savin, 1961). It is, however, possible to base models on the time-dependent strength behaviour of rock, as documented by Bieniawski (1970) and Farmer and Gilbert (1981). Such a model developed by Ladanyi (1974) has been successfully tested against shaft data by Gilbert (1981), as discussed by Gilbert and Farmer (1981).

It is common to concentrate on material models for rock, incorporating them into shaft models. It is, however, less common to question the basic tennents of isotropic stress distribution and lining elasticity. It is this area of shaft behaviour which is discussed in this paper.

THE IN-SITU DATA

Data was obtained from four sets of instruments placed during the construction of No.2 shaft at Riccall mine. Six pressure cells were attached to the bare rock face and the lining cast over them. At the same time two rings of vibrating wire strain gauges cast in concrete briquettes were also incorporated within the lining. After 1 week the shuttering was removed and diameteral extensometer stations installed. Two electro-hydraulic borehole extensometers were also used, however, these proved unreliable and ceased to function before any data was obtained.

INSTRUMENTATION LAYOUT

It was intended to place all instruments on a horizontal plane 326m below ground level, coinciding with the center of a halite band within the middle permian marls. However, the pressure cells required a reasonably flat area of shaft wall. The pressure cells were therefore installed at a depth of 326 $\pm$0.5m. It is not anticipated that this will materially affect the results. Other instruments were placed within 100mm of the 326m plane. The positions are shown in plan in Figure 1.

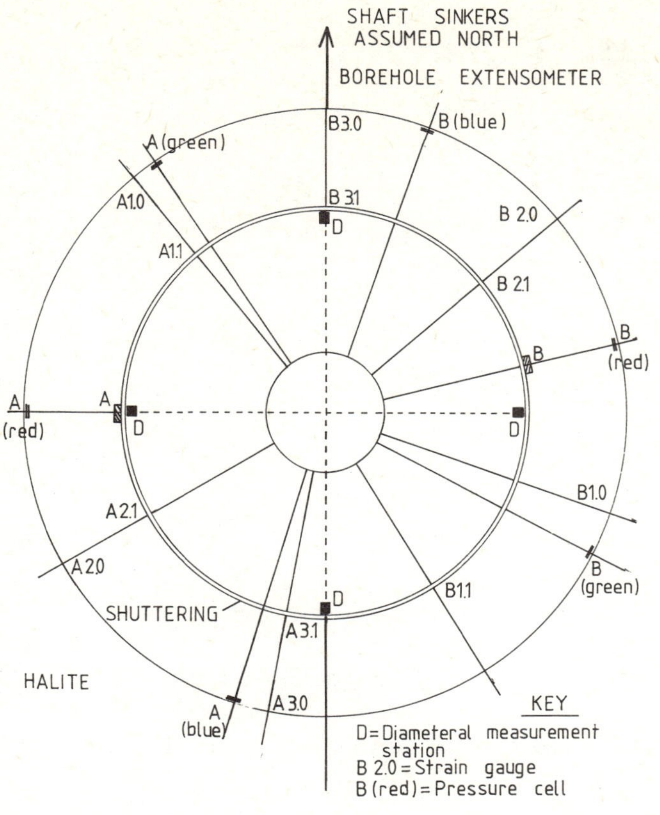

Figure 1. Instrumentation Layout. The instrument leads were brought through the lining in two places, and each instrument is prefixed A or B accordingly.

THE READINGS

(i) The pressure cells

The data obtained from the pressure cells is summarised in Figure 2. It can be seen that after 200 days have elapsed two groups of cells may be identified, cells A (Red) and A (Green) which gave high readings and the others which showed pressures in the low to moderate range. Bloor (1980) has suggested that the high readings may have been caused by kinking of the hydraulic lines, however, the pressure in cell A (Green) is seen to have risen steadily with time.

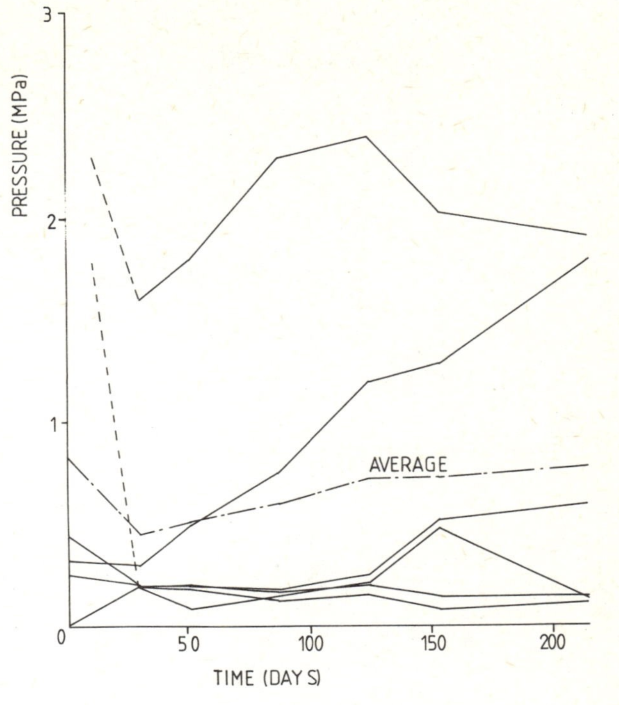

Figure 2. The Pressure Cell data

(ii) The vibrating wire strain gauges

The readings from the inner and outer ring gauges are shown in Figures 3 and 4. All gauges were cast into the lining approximately 150mm from the surface.

Examination of the data shows that the external ring gauges have gone into compression whilst those close to the inside face have generally gone into tension.

Further general trends may be suggested for both sets of gauges. The external gauges appear to have experienced maximum or near maximum compression by the time first readings were taken (27 days). Over this same period the position with the external gauges is much more confused. In the period up to the next reading (49 days) significant tensile strains are observed on the inner ring gauges, no corresponding large movements were recorded on the outer ring of gauges, however. The movements on the external gauges show no apparently logical trend over the next period (49-85 days). Large tensile strain increments were observed on 3 of the gauges, a small tensile strain on gauge B3.0 and a small compressive strain on gauge A2.0. Over this same period large tensile strains were observed on the inner ring gauges which continued up to the next reading (122 days). In contrast the outer ring gauges appear to have stablised by 85 days.

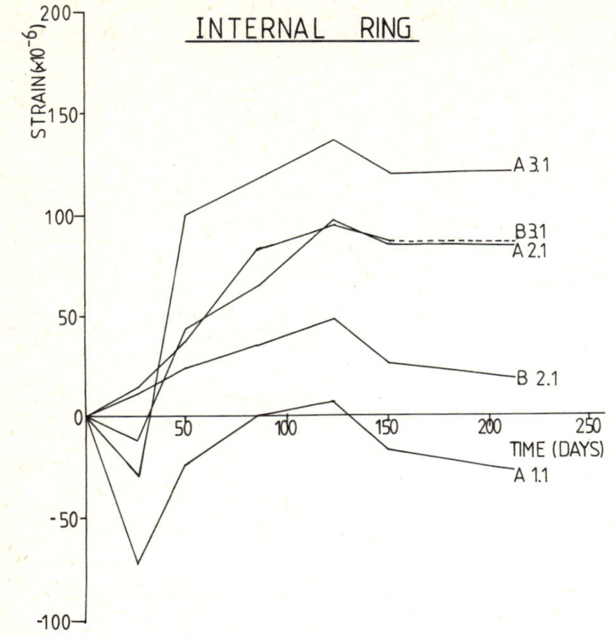

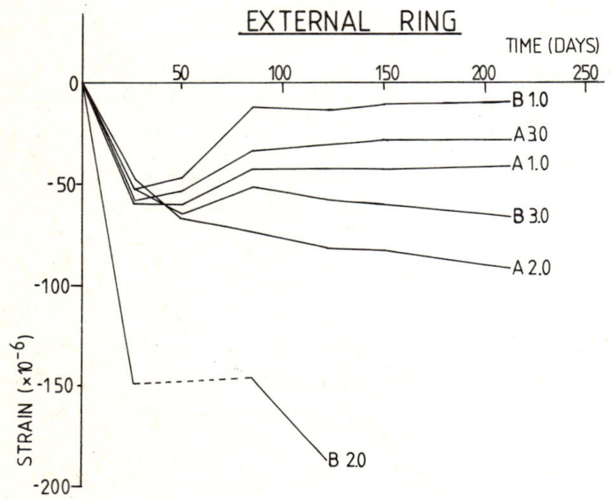

Figures 3 and 4. Data from the vibrating wiere strain gauges (Tensile strains positive sign convention).

(iii) The diametral extensometers

Movements measured across two mutally perpendicular diameters are shown in Figure 5. Unfortunately it was not possible to obtain any readings until a month after the lining was cast.

Interestingly the shaft expanded significantly in the period between the 27 and 49 day readings. This was followed by a period when the shaft diameter remained essentially stable (49-122 days). There then followed a large contraction of the diameter as the shaft began to close for the first time. This movement coincided with back wall grouting. The final movements had opposite sign making it impossible to draw any conclusions from the data.

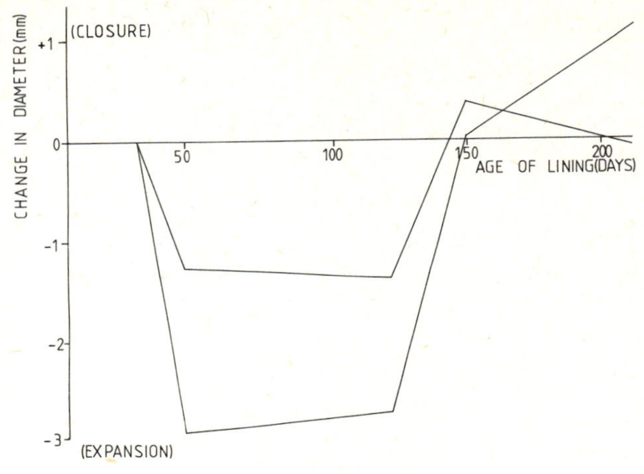

Figure 5. The diameteral extensometer readings

ULTIMATE PRESSURE AND STRAIN DISTRIBUTION

The data discussed above suggests that the shaft was approaching an approximately stable state when the final readings were taken after 213 days. The state of stress and pressure of the lining is shown in Figure 6. It can be seen that the pressure distribution is markedly anisotropic. It is therefore unsurprising that the resulting strain distribution is also uneven. The distribution is discussed further below.

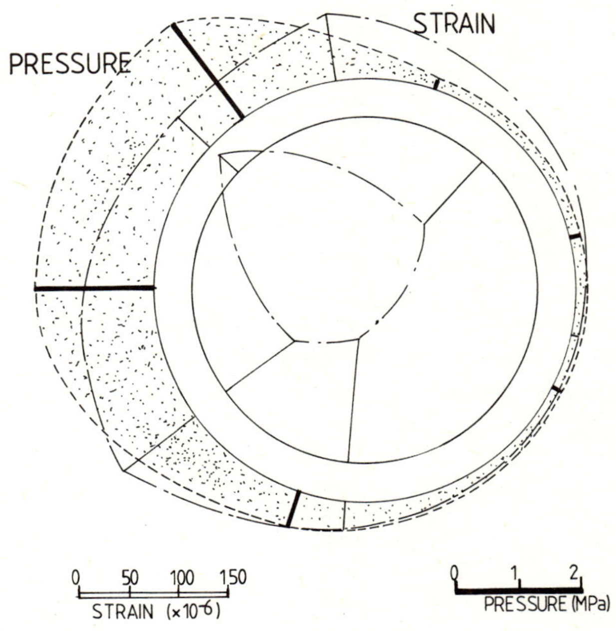

Figure 6. The strain displacement fields established around the shaft after 213 days. The pressures and outer ring strain gauge readings are zeroed on the outside surface of the lining whilst the inner ring strain gauge reading are zeroed on the inside surface. Compression radially outward from datum sign convention. The pressure envelope has been shaded.

DISCUSSION

No one set of data in isolation is sufficient to provide a rational explanation of the processes occurring around the shaft. The diameteral extensometer readings, however, are the best starting point for a discussion of the situation,as they provide a gross measure of all the individual component processes. These show the shaft to be expanding in the period immediately after construction. Whilst this is not predicted by any shaft model it is consistent with concrete shrinkage (see Neville, 1972). Thus concrete shrinkage has been isolated as one factor which is having a marked effect on the structure. It is further likely that even greater shrinkage occurred in the period before the first reading. Over this same period large compressive strains are measured on the outer ring strain gauges, showing that the shaft is in fact closing despite the evidence of the diameteral extensometers. During this period the inner ring gauge reading are relatively confused with an average strain of approximately zero. This is consistent with a shrinkage component manifesting itself as motion towards the outside of the lining being cancelled by a closure movement of approximately the same size. It is also likely that a strong bond has been formed between the shaft and the lining. Thus all shrinkage strains are translated into movement towards the shaft/lining interface.

The pressure distribution as displayed in Figure 6 appears markedly anisotropic. It has been suggested that the high pressures which have been measured are a result of kinking within the hydraulic lines, however, the strain gauge readings are completely consistent with the pressure cell data. In general large compressive strains within the external ring coincide with the region of high pressure and large tensile (shrinkage) strains within the internal ring with the low pressure region. Interestingly a compressive strain is measured on the inside face in only one position, coincident with the centre of the high pressure zone. With this amount of supporting evidence it is reasonable to assume that the pressure distribution is in fact heavily anisotropic, as indicated by the pressure cells. Thus it appears that only half of the lining is being subjected to any appreciable stress on this horizon.

It may therefore be suggested that the behaviour of the shaft is a result of three separate phenomena. A stress field which is markedly anisotropic, concrete shrinkage and a strong bond between the concrete lining and the rock shaft. Obviously all three contravene some basic assumption common to all shaft models.

ACKNOWLEDGEMENTS

The work was supported by the Science Research Council and Cementation Specialist Holdings Ltd. The author particularly acknowledges the valuable contribution to the works of Dr.R.A.Scott of Cementation Specialist Holdings Ltd. and Dr.A.S.Bloor of the National Coal Board, Mining Research and Development Establishment.

REFERENCES

Abel,J.R., Jr. Dowis,J.E. and Richards,D.P. (1979) "Concrete shaft lining design" 20th U.S. Rock Mechanics Symp. Austin, Texas 627-33.

Bieniawski,Z.T. (1970) "Time dependent behaviour of brittle rocks" Rock Mechanics 7.

Bloor,A.S. (1980) "Riccall mine-instrumentation of No.2 shaft lining at the halite horizon" N.C.B. technical memorandum RCM/B (80)4

Farmer,I.W. and Gilbert,M.J. (1981) "Time dependent strength reduction of rock salt" First Conf. on the mechanical behaviour of salt, Pennsylvania State University.

Gilbert,M.J. (1981) "Shafts in squeezing rock" PhD. Thesis University of Newcastle upon Tyne.

Gilbert,M.J. and Farmer,I.W. (1981) "A time dependent model for lining pressure based on strength concepts" Ints. Symp on weak rocks, Tokyo.

Ladanyi,B. (1974) "Use of the long-term strength concept in determination of ground pressure on tunnel linings" Proc. 3rd Int. Conf. ISRM, Denver Pt.B. 1150-56.

Muir-Wood,A.M. (1979) "Ground behaviour and support for mining and tunnelling" 14th Sir Julius Wernher memorial lecture of the Inst. Min. and Metallurgy, London.

Neville,A.M. (1972) "Properties of concrete" Pitman, London.

Savin,G. (1961) "Stress concentrations around holes", Pergamon Press, London, 296-301.

CREEP CLOSURE OF CAVERNS IN ROCKSALT

EVAN K.S. PASSARIS
Department of Geotechnical Engineering,
University of Newcastle upon Tyne.

STEPHEN T. HORSEMAN
Engineering Geology Unit,
Institute of Geological Sciences, London.
(Formerly Research Associate, Department of Mining Engineering,
University of Newcastle upon Tyne).

Summary: The time dependent closure that occurs in solution mined caverns in salt which are to be utilized for storage purposes is a significant design problem. The paper deals with the development of analytical solutions which can be employed in predicting closure rates of internally pressurized underground caverns in rocksalt. Solutions are derived for cylindrical and spherical caverns using a power law expression for the creep constitutive equation. Since the stresses around the caverns are statically indeterminate the derived solutions are only applicable to problems examined at a time remote from the time of the construction of the openings. A comparison of the predictions of the analytical solution with in-situ measured convergence rates in a shaft sunk through a thick salt bed at a depth of 1056m, resulted in an excellent agreement.

INTRODUCTION:

Storage in solution mined caverns has emerged as a rapidly developing improvement and extension of the basic underground storage concept. Storage in solution mined salt caverns is particularly important in reduction or shaving peak load demands. This peak load shaving reduces the huge capital investment that otherwise must be made to accommodate the intensive energy demands by industry and by the private consumer.

Gas storage caverns in salt present a particular and somehow demanding rock mechanics problem because salt tends to flow or "creep" under high deviatoric stresses, which if unchecked, may result in a progressive closure of the cavern and an unacceptable loss of storage volume. Rocksalt is a highly non-linear time dependent material and this non-linearity may be incorporated in a creep model in a number of different ways using empirical creep equations (Lomenick and Bradshaw, 1962), step-wise non-linear rheological models (Serata, 1966) or rheological models with stress-dependent parameters (Schuppe, 1963).

Non-linear models have one thing in common, they predict stress redistribution around a cavern. Initially the peak tangential and peak deviatoric stresses are situated close to the cavern surface. The salt in this area responds to the high deviatoric stress by very rapid creep. This causes the peak tangential stress to move away from the cavity surface. The volume of rock influenced by the creation of the opening becomes progressively larger until eventually a "steady-state" condition is achieved. In this condition the stress distribution remainds constant with time but the cavern may continue to close. Whilst the dynamic effects of stress distribution can only be quantitatively investigated by using numerical stress analysis techniques (Boresi and Deere, 1963), the steady-state condition around openings of simple geometry can be predicted analytically as shown in this paper.

Analytical solutions imply the use of idealized cavern shapes such as spheres and long cylinders and only simple material models lead to closed form solutions. However, analytical methods enable a clear understanding of the essential features of the time-dependent cavern responce and these are emphasised in the following sections.

THE SELECTION OF A CREEP CONSTITUTIVE MODEL

The assessment of the stability of any underground opening is possible only if the mechanical behaviour of the rock is mathematically described with sufficient accuracy. The selection of a suitable mathematical model to represent the time dependent behaviour or rocksalt is an essential and important step in the solution of the creep problem related to storage caverns. The approval of a constitutive model depends on its ability to predict the behaviour of the real material. The model should be complex enough to simulate the real behaviour and yet simple enough to be handled mathematically.

The theory of linear viscoelasticity has been utilized to analyse time dependent problems by means of rheological creep models. In their simplest form such models consist of a differential equation relating stresses and strains at a specific point (Langer, 1981). However, the limitation of linearity of the stress dependence precludes the effective use of such methods for deep salt caverns, since the corresponding time-dependent response of rocksalt is highly non-linear (Lindner and Brady, 1981).

Triaxial creep tests performed on salt cores by various researchers (see table 1) suggest that the creep response can be described by the power law equation of the form:-

$$\varepsilon_c = A\sigma^n t^m \tag{1}$$

where ε_c is the creep strain, σ and t are stress and time, respectively, whereas A, n and m are material parameters. Table 1 is a compilation of the reported values of these material parameters for rocksalt as determined from laboratory triaxial creep tests, in which case the term σ in eq. 1 is equal to the stress difference $\sigma_1 - \sigma_3$.

Table 1. Experimentally determined creep parameters in eq. 1. Stress and time measured in MPa and days respectively.

A ($\times 10^{-7}$)	n	m	Source
16.216	2.98	0.360	Boresi and Deere (1963)
68.763	2.52	0.377	Hansen (1977)
0.675	3.70	0.400	Hansen (1978)
1.893	3.83	0.360	Horseman and Passaris (1981)

The total strain ocurring at any point around a salt cavern is the sum of the following strain components:

$$\varepsilon = \varepsilon_e + \varepsilon_c + \varepsilon_t \qquad (2)$$

where ε_e isthe reversible elastic strain and ε_t is the thermal strain caused by a temperature rise T. Eqn (2) in its expanded form becomes:

$$\varepsilon_{ij} = \left(\frac{1+\nu}{E}\sigma_{ij} - \frac{\nu}{E}\sigma_{kk}\delta_{ij}\right) + \left(At^m\left(\sigma_{ij} - \frac{1}{3}\delta_{ij}\sigma_{kk}\right)^n\right) + \left(\alpha T\delta_{ij}\right) \qquad (3)$$

where ν, E and α denote Poisson's ration, Young's modulus and coefficient of thermal expansion (δ_{ij} is the Kronecker delta).

In order to determine the total closure of a cavern from the time of its construction it is necessary to trace the progressively decreasing rate of closure throughout the transient phase. Since the stresses are statically indeterminate, that is, they are dependent on the creep strains, the complete solution is not amenable to analytical techniques.

However, if as a first approximation the state of stress around a cavern can be considered, at a time remote from the time of its construction as time independent, then eqn.(3) may be differentiated with respect to time and by ignoring the thermal effects we get:

$$\dot{\varepsilon}_{ij} = mAt^{m-1}\left(\sigma_{ij} - \frac{1}{3}\delta_{ij}\sigma_{kk}\right) \qquad (4)$$

CYLINDRICAL CAVERNS

For a cylindrical cavern and using cylindrical coordinates

$$\varepsilon_{zz}^{(creep)} = At^m\left(\sigma_{zz} - \frac{1}{3}\left(\sigma_{zz} + \sigma_{rr} + \sigma_{\theta\theta}\right)\right)^n \qquad (5)$$

Although cylindrical caverns are idealized as infinite cylindrical holes, the physical problem of interest involves a finite length. However, if this length is sufficiently large compared to the diameter and since a time dependent displacement along the axial direction is not probable, plane strain conditions can be assumed hence

$\varepsilon_{zz} = 0$ from which e.g. 5 yields $\sigma_{zz} = \frac{1}{3}(\sigma_{zz} + \sigma_{rr} + \sigma_{\theta\theta})$

or $$\sigma_{zz} = \frac{\sigma_{\theta\theta} + \sigma_{rr}}{2} \qquad (6)$$

Consequently, the first stress invariant σ_{kk} becomes

$$\sigma_{kk} = \sigma_{zz} + \sigma_{rr} + \sigma_{\theta\theta} = \frac{3}{2}\left(\sigma_{\theta\theta} + \sigma_{rr}\right) \qquad (7)$$

Combining eqns (4) and (7) for the radial strain component we get:

$$\dot{\varepsilon}_{rr} = mAt^{m-1}\left(\frac{\sigma_{rr} - \sigma_{\theta\theta}}{2}\right)^n \qquad (8)$$

The cross section which corresponds to the maximum stress concentration in a cylindrical cavern with vertical axis is located at a distance of approximately 2/3 of the height of the cavern measured from the roof (Ottosen and Krenk,1979). At that depth it is acceptable to assume that the Poisson's ratio of the salt surrounding the cavern will be approximately 0.5 (implying incompressibility) in which case in the absence of temperature effects we get:

$$\varepsilon_{kk} = \frac{\Delta V}{V} = 0$$

consequently $\varepsilon_{rr} + \varepsilon_{\theta\theta} + \varepsilon_{zz} = 0$ and since $\varepsilon_{zz} = 0$ the relationship becomes:

$$\varepsilon_{rr} = -\varepsilon_{\theta\theta} \qquad (9)$$

Since no rotation of the principal strain axes is expected during deformation we can differentiate eq (9) with respect to time to get:

$$\dot{\varepsilon}_{rr} = -\dot{\varepsilon}_{\theta\theta} \qquad (10)$$

The strain-displacement equations in cylindrical coordinates are:

$$\varepsilon_{rr} = \frac{\partial u}{\partial r} \qquad \varepsilon_{\theta\theta} = \frac{u}{r} + \frac{1}{r}\frac{\partial v}{\partial \theta}$$

and because of the geometrical symmetry the above equations are simplified to:

$$\varepsilon_{rr} = \frac{\partial u}{\partial r} \qquad \varepsilon_{\theta\theta} = \frac{u}{r} \qquad (11)$$

It may be shown therefore by differentiation of the latter of eqns (11) that the equation of strain compatibility is

$$\frac{\partial}{\partial r}\varepsilon_{\theta\theta} = \frac{\partial}{\partial r}\left(\frac{u}{r}\right) = \frac{1}{r^2}\frac{\partial u}{\partial r}r - u = \frac{1}{r}\left(\frac{\partial u}{\partial r} - \frac{u}{r}\right)$$

and taking into account eqns (11) we have:

$$\frac{\partial}{\partial r}\varepsilon_{\theta\theta} + \frac{1}{r}\left(\varepsilon_{\theta\theta} - \varepsilon_{rr}\right) = 0 \qquad (12)$$

By comparing eqns (10) and (12) we get

$r\frac{\partial}{\partial r}\dot{\varepsilon}_{\theta\theta} = -2\dot{\varepsilon}_{\theta\theta}$ which upon integration gives

$$\dot{\varepsilon}_{\theta\theta} = \frac{B}{r^2} \qquad (13)$$

where B is the integration constant. Furthermore, by combining eqns (10) and (13) we have

$$\dot{\varepsilon}_{rr} = -\frac{B}{r^2} \qquad (14)$$

The stress equilibrium equation for cylindrical coordinates is given by:

$$\frac{\partial}{\partial r}\sigma_{rr} + \frac{1}{r}\frac{\partial}{\partial \theta}\sigma_{rz} + \frac{\partial}{\partial z}\sigma_{r\theta} + \frac{\sigma_{rr} - \sigma_{\theta\theta}}{r} = 0$$

however, since the shear stresses are zero on account of the symmetry of the deformation the above equation is simplified to:

$$r\frac{\partial}{\partial r}\sigma_{rr} = \left(\sigma_{\theta\theta} - \sigma_{rr}\right) \qquad (15)$$

Substituting eqns (14) and (15) in eqn (8) we get

$$\frac{B}{r^2} = mAt^{m-1}\left(\frac{r}{2}\right)^n\left(\frac{\partial}{\partial r}\sigma_{rr}\right)^n$$

or $$\frac{\partial}{\partial r}\sigma_{rr} = 2\left(\frac{B}{mA}\right)^{\frac{1}{n}} t^{\frac{1-m}{n}} r^{-\left(\frac{2}{n}+1\right)} \qquad (16)$$

Since σ_{rr} was assumed at a time remote from the time of the construction of the cavern as time independent, the differential equation (16) may be solved by separation of variables and integration, yielding:

$$\sigma_{rr} = -2\left(\frac{B}{mA}\right)^{\frac{1}{n}} t^{\frac{1-m}{n}} \frac{n}{2}\frac{1}{r^{(2/n)}} + C \qquad (17)$$

The integration constants B and C of the above equation

may be determined from the boundary conditions i.e.

$$\sigma_{rr} = -P_i \quad \text{at} \quad r = r_i$$

and (18)

$$\sigma_{rr} = -P_o \quad \text{at} \quad r \to \infty$$

where P_i and P_o are the internal and external pressure respectively and r_i is the radius of the cylindrical cavern. By combining eqns (17) and (18) we get:

$$-P_i = -\left(\frac{B}{mA}\right)^{\frac{1}{n}} t^{\frac{1-m}{n}} \, n \, \frac{1}{r_i^{(2/n)}} + C$$

and $-P_o = C$

Consequently, by solving for B and C we have

$$C = -P_o$$

and $$B = -\frac{(P_o - P_i)^n r_i^2 \, m \, A}{n^n \, t^{(1-m)}} \tag{19}$$

Substituting eqns (19) in eqn (17) we get:

$$\sigma_{rr} = (P_o - P_i)\left(\frac{r_i}{r}\right)^{2/n} - P_o \tag{20}$$

and differentiating eqn (20) with respect to r we have:

$$\frac{\partial}{\partial r}\sigma_{rr} = -\frac{2}{n}(P_o - P_i)\left(\frac{r_i}{r}\right)^{\frac{2}{n}}\frac{1}{r} \tag{21}$$

Furthermore, substituting eqns (20) and (21) in eqn (15) and solving for $\sigma_{\theta\theta}$ we get:

$$\sigma_{\theta\theta} = (P_o - P_i)\left(\frac{r_i}{r}\right)^{\frac{2}{n}}\left(1 - \frac{2}{n}\right) - P_o \tag{22}$$

Similarly, by introducing eqns (20) and (22) in eqn (6) and solving for σ_{zz} we can have:

$$\sigma_{zz} = (P_o - P_i)\left(\frac{r_i}{r}\right)^{\frac{2}{n}}\left(1 - \frac{1}{n}\right) - P_o \tag{23}$$

For the calculation of the rate of the radial strain we substitute eqns (20) and (22) in eqn (8) and we get:

$$\dot{\varepsilon}_{rr} = m\,A\,t^{m-1}\left(\frac{P_o - P_i}{n}\right)^n\left(\frac{r_i}{r}\right)^2 \tag{24}$$

The radial displacement rate at a radial distance a is given by:

$$\dot{u}_a = \int_a^\infty \dot{\varepsilon}_{rr}\,dr \tag{25}$$

Therefore, by substituting eqn (24) in eqn (25) we get:

$$\dot{u}_a = mAt^{m-1}\left(\frac{P_o - P_i}{n}\right)^n r_i^2 \int_a^\infty r^{-2}\,dr$$

which upon integration yields:

$$\dot{u}_a = mAt^{m-1}\left(\frac{P_o - P_i}{n}\right)^n \frac{r_i^2}{a} \tag{26}$$

To calculate the volumetric closure rate of a cylindrical cavern, consideration is given to the following steps:

The change in volume caused by creep closure is equal to:

$$\Delta V = V_{(\text{initial})} - V_{(\text{after convergence})} = h\pi\left(r_i^2 - (r_i - u)^2\right)$$

where h is the height of the cylindrical cavern. The above equation when expanded and simplified (by assuming that u^2 is negligible) yields:

$$\Delta V = 2h\pi r_i u$$

Consequently, the volumetric strain will be

$$\frac{\Delta V}{V} = \frac{2h\pi r_i u}{h\pi r_i^2} = 2\frac{u}{r_i} \tag{27}$$

Differentiation of eqn (27) with respect to time gives the volumetric closure rate as being:

$$\left(\frac{\dot{\Delta V}}{V}\right)_{r=r_i} = 2mAt^{m-1}\left(\frac{P_o - P_i}{n}\right)^n$$

and substituting eqn (26) in the above relationship for a = r_i, we get:

$$\left(\frac{\dot{\Delta V}}{V}\right)_{r=r_i} = 2mAt^{m-1}\left(\frac{P_o - P_i}{n}\right)^n \tag{28}$$

SPHERICAL CAVERNS

By considering a spherical cavern with radius r_i and internal pressure P_i while at infinity the state of stress approaches the hydrostatic pressure P_o, the corresponding solution follows a similar pattern to the one employed for the cylindrical caverns.

On account of the symmetry there will be three non-zero stress components, the radial component and two tangential components $\sigma_{\theta\theta}$, $\sigma_{\phi\phi}$ which are identical (Timoshenko and Goodier, 1970), hence:

$$\sigma_{\theta\theta} = \sigma_{\phi\phi} \quad \text{and} \quad \varepsilon_{\theta\theta} = \varepsilon_{\phi\phi} \tag{29}$$

Consequently, the first stress invariant σ_{kk} becomes

$$\sigma_{kk} = \sigma_{rr} + \sigma_{\theta\theta} + \sigma_{\phi\phi} = \sigma_{rr} + 2\sigma_{\theta\theta} \tag{30}$$

Combining eqns (4) and (30) for the radial strain component we get:

$$\dot{\varepsilon}_{rr} = mAt^{m-1}\left(\frac{2}{3}(\sigma_{rr} - \sigma_{\theta\theta})\right)^n \tag{31}$$

It is assumed that at the depth where the spherical cavern is located the Poisson's ratio of rocksalt is approximately 0.5, that is that the rock material is considered incompressible, then in the absence of temperature effects:

$$\varepsilon_{kk} = 0 \quad \text{or} \quad \varepsilon_{rr} + \varepsilon_{\theta\theta} + \varepsilon_{\phi\phi} = 0$$

and using the second of eqns (29) the above relation becomes

$$\varepsilon_{rr} = -2\,\varepsilon_{\theta\theta} \tag{32}$$

By assuming no rotation of the principal strain axes during deformation eqn (32) can be differentiated with respect to time giving:

$$\dot{\varepsilon}_{rr} = -2\,\dot{\varepsilon}_{\theta\theta} \tag{33}$$

The strain displacement equations in spherical coordinates are:

$$\varepsilon_{rr} = \frac{\partial u}{\partial r}, \quad \varepsilon_{\phi\phi} = \frac{1}{r\sin\theta}\frac{\partial v}{\partial\phi} + \frac{u}{r} + \frac{\omega}{r}\mathrm{ctg}\,\theta \quad \text{and} \quad \varepsilon_{\theta\theta} = \frac{1}{r}\frac{\partial\omega}{\partial\theta} + \frac{u}{r}$$

and because of the spherical symmetry the above expressions are simplified to:

$$\varepsilon_{rr} = \frac{\partial u}{\partial r} \quad \text{and} \quad \varepsilon_{\theta\theta} = \varepsilon_{\phi\phi} = \frac{u}{r} \tag{34}$$

Consequently, following the same method employed for the determination of eqn (12) we can prove that the strain compatibility equation for spherical symmetry is:

$$\frac{\partial}{\partial r}\varepsilon_{\theta\theta} + \frac{1}{r}\left(\varepsilon_{\theta\theta} - \varepsilon_{rr}\right) = 0 \tag{35}$$

By comparing eqns (33) and (35) we get:

$$r\frac{\partial}{\partial r}\dot{\varepsilon}_{\theta\theta} = -3\,\dot{\varepsilon}_{\theta\theta}$$ which upon integration gives

$$\dot{\varepsilon}_{\theta\theta} = \frac{D}{r^3} \tag{36}$$

where D is the integration constant. Furthermore, by combining eqns (33) and (36) we have:

$$\dot{\varepsilon}_{rr} = -\frac{2D}{r^3} \tag{37}$$

The stress equilibrium equation for spherical coordinates is given by:

$$\frac{\partial}{\partial r}\sigma_{rr}+\frac{1}{r\sin\theta}\frac{\partial\sigma_{r\phi}}{\partial\phi}+\frac{1}{r}\frac{\partial\sigma_{r\theta}}{\partial\theta}+\frac{2\sigma_{rr}-\sigma_{\phi\phi}-\sigma_{\theta\theta}+\sigma_{r\theta}\operatorname{ctg}\theta}{r}=0$$

however, since the shear stresses are zero on account of the symmetry of the deformation, the above equation reduces to:

$$\frac{r}{2}\frac{\partial}{\partial r}\sigma_{rr}=(\sigma_{\theta\theta}-\sigma_{rr}) \qquad (38)$$

Substituting eqns (37) and (38) in eqn (31) we get

$$\frac{2D}{r^3}=mAt^{m-1}\left(\frac{r}{3}\frac{\partial}{\partial r}\sigma_{rr}\right)^n$$

or $$\frac{\partial}{\partial r}\sigma_{rr}=3\left(\frac{2D}{mA}\right)^{1/n}t^{(1-m)/n}\,r^{-\left(\frac{3}{n}+1\right)} \qquad (39)$$

Following the same arguments employed during the derivation of eqn (17) we can integrate eqn (39) to get:

$$\sigma_{rr}=-\left(\frac{2D}{mA}\right)^{\frac{1}{n}}t^{\frac{1-m}{n}}\,n\,\frac{1}{r^{(3/n)}}+E \qquad (40)$$

The integration constants D and E of eqn (40) may be determined from the boundary conditions i.e.

$$\sigma_{rr}=-P_i \quad \text{at} \quad r=r_i$$

and $$\sigma_{rr}=-P_o \quad \text{at} \quad r\rightarrow\infty \qquad (41)$$

By combining eqns (40) and (41) we get:

$$-P_o=E$$

and $$-P_i=-\left(\frac{2D}{mA}\right)^{\frac{1}{n}}t^{\frac{1-m}{n}}\,n\,\frac{1}{r_i^{(3/n)}}+E$$

Consequently, by solving for D and E we have

$$E=-P_o \qquad (42)$$

and $$D=-\frac{(P_o-P_i)^n r_i^3 mA}{2n^n t^{(1-m)}}$$

Substituting eqns (42) in eqn (40) we get:

$$\sigma_{rr}=(P_o-P_i)\left(\frac{r_i}{r}\right)^{\frac{3}{n}}-P_o \qquad (43)$$

and differentiating eqn (43) with respect to r we have:

$$\frac{\partial}{\partial r}\sigma_{rr}=-\frac{3}{n}(P_o-P_i)\left(\frac{r_i}{r}\right)^{\frac{3}{n}}\frac{1}{r} \qquad (44)$$

Furthermore, substituting eqns (43) and (44) in eqn (38) and solving for $\sigma_{\theta\theta}$ we get:

$$\sigma_{\theta\theta}=(P_o-P_i)\left(\frac{r_i}{r}\right)^{\frac{3}{n}}\left(1-\frac{3}{2n}\right)-P_o \qquad (45)$$

For the calculation of the rate of the radial strain we substitute eqns (43) and (45) in eqn (31) and we get:

$$\dot{\epsilon}_{rr}=mAt^{m-1}\left(\frac{P_o-P_i}{n}\right)^n\left(\frac{r_i}{r}\right)^3 \qquad (46)$$

The radial displacement rate at a radial distance a is given by

$$\dot{u}_a=\int_a^\infty \dot{\epsilon}_{rr}\,dr \qquad (47)$$

Therefore, by substituting eqn (46) in eqn (47) we get

$$\dot{u}_a=mAt^{m-1}\left(\frac{P_o-P_i}{n}\right)^n r_i^3\int_a^\infty r^{-3}\,dr$$

which upon integration yields:

$$\dot{u}_a=\frac{m}{2}At^{m-1}\left(\frac{P_o-P_i}{n}\right)^n\frac{r_i^3}{a^2} \qquad (48)$$

To calculate the volumetric closure rate of a spherical cavern, consideration is given to the following steps:

The change in volume caused by creep closure is equal to:

$$\Delta V=V_{(\text{initial})}-V_{(\text{after convergence})}=\frac{4}{3}\pi\left(r_i^3-(r_i-u)^3\right)$$

The above equation when expanded and simplified (by assuming that the expression , and are negligible) yields:

$$\Delta V=4\pi r_i^2 u$$

Consequently, the volumetric strain will be

$$\frac{\Delta V}{V}=\frac{4\pi r_i^2 u}{(4/3)\pi r_i^3}=3\frac{u}{r_i} \qquad (49)$$

Differentiation of eqn (49) with respect to time gives the volumetric closure rate as being:

$$\left(\frac{\dot{\Delta V}}{V}\right)=\frac{3}{r_i}\dot{u}$$

and substituting eqn (48) in the above relationship for $a = r_i$ we get:

$$\left(\frac{\dot{\Delta V}}{V}\right)_{r=r_i}=\frac{3}{2}mAt^{m-1}\left(\frac{P_o-P_i}{n}\right)^n \qquad (50)$$

A COMPARISON OF IN-SITU MEASURED CONVERGENCE RATES WITH THE PREDICTIONS OF ANALYTICAL SOLUTIONS

A cirular 7.6m diameter shaft was sunk through Permian upper evaporite deposits for the Boulby Potash Mine, near Staithes, in Cleveland, U.K. At a depth of 1056m four radial borehole extenseomter were installed in the walls of the underground opening in a thick bed of upper halite of the Permian Zechstein IV. Within each hole there were four anchors located at radial distances of 4.4m, 5.3m, 6.8m and 8.3m, with rods attached for reading with a dial gauge probe at the mouth of station. A detailed description of the measuring instruments and their installation is given elsewhere (Cleasby et al 1975).

After readings have been taken for almost three years, sufficient in-situ data had been obtained to assess the creep rates at each anchor depth and an extremely good fit was obtained (Hebblewhite et al, 1977) using the following expression for the rate of displacement:

$$\dot{u}=F\,t^b \qquad (51)$$

where the parameters F and b are given in Table 2 together with their corresponding correlation coefficient of the curve fitting technique employed.

Table 2 Parameters of eqn (51) after Hebblewhite et al (1977).

Radial distance of anchors (m)	F	b	Correlation coefficient
4.4	1.972	-0.561	0.999474
5.3	1.452	-0.542	0.999670
6.8	1.305	-0.556	0.999841
8.3	0.941	-0.516	0.999320

In eqn (51) the displacement rate is expressed in (mm/day) and time is expressed in days.

For the purpose of the analysis the average temperature around the shaft excavation is required. The average geothermal gradient for the rocks overlying the coal measures in Great Britain (including the Permian) is usually quoted as 1°C per 38.4m. The temperature at a depth of 15.2m is considered to be constant at 10°C. Thus the temperature T at a depth Z in metres may be estimated (Sinclair, 1958) as follows:

$$T\,(^{\circ}C)=10+\frac{Z-15.2}{38.4}$$

The above equation, for a depth of 1056 m corresponding to the instrumented cross section of the shaft, gives a temperature of 37°C.

The assumptions made during the derivation of the

equation for the radial displacement rate (eqn 26), restrict its application to a time remote from the time of the construction of the opening.

Taking into account that any time after a year from the excavation of the shaft is, in real terms, adequately remote, a period of 400 days was considered acceptable for the purposes of this analysis. Considering that it is not feasible to calculate the exact heat losses to shaft air over a period of 400 days it was decided to employ reasonable assumptions based on practical considerations. On that basis a value of 34°C was employed as being a representative average equilibrium temperature of the salt around the shaft at the depth where the in-situ measurements were taken.

The creep parameters shown in table 1 were defined by laboratory tests performed at an ambient temperature of approximately 20°C (with the exception of the last reference where the creep tests were conducted at 57°C), implying that a certain adjustment must be made if they are to be used in conjunction with the radial displacement data from Boubly Shaft. Several investigators (Lomenick 1971, Thoms et al 1972, Hansen 1979, Hansen and Mellegaard 1979) employed an empirical relationship to express the temperature dependency of the creep constitutive equation for salt, in the form of:

$$\varepsilon_c = K T^s \sigma^n t^m \qquad (52)$$

where K and s are constants and T is measured in °K.

By direct comparison of eqns (1) and (52) it may be shown that eqn (1) can be adjusted so that it could be used for temperatures different from the one used during the tests for the derivation of its constants. This can be achieved by multiplying the creep parameter A of eqn (1) by a factor equal to:

$$\left(T_{(\text{in situ})} / T_{(\text{laboratory})} \right)^s$$

where the temperature exponent s is equal with 9.5725, this being the mean of the values quoted by the aforementioned references. Consequently, table 1 can be reproduced with the adjusted values of the parameter A as follows:

Table 3 Creep parameters from table 1 adjusted for a temperature of 34°C.

$A\ (\times 10^{-7})$	n	m	Source
25.350	2.98	0.360	Boresi and Deere (1963)
107.497	2.52	0.377	Hansen (1977)
1.055	3.70	0.400	Hansen (1978)
0.948	3.83	0.360	Horseman and Passaris (1981)
33.7125	3.26	0.374	Mean values

Employing a vertical geostatic stress gradient of 22.6 kPa/m depth and taking into account that a hydrostatic state of stress exists in the salt mass at infinite radial distance from the shaft, P_o = 23.9 MPa at 1056m depth. The internal pressure in the shaft may be taken as zero since all pressures are above atmospheric datum.

Consequently, the radial displacement rate of 400 days for the Boulby Shaft (with r_i = 3.8m) at a depth of 1056m (corresponding to P_o = 23.9 MPa) using the mean values of the creep parameters shown in table 3 and employing eqn (26) is given by

$$\dot{u}_a = 0{\cdot}374 \times 33{\cdot}7125 \times 10^{-7} \times 400^{(0{\cdot}374-1)} \times \left(\frac{23{\cdot}9}{3{\cdot}26}\right)^{3{\cdot}26} \times 3{\cdot}8^2 \times \frac{1}{a} \quad \text{m/day}$$

or

$$\dot{u}_a = \frac{0{\cdot}283}{a} \quad \text{mm/day} \qquad (53)$$

where a is the radial distance expressed in m.

In table 4 the calculated anchor displacement rates using eqn (53) are compared with the measured displacements as given by eqn (51) for t = 400 days.

Table 4 Comparison of measured radial displacement rates at 400 days with the predictions of eqn (26).

Radial distance of anchor (m)	Measured displacement rate $(\text{mm/day})\times 10^{-2}$	Predicted displacement rate $(\text{mm/day})\times 10^{-2}$
4.4	6.842	6.432
5.3	5.645	5.340
6.8	4.665	4.162
8.3	4.275	3.410

The correlation between the measured closure rates and the predicted values based on laboratory testing results reported by various workers is very good. However, the results presented here must be viewed in a realistic manner. The shaft problem was analysed in terms of plane strain. This is only a first approximation to the conditions which will prevail in the walls of the opening.

Both the in-situ temperature and geostatic stresses were assumed. Both assumptions may be subject to considerable error. Finally, it appears from table 3 that large variations exist in the creep parameter values, this being particularly true for the Constant A. As a result, the average values employed in the analysis are subjected to a certain degree of scatter. Consequently, if the analysis were to be repeated for some more extreme laboratory results, then the correlation with in-situ behaviour might be less impressive.

Despite these reservations, the results presented here are sufficiently encouraging to allow the tentative use of the developed solutions for the closure rate of underground caverns in rocksalt.

This paper is published with the permission of the Director, Institute of Geological Sciences (NERC).

REFERENCES

Boresi, A.P. and Deere, D.U., 1963, Creep closure of a spherical cavity in an infinite medium (with special application to project Dribble, Tatum salt dome, Mississippi) Report to Holmes and Narver, Inc., Las Vegas Division TID-20083.

Cleasby, J.V, Pearse, G.E., Grieves, M. and Thorburn, G. 1975, Shaft-sinking at Boulby Mine, Cleveland Potash Ltd., Trans. Inst. Min & Metal./Sect A 84(824):A7-28.

Hansen, F.D., 1977, Case history rock mechanics examination of the Jefferson Island salt mine: II, (RE/SPEC, Inc., Rapid City, SD, USA), Y/OWI/SUB-77/2203/5, June, 1977.

Hansen, F.D., 1978, Quasi-static strength and creep deformational characteristics of bedded salt from the Carey mine near Lyons, Kansas, (RE/SPEC, Inc., Rapid City, SD, USA), Y/OWI/SUB-78/22303/13, June, 1978.

Hansen , F.D., 1979, Triaxial quasi-static compression and creep behaviour of bedded salt from south eastern New Mexico, SAND-79-7045, Sandia Laboratories, Alberquerque, New Mexico, Nov. 1979.

Hansen, F.D., and Mellegaard, K.D., 1979, Creep behaviour of bedded salt from south eastern New Mexico at elevated temperature, SAND-79-7030, Sandia Laboratories, Alberquerque, New Mexico, Nov. 1979.

Hebblewhite, B.K., Miller, H.D.S. and Potts, E.L.J., 1977, A method for predicting time dependent deformations in evaporites around a vertical shaft, Proc.

Conf. Rock Eng., Brit. Geotech. Soc: 529-539.
Horseman, S.T. and Passaris, E.K.S., 1981, Creep tests for storage cavity closure predictions, Proc. 1st Conf. on the Mech. Behaviour of Salt, Pennsylvania State Univ.
Langer, M., 1981, The rheological behaviour of rocksalt, Proc. 1st Conf. on the Mech. Behaviour of Salt, Pennsylvania State Univ.
Lomenick, T.F., 1971, Laboratory pillar model experiments, Chapt. 12 in "Project Salt Vault: A demonstration of the disposal of high-activity solidified wastes in underground salt mines" edited by Bradshaw, R.L. and McLain, W.C., Oak Ridge National Laboratory, Oak Ridge, Tenessee, April, 1971.
Lomenick, T.F. and Bradshaw, R.L., 1969, Deformation of rocksalt in openings mined for the disposal of radioactive wastes, Rock Mech. 1: 5-30.
Lindner, E.N. and Brady, B.H.G., 1981, Memory aspects of salt creep, Proc. 1st Conf. on the Mech. Behaviour of Salt, Pennsylvania State Univ.
Ottosen, N.S. and Krenk, S., 1979, Non-linear analysis of cavities in rocksalt, Int. J. of Rock Mech & Min. Sc. & Geomech. Abstr 16 (4): 245-252.
Schuppe, F., 1963, Ein rheologisches Modell fur das Salzgebirge, Bergakademie , 15: 583-586.
Serata, S., 1966, Continuum theory and model of rocksalt structures, 2nd Symp. Salt 2: 1-17, North Ohio Geological Soc., Cleveland, Ohio.
Thoms, L.R., Char, C.V. and Bergeron, W.J. 1972, Finite element analysis of rocksalt pillar modes, Rock Mech, 1972.
Timoshenko, S. and Goodier, J.N., 1970, Theory of elasticity (3rd edit). N. York, McGraw-Hill Book Company.

SMALL SCALE BRITTLE MODEL STUDIES

OF MINE ROADWAY DEFORMATION

Norman Brook

Department of Mining and Mineral Engineering
University of Leeds LS2 9JT

SUMMARY. A field investigation of the roadway behaviour in the tailgate of a retreat face at Fryston Colliery is reported. The measurements of load, rock properties and deformation are compared with the behaviour of 1 to 150 scale models, designed according to a comparative behaviour criterion, based on tensile strength measurements. The importance of modelling the dominant rock structure, in addition to stress and strength, is considered.

INTRODUCTION

The available methods of studying the behaviour of mine roadways can be considered in four categories. The first category is the "engineering analysis" which seeks to simplify the mathematics, and physics, of the problem by making suitable assumptions as to the properties and behaviour of the materials involved. Because of the still complex applied mathematics remaining after the assumptions, this approach is mainly restricted to the analysis of circular roadways (Labasse 1950, Wilson 1977) A second category can be recognised as 'numerical methods' such as the finite element method, (Zienkiewicz 1971) which although it requires extensive computer time can, in its advanced forms, consider some differences in rock properties, discontinuities and three dimensional situations. The boundary integral method is more efficient in computer time, but is limited to linear elastic material properties. Both these computer based methods are obviously difficult to use for situations where the rock is stressed to failure, and post failure behaviour is generally outside their scope.

The third method of considering roadway behaviour is by field investigations, (Whittaker 1976) often with generalised geological information and the analysis mainly of statistical trends in the performance. A fourth method uses scale models, where brittle fractures can occur, and loading is possible after considerable 'failure' of the constituent parts. The use of models is commonplace in some branches of engineering, such as hydraulics and aerodynamics, where generally accepted scaling criteria are used. The use of models in rock mechanics is considerably less well-established, due in some part to the large sizes sometimes used (Everling, 1964) or the apparently expensive and complex rigs required (Hobbs, 1966), but also due to the difficulties of using dimensional analysis methods in a system where most of the variables involved have the same dimensions, that of stress, so that well established theories of real value in other disciplines such as Buckingham's 'Pi' theorem, are not as directly applicable. This paper is an account of the development of a conveniently small sized scale method, depending on the modelling of selected behaviour, and its comparison with field observations.

SITE INVESTIGATIONS

A comprehensive investigation was conducted at Fryston Colliery in the tailgate of the 77's retreat face. Three detailed observation sites were selected, at 130, 270 and 400 m inbye from the entrance, with the face start line 510 m inbye. One objective of the investigation was to consider the effect of a sandstone layer, known to be near the coal seam but at a varying height above it. Diamond drill cones were obtained at each site to a depth of 10 m below the roadway and to a height of 10 m above or until the sandstone layer had been intersected. Each core was logged for recovery, intact core length (average length of pieces up to 100mm in length) and the indirect tensile strength as determined by the T_{500} point load test (Brook 1980).

Table 1 summarises the main geological observations

Site	Height to Sandstone	Sandstone Thickness	Height to Beeston Rider	Depth to thin floor sandstone
1	13.9m	2.1m	9.9m	3.5m
2	11.5	1.6	6.3	1.5
3	7.5	3.2	2.2	1.9

	Roof Strength Range T_{500}..MPa	Floor Strength Range	Seam Strength Range
1	4.25 - 10.38	4.63 - 11.38	1.17 - 1.79
2	2.75 - 10.25	4.13 - 12.5	1.17 - 1.79
3	2.13 - 10.75	4.75 - 12.5	1.17 - 1.79

TABLE 1 Borehole core data

The initial rock stress in the vicinity of the roadway was attempted to be measured by a modified "doorstopper" technique, summarised in figure 1. The success rate of this was disappointingly low, but two good attachments of the strain gauge carrying device to a subsequently detached but intact core were attained. The measured strains were used to calculate the assumed horizontal and vertical principal stresses using a method which links strain directly to field stresses (Seymour 1979), without the usual intermediate step of considering plane stress at the borehole end, with stress concentration factors sensitive to the value of Poisson's ratio. For field stresses of σ_1, σ_2 in the plane of the borehole end, and σ_3 along the borehole. Young's Modulus E, Poisson's ratio ν and principal strains ε_1, ε_2 at the flattened borehole end, a finite element analysis, which agrees with published fully three dimensional solutions (Seymour 1977) can be used to derive the following equations

$$E\varepsilon_1 = 1.36\sigma_1 - (0.13 + \nu)\sigma_2 - 0.48\sigma_3$$

$$E\varepsilon_2 = 1.36\sigma_2 - (0.13 + \nu)\sigma_1 - 0.48\sigma_3$$

for $0.2 < \nu < 0.4$

- although these equations do depend on Poisson's ratio, its effect is apparently much smaller than in other solutions.

Table 2 summarises the results obtained from the two successful tests, one at Site 2 and one at Site 3. Despite the apparent success of the measurement at Site 2, the more preferable figure for initial stress would seem to be that for Site 3.

Site	2	3
Measured Strains (60° Rosette A vertical)	A = 706 B = 520.5 C = 360.5	A = 1845.5 B = 1261 C = 1155 micro strain
Youngs Modulus	3.45 GPa	3.45 GPa
Poisson's Ratio	0.31	0.31
Principal Strains at Borehole end	729 329	1850 991
Angle of major strain to vertical	13° 46'	4° 6'
In situ stresses for zero stress along borehole	σ_ν = 2.38 MPa σ_η = 1.63 MPa	σ_ν = 6.15 MPa σ_η = 4.50 MPa
Calculated depth pressure	14.96 MPa	14.96 MPa

TABLE 2. Doorstopper stress determination results.

After this determination of the initial stress, the subsequent increase in stress at each site due to the additional loading caused by the approach of the coal face was measured by MRDE stiff inclusion "stress plugs" installed in pairs one to measure vertical stress and the other horizontal stress (Ravon 1980). As this method of stress measurement is sensitive to stress both in line with the measuring element and at right angles to this direction, each plug was first individually calibrated in both directions to determine the coefficients α and β defined as:-

$\alpha = E_b \frac{\varepsilon}{\sigma}$ for the "sensitive" axis of the plug, and

$\beta = - E_b \frac{\varepsilon}{\sigma}$ for loading perpendicular to the axis

where E_b is Young's Modulus for brass, the plug material, ε the measured strain and σ the applied stress. If the principal stresses are assumed to be σ_{vr} and σ_{hr} for the rock with v vertical and h horizontal, these induce stresses in the plug of σ_{vp} and σ_{hp} . The measured strain for the sensitive axis vertical is ε_v , and for the horizontal ε_h, and if α_1 β_1 are the calibration constants for the "vertical" stress plug, and α_2 β_2 those for the plug with the sensitive axis horizontal, then

$$E_b\varepsilon_v = \alpha_1\sigma_{vp} - \beta_1\sigma_{hp} \quad \text{and}$$

$$E_b\varepsilon_h \quad \alpha_2\sigma_{hp} - \beta_2\sigma_{vp}$$

from which $\sigma_{vp} = E_b \frac{(\alpha_2\varepsilon_v + \beta_1\varepsilon_h)}{(\alpha_1\alpha_2 - \beta_1\beta_2)}$

and $\sigma_{hp} = E_b \frac{(\alpha_1\varepsilon_h + \beta_2\varepsilon_v)}{\alpha_1\alpha_2 - \beta_1\beta_2}$

The corresponding stress increases in the rock were then calculated from

$$\sigma_{vr} = K \ \sigma_{vp} + k \ \sigma_{hp}$$

and $\sigma_{hr} = K \ \sigma_{hp} + k \ \sigma_{vp}$

as quoted by Wilson (1960), using the constants K and k as determined by Continho. All the stress and load measurements were made in boreholes 3m deep in the ribside of the roadway, in solid ground, the two sets of measurements being combined to establish values of vertical and horizontal stress in the immediate vicinity of the measuring sites. Not all the load measurements were successful, those at site 1, the first attempted, being the least satisfactory, and for site 2 the assumption that $\sigma_{hr} = 0.64 \ \sigma_{vr}$, obtained from site 3 observations, had to be made.

The measurements of roadway closure, for the approximate original 3.5 m width by 4.9 m height, attempted at each site were by the installation of short rock bolts (0.9m) fully resin grouted, with locked nuts adapted for use with a steel tape. Three measurements were made at each station, the roadway width at a height of about 1m, and the distance up and down from this line to the measuring pins in roof and floor, so that some separation of roof and floor movement could be considered. Two sets of measurements were taken at each of the three sites. One set of rock separation devices was installed at each site, these comprising 3.7m steel bolts, point anchored with resin, for the roof and ribside, a 1.85 m steel bolt in the floor and a 3.7 m wood dowel in the faceside, each with 0.5m steel pipes resin grouted in the hole mouths. The apparent movement of the bolts into the pipes was taken as a measurement of rock separation over the length of the bolt. The floor separation meter at site 3 was unsatisfactory as the lower end could not be anchored.

The taking of the various measurements for the sites was performed at regular intervals, generally weekly, for 10 months at site 1 which was then abandoned due to the passage of the face, for 13 months at site 2 and for 16 months at site 3. At the end of the monitoring period the face had progressed 15 to 20 m past each station, the MRDE stress plugs readings being still possible with the use of the extended cables, but closure measurements being ended on the erection of "no road" signs. Figure 2 summarises the results for the vertical percentage closure for site 2, plotted against the vertical stress, similar results being obtained for site 3, but with the closure limited to about 12% at a stress of 12.5 MPa, as shown in Figure 3.

SCALE CRITERIA FOR MODELLING

Previous methods of deciding the geometrical and strength scales for models include work by Everling (1964) who used 1/10 for size, and 1/10 for strength, based on the compressive strength. Hobbs (1966) attempted to obtain scale factors for the fundamental dimension of "length", chosen at 1/50, "mass" determined by measured density and geometry, and "time", determined by equating the acceleration due to gravity for 'model' and 'prototype'. This unfortunately leads to a fixed requirement of strength scale, and the modelling of 'g' in a static system, seems unnecessary. The scaling systems used in hydraulics generally consider one specific aspect of performance, such as flow pattern in Reynold's number, and for the failure of rock due to compressive loading the critical performance could be the tensile strain, induced by compressive stress. This can be considered a failure criterion for both strength tests and model or rock failure. For a material of tensile strength T_o and Young's Modulus E, the critical strain e_f at failure is $e_f = \frac{T_o}{E}$

For a rock mass, or model, compressed by force F, general linear dimension L, Poisson's ratio ν, the tensile strain is

$$e_t = \frac{F\nu}{L^2 E}$$

For similar behaviour in model and prototype the ratio of these should be the same, i.e.

$$\frac{F\nu}{L^2 E} \div \frac{T_o}{E} = \frac{F\nu}{L^2 T_o} = \text{a constant}$$

As the constant is dimensionless, this is called the Rock Strength Similarity Number, N_R. Its use implies that a model made of stronger material would simply require a bigger force to cause it to fail. It is important to use the same method of finding the tensile strength for both 'prototype' rock, in the mine, and the model material. The T_{500} method of point load testing, using sufficiently sensitive apparatus, is capable of this wide range of strength testing.

A similar method can be used to determine the characteristics of the model arch supports for the roadway. These can be deemed to fail when the stress due to bending is the yield stress of the material. The bending moment will depend on F x L, and if the arch section modulus is Z, the stress due to bending is $\frac{FL}{Z}$. The ratio of this to the yeld stress T_y should be constant, i.e.

$\frac{FL}{ZT_y}$ = a constant, the Support similarity number, N_S

and with F as determined by the Rock Strength Similarity number.

this is $N_S = N_R \left(\frac{L^3 T_o}{Z \nu T_y} \right)$. The use of this "number" allows the section modulus required to be determined from the properties of the mine arches and the selected model material. If Z is the section modulus, with subscript m for the model, then

$$Z_m = Z \left(\frac{L_m}{L}\right)^3 \frac{T_{om}}{T_o} \frac{\nu}{\nu_m} \frac{T_y}{T_{ym}}$$

MODEL CONSTRUCTION AND TESTING

The model test rig, as shown in figure 4, is basically designed to obtain a condition of plane strain for a section of the rock mass around the roadway. It is important to maintain this condition, as although the stress distributions for elastic material is the same for both plane stress and plane strain, the strains are different, especially in the post failure condition. The load was applied via platens and a spring loading system, which allowed some relaxation of the load over the roadway as this yielded. Vertical loading only was directly applied, and the horizontal loading induced entirely by the confinement of the model. To maintain a plane strain situation the model material must be weak enough to fail without inducing high bursting loads on the front and back plates of the test rig.

A mixture of sand, barytes and dental plaster was used to obtain a material with a strength of 1/20 of the equivalent rock, the actual strength being controlled by varying the amount of plaster added to a standard 6 to 1 mix of sand/barytes. The strength range used was 0.06 to 0.5 MPa on the T_{500} scale. The various sections of the model were cast in layers, for corresponding strength, of 3 or 6 mm thickness, with "bedding planes" being modelled by sprinkling a layer of mica flakes on each layer of plaster mixture. The roadway shape was cut out of the cast beams by a saw and jig. The model arches were made from soft aluminium rod of 0.73 mm diameter to model the steel arches of assumed yield strength of 460 MPa . The convergence of the sides and roof to floor was measured by steel wire pins glued into 7 mm deep 0.5 mm diameter holes in sides roof and floor. The ends of the pins were allowed to protrude past the model arches, and the subsequent closure determined by measuring relative positions with a vernier scale travelling telescope. After construction the models were installed in the rig which was then clamped together by the end nuts and screws, and by the front and back wedges, and the whole assembly fitted into an Instron compression testing machine. The models were preloaded for a few hours at the load equivalent to the initial rock stress of 6.15MPa (3kN), before unloading and testing in regular 1 kN or so increments up to about 20 kN. The load equivalent to the full depth pressure would be about 7.5 kN. The vertical closures obtained for the models for sites 2 and 3, these being the average values for these models for each site, are shown in figures 2 and 3. The general correspondence of the graphs for site 3 is particularly good, and indicates that the model system is capable of similar effects of closure and ground movement as the mine situation. The strength similarity number system does not require any gross "correction factor" to allow for the different strength of rock mass and sample, and has provided a reasonably accurate method of scaling the loads involved. A similar conclusion was reached with an earlier investigation (Brook 1977) when the size scale was 1/30, but this scale of model, about 1 m by 0.75 m, was very time consuming to construct and test. With similar results possible using the quickly made models of 1/150 scale (each model taking about 1 day or less to make) it was decided to investigate the role of discontinuities on the deformation of the roadway.

THE EFFECT OF ROCK STRUCTURE

To investigate the effects of varying the structural features of the rock mass on the roadway, a series of twelve model tests was made, with each model made according to the geological data of site 3. Table 3 summarises the various constructions and results. Whilst better results could perhaps have been obtained by repeating each model 3 times or so, as for the three main test sites, the results do show that the dominant control over the floor lift created is the "bedding plane" modelling in 3 mm thickness, (equivalent to 0.45 m) in the floor.

Model No.	Hor.joints/ Vertical joints	Bedding Planes	Type of Floor Failure	Percent.vertical closure at 5 and 10 KN	
1	4/0	No	Plug	4.2	14.3
2	8/0	No	Plug	10.9	21.8
3	12/0	No	Plug	7.8	21.1
4	12/0	3 Floor	Slight Lift	11.9	26.9
5	12/0	6 Floor	Lift	10.6	21.3
6	12/0	3 Roof	Slight Lift	11.2	22.9
7	12/0	6 Roof	Slight Lift	8.9	20.1
8	12/0	6 Roof 6 Floor	Massive Lift	9.2	26.5
9	4/3	No	Side cracks	7.3	13.4
10	4/1	No	Side cracks	5.8	12.8
11	4/7	No	Plug	7.6	20.0
12	12/7	6 roof 6 floor	Massive lift	10.6	26.5

TABLE 3. Model Tests on Rock Structure

CONCLUSIONS

The comparisons of vertical closure in the mine situation and for the corresponding model, shown in figures 2 and 3, indicate that for low loads the model closure exceeds that for the mine. This is presumably due to the uncompacted nature of the model material, although all models were pre-loaded to take up any gross crushing of this material. At higher values of stress and closure the model would require a larger load than used to obtain accurately comparative closure. The most encouraging aspect of the model work was the simple visual result of floor lift occurring at moderate stresses, without the application of horizontal pressure, and the general correspondence of the measured vertical stresses and applied model loads determined from the rock strength similarity number for two different size scales. The importance of modelling the rock structure was emphasised by the separate tests of varying structures. The lack of model closure at the higher loads may well be due to the great difficulty of modelling the rock mass structure in sufficient detail, and a size scale of about 1 to 75 may be much better from this point of view. Some problems are encountered in measuring the strength of the very weak model material by the point load test, as in some cases considerable penetration of the test piece by the loading points occurred before failure, giving low values of strength. Some early tests (Hanif 1974) used the modulus of rupture as the tensile strength, but as this did not at the time seem possible with badly fractured and heavily bedded rock cores from the vicinity of a coal seam, the point load test was adopted and developed. This does have a "size effect", and either standard sizes, or size correction, have to be used (Brook 1980). A simple method of measuring the tensile strength, by bending, has now been developed as the "three point test", which is substantially free from size and shape effects and is not critically affected by loading point penetration. In this test, a thin disc of rock, or other material, obtained in some cases by chisel splitting of a core across bedding planes, is supported on two points (10 mm diam steel balls) and loaded centrally by the third. Provided that the "span to thickness" ratio exceeds 2, reliable values of bending strength are obtained using the classical "beam theory", which correspond very closely to values obtained by more conventional beam tests using cores. The big advantage of this three point test is that it does not require carefully cut or lapped test specimens, especially in bedded rocks. The pieces often trimmed from the ends of compressive strength specimens are in fact readymade test pieces with any rock anisotropy features automatically in the correct direction, and any convenient size can be used. It is hoped that in the future the use of a slightly larger model with the capability of more detailed representation of the structural features, together with the revised method of obtaining tensile strength and perhaps the option of side loading of the test rig will lead to an even more realistic and quantitative modelling system.

ACKNOWLEDGEMENTS

The author would like to gratefully acknowledge the help of Professor P.A. Young for his constant encouragement, Mr. K. Moses, the then Chief Mining Engineer, North Yorks N.C.B. for initiating the field investigation, Mr. E.R. Wastell, agent manager at Fryston Colliery, and many members of his staff, for supporting the field work, the staff of the North Yorks N.C.B. area headquarters, especially the geologists and area boring team for obtaining the geological cores, Dr. A.H. Wilson and Dr. D.P.Ashwin, and other members of MRDE for the provision of the stress plugs and strain meter, colleagues and staff, especially Dr. M. Hanif, Dr. K.P. Seymour, Dr.R. Krishna and Dr. A.A. Rayan who worked on various aspects of research projects connected with the paper at the University of Leeds.

REFERENCES

Brook, N. 1977 "Model Studies of mine roadway deformation. Trans.I.Min.E. ph 375-384. Vol. 136.

Brook, N. 1980 "Size correction for point-load testing. Int.J.Rock Mech.Min. Sci. & Geomech. Abs. 17, No. 4 pp 231-235.

Everling, G. 1964 "Model tests concerning the interaction of ground and roof support in gate roads" Int J.Rock Mech.Min.Sci & Geomech Abs. 1 pp 319-326.

Hanif, M. 1974 "Support and deformation of underground openings". Ph.D. Thesis Dept. Mining and Mineral Engineering, University of Leeds.

Hobbs, D.W. 1966 "Scale model studies of strata movement around mine roadways : apparatus, technique and some preliminary results". Int J. Rock Mech.Min.Sci & Geomech. Abs. 3 pp 101 - 127

Labasse 1950. as quoted by Woodruff, 1966 "Methods of working coal and metal mines" Vol.1 (Pergamon Press)

Rayan, A.A. 1980 "Application of Rock Strength Tests to Physical Models". Ph.D. Thesis. Dept of Mining and Mineral Engineering, University of Leeds.

Seymour, K.P. 1977 Discussion on "Three-dimensional elastic stress distribution around the flat end of a cylindrical cavity" by G. Hocking. Int J. Roch Mech.Min.Sci & Geomech. Abs 14 pp 295 - 296.

Seymour K.P. 1979 "Aspects of underground rock stress determination" Ph.D. Thesis. Dept of Mining and Mineral Engineering, University of Leeds.

Whittaker,B.N. 1976 "Recording, treatment and interpretation of roadway deformation surveys". Tran.I.Min.E. pp 607-617 Vol 135 No. 184.

Wilson, A.H. 1960 "A laboratory investigation of a high modulus borehole plug gauge for the measurement of rock stress". M.R.E. Report No. 2181.

Wilson, A.H. 1977 "The stability of tunnels in soft rock at depth". Conf. Rock Engineering. University of Newcastle-upon-Tyne.

Zienkiewicz, O.C. 1971 "The finite element method in Engineering Science". McGraw Hill

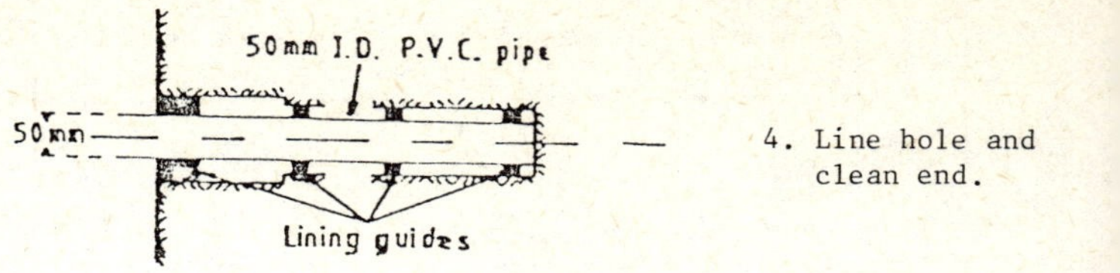

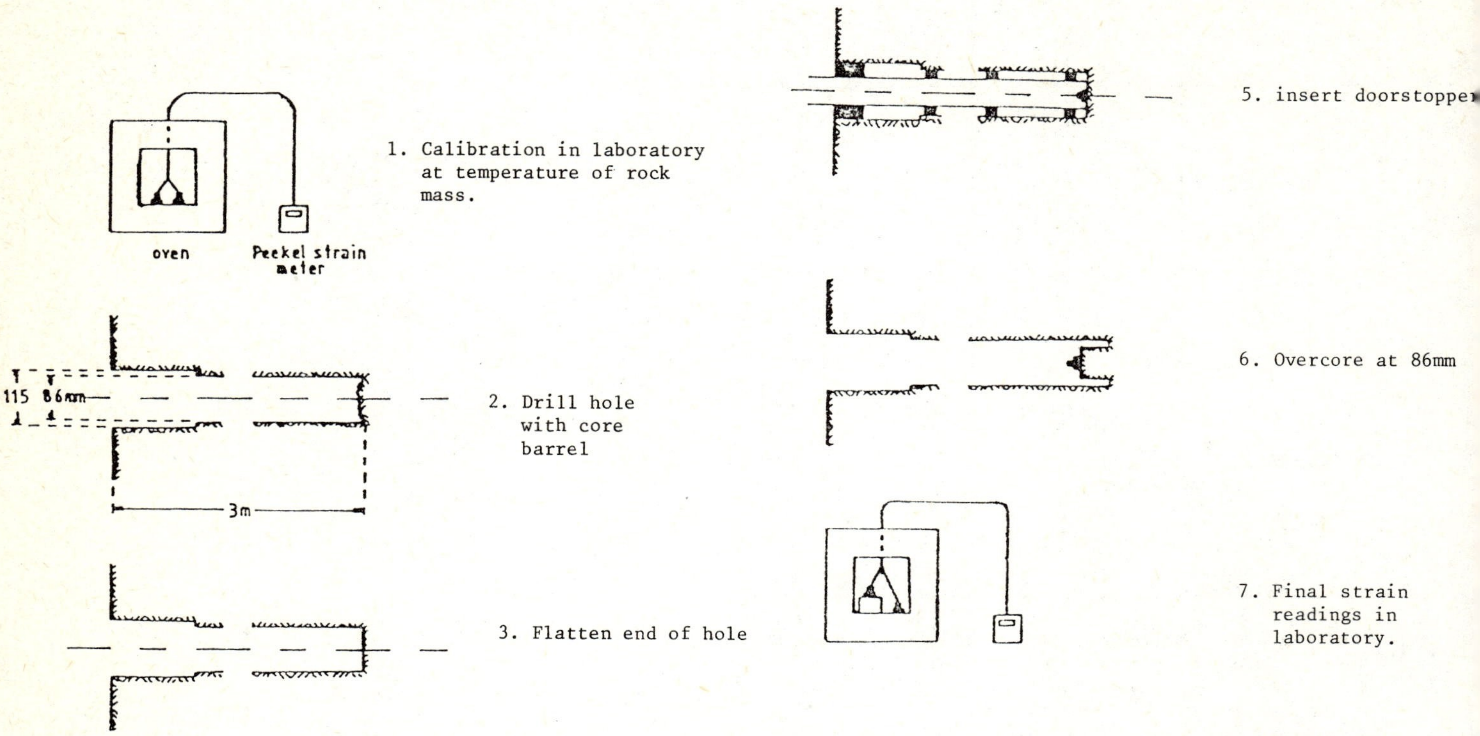

Figure 1. "Doorstopper" stress measurement method

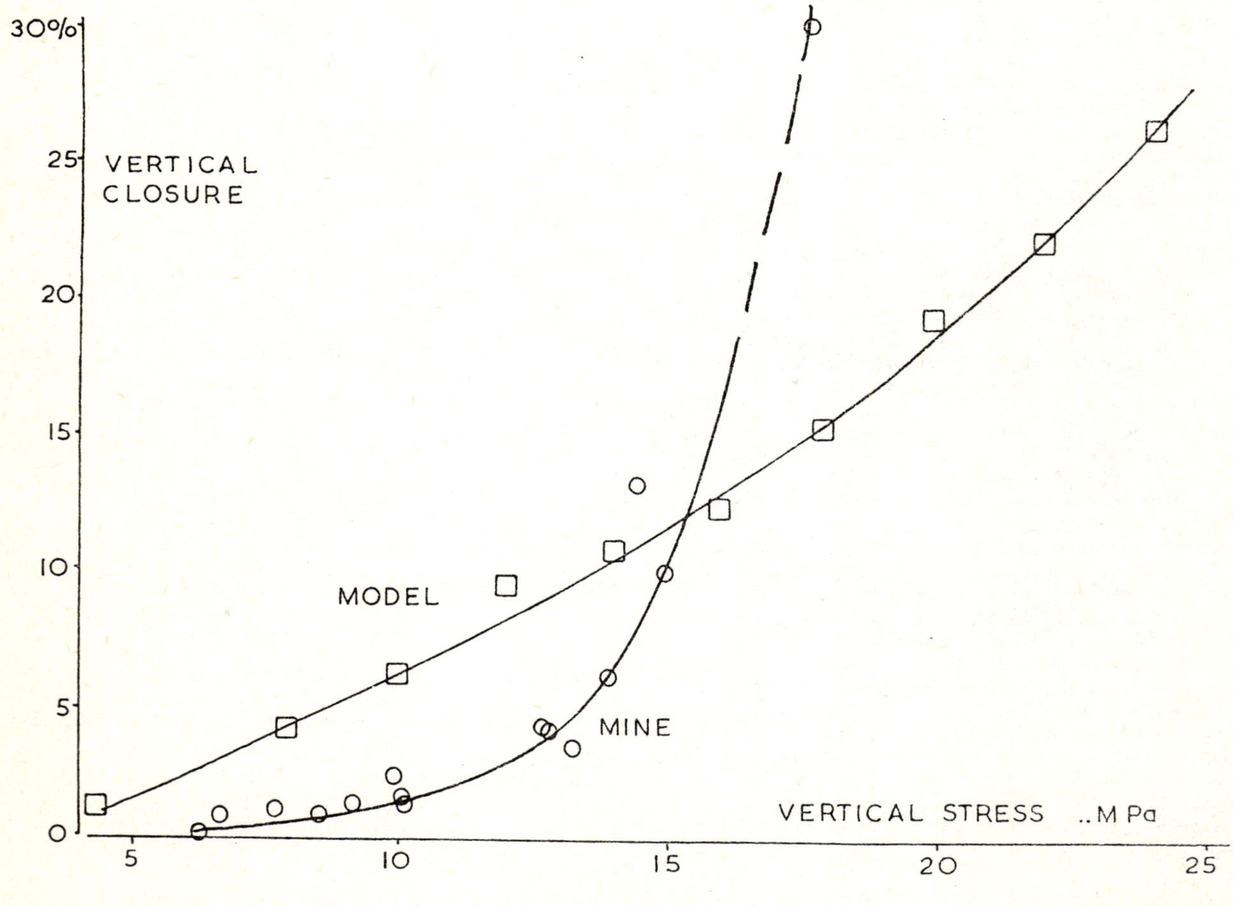

FIGURE 2. VERTICAL CLOSURE AT SITE 2

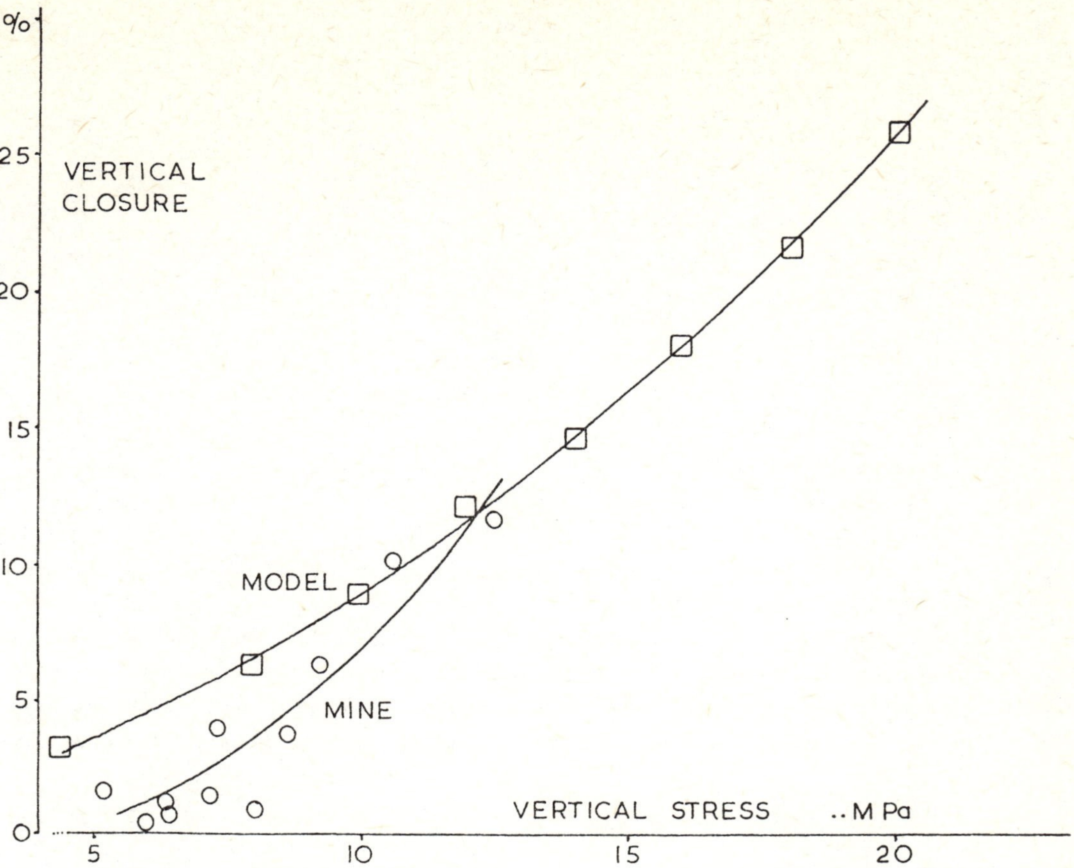

FIGURE 3. VERTICAL CLOSURE AT SITE 3

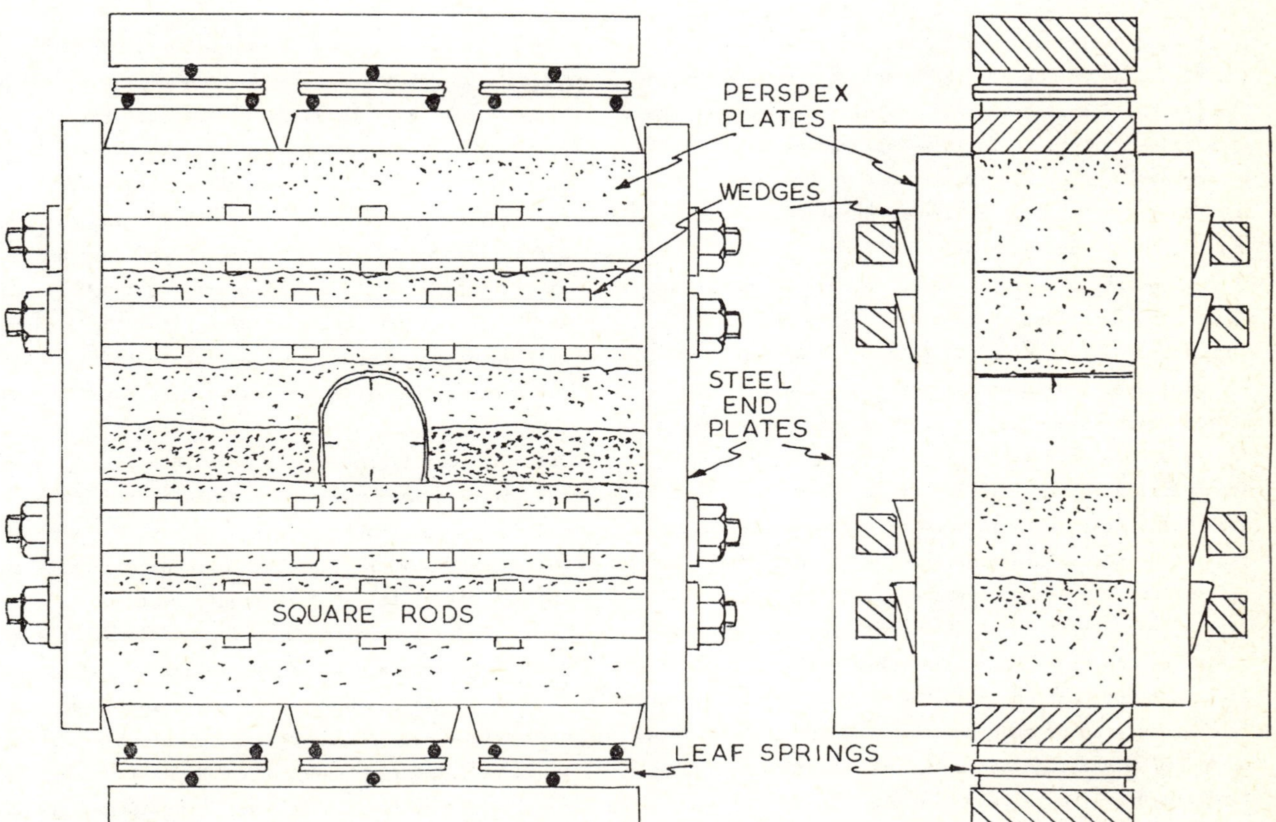

FIGURE 4. DIAGRAM OF 1 TO 150 TEST RIG

MODEL STUDIES OF STABILITY OF MINING TUNNELS

C J Bonsall, S F Smith and B N Whittaker

Department of Mining Engineering,
University of Nottingham,
Nottingham, NG7 2RD, England

SUMMARY. The paper is concerned with the stability of coal mining tunnels and briefly reviews the different forms of deformation and instability which occur in such tunnels. The use of small scale model studies involving sand-plaster materials is shown to provide a valuable form of general study of tunnel deformational behaviour under different stress fields. Results are presented of common shapes of tunnel excavated openings subjected to different combinations of vertical to horizontal stresses. Tunnel opening deformational comparisons have been made together with general observations of the extent of the yield zones immediately surrounding the particular tunnel shapes investigated.

INTRODUCTION

Investigations into the stability of coal mining tunnels can be performed mainly by field studies of instrumented sites and by suitable modelling. There is no doubt that field investigations perhaps provide the most useful detailed behavioural characteristics of tunnels under specific geological and mining conditions and this will continue to be favoured as the most realistic basis for establishing detailed information. However, modelling of mining tunnels can provide an important indication of likely deformational behaviour. Theoretical models have been useful to examine the role of various factors in influencing support requirements. A combination of theoretical and empirical principles has been successfully applied to obtaining a formula which gives a good assessment of support requirements in Coal Measures' strata; the same formula takes into account post-failure characteristics of the rocks immediately surrounding the excavation, Wilson (1980).

The use of sand and plaster small scale physical models has been applied to mine tunnels and roadways fairly extensively particularly by Hobbs (1966, 1968) and Everling (1962). This form of research tool is especially attractive in view of convenience and time scale over which investigations can be performed. It is relatively inexpensive and providing the models are fairly representative of the actual conditions, the technique can prove of considerable value in providing important information which can supplement field studies. The convenience of changing the geometry and stress conditions is recognised as an important factor supporting the use of physical modelling.

The research reported here has directed specific attention towards the development of yield zones around different shapes of mining tunnel under different combinations of vertical to horizontal stress.

STABILITY ASPECTS OF MINING TUNNELS

The strength and relative thickness and dispositions of the rocks immediately surrounding a mining tunnel, and the magnitude of the strata stresses greatly govern the deformational response of the tunnel immediately after excavation. The subsequent support operations are required to control the degree of strata movements immediately around the excavation. The strength characteristics of the chosen support system are influenced by mining economics in addition to technical factors. Gate roadways serving longwall coal faces are supported with emphasis on controlling deformation with supports which can be rapidly installed and whose costs are comparatively low when contrasted to major tunnelling projects. The latter are designed not to tolerate significant strata movements, since the method and intensity of support are aimed at minimising subsequent movements to achieve a high level of stability. There are a number of design methods currently available for estimating the support requirements of a circular tunnel in various geotechnical conditions but assumptions need to be made and a certain amount of judgement applied. The use of small scale models can help to verify certain assumptions made in the case of particular deformational features of mining tunnels. Such models have been applied widely in research into mine tunnel stability and there is no doubt that they will continue to provide an attractive means of investigating various stability aspects.

MODELLING OF MINE TUNNELS

There are numerous modelling techniques for investigating mine tunnel stability ranging from photoelasticity and related methods, base friction methods, brittle low-strength models involving such materials as sand and plaster, other physical models involving highly deformable materials but without brittle fracture to mathematical and computer based models. Each technique has particular merits. The most successful of the commonly applied modelling techniques has probably been the sand-plaster small scale models which have been used to investigate a wide range of excavation stability problems in mining in several countries. The results are especially meaningful to likely deformational behaviour of the particular situation being modelled and important conclusions are frequently drawn which are of special value to the tunnel support designer. The models described in this paper have been entirely of sand and plaster mixed and cured to give suitable properties for meaningful tests to be carried out. The biaxial compression rig used to test all the models enabled a suitable combination of stresses, both vertically and laterally, to be used for making observations of deformational behaviour of the mine tunnel yield zone development and closure characteristics.

DIMENSIONAL ANALYSIS AND SCALE FACTORS

The model studies were preceded by a detailed examination of the required choice of model scale factors by performing a dimensional analysis of the problem. Details of the dimensional analysis method are given in Obert and Duvall (1967) and the reader is referred to this standard text for basic theory and its application to modelling. Applying the analysis to a model of a mine tunnel it was established that the fundamental quantities would be length, stress and time. For geometrical similarity the distance between any two points in the model required to bear a constant ratio between the corresponding two points in the original system.

The geometrical scale factor (λ) is expressed by L/L', whilst the stress scale factor (ξ) is the ratio of stress in the model material to the stress in the rock around the tunnel. The factor ξ applies to any quantity with units of stress i.e. $\xi = \sigma_o/\sigma'_o = E/E' = \sigma_v/\sigma'_v = \sigma_t/\sigma'_t$ where σ_o is uniaxial compressive strength, E is Young's modulus, σ_v is the vertical component of geostatic stress, σ_t is tensile strength. Dimensional quantities such as strain (ϵ) and Poisson's ratio (ν) and angle of internal friction (ϕ) of any material should be the same for both model and original system.

The geometrical scale factor was essentially governed by the size of the biaxial test rig and was set at 1/50. The density of the model material was approximately 15/23 of that encountered for typical Coal Measures rocks. The stress scale factor was determined to be approximately 1/76. Typical rock strength data are given in Table 1.

	σ_o MN/m^2 Actual	(Model)	E GN/m^2 Actual	(Model)
Sandstone	80	(1.05)	25	(0.33)
Siltstone	60	(0.80)	20	(0.26)
Mudstone	30	(0.40)	10	(0.13)
Coal	20	(0.26)	10	(0.13)
Seatearth	25	(0.33)	8	(0.10)

Table 1 Rock strength data (actual and model)

Figure 1 Biaxial stress field model

MODEL MATERIALS AND PREPARATION

The materials used in the construction of the models used for the tests described in this paper were a combination of sand, fine casting plaster, and water. The models were oven cured at 120°C for a period of 48 hours. The average size of the sand was 0.28 mm. Initially cylindrical specimens 51.5 mm dia x 103 mm long were moulded and tested using various combinations of sand-plaster in order to obtain a material with a Young's modulus value of around 48 MN/m^2. Eventually the combination of 535/215/250 sand/plaster/water produced the strength value required. Table 2 gives

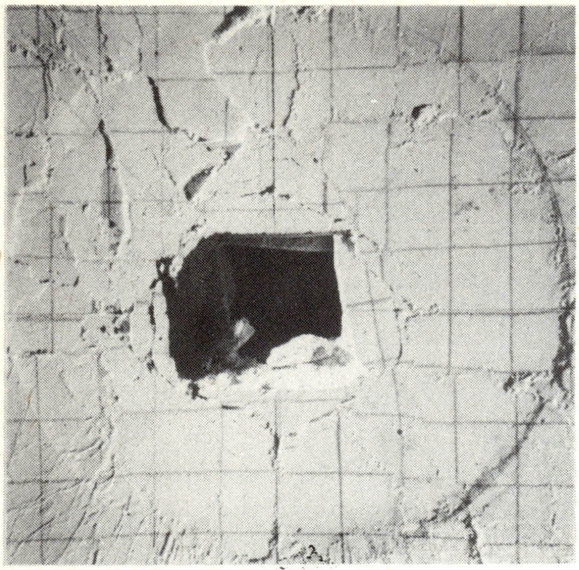

<u>Figure 2</u> Model tunnel deformation: originally square profile subjected to hydrostatic stress field

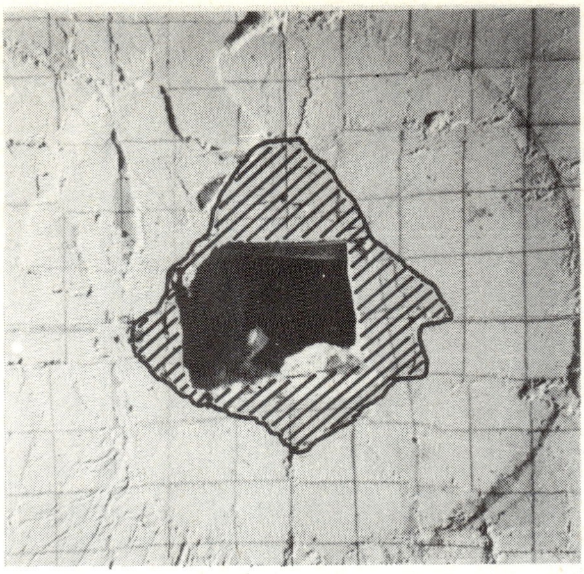

<u>Figure 3</u> Model tunnel deformation: extent of fractured zone around square excavation and hydrostatic stress field

strength property details of the model material used and other mixes tested.

Model material mix Sand/Plaster/Water			Density kg/m^3	σ_o MN/m^2	E MN/m^2	ν
450/	300/	250	1490	0.959	112.4	0.18
535/	215/	250	1485	0.550	48.0	0.19
560/	190/	250	1483	0.201	32.4	0.25
600/	150/	250	1499	0.131	11.9	0.18
625/	125/	250	1480	0.098	10.0	0.16

<u>Table 2</u> Model material properties

After curing, a simulated roadway profile was cut in the centre of the model using a 65 mm dia hand held core barrel for a circular drivage. For extending a circular profile into an arched shape the bottom corners were cut using a saw blade. In the case of simulated square and rectangular drivages a 16 mm twist drill was used to bore a hole at each corner of the outline of the roadway and the remainder was again cut out using a saw blade. Before being placed in the model testing rig each model was marked in various patterns with either small pins or etched lines to show movement of the material.

The model rig, as shown in Figure 1 was manufactured in the Mining Engineering Department at the University of Nottingham. The model was placed in the rig between sets of hydraulic rams (4 sets of 5), which could be operated independently in the vertical and horizontal directions. The model was held in position by front and back plates. Each model had dimensions 450 mm x 450 mm x 75 mm.

Regarding preparation of the model, this was carried out as a monolithic structure, which could be built as a layered model by casting horizontal sections independently. In the latter case, subsequent layers were poured directly onto a previously hardened layer. All the models described in this paper were uniform throughout and represented a mine tunnel in medium Coal Measures strata conditions.

Distortion aspects of the model required consideration. If two quantities in the original system, which have the same dimensions, are not equally scaled, the model is said to be distorted. When modelling underground excavations a certain amount of distortion is inevitable and in many situations is acceptable. Scale model studies are essentially qualitative and should be used to observe trends rather than obtaining absolute values.

In producing the model material distortion has occurred as a result of the modulus of elasticity with units of stress and the dimensionless parameter of strain induced at failure. For typical Coal Measures rocks the ratio of uniaxial compressive strength to modulus of elasticity is of the order of 250 while for the sand/plaster/water mix this ratio is around 118. This distortion can be explained and balanced by the fact that the strain at failure of a typical Coal Measures' rock is around 10×10^{-3} while for the plaster model the strain was around 20×10^{-3}. This will give a magnification of the movements around the excavation; this is acceptable in view of investigating the shape of the yield zone and the movement of material into the excavation.

TESTING PROCEDURE

The models were loaded into the biaxial stress field loading rig and constrained in the tunnel axis direction. Vertical and horizontal loading was carried out either together at the same increases in stress or

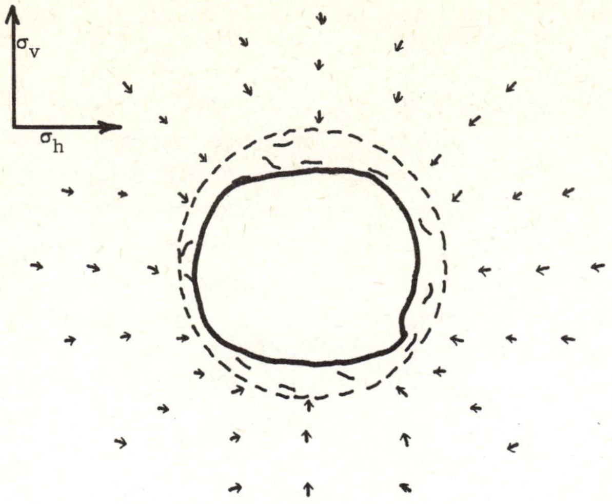

Figure 4 Deformation of circular opening ($\sigma_v = \sigma_h$)

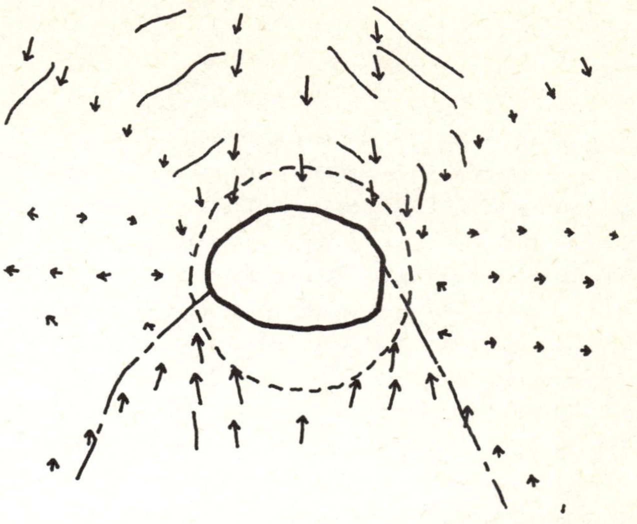

Figure 7 Deformation of circular opening ($\sigma_v = 3\sigma_h$)

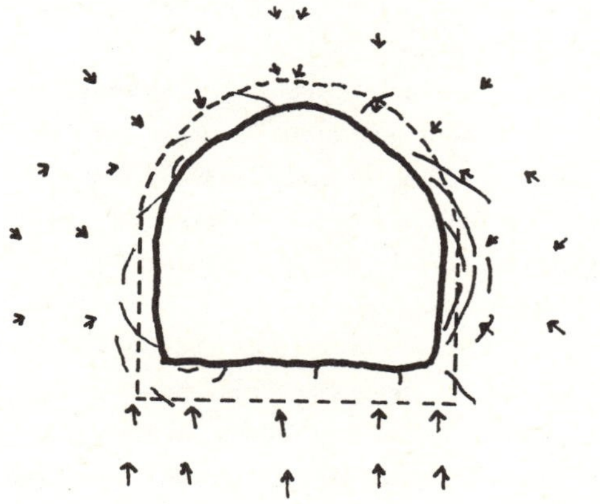

Figure 5 Deformation of arch opening ($\sigma_v = \sigma_h$)

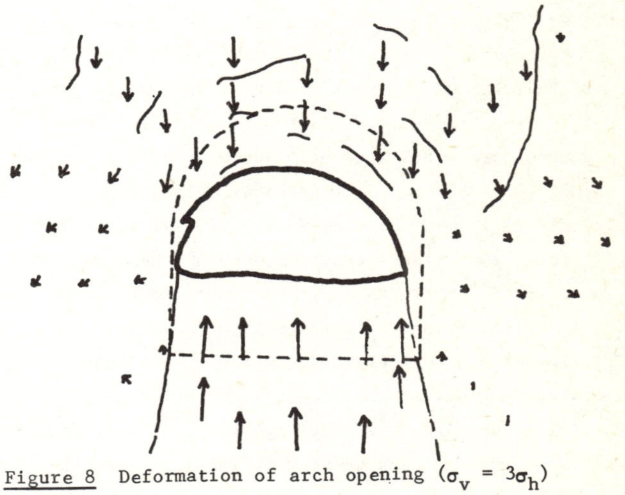

Figure 8 Deformation of arch opening ($\sigma_v = 3\sigma_h$)

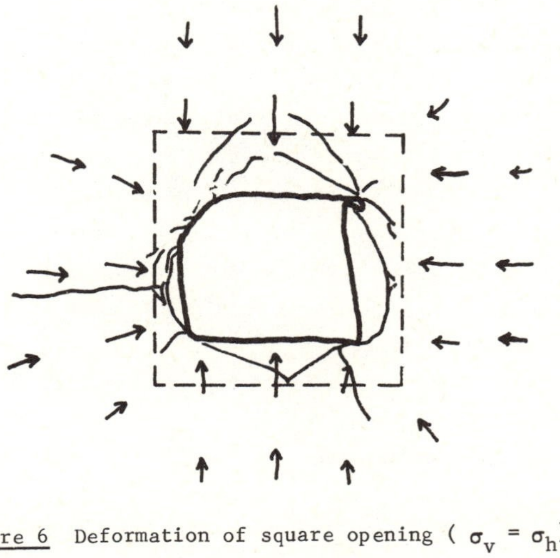

Figure 6 Deformation of square opening ($\sigma_v = \sigma_h$)

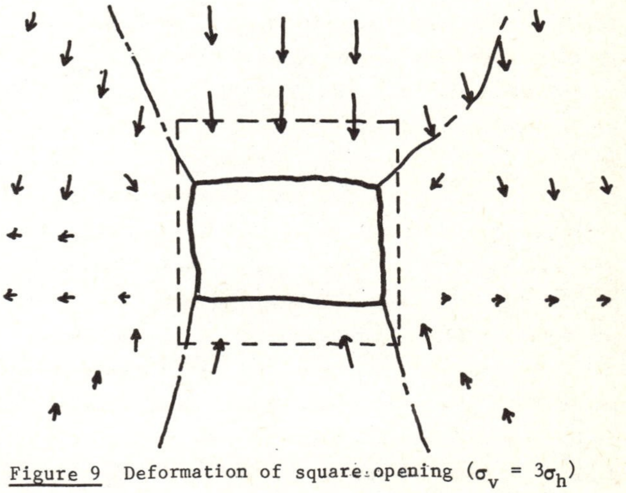

Figure 9 Deformation of square opening ($\sigma_v = 3\sigma_h$)

at a specific ratio depending upon the test. Deformation of the model could be observed through a perspex window and photographs were taken at particular increments of vertical loading. The stresses were increased at the sides of the block containing the model tunnel up to a maximum vertical stress of 7.8 MN/m^2. A grid pattern was used on the model to assist displacement measurements which were carried out at the end of each complete test.

Three particular opening shapes were examined (i) circular, (ii) arch, and (iii) square, the width being kept constant in all cases. Two combinations of applied stress were used (a) hydrostatic, and (b) vertical stress = 3 x horizontal stress. The average vertical loading rate adopted was 2 $KN/m^2/s$.

RESULTS

The type of deformation observed following each hydrostatic test was similar to that illustrated in Figure 2 which shows the result of compression of the square tunnel profile. The form of grid pattern (originally 2 cm squares) used readily allowed the in-plane displacements to be determined. The fracture patterns associated with each shape were also easily identified. Figure 2 demonstrates a rotational shear failure in the tunnel side wall. The extent of the fractured zone immediately surrounding the model tunnel is indicated in Figure 3 by cross-hatching.

Results from the main tests carried out are given in Figures 4 to 9 inclusive. Figures 4, 5 and 6 present results from hydrostatic loading on the three excavation profiles considered, whilst Figures 7, 8 and 9 show the results for the tests where vertical stress = 3 x horizontal stress. The results are discussed according to profile.

Circular tunnel shape

For hydrostatic loading, this shape proved to be the most stable with relatively minor closure. The circular shape (see Figure 4) was still evident at the end of the test. Movement of the material was generally of limited extent and fairly close to the excavated surface.

In the case of where the lateral stress was ⅓ x vertical stress, fairly large displacements were observed especially vertically with small and often negative displacements horizontally (see Figure 7). The crown and floor essentially retained their circular shape; the floor tended to move as a block with almost vertical major shear planes below the sides. This shearing phenomenon was not so evident in the roof material where fracturing was observed to within 1 tunnel diameter of the crown. The major proportion of the tunnel closure was due to the floor uplift.

Arch tunnel shape

The hydrostatic loading case indicated that an appreciable degree of stability ensued. There were small displacements in the roof, with increasing amounts in the sides and the largest amount being encountered in the floor. Figure 5 indicates that movement was generally radially inwards and failure appeared to be due to over-compression although some side slabbing occurred. The crown was relatively free of fractures.

For the lateral stress being ⅓ x vertical stress, the results (see Figure 8) indicate large vertical displacements. Predominantly vertical movements of roof and floor occurred although the latter was greater than the former. Shear failure readily occurred and promoted excessive movement of floor material in particular, although the same was true for the roof to a lesser extent.

Square tunnel shape

The hydrostatic loading results of the square tunnel shape are presented in Figure 6. This geometry gave the least stability of all those tested. Deformation was significantly high both horizontally and vertically when compared to the other tunnel shapes tested under similar conditions. The major fractures extended to 0.75 x excavated diameter into the sides, roof and floor. There was a tendency for the fracture pattern to become approximately circular. Side wall slabbing was strongly evident, and exhibited rotational shear failure (see Figure 6).

Figure 9 shows the results for the lateral stress being ⅓ x vertical stress. Major shearing occurred in the roof and floor which produced a block-type of movement mainly in the vertical plane above and below the excavation. A very low order of stability was indicated by comparison to the other shapes tested.

CONCLUSIONS

The main conclusions drawn from this experimental work to-date are listed as follows:

1. The experimental technique involved using a monolithic block within which the model mining excavation was formed. This method of using a complete block has been found to give more realistic results than the form of model construction which employs an assembly of independent layers with no positive bonding between the slabs, owing to the latter giving a specific form of strut-type of failure.
2. The form of stress field adopted has been found to be of major significance to the resulting stability of the mine opening. The hydrostatic test models indicated appreciably more stability than those cases where non-hydrostatic loading was employed. It would appear that knowledge of the nature of the magnitude of lateral to vertical stresses is of major importance to stability. Also it would appear that assuming the existence of a hydrostatic

stress field might lead to underestimating the magnitude of resulting tunnel closure and the required support resistance.

3. Of the model tunnel shapes investigated, the square opening was considerably less stable than the circular and arch ones; shear failure on a large scale appeared to predominate.
4. The extent of the fracture envelope around each model was clearly identified in all tests; this envelope was mainly responsible for the closure encountered in the hydrostatic loading condition. The extent of this fracture envelope from the excavation was greatly affected by the shape of the opening.

ACKNOWLEDGEMENT

The authors express appreciation to the European Coal and Steel Community, and the Mining Research and Development Establishment of the National Coal Board for financial assistance given to this research project.

REFERENCES

Everling, G., 1962, "Model test study on the effects of strata and supports in gate roadways", Gluckauf, Vol. 98, 1471-1479

Hobbs, D.W., 1966, "Scale model studies of strata movement around mine roadways. Apparatus, technique and some preliminary results", Int. Journal Rock Mech., Vol. 3, No. 2, 101-127

Hobbs, D.W., 1968 "Scale model studies of strata movement around mine roadways. I The dependence of roadway closure on rock strength", Int. Journal Rock Mech., Vol. 5, No. 3, 219-235

Obert, L and Duvall, W.I., 1967, "Rock Mechanics and the Design of Structures in Rock", New York, Wiley

Wilson, A.H., 1980, "The stability of underground workings in the soft rocks of the Coal Measures" PhD Thesis, University of Nottingham.

RETREAT WORKING USING AN ANHYDRITE PACKED GATEROAD TWICE

R Mallory

Rock and Coal Mechanics Branch
Mining Research and Development Establishment
National Coal Board
Burton-on-Trent, Staffordshire. DE15 0QD

SUMMARY. An investigation was undertaken in the anhydrite packed maingate of T24's advancing face in the Trencherbone Seam at Parkside Colliery. The roadway was used a second time as the tailgate for the adjacent T25's retreating face. The results indicate that:

(a) the system of packing and roadway support achieved a stable roadway capable of being used a second time,
(b) during the advancing phase, load in the anhydrite pack yielded when it reached 4 MPa. It also reduced as the retreating face approached. In both cases it is probable that the fireclay floor yielded. However, when the floor was restrained pack load exceeded 12 MPa.

INTRODUCTION

The first recorded attempt of using anhydrite successfully to form a gateside pack was Holland Colliery West Germany in 1964, Genthe (1970). The reasons for using it were:

1) to reduce the volume of chock wood required for the gateside packs, the quantities of which impeded other essential face supplies,

2) to create a better seal at the gateside to reduce the possibility of spontaneous heatings.

Within the UK mining industry the formation of gateside packs utilising anhydrite is rare. To date only two mines have operated the system for a reasonable length of time. The first was Easington Colliery where it worked successfully in production districts, Smith (1976) and Dudley (1977). Parkside Colliery operated the equipment in two districts, Mallory (1980).

The primary purpose of the Anhydrite Packing System in the UK was to increase roadway support, retaining a stable roadway with a larger cross-sectional area. Other methods exist for increasing roadway support, eg increasing the cross-section of the arch support or decreasing the intervals at which the arches are set. However, both methods are subject to a law of diminishing returns and could be grossly uneconomic. Supports set at too frequent intervals may even weaken the floor. Results from Cotgrave Colliery indicated that increasing arch girder spacing from 0.61 m to 1.07 m caused roof lowering to increase from 0.59 to 0.96 m whereas floor lift decreased from 0.86 m to 0.42 m, Mallory (1975).

An investigation was undertaken in the anhydrite packed maingate of T24's advancing face in the Trencherbone Seam at Parkside Colliery. The roadway was used a second time as the tailgate for the adjacent T25's retreating face.

Anhydrite packing started in the maingate at the face starting line. Roadway investigation techniques were used to find closure characteristics. Pack load and convergence were also determined. Three sections were installed in the maingate (Figure 1).

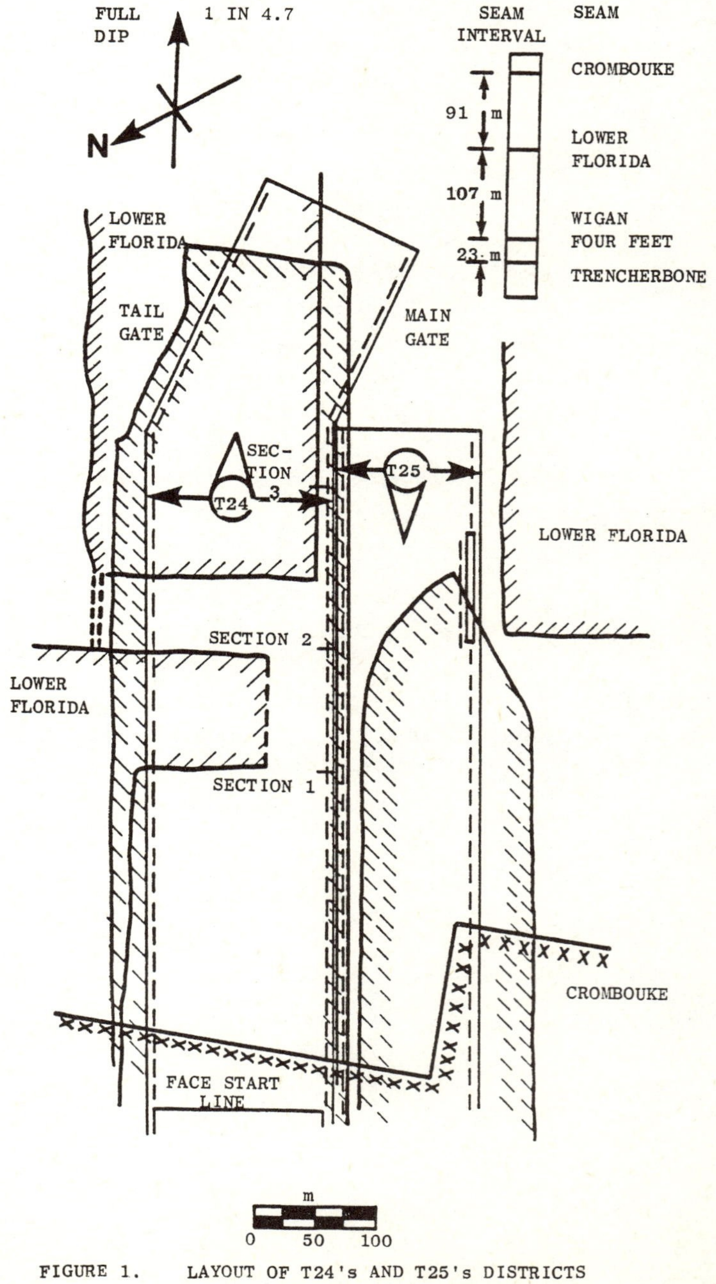

FIGURE 1. LAYOUT OF T24's AND T25's DISTRICTS AND POSITION OF THE SECTIONS

DESCRIPTION OF SITE

Seam details and geology

T24's District lay at a depth of about 690 m at the face starting line. When the face ceased production the depth was 835 m. The seam was 1.6 m thick and extraction height was 1.5 m (Figure 2). The seam was overlain by mudstone over 3 m thick and underlain by 0.35 m fireclay and the 1.93 m Peacock Seam.

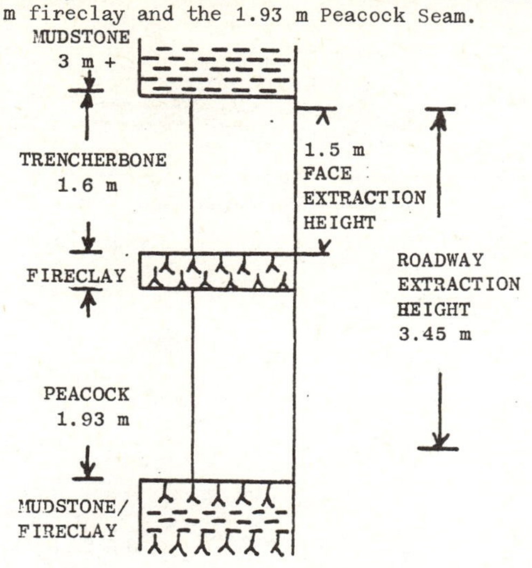

FIGURE 2. SECTION OF STRATA IN THE VICINITY OF T24's/T25's DISTRICTS

In the vicinity of T24's the seam dipped at 1 in 4.7 to the South East. The Wigan Four Feet, Lower Florida and Crombouke Seams, situated at distances of 23 m, 130 m and 221 m respectively above the Trencherbone Seam had been partially extracted (Figure 1). A 30 m down throw fault was present at T24's tailgate.

Face details

T24's advancing face was initially 153 m long, and was situated directly beneath the old waste of the superjacent Wigan Four Feet Seam, W24's. The tailgate contacted the 30 m fault after a face advance of 574 m, eventually both roadways were turned and the face length had reduced to about 137 m when coal production stopped. The coal was won by two SERDS. The shearer at the main end formed a 0.9 m dint in-line with the proposed maingate. The face was supported by Dowty 6-240 Tonne powered supports.

T25's face line was formed 125 m behind T24's face finishing line and was 118 m long. It was separated from T24's maingate by a 3 m wide pillar. When heavy weighting affected the maingate solid drivage the face length was reduced to about 109 m over a 99 m section (Figure 1) and was then increased to its original length. The face was supported by Gullick 5 leg-200 tonne supports, and the coal was won by a Double Ended Ranging Drum Shearer.

Roadway details

T24's maingate

Support in the maingate was by 4.57 m (4.88 m) x 3.35 m (0.152 m x 0.127 m), three piece arches set on 0.4 m Hayden Nilos stilts and spaced at 1 m intervals. The supports were tied with seven 50 mm diameter tubular steel struts and lagged with 'Strutlag'. The roadway was formed at the conventional rip position, a 0.9 m dint being taken by the shearer (Figure 3). The rip was normally drilled and fired. When conditions were poor at the roadhead it was ripped by pneumatic pick. Part of the ripping debris was loaded out onto the stage-loader by an Eimco side loader.

The maingate pack was formed by a 0.4 m wide wall of 'Stranlite' blocks at the waste breaking off line, a 2.8 m wide section containing debris filled hessian sacks and, at the gate side, 1.5 m wide anhydrite contained by waterproof brattice (Figure 3). The rib was maintained by the solid coal; when the profile was uneven the gaps were back-filled with debris filled hessian sacks.

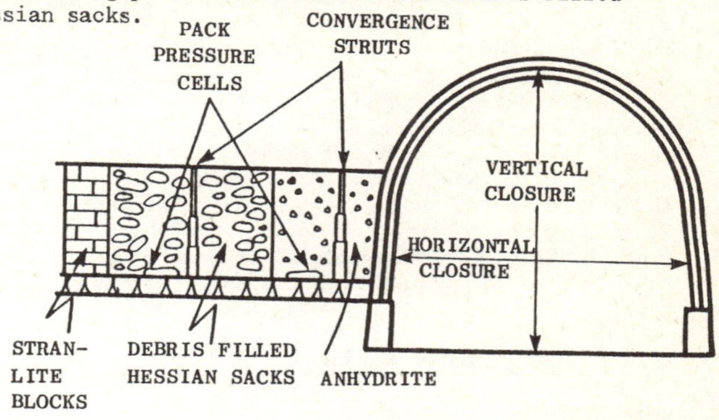

FIGURE 3. ROADWAY MEASURING SECTION USED IN T24's/T25's COMMONGATE

When the maingate was turned the methods of roadway formation, roadway support and pack support were retained. Conditions in the maingate were very good.

T25's tailgate (common gate)

T25's tailgate was originally T24's maingate. Support used was as described for T24's maingate, however, the roadway cross-section was smaller because of the closure associated with T24's face extraction. The tailgate was separated from T25's face-line by a 3 m wide coal pillar. Cross-cuts connecting the face and the gate were driven at about 30 m intervals. The cross-cuts were 1.5 m high by 2 m wide and supported by H-section RSJ's and 0.15 m diameter wood props. A 2 m wide by 1.5 m high access road was formed at the face end, also supported by H-section RSJ's and wood props. A face breaking off line was formed of hardwood chocks. Brattice cloth was attached to the chocks to enable the ventilation to circulate the working places. Both the face access road and the commongate had stoppings built at about 30 m intervals to stop air leakage.

T25's maingate

T25's maingate was a purpose driven solid drivage formed by a Dosco Dintheading machine. It was 4.2 m wide by 2.2 m high, supported by straight H-section, (0.127 m x 0.127 m) beams and H-section (0.127 m x 0.115 m) legs connected by single bolt corner brackets. The supports were tied with seven Fastlock tie bars, lagged with strutlag and were set at 0.92 m intervals, the beams centre propped with 0.23 m diameter wood props.

General conditions in the rectangular heading were very good where it was overlain by the previously worked Wigan Four Feet Seam. However, the final 160 m of the heading and 108 m of the face-line were in virgin strata and conditions were poor even before production started. Face conditions continued to deteriorate during the installation of the face equipment. They became so difficult it was necessary to re-drive a 108 m length of the face-line 5 m ahead of the original position. A 10 m length adjacent to T25's tailgate and overlain by Wigan Four Feet Seam workings was unaffected.

When production started conditions in the maingate became extremely difficult and it was necessary to stop production and drive a twin maingate. The face length was reduced to 109 m and six face supports were withdrawn. A 2.7 m wide by 1.8 m high heading was mined in two directions. It was 99 m long and supported by H-section RSJ's and wood props set at 1 m intervals.

Production was delayed about six weeks. When it re-started four slips occurred in the new heading below the Wigan Four Feet Seam pillar edge. Conditions in the heading varied; the part in virgin strata was affected by floor lift, whereas the outbye, stress relieved length was quite good. The face retreated the 99 m with little trouble, and thereafter the six face supports were returned to the face-line and retreat was continued using the original maingate. Conditions in the maingate were good until it was overlain by Crombouke waste in addition to the Wigan Four Feet waste. Then at a distance of about 40 m ahead of the face considerable floor lift and side closure occurred. A dint up to 1 m was taken. The possible cause was that the effect of two superjacent workings could have been to relieve the intervening strata to such an extent that negligible vertical restraint was being provided to the buckling effects of the horizontal pressure, Lawrence (1973).

MEASURING TECHNIQUES AND RESULTS

Measurements

Marks were placed on the inside flanges of the arch girders. The roof mark was at the centre of the arch crown and side marks were at a point 1 m from the bottom of the arch legs (Figure 3). Roadway closure was determined by steel tape measurements (reading to 1 mm) of the distance between the roof mark and the true floor, and the horizontal distance between the marks at the sides.

Pack load and convergence was determined by installing Soil Instrument hydraulic pressure cells and pack convergence struts. Load was read on a portable "Read Out" instrument and closure was read directly by a wire from the mechanical, telescopic, pack convergence strut.

Samples of strata and anhydrite were tested for compressive strength and triaxial stress factor in the laboratory (Table 1).

Table 1 - Strength of Rock Samples Around the Roadway and Anhydrite Pack Samples

Sample	Uniaxial Compressive Strength (MPa)		Triaxial Stress Factor 'K'
	Actual	Equivalent NCB Standard 2:1 H/D Ratio	
T24's Returngate Pack-Anhydrite	34.3	33.2	2.3
T24's Maingate Pack-Anhydrite	28.0	25.8	3.2
T24's Roof Mudstone	46.6	44.2	2.1
T24's Coal Trencherbone Seam	30.0	29.4	3.2
T24's/T25's Floor Coal Peacock Seam	21.4	Not Applicable	-
T25's Coal Trencherbone Seam	41.3	39.4	3.4

NOTE 1) 10 MPa = 1450 lbf/in^2
2) Triaxial Stress Factor 'K' is the relationship between stress at failure and confining pressure.

Roadway closure

The pattern of roadway closure in T24's/T25's common gate during the advancing and retreating phases are shown in Figures 4, 5 and 6.

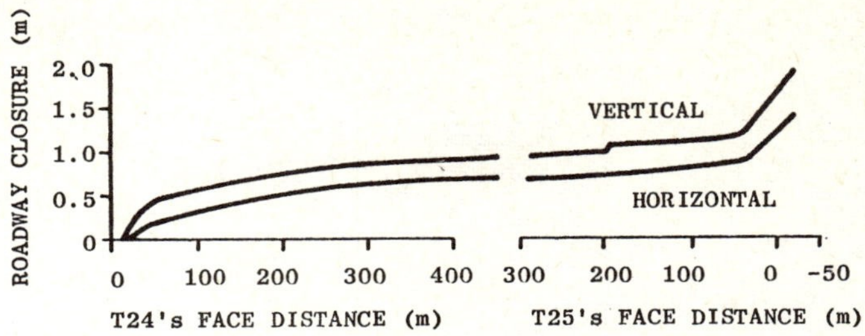

FIGURE 4. ROADWAY CLOSURE: SECTION 1

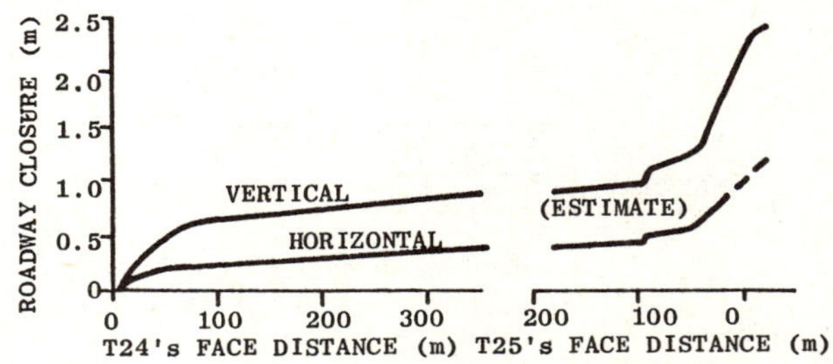

FIGURE 5. ROADWAY CLOSURE: SECTION 2

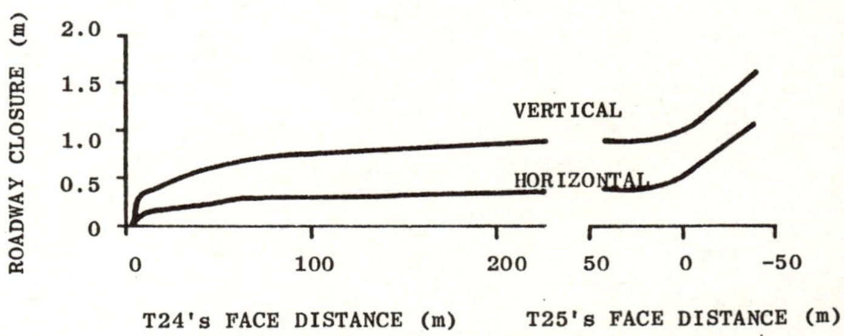

FIGURE 6. ROADWAY CLOSURE: SECTION 3

Both vertical and horizontal closure curves are typically S-shaped except for the jump caused by the six week delay when T25's production was stopped while the maingate was partially redriven.

Vertical closure at each of the three sections was about 0.9 m before T25's retreating face started production. The greatest horizontal closure before T25's commenced production was 0.7 m at Section 1. In all cases the horizontal and vertical closure adapted a similar trend. Roadway closure during the advancing stage was not excessive indicating that the system of packing and roadway support was achieving a stable roadway capable of being used a second time.

Roadway closure at Section 3 during the retreating phase was less than at Sections 1 and 2. This was because it derived greater benefits from end, Mallory (1978), and time, Wilson (1964) effects than Sections 1 and 2. The change of stress ahead of a retreating face is dependent upon the formation of the maximum front abutment. The abutment is unlikely to reach its maximum value until the face has advanced or retreated a certain distance. This distance determines the length of roadway end effect and is probably between 0.2 and 0.3 times depth. When this distance is reached a 'stress balance' must exist between total stress above cover load on the solid coal and a compensating stress reduction over the waste, Wilson (1980). T25's had retreated about 45 m when it passed Section 3, this would not be far enough for the abutment to be fully formed. Johnson (1967) considered the roadway end effect to be negative exponential in form (probably only slightly different in practice).

$$c = C\,(1 - a^{-bx}) \qquad \text{............} \quad (1)$$

where c = basic closure
C = total closure
a,b = constants
and x = distance behind face

such that

$$a^{-bx} \rightarrow 1 \text{ as } x \rightarrow 0$$

$$a^{-bx} \rightarrow 0 \text{ as } x \rightarrow \infty$$

Variation of closure with time benefits Section 3 as it was read over a shorter time than Sections 1 or 2.

However, roadway closure would be less throughout the gateroad if greater rates of advance and retreat were possible.

Final closure results at Section 1 and 2 (Figures 4 and 5) were consequently greater than at Section 3. The greatest closure, 2.48 m, was recorded at Section 2, 18 m behind T25's face. The longwall abutment effect was apparent at Sections 1 and 2 at distances of about 60 m ahead of the face.

Pack load and closure

Graphs of pack load and closure are shown in Figures 7, 8 and 9.

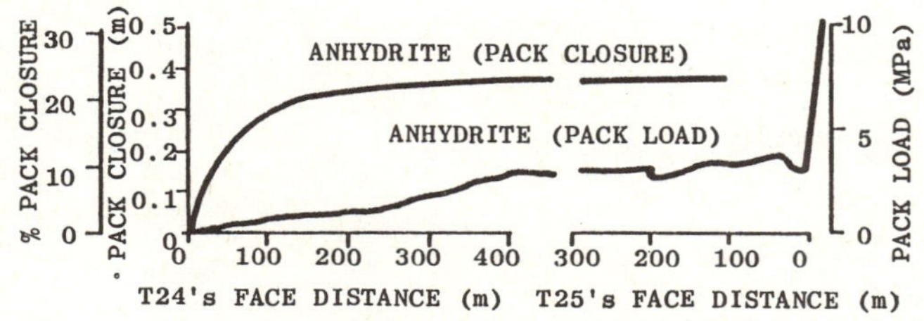

FIGURE 7. PACK LOAD AND CLOSURE: SECTION 1

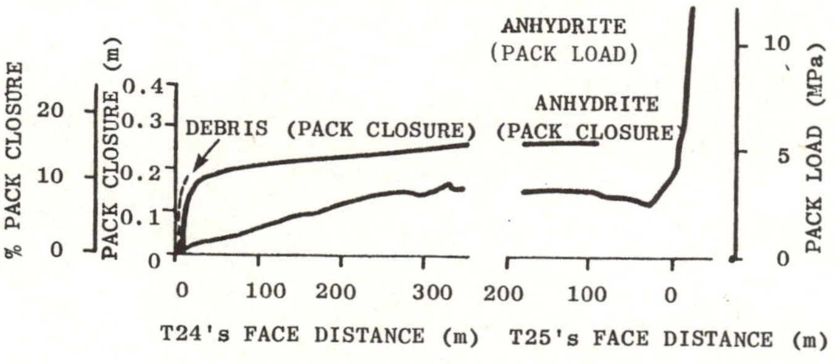

FIGURE 8. PACK LOAD AND CLOSURE: SECTION 2

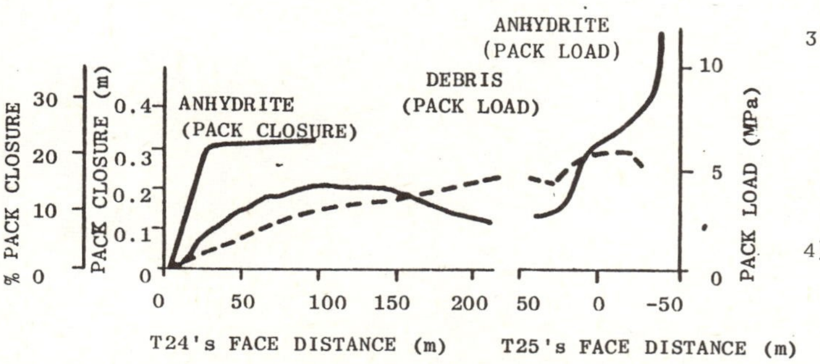

FIGURE 9. PACK LOAD AND CLOSURE: SECTION 3

During the advance of T24's the greatest load recorded in the anhydrite was 4 MPa at Section 3. Thereafter, it tended to shed load, reducing to 2.4 MPa. A possible reason for this fall off in load was that the floor beneath the pack yielded into the roadway. It is possible in weak strata where the roadway is formed wholly or partially by a dint for this to occur as the weak bed would have negligible lateral constraint. However, in this example the load continued to be taken by the adjacent debris pack further up the face. This indicates the importance of adequate pack widths in roadways formed by dints where the strata underlying the seam is believed to be weak.

The tendancy for the fireclay floor to yield was eventually restrained by floor heave in the gateroad. At Sections 1 and 2 the lowest readings were in front of T25's face at face distances of 17 m and 32 m respectively. However, Section 3 indicated that load continued to be taken up by the debris pack. The greatest loads occurred in the anhydrite pack behind T25's face, possibly as lateral restraint was built up by floor heave in the gateroad. It seems probable that the 4.7 m wide pack plus the 3 m wide pillar acted as a 12.7 m wide split pillar separating the two worked out faces and was strong enough initially to take up the combined load from the side abutment.

Convergence of the packs is shown in Figures 7, 8 and 9. Percentage closure in the anhydrite packs at Sections 1, 2 and 3 was 23.49%, 17.64% and 23.16% respectively. It is probable that the variation in the triaxial stress factor of the anhydrite between 2.3 and 3.2 (Table 1) was responsible for the difference in pack closure. This was probably due to variations in the proportion of either the water or the accelerator. The convergence strut at Section 2, debris pack must have failed after the first reading was taken since it is unlikely that the closure in the debris pack would be less than in the anhydrite.

Closure was reactivated at Sections 1's anhydrite pack when T25's face started to retreat. It was not possible to determine final pack closure or the effect of the longwall abutment as all the strut wires were damaged before the high stress zone was reached. However, general observations made almost throughout the life of the roadway gave no indications of instability or excessive yield in the anhydrite packs themselves.

CONCLUSIONS

The Parkside Colliery in T25's anhydrite packed second usage (common) gate it was found that:

1) Roadway closure during the advancing stage was not excessive. This indicated that the system of packing and roadway support achieved a stable roadway capable of being used a second time.

2) During the advance of T24's the greatest load in the anhydrite pack was 4 MPa, thereafter it shed its load and reduced to 2.4 MPa. However, the load continued to be taken up by the adjacent debris pack. A possible reason for the fall off in load was that the floor beneath the pack load cell yielded laterally into the roadway.

3) As T25's retreating face approached, a reduction in pack load occurred in the anhydrite. It is possible that the fireclay floor yielded until it was eventually restrained by floor heave in the gateroad. The greatest loads in the anhydrite occurred behind the face possibly as lateral restraint was built up in this manner.

4) Closure was reactivated in the anhydrite pack when T25's face started to retreat. However, there were no signs of excessive yield or instability in the anhydrite packs themselves at any time.

ACKNOWLEDGEMENTS

The author wishes to thank the management and personnel at Parkside Colliery, in particular Mr B Dyer, whose co-operation made the work possible. Thanks are also due to members of MRDE Roadway Stability Group for advice and practical assistance.

The author is grateful to Mr P G Tregelles, Director of Research and Development, for permission to give this paper. The opinions expressed are those of the author and not necessarily those of the National Coal Board.

REFERENCES

Dudley, W. R., 1977, "A Field Investigation into the Influence of Anhydrite Packing on Longwall Gate Road Stability", PhD Thesis, University of Newcastle upon Tyne, August 1977.

Genthe, M., 1970, "Abbaustreokenbegleitdamme aus hydraulisch abbindenden Materialien". Glückauf, Yr 106, No 2, 22nd January 1970, pp 67-74. NCB Translation Trans 14.2802/HSA Gateroad Side Packs made of Hydraulically Setting Material.

Johnson, G., 1967, "The Closure of Gate Roads related to the Method of Goaf Control in the Low Main Seam at Bilsthorpe Colliery". NCB MRE Report No 2309

Lawrence, D., 1973, "Scale Model Studies of Strata Movements around Mine Roadways - VII. Effects of Horizontal and Vertical Pressure". Int. J Rock Mech. Min. Sci. and Geomech. Abstr. Vol 10, pp 173-182.

Mallory, R., 1980, "Anhydrite Packing in the Wigan Four Feet Seam at Parkside Colliery". MRDE Internal Report No 80/37.

Mallory, R., 1975, "Effect of a Ribside on Subjacent Workings in Close Proximity at Cotgrave Colliery". MRDE Report No 61.

Mallory, R., 1978. "The Effect of Adjacent and Subjacent Workings on a Gateroad in the Seven Feet Seam at Birch Coppice Colliery. MRDE Internal Report No 78/16.

Smith, I., 1976, Experience with Anhydrite Packing (ANPAC) at Easington Colliery", Mining Engineer, Vol 136, October 1976, pp 47-51.

Wilson, A. H., 1964, "Conclusions from Recent Strata Control Measurements made by the Mining Research Establishment". Mining Engineer, Vol 123 No 43, April 1964, pp 367-380.

Wilson, A. H., 1980, "The Stability of Underground Workings in the Soft Rocks of the Coal Measures". PhD Thesis, University of Nottingham, July 1980.

RATIONAL LAYOUT OF ROADWAYS IN FLOOR STRATA AFFECTED BY LONGWALL EXTRACTION

Lu Shiliang

Associate Professor
Chairman of the Mining Section
Department of Mining Engineering
China Institute of Mining

SUMMARY. On the basis of the rules of displacement concerning the surrounding rocks of roadways in floor strata, this paper analyzes the factors influencing the maintainence of these roadways, i.e. the extraction of the overlying coal seams, the mechanical properties of surrounding rocks, the vertical distance between the roadways and the overlying seams, as well as the horizontal distance between the roadways and the edge of coal pillars left in overlying seams. The amount of the displacement of surrounding rocks of various roadways are derived from nomogram. The rational layout of roadways in floor strata is also described.

1. THE INFLUENCE OF THE SEAM EXTRACTION ON THE ROADWAYS IN FLOOR STRATA

According to its relative position to opening in overlying seam, roadway in floor strata or roadway close to coal seam can be classified into six categories as shown in Fig.1. Table 1 indicates the loading of the roadways.

The closure of the roadways caused by seam extraction(u) can be expressed by the following equation:

$$u=v_o t_o+u_1+v_1 t_1+u_2+v_2 t_2 \qquad (1)$$

where: t_o - service time of the roadway in front of the working face I, in overlying seam (days)

t_1 - service time of the roadway behind the working face I in overlying seam (days)

t_2 - service time of the roadway behind the working face II in overlying seam(days)

v_o - closure rate of the roadway before mining of the longwall face I,(mm/day)

v_1 - closure rate of the roadway after mining of working face I, (mm/day)

v_2 - closure rate after mining of working face II,,(mm/day)

u_1 - additional closure of the roadway during the mining of working face I,(mm)

u_2 - additional closure of the roadway during the mining of the working face II,(mm)

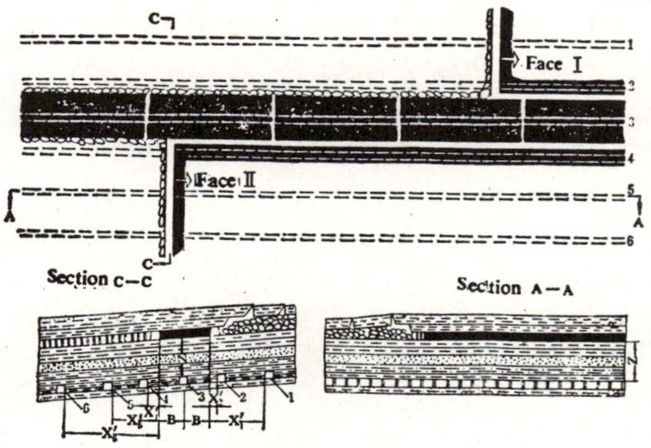

Fig.1 Layout of roadways affected by the extraction of overlying seam.

The field observations made in coal mines of Yangquan, Pingdingshan and Fuxin indicate that the mechanical properties of surrounding rocks of roadways are comparatively stable when mining depth is around 300 m. Fig.2 and Table 2 shows the rules and amount of the deformation of aforecited six categories of roadways when vertical distance between roadways and overlying seams is 10 m. We can see from the table that each roadway has the same value of V_o, but due to the difference of the relative position of the roadway to the pillar between two contiguous working faces and thus the defferent loading conditions, the values of u_1, v_1, u_2 and v_2 vary greatly. In the view point of the amount of roadway deformation throughout its service time, the location of the roadway 1 and 6 is certainly better than others. But it should be pointed out that when the coal face is advancing above the roadway, the overlapping abutments of the working face and the adjacent goaf on roadway should be avoided. After finishing of the face, the roadway should be located in stress relief area, and not be affected by high stress transmitted from overlying coal pillars. This is a very important condition for keeping the roadway in good shape.

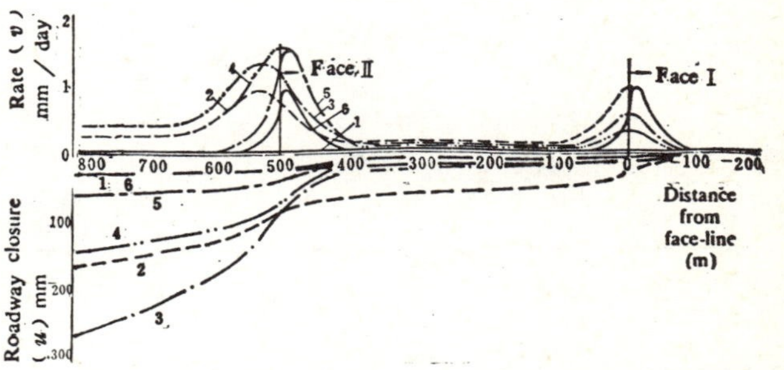

Fig.2 Roadway closure affected by the extraction of overlying seam.

Table 1. The influence of the mining of overlying seam on roadway closure

<table>
<tr><th colspan="2" rowspan="3">Category of roadway</th><th colspan="3">Face I</th><th colspan="2">Face II</th></tr>
<tr><th>before mining</th><th>during mining</th><th>after mining</th><th>during mining</th><th>after mining</th></tr>
<tr><th>rate of closure, v_o (mm/day)</th><th>additional closure, u_1 (mm)</th><th>rate of closure, v_1 (mm/day)</th><th>additional closure, u_2 (mm)</th><th>rate of closure, v_2 (mm/day)</th></tr>
<tr><td rowspan="2">The face I of overlying seam is advancing just above the roadway</td><td>1.roadway with large x' value (x', the horizontal distance between roadway and the margin of coal pillar above) (roadway No.1 in Fig.1)</td><td rowspan="6">all the roadways are driven in rock strata</td><td rowspan="2">affected by the abutment caused by the advancing of the working face I</td><td colspan="3">situated beneath the goaf and in the stress relief area</td></tr>
<tr><td>2. roadway with small x' value. (roadway No.2 in Fig.1)</td><td rowspan="2">affected by the pillar being mined at one side</td><td rowspan="3">mining taking place at both sides of the pillar above, subjected to the overlapping stress caused by the mining of the working faces No.I and II</td><td rowspan="3">affected by the abutment of the pillar being mined out at both sides</td></tr>
<tr><td colspan="2">3. roadway situated beneath the coal pillar after the extraction of the overlying seam (roadway No.3 in Fig.1)</td><td>affected by the abutment caused by the mining of the working face I</td></tr>
<tr><td rowspan="3">The faceII of overlying seam is advancing just above the roadway</td><td>4. roadway with small x' value (roadway NO.4 in Fig.1)</td><td colspan="2">compared with roadway No.3 the distance to face I is larger, so it is less affected by the mining of face I</td></tr>
<tr><td>5. roadway with large x' value (roadway No.5 in Fig.1)</td><td colspan="2">located far from working face I, being little affected by the mining of face I</td><td>little affected by the overlapping stress caused by the mining of faces I and II</td><td rowspan="2">situated beneath the goaf, and in the stress relief area</td></tr>
<tr><td>6. roadway with x' value greater than 60-70 m. (roadway No.6 in Fig.1)</td><td colspan="2">located far from working face I, not affected by the mining of face I</td><td>only affected by the abutment caused by the advancing of face II</td></tr>
</table>

Table 2. Closure of various roadways at different mining stages

Roadway category (numbered corresponding to Fig.1)	Rate of closure, v_o (mm/day)	Additional closure, u_1 (mm)	Rate of closure, v_1 (mm/day)	Additional closure, u_2 (mm)	Rate of closure, v_2 (mm/day)
1	0.005	8	0.005	0.005	0.005
2	0.005	20	0.015	20	0.025
3	0.005	0-10	0.005-0.01	40-80	0.05-0.1
4	0.005	0-6	0.005-0.007	20-40	0.025-0.05
5	0.005	0	0.005	20	0.005
6	0.005	0	0.005	8	0.005

2. THE INFLUENCE OF MECHANICAL PROPERTIES OF SURROUNDING ROCKS ON THE MAINTENANCE OF ROADWAYS

The mechanical properties of surrounding rocks have great influence on the maintenance of roadways. The followings are some examples showing the relationship between the mechanical properties of surrounding rocks and roadway deformation under different mining stages.

2.1 Relationship between the mechanical properties and displacement of surrounding rock during roadway driving.

Due to the different mechanical properties of the surrounding rocks of the raise in floor strata in Colliery No.7, Pingdingshan Coal Mines, closure of different section of the raise varied greatly during driving. When the surrounding rock was massive sandy shale, the closure of the raise (u_o) after 40 days was only 5 mm (Fig.3-1); u_o amounted to 35 mm when the surrounding rock was grey shale with bands of carbonaceous shale at both sides(Fig.3-2); u_0 reached 65 mm when carbonaceous shale occurred on roof and both sides (Fig.3-3); and u_0 increased to 105 mm when both sides and floor was aluminous shale which changed into powder when wetted. The ratio of the closures was 1:7:13:21, and the closure rate of two sides (v_0) was 0.03, 0.3, 0.4 and 0.7 mm/day respectively after surrounding rock tended to stability.

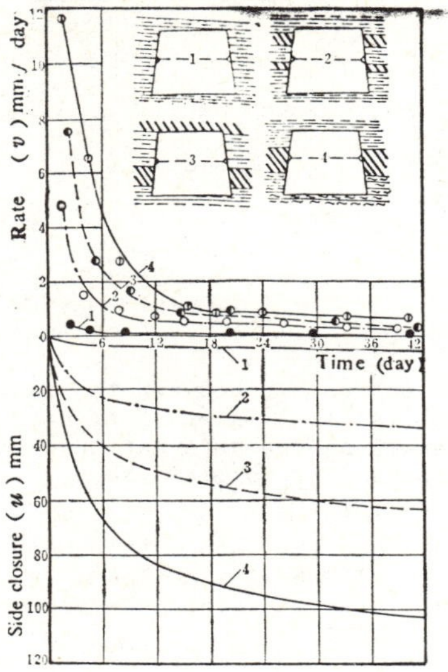

Fig.3 The deformation of the raise in Colliery No.7, Pingdingshan Coal Mines during driving.

2.2 Relationship between mechanical properties and the displacement of surrounding rocks during the extraction of overlying seam.

The field measurements in Pingdingshan, Yangquan Coal Mines indicate that during the extraction of overlying seam, with value Z (vertical distance between the roadway and overlying seam) being about 10 m, the additional closure u_1 for roadway 1 and 6 (as shown in Fig.1) is as follows:

about 10 mm for stable sandstone or sandy shale;
about 30-50 mm for medium stable interbedded shale and sandy shale;
about 150-200 mm for soft strata with thick bands of carbonaceous shale or clay shale.

In China, rock roadways are divided into three categories in terms of the properties of surrounding rocks, namely: stable, medium stable and soft. The ratio of the closures of these roadways during the extraction of overlying seams is 1:3-5:15-20.

3. THE INFLUENCE OF THE VERTICAL DISTANCE BETWEEN ROADWAY AND THE OVERLYING SEAM ON THE STABILITY OF THE ROADWAY

The determination of the vertical distance between the roadway and the overlying seam (Z) mainly depends on the properties of surrounding rocks and the extraction of overlying seam. The field observations reveal the following laws.

3.1 The effect of the properties of surrounding rocks on the determination of value Z.

The positions of all the roadways shown in Fig.4 is corresponding to that of the roadway No.1 in Fig.1. The difference in value Z and surrounding rocks results in different roadway closure. From Fig.4, we can see that:

(1). Relationship between (u) of various roadways and Z can be expressed by:

$$u = a\,Z^{-b}$$

where: a,b, - coefficients, determined by rock properties and mining status of the overlying seam as show in table 3.

Table 3. The value a,b for each kind of roadways

Roadway	a	b
1	33.1	0.442
2	124.5	0.490
3	372.6	0.457
4	1083.6	0.502

(2). The influence of rock properties on roadway closure (u) is much greater than the influence of Z. Under a given rock condition, when value Z is increased from 5-10 m to 20-25 m, value u is only lowered by about 50%. But at a given Z, value u varies greatly as the properties of surrounding rock varies. As shown in Fig.4, the ratio of value u of four categories of roadways is 1:3.6:11:30.

(3). Above a certain value, the increase of Z creates little effect on the reduce of u; but below which, the value u increases significantly with the decrease of Z. That means there exists an optimal value of Z for each kind of surrounding rocks. For the roadway with stable surrounding rocks, the optimal value of Z is about 6 m, and 10 m for medium stable rock, 15-20 m for soft strata.

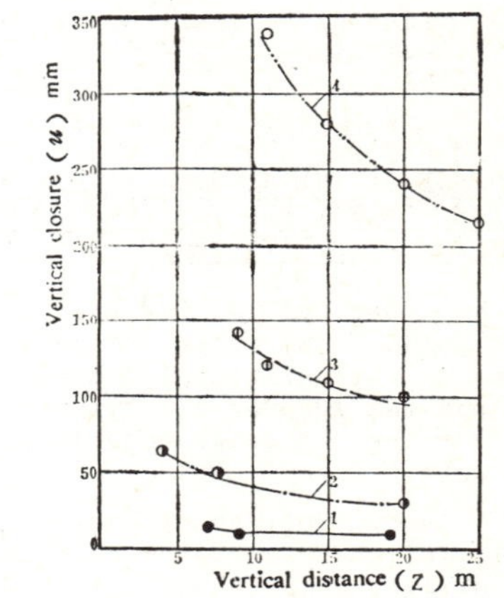

Fig.4 The relationship between roadway closure and value Z.

1 - rock roadway with stable surrounding rocks
2 - rock roadway with medium stable surrounding rocks
3 - roadway in seam with medium stable surrounding rocks
4 - roadway in seam with unstable surrounding rocks

3.2 The effect of the mining status on the determination of value Z.

Two rock roadways shown in Fig.5 were driven in medium stable rocks, and the working faces were advancing just above the roadways. The displacements of the surrounding rocks vary with the mining status and Z values. Fig.5 also indicates that:

(1) The mining status of overlying seam exterts a greater influence on the roadway closure than value Z does. The closure of the roadway subjected to overlapping stress of goads on both sides before the extraction of the straddle working face is four times larger than that of the roadway influenced only by the extraction of the straddle working face (Fig.5-1).

(2) In the case of the medium stable surrounding rock, the optimal Z value for roadway No.1

(Fig.5) is about 10 m. But even when the Z value for roadway No.2 comes up to 20-25 m, the closure of roadway No.2 is still much greater than that of roadway No.1. If the roadway is located beneath the goaf of the overlying seam all the time, 5-6 m of value Z is enough.

In brief, the determination of value Z should be in accordance with the mechanical properties of the surrounding rocks and the mining status of the overlying seam.

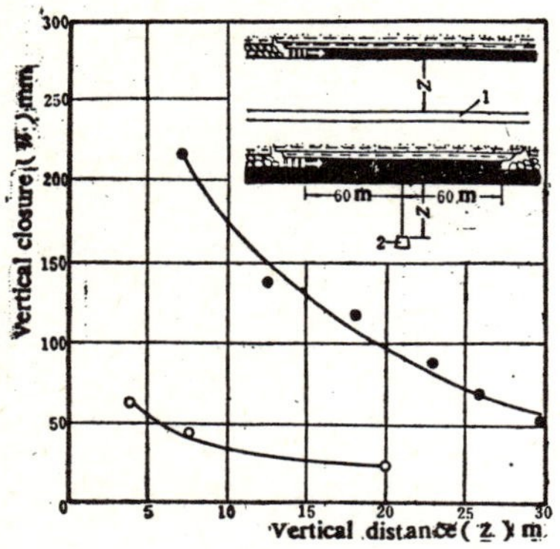

Fig.5 Roadway closure during the advancing of overlying working face.

1 - district gathering roadway of Colliery No.3, Pingdingshan Coal Mines
2 - raise of Colliery No.5, Fengfeng Coal Mines

4. THE INFLUENCE OF THE HORIZONTAL DISTANCE BETWEEN THE ROADWAY AND THE EDGE OF OVERLYING PILLAR ON THE STABILITY OF THE ROADWAY.

The rock properties, mining status of overlying seam and value Z has great influence on the determination of the horizontal distance between the roadway and the edge of overlying pillar(x'). The following rules had be deduced from field observation data.

4.1 Effect of the extraction of overlying seam.

The deformation data of the main haulage roadway of Colliery No.5, Pingdingshan Coal Mines show that the section of the roadway beneath the pillar being mined out on both side (Fig.6) or the section of roadway beneath the coal pillar being mined out only on one side has the following features:

(1). Taking v as the closure rate of the roadway, and x=x'+B as the horizontal distance between roadway and the center of the pillar above, v-x relationship can be expressed as:

$$v = 0.04 + 1.47\ e^{-0.00436\ x^2}$$

In Fig.7, the maximum value of v occurs at the place about 6 m away from the edge of the pillar, and in Fig.6, at the middle part of the pillar.

(2). The extraction of the overlying seam only exerts influence on a limited section of the roadway, with the spread angle β being about 32-35° both in Fig.6 and Fig.7 (In Fig.6, the angle included between the vertical line and the connecting line of the points A and B to the edge of pillar above). But the extraction of overlying seam has significant influence on rate of roadway closure within the section.

(3). When the overlying pillar was destroyed, the v values measured at all stations decreased rapidly and then tended towards stability(Fig.6-2).

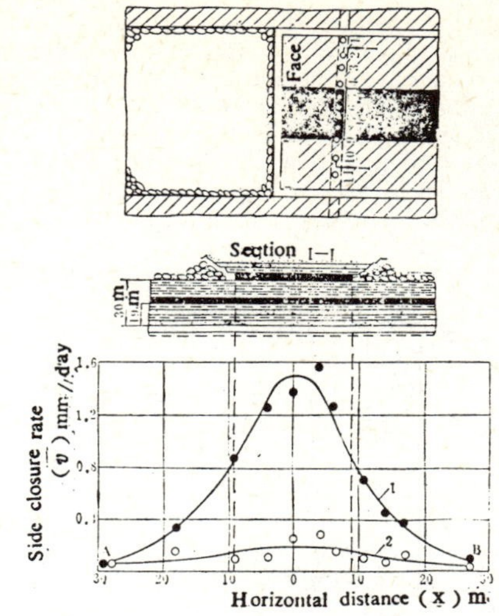

Fig.6 Relationship between the position of pillar left in mined area in overlying seam and the deformation of main haulage roadway of Colliery No.5, Pingdingshan Coal Mines.

1 - average rate within 0-30 m in front of the working face.
2 - average rate, 30 m behind the working face.

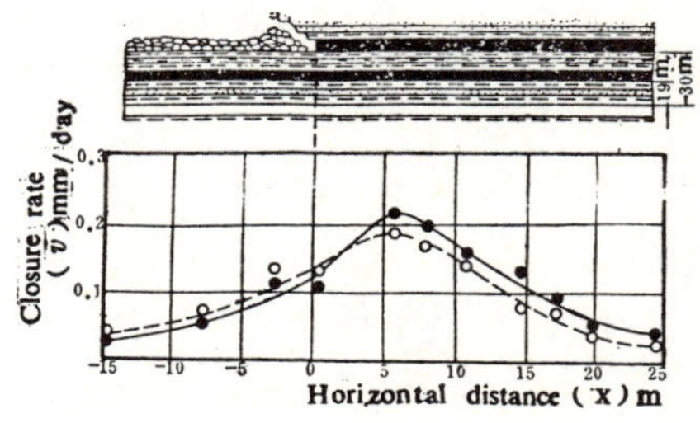

Fig.7 Relationship between the position of the goaf of overlying seam and the deformation of the main haulage roadway of Colliery No.5, Pingdingshan Coal Mines.

black line - roof-to-floor convergence.
dotted line - rate of closure of two sides.

4.2 Effect of rock properties

The data measured from rock roadway with soft clay shale surrounding rock in Furong Coal Mines show that the v versus x still assumes the probability curve distribution, with spread angle β being about 35°(Fig.8). But the value v beneath the center of the pillar differs greatly with that beneath the edge of the pillar. The value v in the roadway of Furong Coal Mines is 6-10 times larger than that in main haulage roadway of Pingdingshan Coal Mines.

4.3 Effect of rock properties and mining status of overlying seam on the determination of x'.

Fig.9 shows the effect of value x' on roadway closure under varying mining status and rock properties in Pingdingshan Coal Mines.

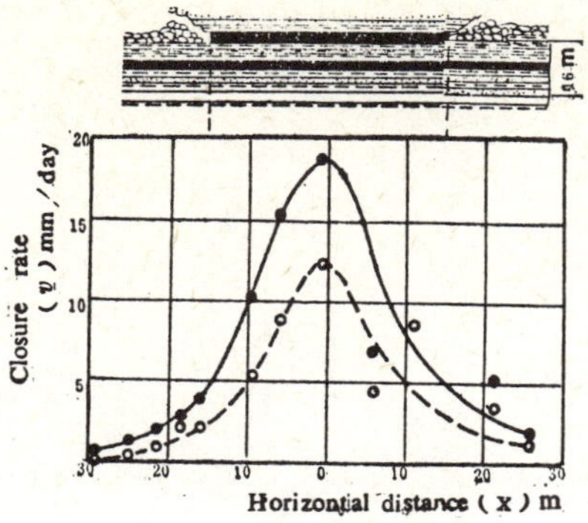

Fig. 8 Relationship between the position of pillar and the deformation of raise of Baijiao Colliery, Furong Coal Mines.
black line - rate of roof-to-floor convergence
dotted line - rate of sides closure

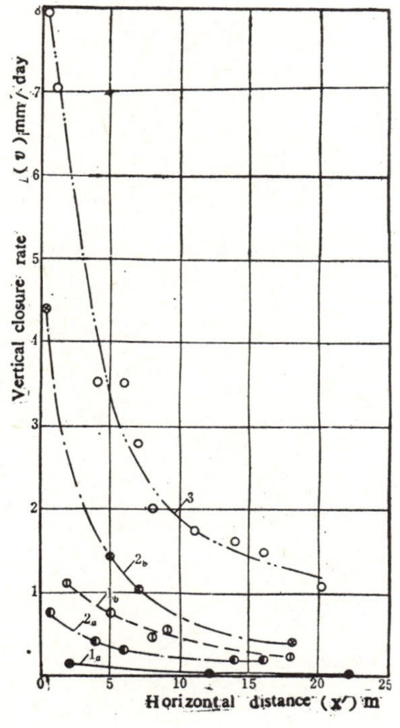

Fig.9 Relationship between x' and roof-to-floor convergence of roadway in Pingdingshan Coal Mines

No.	rock	$\left(\frac{Z}{m}\right)$	Status of pillar above
1a	sandstone, shale	19	mined out at one side
1b	snadstone, shale	19	mined out at both sides
2a	roof being shale, sandy shale; medium hard coal	10	mined out at both sides, tending to stability
2b	roof being shale, sandy shale; medium hard coal	10	mined out at one side, the other side is under operation
3	roof being carbonaceous shale; soft coal	8.5	mined out at both sides, tending to stability

(1). Relationship between v and x' can be expressed as:

$$v = c + a\,e^{-bx'}$$

where: a,b,c, - coefficients, determined by the rock properties and mining status (as shown in Table 4)

Table 4. Values of a,b,c, for various roadways

Roadway	a	b	c
1a	0.211	0.114	0
1b	1.145	0.118	0.23
2a	0.799	0.347	0.2
2b	3.785	0.244	0.37
3	6.649	0.201	0.95

(2). The mining status of overlying seam exerts a great influence on v-x' relation. The v value of the roadway close to pillar being mined out at both sides (Fig.9-1b) is six times greater than that of the roadway close to the pillar being mined out at one side (Fig.9-1a). The v value of the roadway near mining influenced area (Fig.9-2b) is 3-4 timesgreater than that of the roadway near stress relief area (Fig.9-2a).

(3). Rock properties exert a tremendous influence on relationship between v and x'. The v value of the roadway with soft surrounding rocks(Fig9-3) is 7-10 times higher than that of the roadway with medium stable surrounding rocks (Fig.9-2a).

(4). As on the u versus z curve shown in Fig.4, there also exists an optimal value of x' on v versus x' curve. The field observations indicate that the x' to z ratio is about 1(0.7-1.4), β ranging from 35° to 55°, regardless of the rock properties andthe mining status of overlying seam. During actual operation, in the case of soft surrounding rock, intense stress caused by extraction of the overlying seam or smaller value Z, the value x' chosen can be greater than the value Z so as to control roadway deformation, and vice versa.

On the whole, the rock deformation rules revealed in Fig.6,7,8,9 are very important for the selection of the layout of rock roadway and the determination of the rational parameters such as x' etc. It is obvious that the progress of the mining operation from one side of the pillar to both sides isthe main reason causing rapid deformation of the roadway No.2,3,4 as shown in Fig.1. The positions of roadway No.1 and 6 in Fig.1 have more advantages over roadway No.2,4 and 3.

5. ESTIMATION OF THE CLOSURE OF ROADWAYS AFFECTED BY MINING OPERATION

In this paper, all the factors guiding the selection of rock roadway position are closely related to and interact on each other. Here, a nomogram is used for summing up the inner link of the factors and estimating roadway closure, so as to provide informations for selection of the layout of roadways and support type (Fig.10). Roadway positions shown in Fig.10 are identical with those in Fig.1. The mining depth is about 300 m. All the roadways are supported by arch-setting of cut-stone or precast reinforced concrete. Roadway closure are calculated according to Table 2 and formula 1. The width of coal pillar (B) above roadways No.2,3,4 is either 15-20 m or 40 m, but it does not affect roadways No.1,5,6. Three kinds of value x' (0, Z/4, Z/2) are chosen for roadways No.2 and 4. Rock properties fall into three categories: stable, medium stable and soft. There are three kinds of Z values: 10 m, 20 m, and 30 m. Based on above conditions, it is able to find out roadway closure from the nomogram. For example, in Fig.10, service time of roadway No.2 is 5 years; B, 15-20 m; x', Z/4; surrounding rock, medium stable; Z, 10 m; then indicated by the arrow with dash line, roadway closure is about 260 mm.

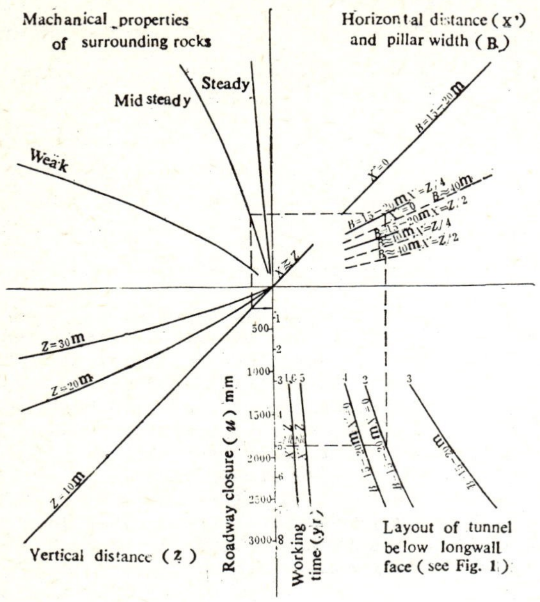

Fig.10 Estimation of the closure of rock roadways affected by mining operation

6. CONCLUSION

Of all the factors affecting the deformation and maintenance of rock roadways, the properties of the surrounding rock is the most important, for the displacement caused by rock properties can be increased as high as 20 times, The next is the mining status of overlying seam and value x', about 5-10 times; then the value Z, 2-5 times. In the coal mines in China, due to the difference of the above mentioned factors, the amount of the displacement of the existing rock roadways in floor strata during service time ranges from 10 mm to 2000-3000 mm. Hence the proper and rational selection of these four factors is of great immediate significance.

DEFORMATION OF ACCESS ROADWAYS AND ROADSIDE PACKS IN COAL MINES

Ian W. Farmer

Department of Mining Engineering
University of Newcastle upon Tyne
Newcastle upon Tyne, NE1 7RU, England

SUMMARY. Although empirical relations between various parameters have been used in predict deformation of access roadways to longwall faces, the mechanism of deformation is very complex. Observations of roadway closure, pack closure and pack stress which were obtained from various sources, including recent investigations by MRDE have been examined.

The principal conclusions are:

(a) Roadway closure has two major components. Initial closure is closely related to pack closure, which occurs over the first 50 - 100m of face advance and rapidly reaches an ultimate level. Secondary convergence continues for sometime and is caused mainly by penetration into or uplift of weak or plastic floor clays.

(b) Pack closure can be defined in terms of pack stress, initial void ratio of the pack material and the compression index of the pack material.

(c) In order to estimate pack closure and secondary convergence due to floor penetration an accurate estimate of pack stress in necessary. Measurements indicate that this is unlikely to exceed by a large amount, half the vertical geostatic stress.

INTRODUCTION

A major factor determining the efficiency and economy of longwall mining operations is the stability or degree of deformation of the access roadways. Although several systematic attempts have been made to isolate the factors affecting roadway deformation (see for instance, Johnson, 1973; Götze and Kammer, 1976; and Whittaker, 1976), these have not always been based on the Mechanics of roadway deformation. It is generally agreed that the main factors determining the magnitude of roadway deformation are:

(a) The stresses acting on the roadway, resulting from redistribution of geostatic stresses during excavation of coal from the longwall working or from adjacent workings in the same or other seams.

(b) The seam height and the compressibility of the face-side pack.

(c) The strength of roof and floor rocks, and the effect of water on strength.

(d) Proximiate geology including the presence of major discontinuities, direction of cleat and strong bands in the roof strata, resisting caving.

(e) The method of forming the roadway, particularly between drill and blast and a machine cut profile; the degree of roadway support and the treatment of the ribside.

(f) The layout of faces and the presence of ribside pillars.

Of these, none will be important unless the stresses acting on the roadway are sufficiently high to cause convergence and/or floor lift to a significant degree.

During longwall working, although some form of waste or goaf support is feasible, caving is normally adopted. During caving, the height of collapse of the roof rocks increases with the area of roof exposed. Although this will be affected by the dilational limits imposed by the seam height (see Farmer and Altounyan, 1980) the maximum height of collapse is usually about 0.8 times the face width. The collapsed rocks then compact, partly under self weight and partly by the action of a surcharge load exerted by the overlying uncollapsed strata. Wilson and Ashwin (1972) have, however, shown that pillars adjacent to the caved area retain redistributed stresses.

Caving is a complex process and prediction of redistributed stresses around the periphery of the excavation, where they will affect the working face and the access roadways, can be extremely difficult. Probably the most satisfactory estimates of stress redistribution in the plane of an advancing longwall face have been obtained by Harris (1974) from a sandbox model. They indicate that a relatively low rib-side abutment stress at the gate end increases as the face advances, reaching a level of about twice the over-burden stress after a face advance roughly equal to the face length. This level results from yielding or loosening of the abutment rocks and as in the case of the coal face, the abutment compression zone can be spread over several metres. The stress in the centre of the goaf increases to about three-quarters of the overburden stress during face advance, but remains low at the edges of the caved area.

OBSERVATIONS OF ROADWAY DEFORMATION

Because of the wide variability of geology and face and roadway geometry it is difficult to isolate the exact mechanism of access roadway deformation. In an attempt to examine some of the more obvious variables, a series of investigations including roadway and pack deformation measurements were carried out and compared with published and unpublished case histories available from MRDE reports and other sources. In the case histories studied as part of the project, a new approach to pack instrumentation was used. To measure stresses a Glötzl or Soil Instruments hydraulic pressure cell was used instead of the NCB pack load cell. These cells were much slimmer than traditional pack load cells and were filled with mercury in an enclosed system connected to a sampling by-pass valve. This construction means that the cells are less subject to bridging and because of their low compressibility are able to sample more accurately pack stresses. A modified strut was used to measure pack closure, and in addition roof lowering and floor heave were measured.

Typical examples of roadway closure with face advance are given in Figure 1. These were all obtained by Mallory (1980).

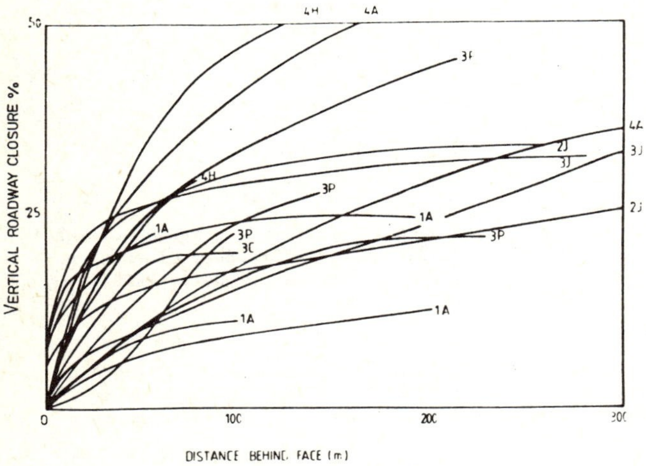

Figure 1. Plot of vertical roadway closure expressed as a percentage of original roadway height against distance behind face (after Mallory 1980). The data is from:

(1) Parkside Colliery, Trencherbone Seam
(2) Kellingley Colliery, Beeston Seam
(3) Daw Mill Colliery, Warwickshire Thick Seam
(4) Parkside Colliery, Wigan Four Feet Seam

The pack types are designated A - Anhydrite, J - Jet stowed, C - Coal, H - Hand built rubble, P - Pump pack.

The basic geometry was:

Parkside Colliery - Trencherbone seam: 1.6m seam with mudstone roof above fire clay on coal floor at a depth of 690m and advancing to full dip at 1 in 4.5.

Kellingley Colliery: 3.1m (2.8m extraction) seam with mudstone roof above fire clay on siltstone floor at a depth of 653m. The seam was nearly horizontal.

Daw Mill Colliery: 2.2m extraction in the seam section of about 8m with 1m of roof coal at a depth of about 600m. The dip was not significant.

Parkside Colliery - Wigan Four Foot Seam: 17m seam with silty mudstone roof and 0.9m coal floor over 2m seatearth. The seam was at a depth of about 750m dipping at 1 in 4.5.

The results, whilst showing variation, illustrate a point which is intuitively acceptable - namely that a stiffer pack will reduce roadway closure. They also show that pack closure is responsible for only a limited proportion of roadway closure. Figure 2 plots Mallory's (1980) data for cases 1, 3, 4 of pack closure with roadway advance and it is evident that this closure although again variable covers a relatively small proportion of the vertical settlement in Figure 1.

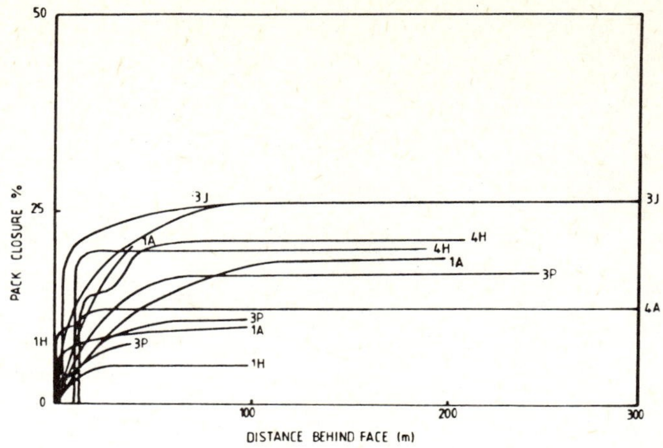

Figure 2. Plots of pack closure against distance behind face (after Mallory 1980). The pack types and locations are indexed under Figure 1.

An indication of the relation between pack convergence and vertical roadway convergence can be obtained from the relations in Figure 3, based on a slightly wider series of case histories listed in Table 10. These show that initially, pack compression is directly related to roadway convergence, but that roadway convergence continues after pack convergence ceases.

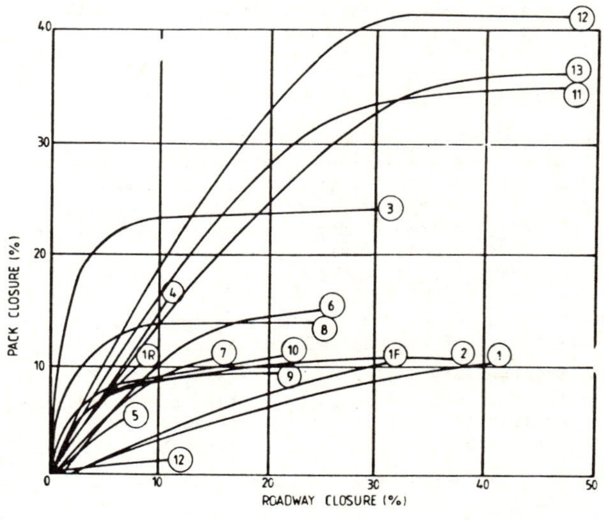

Figure 3. Plots of pack closure against roadway closure against roadway closure for case histories listed in Table 1.

There is evidence (see for instance case 1 in Figure 3) that whilst initial convergence is a direct result of roof lowering following pack convergence, later convergence is caused by floor penetration. It is particularly noticeable from Figures 2 and 3 that pack convergence is relatively quickly completed in less than 100m after face advance. In the case of strong roofs and floors or roadways at shallow depths this is likely to represent the limit of convergence. In weak rock, in the vicinity of discontinuities or shear zones or at depth, considerable additional deformation may occur.

It is interesting to note that roof lowering appears in many cases to mobilise an active shear break close to the ribside and relatively close to the vertical. A possible mechanism for this and for other factors controlling roadway deformation are

Table 1. Access Roadway Deformation - Case Histories

No.	Colliery and seam	Extracted Seam Section (m)	Depth (m)	Type of Pack	Face and Roadway details	Proximate Geology	Reference
1	EASINGTON Low main	1.6	550	Anhydrite	174m advancing face. 30m advance heading 4.5 x 3m arch roadways. Pack 1.8m wide.	8m laminated mudstone roof;thin fireclay and sandstone floor on shale	Dudley (1977)
2	PARKSIDE Wigan four foot	1.7	808	Anhydrite	180m advancing face.4.3 x 3m arch roadway. Pack 1.3m wide on dint.	3m laminated siltstone roof;1m weak seatearth floor on 0.6m coal on shale.	Mallory (1980)
3.	ABERNANT Red vein	1.2	395	Thyssen Pump Pack	160m advancing face.5m x 2m square set roadway. 2.5m wide pack.	9m sandy shale roof. 9m fireclay and shale floor.	Jones (1980)
4	ABERNANT Red vein	2.1	395	Thyssen Pump Pack			
5	MERTHYR VALE Seven foot	2.3	700	High Pressure Stowing with Added Pack Bind			
6	MERTHYR VALE Seven foot	2.3	700	High Pressure Stowing with Added Cement	140m advancing face. 4.3 x 3m half moll arch roadways. 4.5m advance heading. Pack 6.5m wide. Face 40m beneath Nine. foot seam goaf.	3.4m siltstone roof. 2.1m fireclay floor above 8m laminated siltstone.	Jones (1980)
7	METHYR VALE Seven foot	2.3	700	High Pressure Stowing			
8	ABERNANT Red vein	2.1	395	Thyssen Pump Pack	190m retreat face;pack used previously for advancing face. Roadways 5m(8) to 4m (9) wide, 2m high. Pack 2.5m wide.		
9.	ABERNANT Red vein	2.1	395	Thyssen Pump Pack		9m sandy shale roof; 9m fireclay and shale floor	Jones (1980)
10	WOLSTANTON Great Row	1.8	594	Hand Placed Rubble	330m advancing face. Pack 9m wide.	30m weak shale roof; 1m roof coal. Soft fireclay floor.	Evans et al (1941)
11	KELLINGLEY Beeston	2.6	630	Low Pressure Stowing			
12	KELLINGLEY Beeston	2.6	630	Low Pressure Stowing	300m advancing face. 4.5 x 3.5m arch roadways. Pack 3m wide.	3m thick mudstone roof. Thin fireclay floor on siltstone	
13	KELLINGLEY Beeston	2.6	630	Low Pressure Stowing with added cement			
14	LABORATORY	-		Stone Rubble			

discussed in Farmer (1980) and Farmer and Robertson (1975). A schematic approach is illustrated in Figure 4, based on the observations in Figures 1 and 3. This suggests that an estimate of roadway closure could be obtained provided that an estimate of pack stress, pack compressibility and roof and floor strength could be obtained.

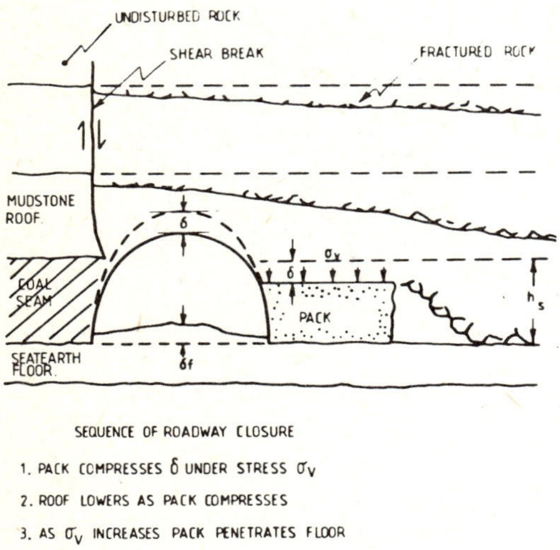

Figure 4. Idealised representation of the mechanics of roadway closure.

COMPRESSIBILITY OF PACK MATERIALS

Large numbers of pack materials (see Blades, 1975) have been introduced in the past ten years in order to give added support to and in some cases to reduce permeability of the face-side of access roadways to advancing longwall faces. Materials have usually been selected arbitrarily for use with various packing systems developed to eliminate manual packing. To date, research has been concentrated on the method of packing, and to a lesser extent, on the interaction of the packing material and the packing method. Less emphasis has been placed on the mechanical reaction of packing materials to the redistribution of forces following mining and the mechanics of side support and its effectiveness in improving roadway conditions

In practice a face or ribside pack is subjected to a degree of confinement and its characteristics can only be correctly defined by taking this into account. A pack is confined longitudinally by adjacent packs, quite rapidly on the goaf side by the caving roof and on the roadway side by the arches and by any lagging between the arches. If it is assumed that the pack material is completely confined, then is should closely follow the fundamental compression characteristics of other drained granular materials. On a plot of void ratio (e) against the logarithm of the vertical pressure (log σ_v) this should (depending on the initial void ratio) follow a horizontal line to between 10 and 100 kN/m^2, a constant slope straight line curve up to about 10,000 to 100,000 kN/m^2 and a horizontal line beyond this. Illustrations of curves for particular granular materials are given in Terzaghi and Peck (1967), page 65. The first part of the curve at constant void ratio represents a pressure too small to compress the material. During the second part grains are crushed and re-arranged and during the third part, the material is too dense for further significant deformation. The constant slope part of the curve defines a fundamental reaction to stress of a granular material in confinement and can be described by the form:

$$e = e_o - C_c \log_{10} \frac{\Delta\sigma_v}{\sigma_{vo}}$$

where σ_{vo} is the pressure at the start of the linear part of the curve,

$\Delta\sigma_v$ is a pressure increment within that part of the curve,

e_o is the initial void ratio

and C_c is a dimensionless ratio known as the compression index.

From the definition of void ratio the compression (δ) of a layer of material of initial thickness h_s can be given by:

$$\frac{\delta}{h_s} = \frac{C_c}{1+e_o} \log_{10} \frac{\Delta\sigma_v}{\sigma_{vo}}$$

Expressed as a percentage, $\frac{\delta}{h_s}$, will give the pack closure for a given $\Delta\sigma_v$ provided satisfactory values of C_c, σ_{vo} can be determined. Terzaghi and Peck quote typical values for sands as:

			C_c	σ_{vo} kN/m^2
(a)	80% sand + 20% mica	-	0.50	10
(b)	90% sand + 10% mica	-	0.29	10
(c)	Loose sand	-	0.10	100
(d)	Dense sand	-	0.08	1000
(e)	Soft Detroit clay	-	0.21	10

A basic question is - do packs deform in a similar way to confined granular materials - and can anything be learnt from using the same criteria - particularly compression index to describe that compression? Case history data is limited at the moment and the cases quoted in Table 1 and plotted in Figure 5 as pack closure % against the logarithm of measured pack stress, designated σ_v, are the best available. Even so some of both the stress data and deformation data must be queried, and generally speaking, data refers to the first 100 days face advance or less. In several cases, the pack, or more probably, the floor, failed.

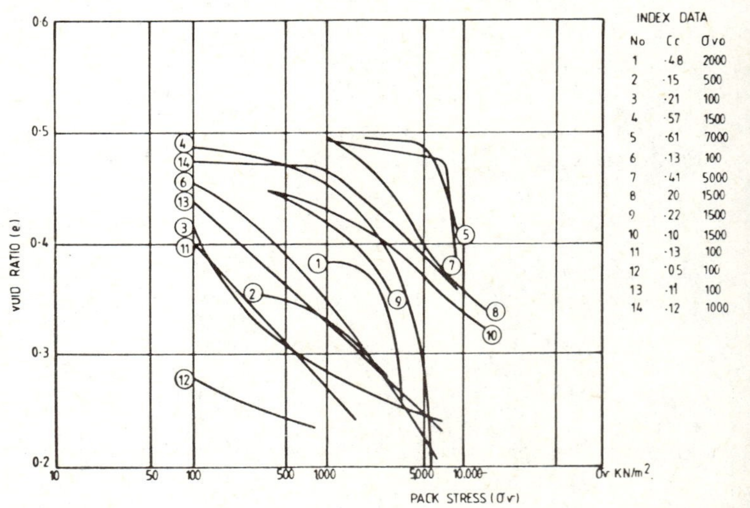

Figure 5. Plot of pack closure (δ/hs) against the logarithm of measured vertical pack stress (σ_v) for case histories in Table 10.

In Figure 6, the data is plotted as void ratio e, against log σ_v. Initial void ratio data was not available and e_o of 0.6 was assumed for rubble packs, 0.5 for monolithic (except in the case of 6) and 0.4 for anhydrite packs. Bearing in mind that 0.3 is a reasonable minimum void ratio, this may have been too low.

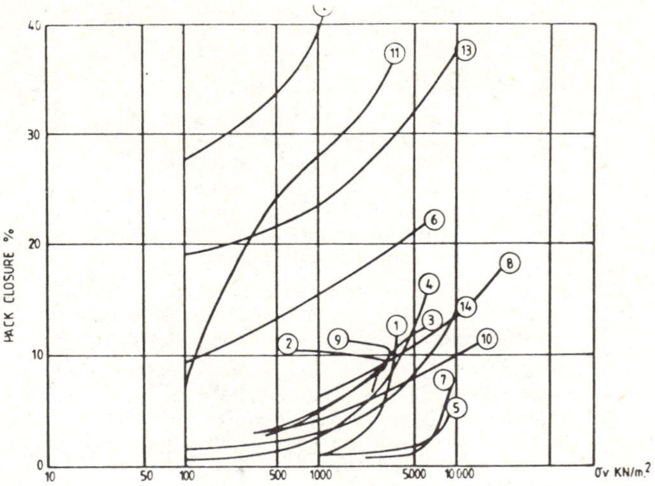

Figure 6. Plot of void ratio (e) against (σ_v) for case histories in Table 1.

Two curves in Figure 6, case 8 and case 10 follow what Terzaghi and Peck 1967 describe as a typical compression curve, but all curves have a significant straight line portion, and many exhibit parts of the typical curve. The conclusion must be that if pressures below 100 kN/m^2 could have been measured and above 10,000 kN/m^2 had been monitored, then all pack materials would have exhibited typical deformation characteristics for confined granular materials.

If this is correct, then the ideal pack material will be one which has a low C_c value at a relatively high pressure threshold. The results are not sufficiently reliable or complete for detailed analysis, but the following general points may be inferred from Figure 6:

(a) One anhydrite pack (case 1) exhibited a high C_c value at a high threshold stress, and case 2 failed at about the same threshold value. Both these results may have been affected by floor penetration.

(b) The monolithic or pumped packed systems (cases 4-7, 11, 12) exhibited both high and low C_c values at a high threshold stress although in two cases (6,11) the threshold stress was low, in the latter case due to poor pack quality.

(c) The low pressure stowing (case 13-15) was much better than the hand packing (case 3), although in the case of pressure stowing, convergence was high.

(d) The high pressure stowing (cases 8-10) even on floor clay gave good results having a high threshold stress and low C_c value. Addition of packbind (8) or cement (9) did not seem to improve characteristics.

From the cases studied, suitable values for substitution in the compression equation can be summarised approximately as:

	e_o	C_c	σ_{vo} kN/m^2
Mechanised pack	0.7	0.15	10
Hand pack	0.6	0.15	50
Stowed pack (low pressure)	0.6	0.15	50
Stowed pack (high pressure)	0.6	0.2	1500
Pump pack (Bentonite)	0.5	0.4	3000
Monolithic pack (Cement or anhydrite)	0.4	0.4	2000

The higher C_c values for the stronger packs indicates a tendency to floor penetration or failure at the onset of rapid convergence. The model is probably less reliable in the case of these packs than in rubble or stowed packs. In these packs, provided σ_v can be predicted accurately, it should be possible to estimate h_s from the consolidation equation and the data summarised from Figure 6. Prediction of σ_v on the pack is, however extremely difficult, because of the proximity of the edge of the excavation and other factors. σ_v was plotted for the available data against seam height, rock strength, face width and seam depth, without finding any significant relations. In Figure 7 it can be seen that in all except one exceptional case (Merthyr Vale) σ_v is less than the vertical geostatic stress, γz. However in all except 8 cases (mainly failed anhydrite packs, and questionable measuring equipment) σ_v is greater than $0.2\gamma z$ predicted by Harris' (1974) model. A reasonable limit would appear to be 0.5 γz and a rule of thumb that the maximum pack stress should not exceed by a large amount, half the geostatic stress in most longwall mining appears to be an acceptable criterion for pack design.

ESTIMATION OF FLOOR HEAVE

Floor heave is not easily predicted and further research is required. If the floor clay when wet behaves in a ductile manner, prediction of closure is difficult. For instance where thick layers of weaker seatearths are confined and subjected to high deviatoric stresses deformation of a strain hardening type will lead to flow of the deforming rock which will exert considerable pressures on any unconfined layers in the immediate floor area. This type of deformation is common in arch supported gate or access roadways. Figure 8, based on measurements from case history 1 in Table 1 shows comparative arch and floor deformation. The continuing deformation of the floor and the virtual stability of the arch is a feature of this type of deformation. The simplest approach to design is to take some of the rules of thumb developed by civil engineers for estimating foundation settlements. The first of these is that for penetration to occur, the pack stress must exceed the bearing capacity of the floor material. Undrained conditions may be used, since consolidation, even of wet floor clays is unlikely to lead to sufficient penetration to cause large scale floor heave.

Then from Terzaghi's modified bearing capacity equation, penetration will occur when:

$$\sigma_v > 9\ Cu$$

where σ_v represents the pack stress
and Cu the undrained shear strength of the floor clay.

Cu values for clays quoted in various sources follow the range

very soft clay	25	kN/m^2
soft clay	25-100	kN/m^2
medium clay	50-100	kN/m^2
stiff clay	100-200	kN/m^2
very stiff clay	200-400	kN/m^2

clay shales	500-1000	kN/m^2
strong seatearths	500-1000	kN/m^2

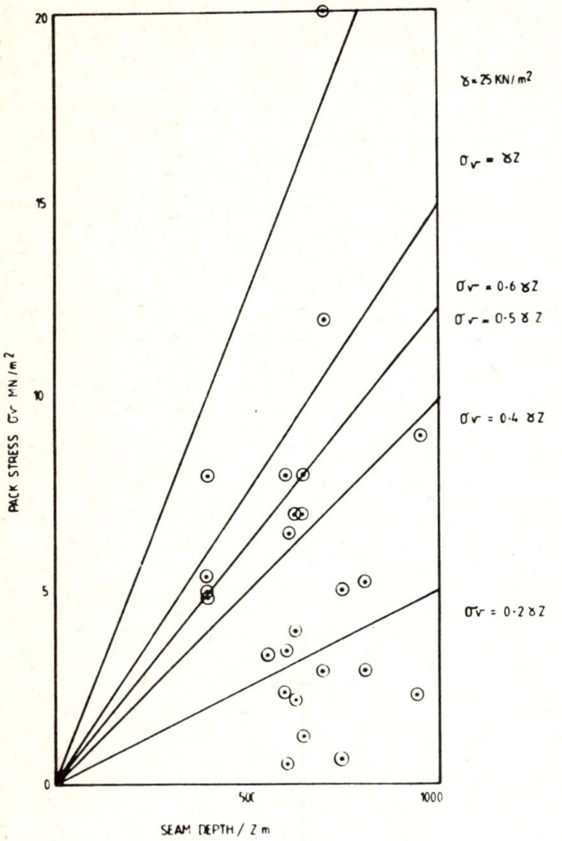

Figure 7. Plot of measured vertical pack stress for all the case histories in Figure 67(a) and Table 1.

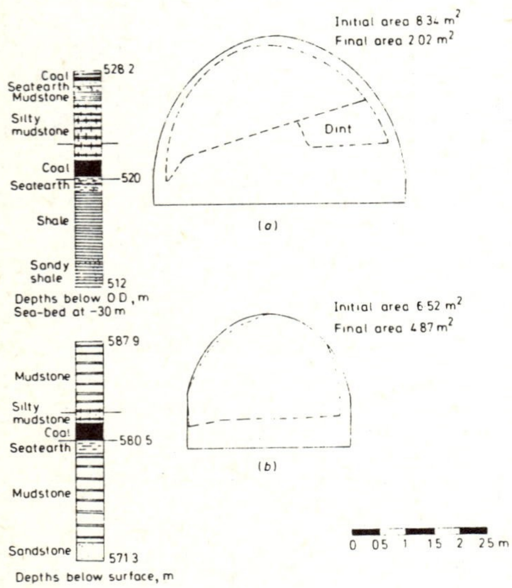

Figure 8. Roof and floor deformation in an access roadway at (a) Dawdon Colliery and (b) Wearmouth Colliery.

These values will be reduced if the floor clay is wet. In the case of active or swelling clays, further information on the plasticity index would be required.

The depth of penetration, once initial failure has occurred will depend on the strength of underlying layers, and will be difficult to predict without a knowledge of local conditions and/or observations and experimentation.

CONCLUSIONS

The principal conslusions are:

(a) Roadway closure has two major components. Initial closure is closely related to pack closure, which occurs over the first 50 - 100m of face advance and rapidly reaches an ultimate level. Secondary convergence continues for sometime and is caused mainly by penetration into or uplift of weak or plastic floor clays.

(b) Pack closure can be defined in terms of pack stress, initial void ratio of the pack material and the compression index of the pack material.

(c) In order to estimate pack closure and secondary convergence due to dloor penetration an accurate estimate of pack stress is necessary. Measurements indicate that this is unlikely to exceed by a large amount, half the vertical geostatic stress.

REFERENCES

Blades M.J. (1975), "Developments in monolithic packing systems". Paper presented to Inst. Min Engrs., South Midlands & South Staffs Branch.

Dudley W.R. (1977), "A field investigation into the influence of anhydrite packing on longwall gate road stability". Ph.D. Thesis, University of Newcastle upon Tyne.

Evans W.H., Hogan M.A. and Vallis E.H. (1941), "An investigation of the load on packs at moderate depths". Trans. Inst. Min. Engrs., 50, 339-377.

Farmer I.W. and Robertson J.T. (1975), "The effect of pack construction on roadway stability behind working faces" Mining Engineer, 134, 599-606.

Farmer I.W. (1980), Face and roadway stability underground in coal mines; geotechnical criterion. Report to ECSC, No. EUR 7298EN.

Gotze W. and Kammer W. (1976), "Die Auswirkungen von Strecken furung und Ausbau au die Querschnitts verminderung von Abbaustrecken". Gluckauf. 112, 846-853.

Harris G.W. (1974), "A sandbox model used to examine the stress distribution around a simulated longwall coal-face." Int. Journal Rock Mech. Min. Sci. and Geomech. Abst. 11, 325-335.

Hobbs D.W. (1968), "Scale model studies of strata movement around mine roadways III". Int. Jl. Rock Mechs. Min. Sci., 5, 245-251.

Johnson G. (1973) "Rock mechanics - a nomograph for the assessment of roadway conditions". Colliery Guardian, 221, 16-20.

Jones R.T. (1980) Internal Report, National Coal Board, Mining Research and Development Establishment.

Mallory R. (1980) Internal reports, National Coal Board, Mining Research and Development Establishment.

Terzaghi K. and Peck R.B. (1967) "Soil mechanics in engineering practice". 2nd Edn. Wiley, New York.

Whittaker B.N. (1976) "Recording, treatment and interpretation of roadway deformation surveys". Mining Engineer, 135, 607-617.

ACKNOWLEDGEMENT

The work of R. Mallory of the National Coal Board, Mining Research and Development Establishment and P.F.R. Altounyan of Newcastle University in obtaining some of the data quoted in the paper is acknowledged.

TUNNELLING INDUCED DEFORMATION AND STRESS CHANGES IN THE LOWER CHALK: CHANNEL TUNNEL PROJECT

ANDREW G. REID *
DOUGLAS J. MACONOCHIE **

* Lecturer, Department of Civil Engineering, University of Strathclyde, Glasgow

** Head of Geomechanics, Peter Hollingsworth and Associates, Brisbane

SUMMARY. Geotechnical studies for the Channel Tunnel Project involved considerable insitu instrumentation and subsequent analysis. This paper examines only deformation and stress changes induced in the Lower Chalk during passage of the mole through the instrumented zones, occurring over a relatively short period of time and distance.

INTRODUCTION

Proposals for a fixed link between the United Kingdom and France date from Napoleonic times; the first serious proposal was made in 1802 and schemes have continued to be propounded, almost unabated, until the present day. The aphorisim that 'more words than work have been expended on the project' (Peyton, 1973) is still true today.

Historical Investigations

The early geological studies, circa mid 19^{c}, comprised surficial examinations and sea bed sampling, later followed by sinking boreholes at St. Margaret's Bay and on the French coast, to depths up to 170 m. In the period 1877 to 1882 the British and French Channel Tunnel Companies undertook shaft sinking and tunnelling trials at Abbot's Cliff and Shakespeare Cliff and at Sangatte, using Beaumont's compressed air driven tunnelling machine. The 2.1 m diameter Beaumont tunnel, extended 1840 m from Shakespeare Cliff towards the Admiralty Pier, was partially resurrected 1974 to 1975, and used to provide access to the instrumentation area (Figure 1).

Recent Investigations and Developments

In more modern times, 1958 to 1960, preliminary geological work conducted by British and French consultants, resulted in a proposal for twin 6.5 m diameter single track rail tunnels at 30 m centres with a 3.8 diameter central lower service tunnel to be favoured (Government White Paper, 1963). Major site investigation work during the years, 1964 and 1965,

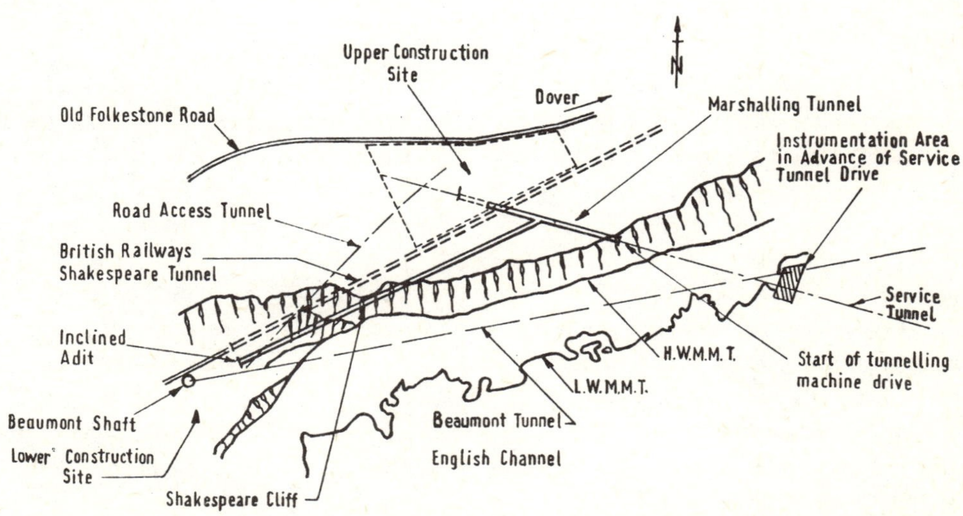

Figure 1 INSTRUMENTATION GENERAL ARRANGEMENT OF PHASE II TUNNELLING WORKS.

followed the U.K. and French Government's accord on tunnelling feasibility and viability. A phased sequence of events was agreed in 1972 viz, Phase I, a Study Report, which was accepted, leading to Phase II, as follows

..Construction of access works and underground marshalling yard near the Shakespeare Cliff, Dover and thereafter driving about 2 km of service tunnel ...Construction of a sloping adit, underground marshalling yard at Sangatte and thereafter driving about 1.5 km of service tunnel.... (additional surveys and tests were referred to, viz.)...boreholes; tests as to the treatment of bad ground and optimisation of construction methods; lining and lining joint tests; tests on rock, soil and concrete ...(Government White Paper, November 1973).

The tunnel layout finally selected for Phase II comprised twin 7.1 m I.D., single track rail tunnels at 30 m centres with a 4.5 m I.D. central lower service tunnel with cross adits at 250 m intervals; submarine tunnelling would be wholly within the Lower Chalk.

ANALYSIS OF INSTRUMENTATION RESULTS

Deformational Behaviour of the Chalk

This section examines only total deformation of the Chalk. The question of creep has been considered by Maconochie (1977) who observed that given the opportunity, creep could amount to some 25 per cent of the total deformation. His studies involved, inter alia, examination of deformation behaviour during stationary periods of the tunnelling machine, and separation of movements into immediate and creep components.

Deformation movements were detected by Linear Variable Differential Transducers, data logged during tunnelling, recording the behaviour of anchors embedded in the Chalk (System resolution 0.001 mm). Space permits only a small sample of the processed results to be shown for Instrument Sections (IS) 1, 2 and 3. The latter IS comprises radially disposed boreholes 3-1 to 3-9; Figure 2 shows layout . Full details are described elsewhere (Maconochie 1977, Reid 1979).

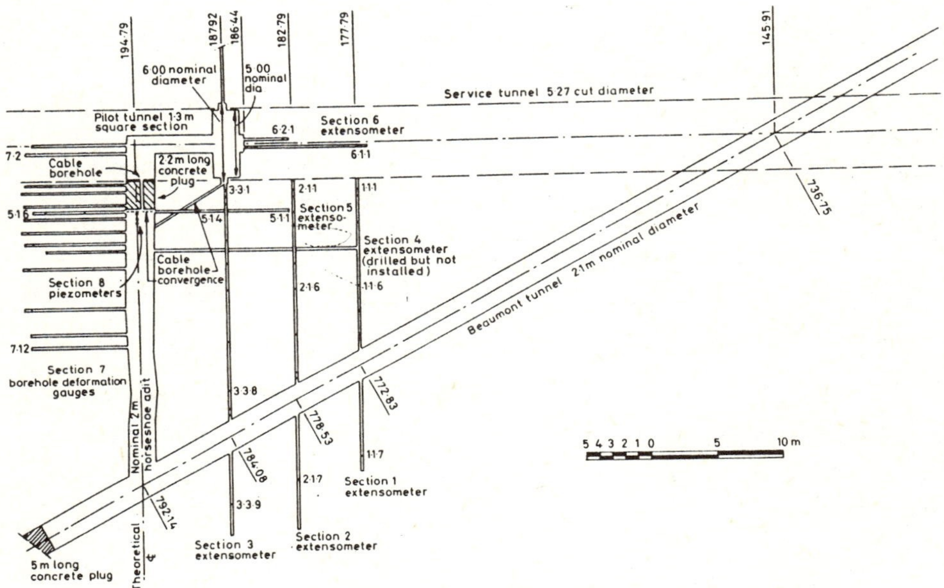

Figure 2 GENERAL LAYOUT OF INSTRUMENTATION AT BEAUMONT JUNCTION

The comprehensive geotechnical programme developed and installed to meet the requirements of the 1973 White Paper has been reported elsewhere (Potts et al,1976 and Curtis et al,1976). The insitu instrumentation scheme referred to in this paper is shown in Figure 2. The scheme comprised, inter alia, extensometers to monitor movement of the chalk due to excavation of the service tunnel and borehole deformation gauges (Talbott Strain Cells). This paper examines the deformational response of and the stress changes occurring within the Chalk due to the limited 260 m tunnelling drive permitted under the curtailed Phase II construction works.

The geotechnical properties (Reid et al, 1982) of the Lower Chalk are referred to as required but not discussed in detail.

Instrument Sections 1, 2 and 3-3

Figure 3 shows the contours of essentially horizontal displacements occurring towards the service tunnel at IS 1, for various positions of the tunnel face Z/D (ratio of distance of face Z from instrumentation section to tunnel diameter D). The results in this figure cover a time span of some 64 hours with stationary mole periods of approximately 7 hours just beyond IS 1 and a 14 hour inactive period at +3 diameters in an otherwise reasonably steady excavation phase.

The contour pattern, while exhibiting reasonable sympathy with expectation i.e. high deformation gradients at ± one diameter from the IS and the tendency to radiate from a low negative Z/D point, the

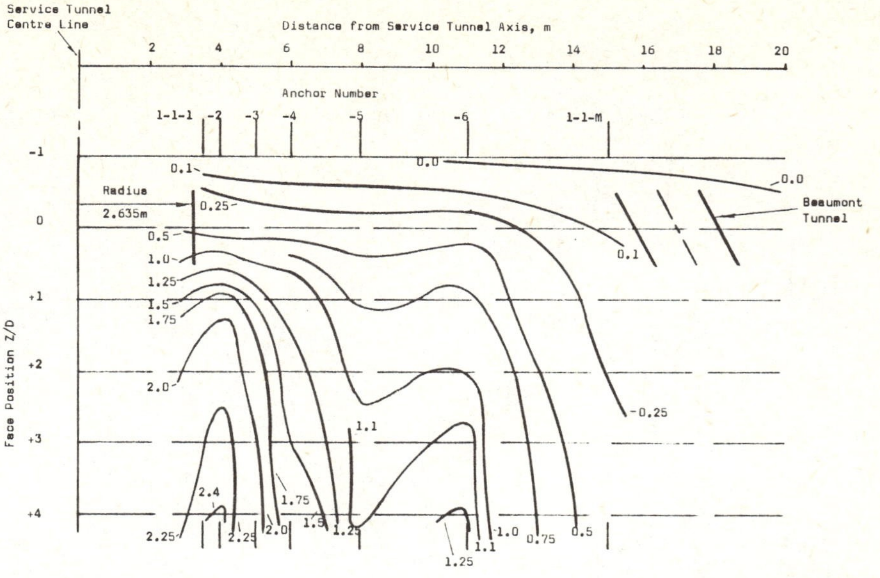

Figure 3 CONTOURS OF TOTAL DISPLACEMENT IN MM, RELATIVE TO FACE POSITION Z/D AT IS 1

feature of peaking at two points 1.1 m and 7.7 m from the excavation wall serves to confirm the non-homogeneity of the Chalk.

At passage of the mole, a wall deformation of 0.5 mm is detected increasing to 1.5 mm at + 1 diameter. At Z/D just over 2 diameters ring construction at the IS has been completed and this is reflected in the increased spacing of the contour interval.

Figure 4 shows a composite of the behaviour of the three essentially planar IS's 1, 2 and 3-3 at Z/D = +4.

The highlights of this figure are clearly (i) the effects of divergence of the Beaumont Tunnel with tendency for low value contours to become parallel to it, and (ii) peaks and hollows indicative of load being attracted to stronger ground and shed by weaker zones.

Another possible explanation of the latter involves the presence of fissures. Features such as zones of softened chalk and fissures were detected while drilling in the chalk.

Instrument Section 3

The section comprises radially disposed IS's 3-1 to 3-9 including IS 3-3 referred to above. The orientation of these IS's are shown in Figures 5 and 6.

Two sets of data have been selected to illustrate deformation behaviour viz. at passage of the mole Z/D = 0, Figure 5, and at Z/D = +3 or +4, Figure 6.

Figure 5 shows maximum deformation of 0.55 mm at the R.H. spring point i.e. IS 3-3. In contrast, at the opposite side of the tunnel IS 3-7 through 3-9,

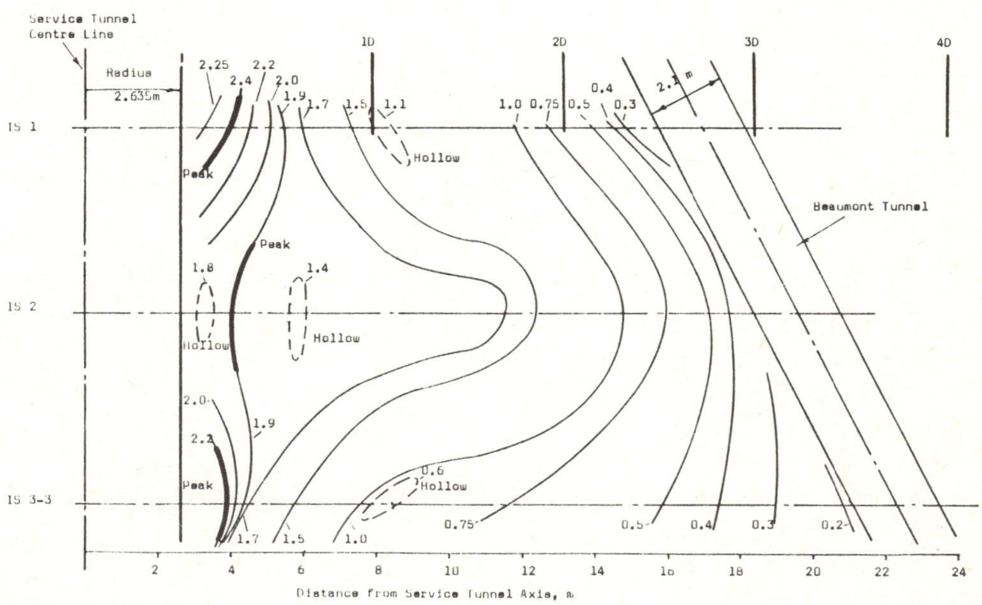

Figure 4 CONTOURS OF DISPLACEMENT AT Z/D = 4 DIAMETERS FOR IS 1, 2 AND 3-3, IN MM.

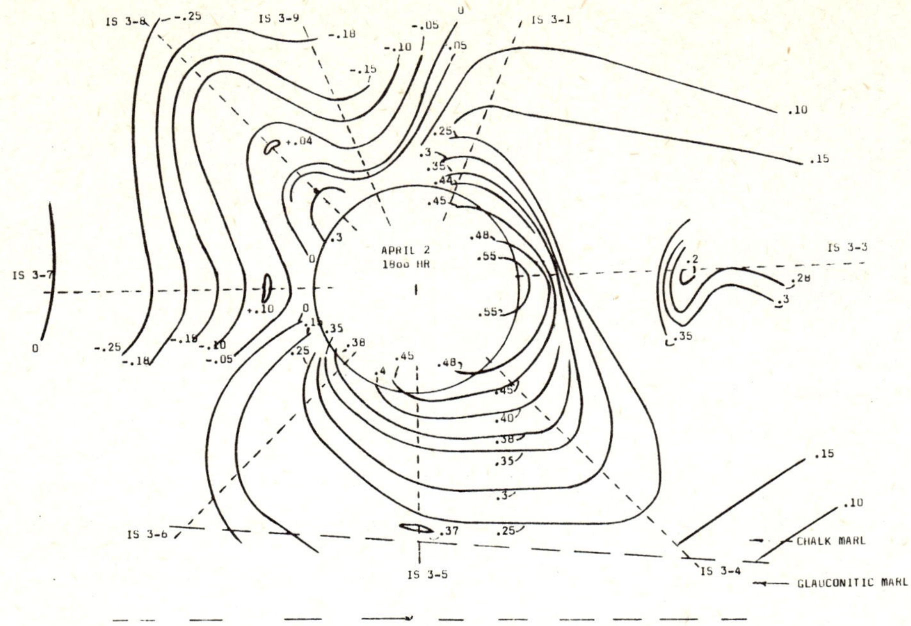

Figure 5 CONTOURS OF TOTAL DISPLACEMENT IN MM RELATIVE TO FACE POSITION Z/D = 0 , IS3

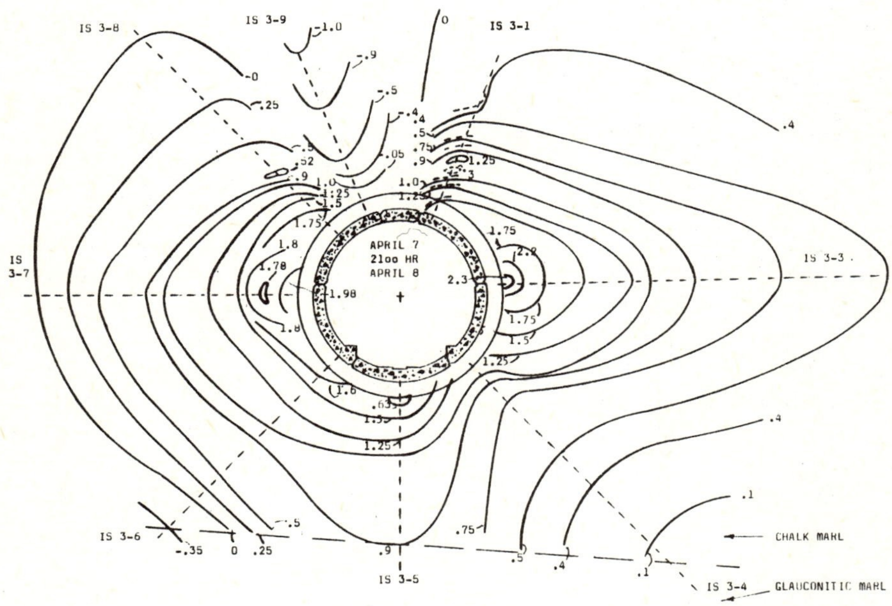

Figure 6 CONTOURS OF TOTAL DISPLACEMENT IN MM RELATIVE TO FACE POSITION Z/D = 3 and 4, IS 3.

the movements are mostly negative namely away from the excavation. Contours of deformation in a homogeneous medium in a hydrostatic stress field would be concentric circles. The elliptical shape with major axis along IS 3-4 and 3-8 suggests a rotated principal stress field implying a higher horizontal virgin stress than vertical stress and, not inconsistently, the 80m high Shakespeare Cliff is only 90 m to the right of the Service Tunnel (the view of Figure 5 is landward). Further, this deformation pattern is not unexpected when considered in relation to the complex geological history of the area viz., the preconsolidation by the landbridge once connecting the U.K. to the Continent and subsequently removed by erosion, and tectonic stresses.

For tunnel face position Z/D = +3 or + 4, Figure 6, the excavation at the IS has been lined and grouting commenced. The maximum deformation has increased to 2.3 mm at IS 3-3 and there are now positive deformations at IS's 3-7 and 3-8. The zero deformation contour has moved outwards some 1.5 diameters as a result of face position Z/D increasing from zero to +3.

The presence of the underlying softer Glauconitic Marl has caused a downward shift of the contours as excavation has progressed i.e. a greater volume of

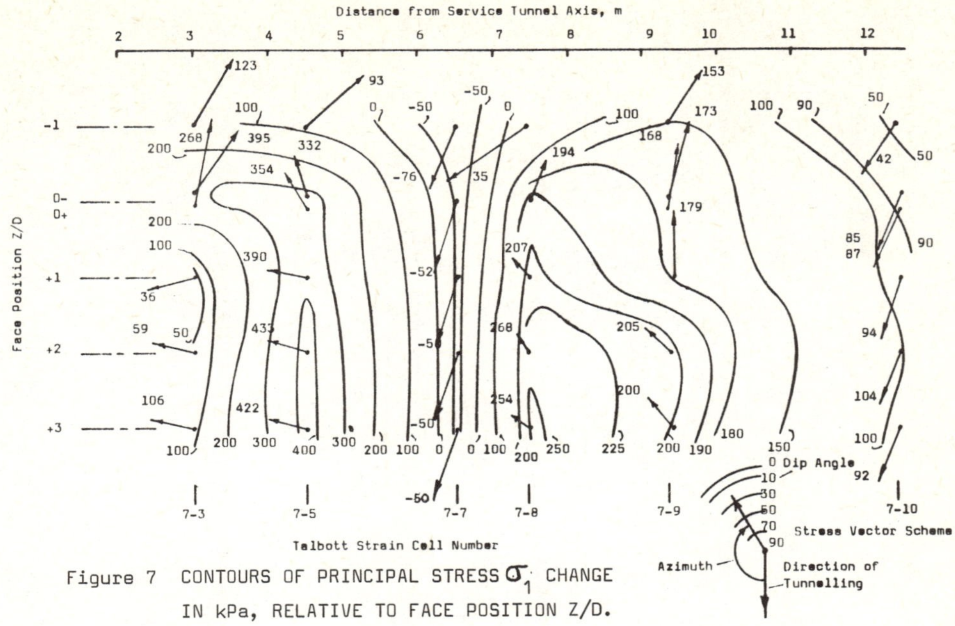

Figure 7 CONTOURS OF PRINCIPAL STRESS σ_1 CHANGE IN kPa, RELATIVE TO FACE POSITION Z/D.

material is involved in the upward deformation process than occurred in the early stage of the excavation.

Stress Changes Induced by Tunnelling

Stress changes in the Chalk were calculated from strain changes detected by the Talbott Strain Cells (or borehole deformation gauges) installed at IS 7, Figure 2, (described by Potts et al 1976), using the analysis of Hirumatsu and Oka (1962) and Leeman and Hayes (1966).

The mathematical analysis required an elastic modulus (E) value for the Lower Chalk. Using the results of laboratory cyclic load tests and an iterative technique, a value of E = 2100 MPa was determined and used in stress calculations. For each of the six strain cells 7-3, 7-5 and 7-7 through 7-10, the six orthogonal normal and shear stress changes

σ_{ij} (i,j = 1 to 3)

were calculated and transformed to principal stress changes

σ_i (i = 1 to 3)

A method was devised to show the spatial orientation of the principal stress changes. This Stress Vector scheme is shown in Figure 7. The vector length relates to dip and the vector direction depicts its azimuth relative to tunnelling direction. The longer the vector implies the nearer the stress to the horizontal i.e. dip zero. The vector is labelled to give the magnitude and sign of the stress, compression being positive. Labelled stress change contours are also given.

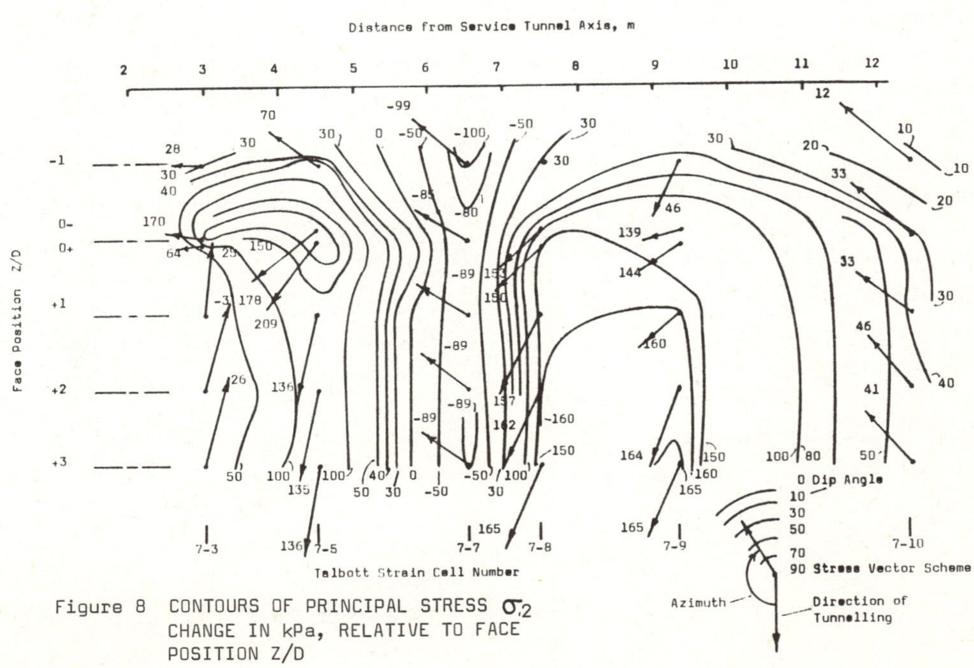

Figure 8 CONTOURS OF PRINCIPAL STRESS σ_2 CHANGE IN kPa, RELATIVE TO FACE POSITION Z/D

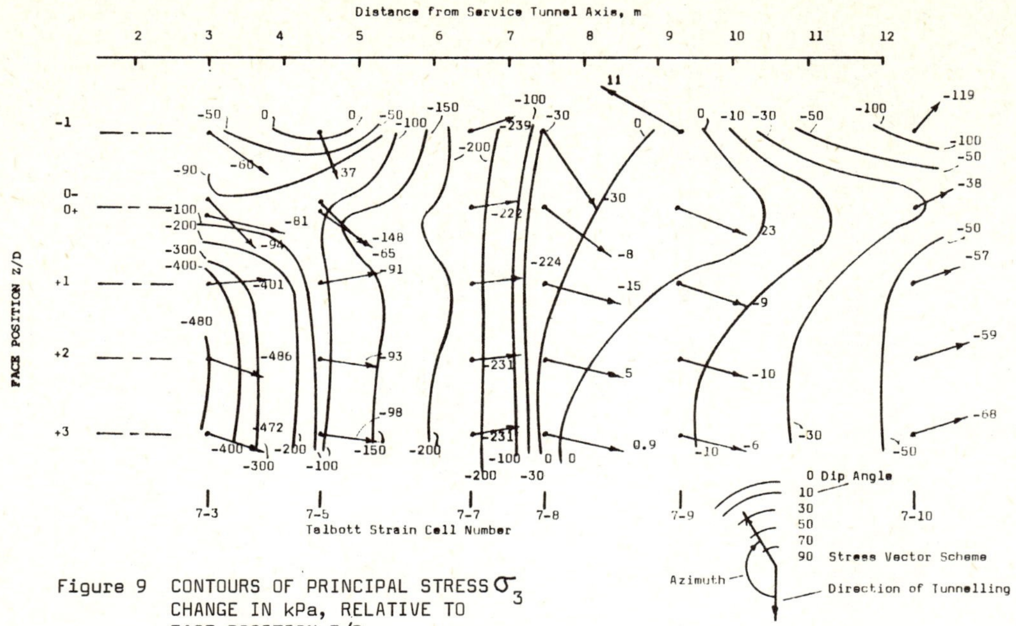

Figure 9 CONTOURS OF PRINCIPAL STRESS σ_3 CHANGE IN kPa, RELATIVE TO FACE POSITION Z/D

The principal stress changes

σ_1, σ_2 and σ_3

are shown in Figures 7, 8 and 9 respectively. These are the stress changes occurring at IS 7 for various tunnel face positions.

Greatest stress changes relative to tunnel excavation are characterised by contours running at right angles to the Service Tunnel. There is a tendency towards this behaviour in the excavation range of Z/D = -1 to +1, thereafter the contours show peaks and hollows consistent with the deformational response of IS's 1 to 3-3 located 25 to 30 m landwards. Talbott Strain Cell 7-7 shows a decompression in all principal stress changes suggestive of weaker ground at 6.5 m from Service Tunnel axis, as interpreted for IS's 1 to 3-3.

The importance of this type of plot lies in that it can quickly depict the three dimensionality of the stress changes, both in direction and in magnitude relative to the dynamics of the tunnelling operation. At IS 7-3, Figure 7, for face position Z/D = - 1 the σ_1 stress change of 123 kPa is acting almost horizontal and at 30° clockwise from the tunnel axis. As excavation proceeds, the vector rotates becoming almost normal to the tunnel wall and the magnitude of the stress change decreases temporarily at Z/D = +2 thereafter increasing.

For strain cells 4 m distant and more from the tunnel axis, all the stress vector plots show little change in direction and magnitude for for face positions beyond Z/D = +1.

Stress changes can be related to other construction activities besides face excavation. For example, at IS 7, the lining was completed at face position +2.5, some 12 to 14 hours after the mole passed the instrument section. The insertion of the lining is seen in its effect on stress increases near the excavation mainly in σ_1 and σ_2 stress changes and to a lesser extent on the σ_3 stress change.

In addition to monitoring stress changes Talbott Strain Cells were also used in the Lower Chalk to determine the insitu stress regime; the overcoring technique was used and stress release values calculated from strain changes. This technique has been used elsewhere, (For example Miller et al 1977).

The use of Talbott Strain Cells thus allows insitu virgin orthogonal stresses and orthogonal stress changes to be determined. Added together they give the absolute insitu state of stress following tunnelling, and can be converted to absolute principal stress, σ_{iA}, at any desired stage in tunnelling.

Lining Design Considerations

Of interest to the tunnel designer is the absolute insitu state of ground stress following the tunnelling operation. If no post lining movement of the ground occurs then the lining carries no ground load. In reality, movements occur due to further advancing of the tunnel face and for reasons related to ground disturbance, creep and the lining insertion technique used.

Adopting the design to yield philosophy results in economic lining design. This requires a ground reaction curve and a knowledge of ground/lining interaction . To show briefly how the results of this study can be used to this end, the state of stress in the ground can be characterised by the first stress invariant

$$J_1/3 = (\sigma_{1A} + \sigma_{2A} + \sigma_{3A})/3$$

or at great distance from tunnel work by

$$(\sigma_v + 2\sigma_h)/3$$

or whatever the designer elects to choose e.g. σ_r radial stress. By plotting the $J_1/3$ stress values against corresponding radial deformations (as here , by ground anchors and LVDT's at IS 3-3)the ground reaction curve can be obtained; related to the lining deformation curve enables a realistic appreciation of lining pressure to be assessed. (Reid, 1982).

CONCLUSIONS

The paper has attempted to show some of the results obtained during the limited advance of the tunnelling mole during Phase II of the Channel Tunnel works.

Ground deformation was monitored by extensometers comprising anchors and LVDT'S installed in boreholes. Talbott Strain Cells (borehole deformation gauges) were used to calculate stress changes occurring in the ground.

Deformation and stress changes are essential for studies involving comparison of measurements with theoretical predictions of stress concentration and parametric studies relating deformation to E values and insitu stresses. The deformation pattern presented shows the dramatic change in deformation near the tunnel wall from Z/D = -1 (0.1 mm) to Z/D = +1 (1.5 mm) i.e. a 15 fold increase compared with a 1.12 increase after the lining was installed at Z/D = +2.

The σ_1 principal stress change acts towards the excavation and therefore should feature in lining design considerations. As ground relaxation occurs between Z/D = 0 to +1, the stress change reduces to approximately one seventh of its value at passage of the mole. The timing of the lining insertion to take advantage of this effect is consistent with the design to yield philosophy.

Finally, extension of stress and deformation data to compile a ground reaction curve relative to lining design has been briefly outlined.

REFERENCES

Curtis, D.J., Lake, L.M., Lawton, W.T., and Crook, D.E., 1976, "Insitu Ground and Lining Studies for the Channel Tunnel Project", International Symposium, Tunnelling '76, London.

Government White Paper, 1963, "Proposals For a Fixed Channel Link".

Government White Paper, 1973, "Treaty Between The United Kingdom of Great Britain and Northern Ireland and the French Republic concerning the Construction and Operation of a Rail Transportation System Under the English Channel".

Hirumatsuy, Y. and Oka, Y., 1962, "Stress Around a Shaft or Level Excavated in Ground With a Three Dimensional Stress State", Mem. Fac. Eng., Kyoto University 24, Part I, 56-75.

Leeman, E.R., Hayes, D.J., 1966, "A Technique for Determining the Complete State of Stress in Rock Using a Single Borehole", 1st Int. Congress SRM, Lisbon, 11, 17-24.

Maconochie, D.J., 1977, "Geotechnical Studies for The Channel Tunnel", Ph.D. Thesis, University of Newcastle Upon Tyne.

Miller, H.D.S., Potts, E.L.J., and Szeki, A., 1977, "Insitu Stress Measurements in Halite in Cheshire Potash in North Yorkshire and Slate at Dinorwic", Proc. Conf. Rock Engineering, Newcastle Upon Tyne, 257-274.

Peyton, J., 1973, "The Channel Tunnel Project", Government White Paper.

Potts, E.L.J., Dunham, R.K., Maconochie, D.J., and Reid, A.G., 1976, "Design and Installation of Ground Instrumentation for the Channel Tunnel", International Symposium Tunnelling '76, London

Reid, A.G., 1979, "Rock Mechanics Investigations Relating to Tunnelling in Chalk", Ph.D. Thesis, University of Newcaslte Upon Tyne.

Reid, A.G., 1982, "Ground Reaction Study Based on Total Deformations and Absolute State of Stress in Lower Chalk; Channel Tunnel Project", to be published.

Reid, A.G., Collins, K. , and Healy, P., 1982, "Geotechnical Properties of the Lower Chalk; Channel Tunnel Project", to be published.

EFFECT OF MINING SUBSIDENCE ON NEAR SURFACE UNDERGROUND STRUCTURES

Leszek Litwinowicz, Dr Eng.
Assist Prof., Technical University of Lublin, Poland.
Kazimierz Kłosek, Dr Eng.
Assist, Silesian Technical University, Poland.

SUMMARY The effect on communication tunnels of horizontal tensile deformations perpendicular to the tunnel axis has been examined and a generalized hypothesis for vertical and horizontal pressures has been developed. The need for further research work has been indicated, particularly on the influence of horizontal specific compressive deformations on construction.

GENERAL CONSIDERATIONS

Underground exploitation of coal deposits can lead to damage of buildings erected above the mined area. The most frequent damage has been caused by deformation of building foundations.

There are serious problems when tunnels are constructed above mined areas (see Litwinowicz, Malcharek, Rosikoń, 1975). In this case the basic problem to solve, is the proper choice of interpretation of ground and tunnel lining interaction. Different tests connected with the use of solutions based on continuous medium models have not provided satisfactory design data.

The magnitude of earth pressures on tunnel lining which occur during ground deformation over mine excavations can be related to horizontal strains ε_L (or ε_c), through the coefficient of lateral pressure which is functionally dependent on these deformations. A procedure based on such interpretation has been successfully used to predict the influence of mining subsidence on the stability of an embankment (Litwinowicz, 1977) and the magnitude of earth pressure on support constructions (Kłosek, 1978) with the help of integration of the equations of Kandaurov-Muller.

In this paper a simpler version has been used to calculate earth pressures on tunnels with the use of Terzaghi's (1948) method modified by Gałczyński (1974).

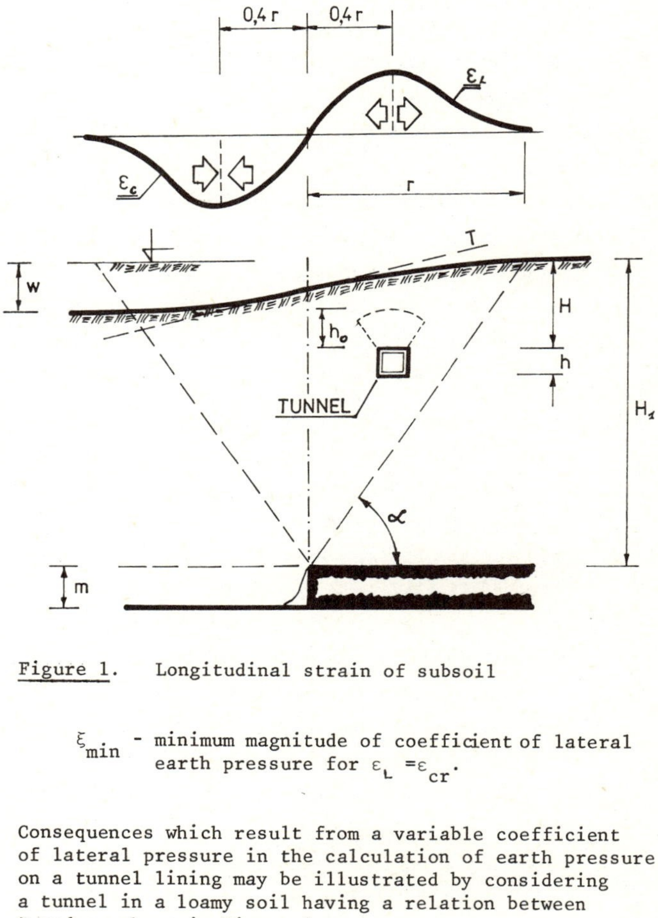

Figure 1. Longitudinal strain of subsoil

THE INFLUENCE OF EARTH LOOSENESS CAUSED BY MINING ON MAGNITUDE OF VERTICAL PRESSURE p_v

Used in engineering practice the method of calculation of vertical earth pressure p_v on tunnel linings assumes (Terzaghi, 1948) a constant value for the coefficient of lateral earth pressure $\xi = 1$. It is impossible to maintain this condition for tunnels influenced by mining subsidence (Fig. 1).

Investigations in different soils suggested that the coefficient of lateral pressure ξ reduces under the influence of increasing looseness (see Glinko, Kwiatek, 1975; Kłosek, Litwinowicz, 1977).

For the general case this relation can be expressed using the following empirical equation:

$$\xi(\varepsilon_L) = \xi_{min} + (\xi_o - \xi_{min})\left(1 - \frac{\varepsilon_L}{\varepsilon_{cr}}\right)^m \qquad (1)$$

where: m is an empirical constant. For the majority of soils good approximation gives a value m = 3.14 (Litwinowicz, 1981),

ε_{cr} - the critical magnitude of horizontal loosening strain.

ξ_o - the coefficient of lateral earth pressure for $\varepsilon_L = 0$.

ξ_{min} - minimum magnitude of coefficient of lateral earth pressure for $\varepsilon_L = \varepsilon_{cr}$.

Consequences which result from a variable coefficient of lateral pressure in the calculation of earth pressure on a tunnel lining may be illustrated by considering a tunnel in a loamy soil having a relation between ξ and ε_L shown in Figure 2.

The angle of internal friction of this clay, $\phi = 25^o$, cohesion c = 160 kN/m^2, unit weight γ = 20 kN/m^3 and coefficient of compressibilty E = 10300 kN/m^2.

The initial magnitude of $\xi_o = 0.96$ was calculated for a depth of tunnel H = 40m and width of tunnel 2.0m.

The magnitude of vertical pressure of the soil on tunnel casing taking into account mining can be written in the following way:

$$p_v = \alpha(\varepsilon_L)\gamma b_1 + \beta(\varepsilon_L) q \qquad (2)$$

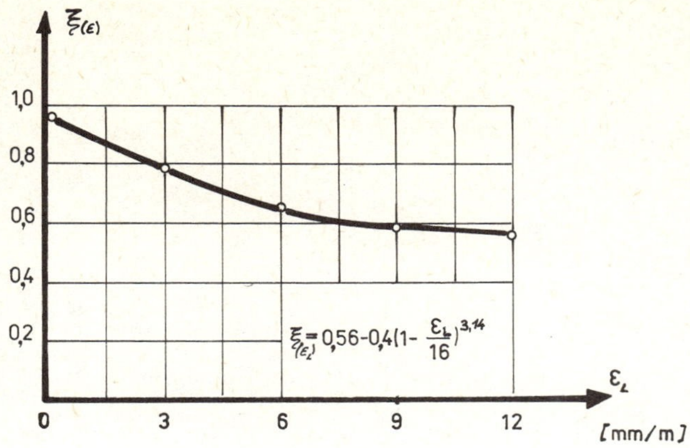

Figure 2. Variation of the coefficient of lateral pressure.

$$\beta(\varepsilon_L) = \exp\left\{-\left[\xi_{min} + (\xi_0 - \xi_{min})\left(1 - \frac{\varepsilon_L}{\varepsilon_{cr}}\right)^m\right] f \frac{z}{b_1} \cdot k\right\} \quad (2a)$$

$$\alpha(\varepsilon_L) = \frac{1 - \beta(\varepsilon_L)}{\left[\xi_{min} + (\xi_0 - \xi_{min})\left(1 - \frac{\varepsilon_L}{\varepsilon_{cr}}\right)^m\right] f \cdot k}$$

where

$$f = tg\,\Phi + \frac{c}{\sigma}; \quad b_1 = b\,tg(\pi/4 + \Phi/2) \cdot f^{-1}; \quad k = 1 - \frac{\varkappa_1 \gamma H}{E}; \quad (2b)$$

In equations (2) $\varkappa_1$ is the coefficient of tension concentration in the tunnel side walls. The remaining symbols are defined in Figure 3.

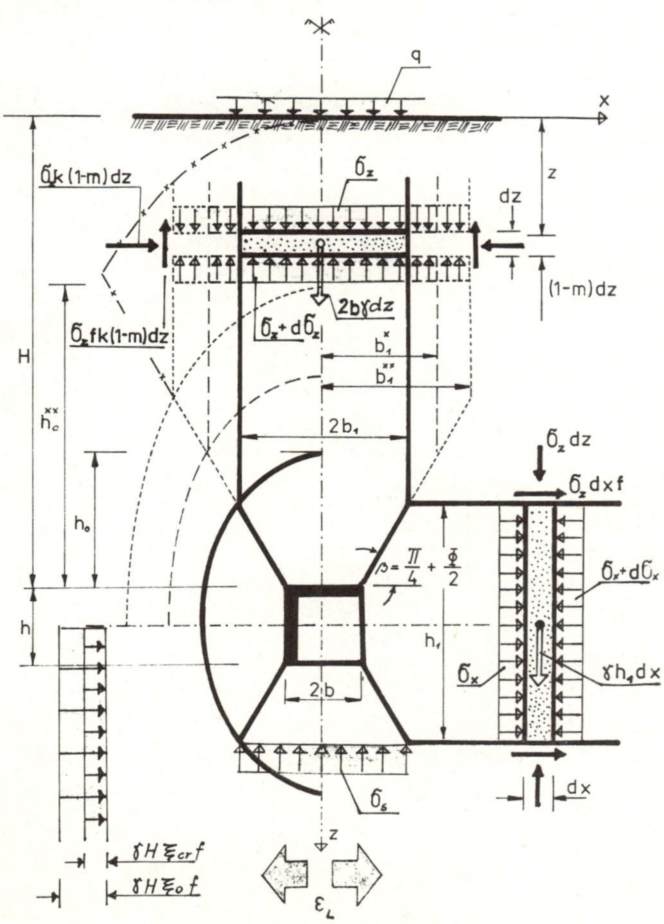

Figure 3. Forces acting on a unit cube of loosened zone of strata above a tunnel.

The relations between the variable functions $\alpha(\varepsilon_L)$ and $\beta(\varepsilon)$ are illustrated in Figure 4 for different values of $\frac{z}{b_1}$, which have been calculated where $\varkappa_1 = 1.5$.

These relations indicate that:

- the loosening of soil due to the influence of mining subsidence increases the influence of external loading q on the magnitude of the vertical earth pressure p_v, in a way which is related to the magnitude of ε_L,
- in the soil where mining subsidence induces looseness the zone h_o reduces from its level when $\varepsilon_L = 0$,
- in areas where mining subsidence affects soil looseness, increases in vertical pressure on tunnel linings must be taken into account in the early design stages.

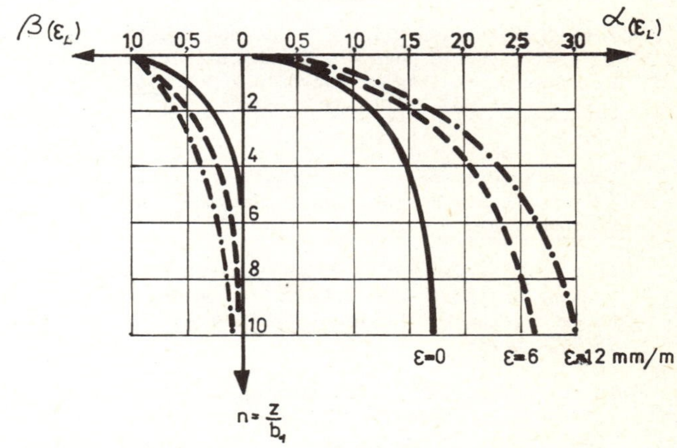

Figure 4. Load curve coefficients $\alpha(\varepsilon_L)$ and $(\beta\ \varepsilon_L)$ for different values $\frac{z}{b_1}$.

THE INFLUENCE OF SOIL LOOSENESS CAUSED BY COAL MINING ON THE MAGNITUDE OF HORIZONTAL PRESSURE p_h

In a similar way it is also possible to evaluate the effect of looseness on the magnitude of horizontal soil pressures in areas affected by mining subsidence. To show the most fundamental elements of this effect the following forumla can be used:

$$p_h = \eta(\varepsilon_L)\,\gamma H; \quad (3)$$

where

$$\eta(\varepsilon_L) = \left[\xi_{min} + (\xi_0 - \xi_{min})\left(1 - \frac{\varepsilon_L}{\varepsilon_{cr}}\right)^m\right]\varkappa_2 \cdot \exp\left\{-\frac{2bf\sqrt{f^2+1}}{\left[\xi_{min} + (\xi_0 - \xi_{min})\left(1 - \frac{\varepsilon_L}{\varepsilon_{cr}}\right)^m\right]\varkappa_2\, h\, tg(\pi/4 + \Phi/2)}\right\};$$

The variation of $\eta(\varepsilon_L)$ with ε_L based on calculations for loamy soil is presented in Figure 5.

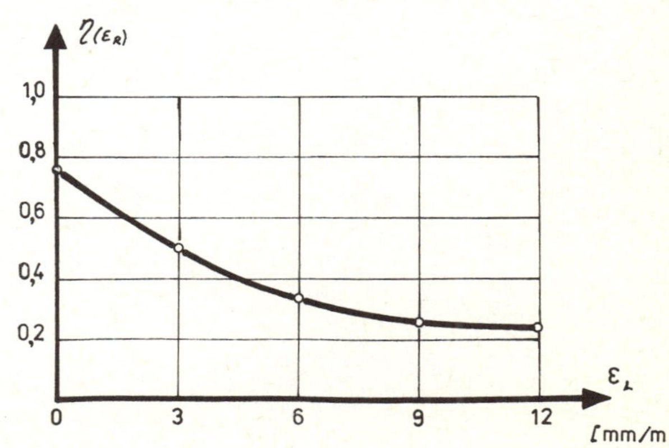

Figure 5. Variation of function.

The considerable fall in the value of this function shows that soil looseness occurs through the loss of horizontal constituent tensions in the soil surrounding the tunnel, thus causing a clear reduction of horizontal earth pressure p_h on the lining.

These calculations have been based on a value for the coefficient of tension concentration $\mathcal{H}_2$ of 1.5.

REFERENCES

Glinko, H. and Kwiatek, J., 1975, "Vlijanije gorizontalnogo rozrychlenija grunta na vielicinu bokovogo davlenija", Proceedings of the First Baltic Conference on Soil Mechanics and Foundation Engineering, Gdansk.

Litwinowicz, L., 1981, "Ocena parametrow geotechnicznych przy obliczaniu nasypow na terenach gorniczych", II Konferencja Naukowa Instytutu Inzynierii Ladowej Politechniki Czestochowskiej.

Litwinowicz, L. and Malcharek, K. and Rosikon, A. "Wplyw na tunele gorniczych odksztalcen terenu o kierunku poprzecznym", Kwartalnik Ochrona Terenow Gorniczych, No. 32, 1976, Katowice.

Klosek, K. and Litwinowicz, L., 1977, "Parcie gruntu na tunele komunikacyjne w warunkach poziomych rozluzniajacych odksztalcen osrodka gruntowego", Kwartalnik Ochrona Terenow Gorniczych, No. 40.

Klosek, K. "Wplyw poziomych odksztalcen gorniczych podloza na parcie w osrodkach dyskretnych", Zbior referatow V Krajowej Konferencji Geotechniki, 1978, Katowice.

Galczynski, S., 1974, "Technicznz ocena cisnienia gorotworu na obudowe wyrobisk chodnikowych i tunelowych", Prace Naukowe Instytutu Geotechniki Politechniki Wroclawskiej, No. 12, Seria: Monografie No. 3.

Litwinowicz, L., 1977, "Oddzialywanie poziomych odksztalcen gorniczych terenu na budowle nasypowe w ujeciu mechaniki osrodkow dyskretnych", Materialy Konferencji Naukowej - Budowle ziemne na terenach gorniczych Komisja Ochrony Terenow Gorniczych PAN, Katowice.

Klosek, K., 1979, "Wplyw poziomych odksztalcen podloza gorniczego na konstrukcje podporowe" Dysertecja doktorska, Politechnika Warszawska.

Terzaghi, K., 1948, "Theoretical soil mechanics" New York, McGraw-Hill.

DEFORMATION OF A CIRCULAR CONCRETE ROADWAY

LINING IN RESPONSE TO STRATA MOVEMENT

A S Bloor

National Coal Board, Mining Research and Development Establishment,
Ashby Road, Burton-on-Trent

SUMMARY. The variation of strain in a circular concrete segmental roadway lining was measured as a function of angular position and time, using vibrating-wire strain gauges. It was found that the eventual failure of the lining was due to excessive bending strains generated by cavities in the roof strata.

INTRODUCTION

A 46 m length of the West Main Return at Cadley Hill Colliery has been repaired and re-lined with pre-cast concrete segments manufactured by C V Buchan (Concrete) Limited. Because of the experimental nature of the lining, MRDE were requested by South Midlands Area to monitor its performance. Vibrating-wire strain gauges, embedded in the concrete, were chosen as the primary instrumentation. Measurements were also made of the relative movement of segments and the rock pressure on the lining.

LOCATION AND HISTORY

The West Main trunk roadways provide access to a number of retreating roadways in the Main Coal at a depth of approximately 500 m. The Main Coal is a composite seam about 5 m thick excluding partings and is overlain by 10 to 15 m of mudstone, including the Molyneux Marine Band. Listricated seatearth mudstones extend to 3 m below the seam. The Colliery has a history of severe unpredictable faulting, often with large local variations in throw, and generally poor roadway conditions.

The West Main Return was driven 10 m above the seam in an attempt to avoid the large closures observed in an adjacent area with in-seam roadways. Figure 1 shows the position of the roadway in relation to the strata proven in a nearby borehole. Inspite of the new drivage horizon, considerable roadway closure was encountered, requiring practically continuous repair work. The major part of the closure was floor-lift, but strata deformation into the roof was also present, resulting in severe arch-distortion.

In view of the success of concrete block linings at other sites (Haunchwood Colliery for example) and the experience of Stassen and Van Duyse (1977) in Belgium with larger concrete panels, it was decided to install an experimental section of segmental concrete lining in the return trunk roadway adjacent to the proposed 71's retreating face.

The site was visited in May 1979 and rock samples obtained for the determination of physical properties. The results of these tests are included in Figure 1. At this time floor lift of 1 - 1.5 m was evident, together with some flattening and distortion of the 4.3 x 3 m steel arches (0.15 x 0.13 m section), and a dinting operation was in progress. By December 1979, when erection of the concrete lining was about to start, it had been necessary to repeat the dinting operation in order to recover the original floor level. The total floor material removed in these two operations was approximately 3 m. Replacement of the steel arches by the concrete rings continued until February 1981, during which time a further 1 m dint was taken in the length of roadway not re-lined. 71's face commenced retreating towards the roadway in March 1981 from a distance of 400 m and finished in November 1981, leaving a 100 m wide pillar beside the re-lined section.

THE CONCRETE LINING

The lining consisted of 75 rings with an internal diameter of 3.66 m and 0.61 m long, giving a total length of 46 m. The concrete thickness was 0.24 m. Each ring comprised 11 segments and a parallel-sided key-piece in the crown. The segments were of lightly reinforced concrete with a minimum 28-day cube strength of 45 MPa. The 12 radial joints between segments were fitted with 14 mm thick wooden inserts, covering the mating surfaces. The segments were bolted together through the inserts from recessed bolt-pockets. Figure 2 illustrates the principal features of the lining.

Erection was by means of a purpose-built two-deck framework fitted with circular guides. A hydraulic winch and two rams were used to maneouvre the segments into position. Removal of a steel arch from the original roadway usually resulted in the formation of a triangular cavity in the roof, typically 2 m wide and 2 m high. The cavity around the lining (mainly in the roof) was filled by pumping in a cement grout after stopping the open end with dirt-filled bags. This operation was carried out at intervals of 5 rings (3 m) advance. Up to 20 tonnes of grout were used for each set of 5 rings.

INSTRUMENTATION

Two rings (Ring 23 and Ring 51), situated 14 m from the outbye and inbye ends, respectively, of the lining were provided with embedded vibrating-wire strain gauges; pressure cells between the concrete and the rock; and reference points to determine the compression of the wood packing between segments.

STRAIN GAUGES

The vibrating-wire strain gauges (Strainstall Limited, Type PC 657) were embedded in 12 of the concrete segments during manufacture. The two instrumented rings each contained 6 such segments. The strain gauges were first cast in pairs into briquette moulds, using production concrete from the segment manufacturing plant. This ensured that the pairs of gauges were aligned with their axes parallel and 100 mm apart. After 1 to 2 days the briquette was cast into the segment mould with the two gauges aligned so as to measure circumferential strain at two radii 50 mm on either side of the central radius of the segment. Figure 3 shows the positions of the strain gauges within the completed ring and the numbering system adopted for the segments.

A reference briquette containing a single strain gauge was not cast into a segment, but remained with the instrumented segments for 12 months during storage and until some time after installation. This briquette provided a zero-load reference and was eventually cored to determine the elastic properties of the concrete.

The vibration frequency of the gauges was nominally 1000 Hz and the time period for 100 oscillations was measured to 10^{-5} sec using an NCB/Strainstall intrinsically safe read-out unit. Changes in strain of the order of 10^{-6} (1 microstrain) were detectable using this system.

Pressure cells

Hydraulic pressure cells of the Glotzl type (Soil Instruments Limited) were positioned at 60° around the instrumented rings between the rock and the concrete lining, with the object of measuring radial pressures. The cells have a stiffness at least an order of magnitude greater than that of coal-measure rock and a range of 0 to 24 MPa. The pressure in the cell is determined by balancing it against hydraulic oil pressure applied through a nylon tube. A return flow of oil is observed through a second nylon tube when the cell pressure is exceeded.

Packing compression

Reference studs were fixed on each side of accessible radial joints in Rings 15, 23 and 51 in order to determine the compression of the wooden inserts as the lining was loaded. The distance between studs was measured with a dial caliper gauge to within 0.05 mm.

RESULTS

Strain gauges

With the chosen configuration of strain gauges it was possible to determine both the circumferential and bending components of strain as a function of time after erection of the lining and also as a function of angular position within the ring. The thickness of the ring is sufficiently small in relation to its radius for a linear distribution of strain between the inner and outer surfaces to be assumed. The strains at the inner and outer surfaces may therefore be calculated from the measured strains at the gauge positions. The circumferential component of strain is then given by the mean of the inner and outer surface strains and the bending component of strain by the difference between them.

Figures 4 and 5 show the mean strains and strain differences for Rings 23 and 51 at six time intervals after erection. The strains are plotted in polar co-ordinates to indicate both magnitude and angular position within the ring and the time intervals have been chosen to be approximately equal for the two rings.

The mean strain, ie the circumferential component, was compressive (positive), practically independent of angular position, and increased continuously during the life of the lining. The maximum recorded values were 940 microstrain for Ring 23 and 1320 microstrain for Ring 51. The maximum design strain for the segments in compression was 540 microstrain and estimated failure strain (compressive) was 1300 microstrain.

The bending component of strain, as indicated by the strain difference plots showed marked variations of both sign and magnitude with angular position and time. The sign convention adopted is that outward bending of a segment, with the outer surface in tension relative to the inner surface, is represented by a positive strain difference. For a segment on the point of failure, with opposite surfaces at the estimated ultimate compressive and tensile strains, the strain difference would be approximately 1500 microstrain, positive or negative.

The most pronounced effect, common to both rings, was an outward bending of Segment 1 (in the crown) which increased with time. In Ring 23 this outward bending increased more rapidly after 220 days, reached a maximum in excess of +5000 microstrain after 345 days and then decreased. In Ring 51 the outward bending decreased again after 192 days having reached +1260 microstrain. Both rings exhibited inward (negative) bending of segments in the floor, the effect being most pronounced and localised in Ring 51, where a strain difference of -2800 microstrain was recorded. Segment 3, nearest the horizontal diameter, showed little bending strain in either ring, the strain differences being in the range -100 to +200 microstrain. By 354 days after erection, bending sign of Segment 5 in Ring 51 had suddenly reversed and the strain gauges in Segment 1 had ceased to function. It was concluded that, as in Ring 23, physical failure of the concrete had occurred and strain gauge readings were therefore discontinued.

Pressure cells

Of the six pressure cells installed at Ring 23, four had failed to an off-scale condition at an early stage, probably due to damage to the nylon supply tubes running behind the lining. The remaining two cells, diametrically opposed at 120° and 300° from the top of the ring, recorded considerably different pressures, as shown in Figure 6. The final readings at 331 days after installation were 1.32 MPa in the invert (120°) and only 0.44 MPa in the crown (300°).

The variation of radial pressure with time and angular position for Ring 51, where only one cell failed, is shown in Figure 7. Severe non-uniform radial loading was apparent from shortly after installation, the pressures on all cells increasing continuously with time. At 297 days the highest and lowest radial pressures were 2.60 MPa and 0.16 MPa on adjacent cells in the crown (0° and 60°).

Packing compression

Figure 8 shows the packing compressions measured on Ring 15 as a function of time. It was not practicable to monitor joints 5 to 8 below the rock infill of the invert. Some reference studs, particularly on Rings 23 and 51, were accidentally dislodged, so the record of packing compression is incomplete. The pattern of compression in Ring 23 however was very similar to that in Ring 15.

The maximum compression occurred on either side of the key-piece (Joints 1 and 12) and amounted to 8 mm - 9 mm in the first 250 days, 60% of the original thickness. The compression in the crown was markedly asymmetric, the corresponding joints 2 and 11 closing by 2 mm and 7 mm respectively. Joint 10 opened initially before closing by 1 to 2 mm, indicating a relative rotation of Segments 9 and 10. The remaining joints above the infill all closed by between 1 mm and 2 mm.

Failure mode

Physical failure of the concrete was not visible until some time after it had been detected by strain gauge readings. The allowable tensile surface strains according to CP114 were first exceeded at 220 days and 192 days after erection for Rings 23 and 51 respectively. In both cases this was due to outward bending of Segment 1. Discontinuities in strain gauge readings indicative of concrete failure occurred at 350 days, but visible disruption of the structure was not observed until 450 days in Ring 23. No disruption was apparent in Ring 51 when readings were discontinued at

507 days.

The general pattern of visible failure commenced with spalling and crushing of the key-piece, followed by upward movement of the key and adjacent crown segments. Shear failure of one crown segment then usually occurred and movement along the shear break forced one crown segment into the roof and the other into the roadway. This was followed by floor-lift of the in-fill, indicating failure of the invert segments. Figure 9 shows the roof area near Ring 23 after physical failure.

DISCUSSION

Estimated roadway closure

The formula due to Wilson (1980) was used to estimate the closure of the roadway when supported

(a) by the original steel arches, and

(b) by the concrete segmental lining.

The values of the parameters used are given in Table 1. The estimated closures on an effective diameter of 4.5 m were as follows:

Original roadway, no support	:	2400 mm
Original roadway, steel arches	:	800 mm
Concrete lined roadway	:	15 mm

From these figures, the expected floor lift and roof lowering in the original arched roadway are approximately 1500 mm and 400 mm respectively.

Tests on the compressible packing between segments indicated that, at the designed working stress in the concrete of 20 MPa, each packing piece would compress by 8 mm. This would allow a reduction in lining diameter of 30 mm, twice the estimated clsoure for the same support conditions.

The concrete lining was therefore expected to be capable of supporting the excavation in a stable manner.

Non-uniform loading

Both the calculation of packing compression and the estimation of roadway closure assumed a uniform radial distribution of load. The fact that the eventual floor lift in the original roadway amounted to about 4000 mm compared with the 1500 mm predicted and the observation of severe bending strains in the concrete indicate that this assumption was not valid.

An obvious cause of non-uniform pressure distribution was the presence of major cavities in the roof. Although cement grout was pumped behind the lining, the possibility of unfilled pockets and bed separation in the immediate roof cannot be discounted. When the signs of severe bending in the crown were observed, secondary grouting was recommended but not implemented. The mode of failure of a model concrete segmental lining with a partially filled cavity in the roof is illustrated in Figure 10. Bending of the crown segments into the roof is clearly visible, as is failure of one of the crown segments in the manner observed at Cadley Hill and shown in Figure 9.

A contributing factor to non-uniform loading may be the existence of intraformational shear zones in the vicinity, as reported by Elson (1979). This phenomenon, described by Salehy, Money and Dearman (1977), is associated with high fault intensity and the strata types present at Cadley Hill. The resulting strata movement could explain the unusually severe floor-lift in the original roadway and the inward bending of invert segments in the concrete lining.

Effect of 71's face

Any additional loading caused by the retreat of 71's face towards the roadway from a distance of 400 m was not considered to be a major factor in the failure of the lining. Visible damage was apparent in Rings 1 to 6 some 220 days before the face started production and severe bending strains were recorded in Ring 23 at about 100 days before the starting date. No effects of the face on Ring 51 were identified and the ring, unlike Ring 23, was still visibly intact when the face ceased production at a distance of 100 m.

CONCLUSIONS

Instrumentation

The use of vibrating-wire strain gauges embedded in a concrete lining provides a sensitive method of monitoring the response of the structure to strata movement.

Cause of failure

The primary cause of lining failure was the inadequate treatment of cavities in the roof strata. This allowed outward bending of the crown segments leading to fracturing of the concrete in this area. A contributory factor was the asymmetric loading of the lining from below, associated with the massive floor-lift previously experienced in the roadway.

Precautions

The success of a block or segmental lining relies on the circumferential compressional forces within it. If overbreak or strata collapse occurs, great care must be taken to fill the space behind the lining in such a way as to avoid excessive bending strains leading to movement and eventual collapse.

ACKNOWLEDGEMENT

Thanks are due to the Director of MRDE, Mr P G Tregelles, for permission to publish this paper. The views expressed are those of the author and not necessarily those of the National Coal Board.

REFERENCES

Elson, R J, 1979. "The Stability of Mine Roadways in Relation to Geology - The Possible effects of intraformational shears in carboniferous rocks", British Geotechnical Society Conference on Rock Engineering, Newcastle upon Tyne, 311-328.

Stassen, P and Van Duyse, H, 1977. "Development of supports in stone-roads in the Campine Coalfield", 6th International Strata Control Conference, Banff, Canda.

Wilson, A H, 1980. "The stability of underground workings in the soft rocks of the coal measures", PhD Thesis, University of Nottingham.

Parameter	Value
Diameter of roadway	4.5 m
Depth of cover	480 m
Cover load	12 MPa
Strata properties:	
Poisson's ratio	0.24
Young's modulus	11.9 GPa
Expansion coefficient	0.2
Compressive strength (laboratory)	38 MPa
Compressive strength (in-situ)	8 MPa
Triaxial stress factor	2.3
Support pressure:	
Steel arches	0.1 MPa
Concrete lining	2 MPa
Cohesive augmentation	0.1 MPa

Table 1 - Input parameters for roadway closure estimates

Figure 2 Principal features of the segmental concrete lining showing erector framework

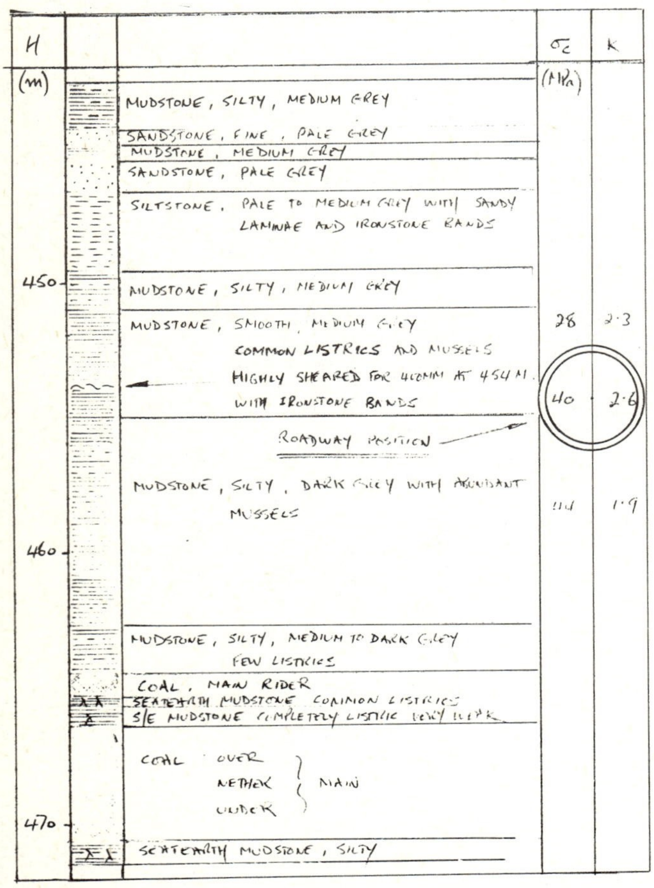

Figure 1 Position of roadway in relation to strata
H = borehole depth;
σ_c = uniaxial compressive strength;
k = triaxial stress factor

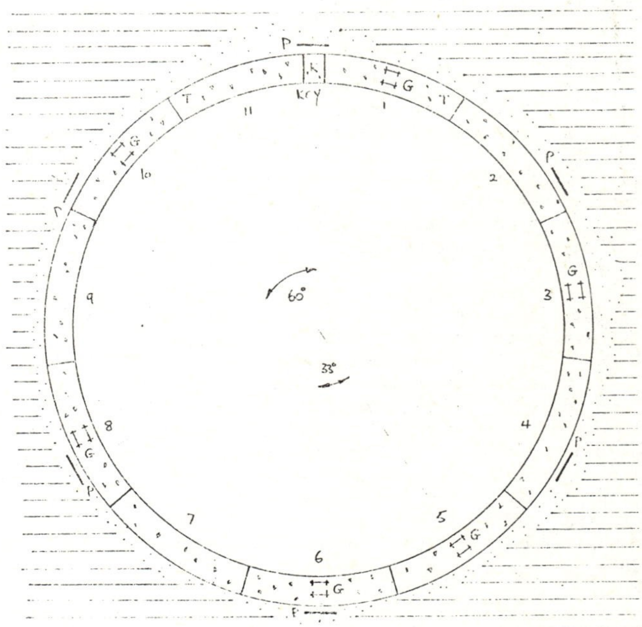

Figure 3 Positions of strain gauges (G) and pressure cells (P) in an instrumented ring

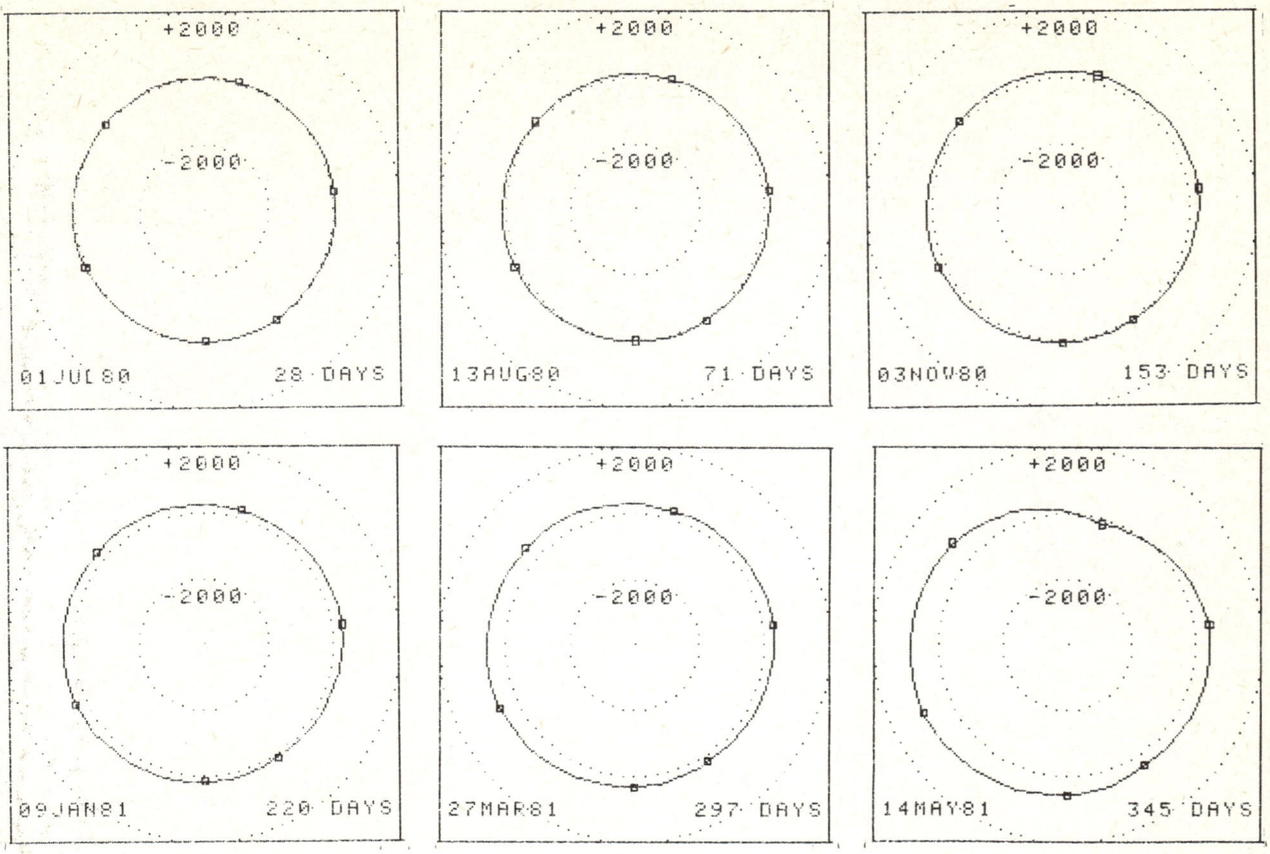

Figure 4(a) Variation of mean strain in Ring 23 with angular position and time since erection

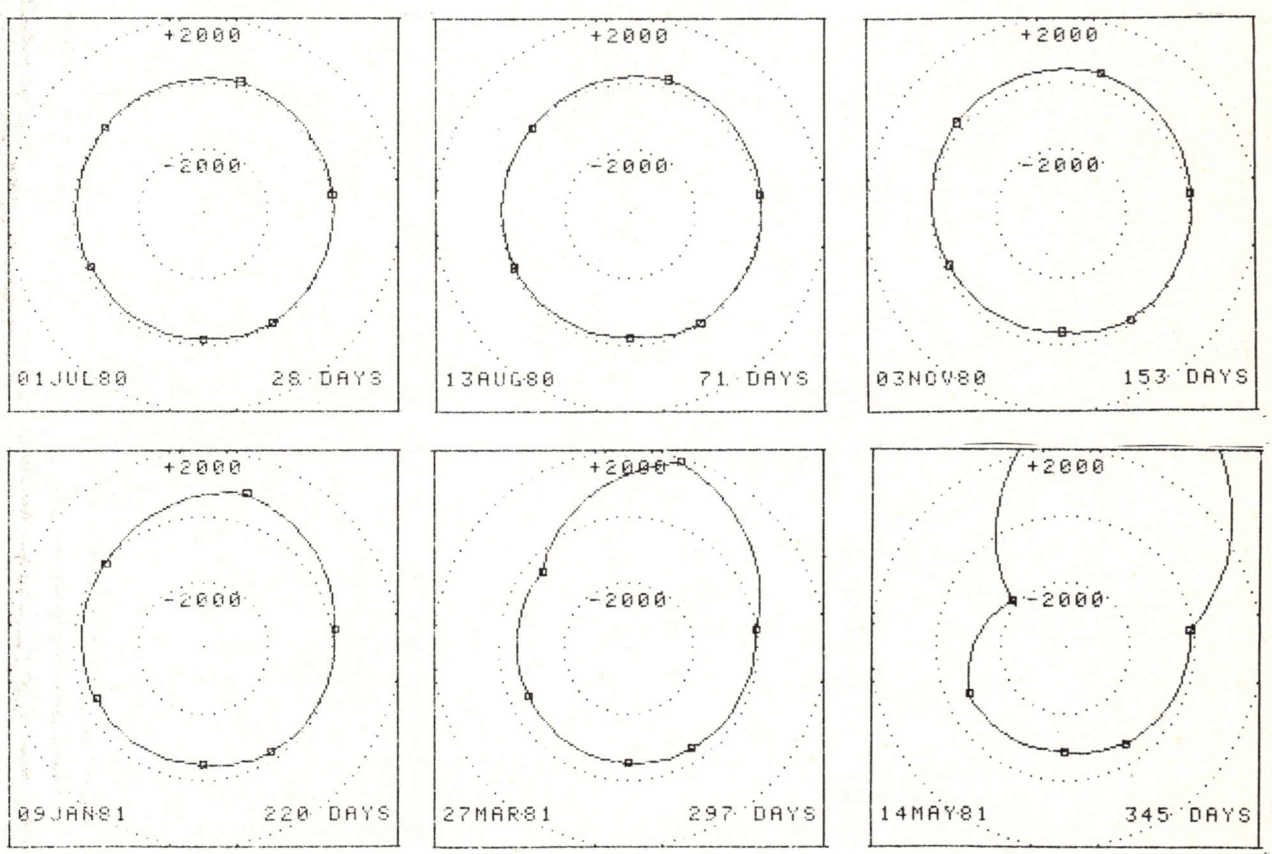

Figure 4(b) Variation of strain difference in Ring 23 with angular position and time since erection

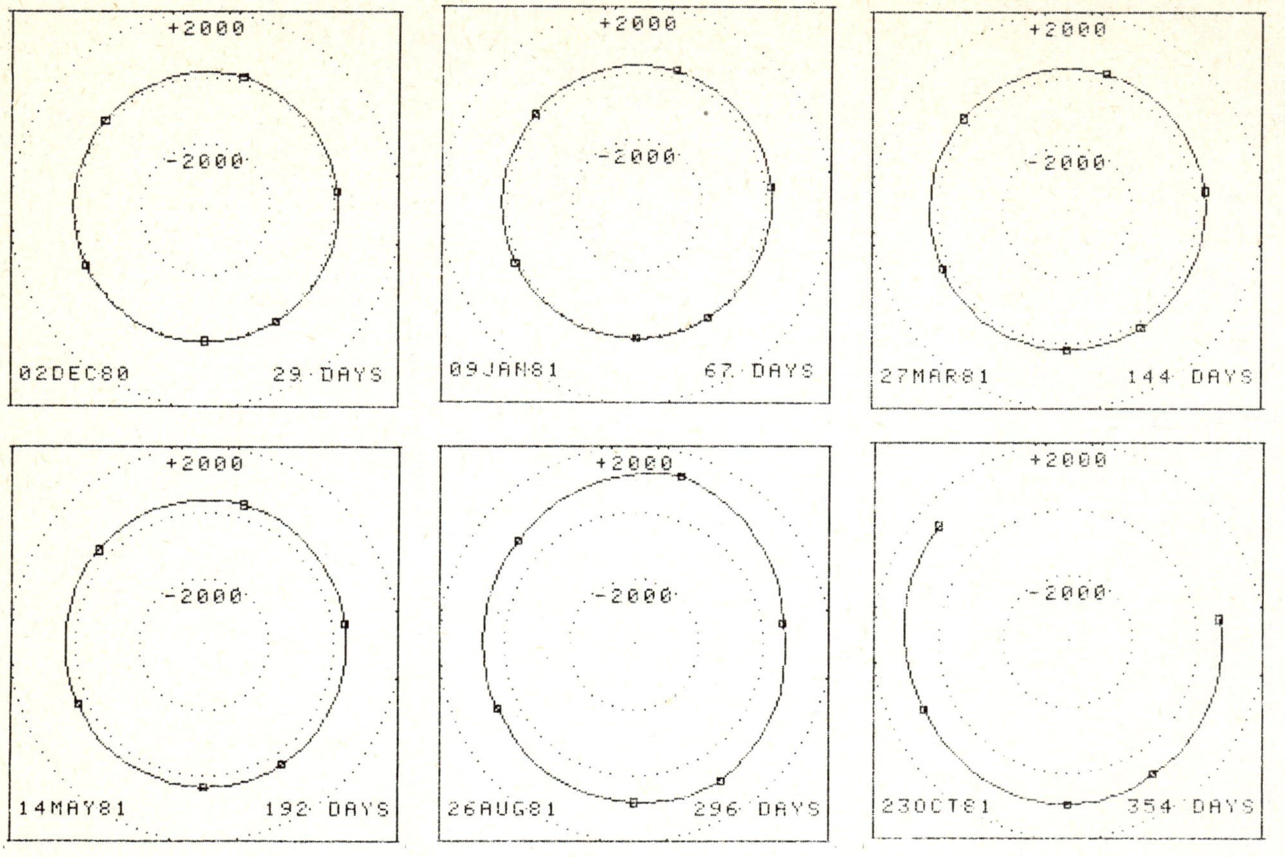

Figure 5(a) Variation of mean strain in Ring 51 with angular position and time since erection

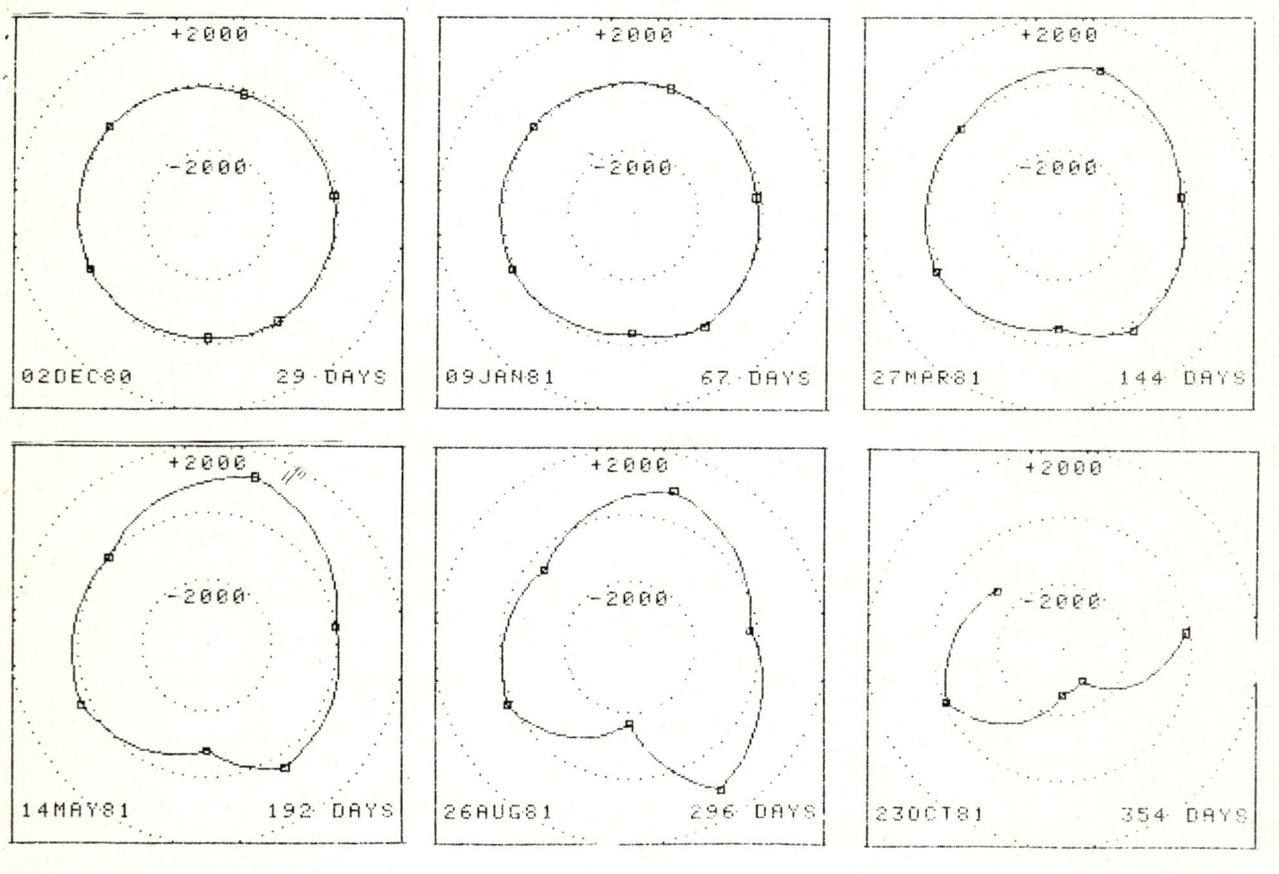

Figure 5(b) Variation of strain difference in Ring 51 with angular position and time since erection

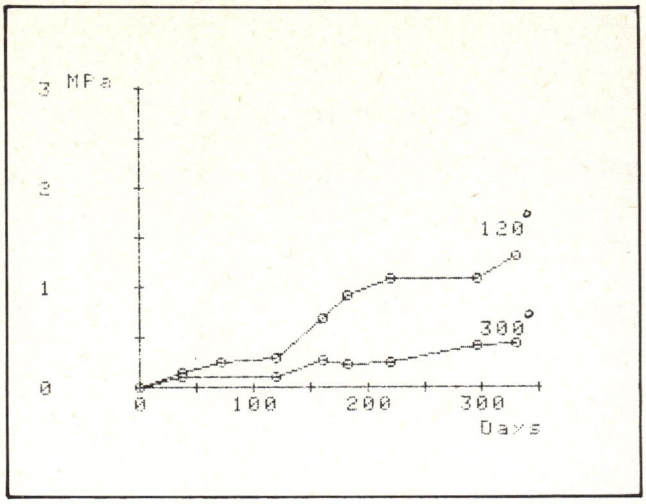

Figure 6 Radial pressures, Ring 23, versus time since erection and angular position

Figure 9 Failure of crown segments near Ring 23

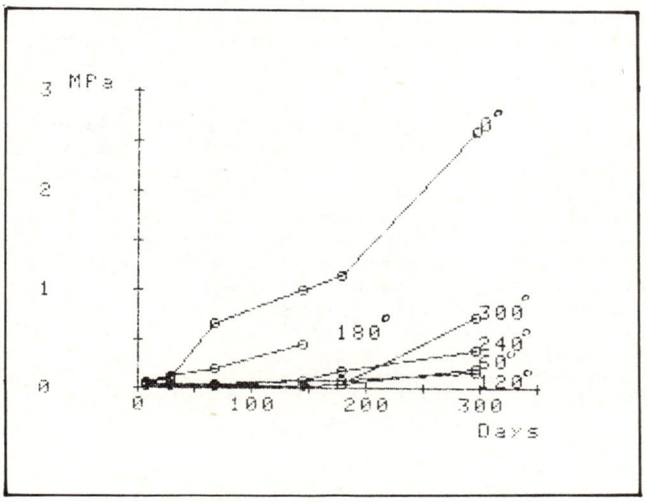

Figure 7 Radial pressures, Ring 51, versus time since erection and angular position

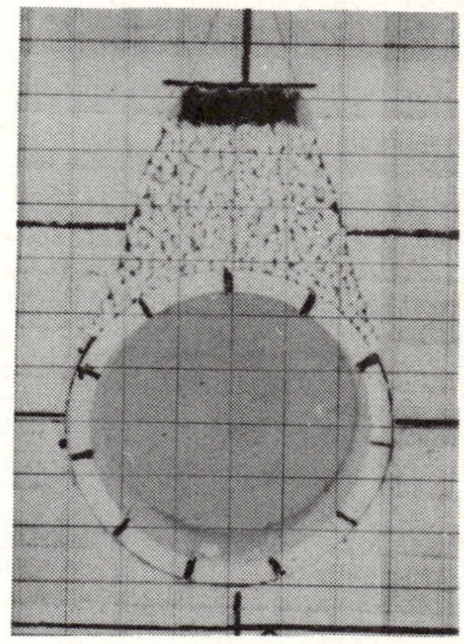

(a)

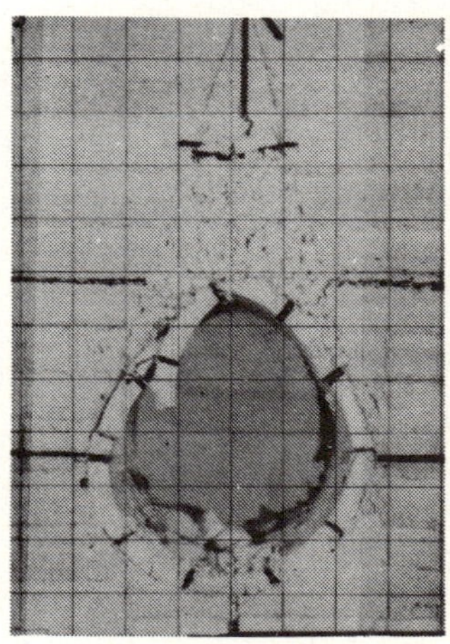

(b)

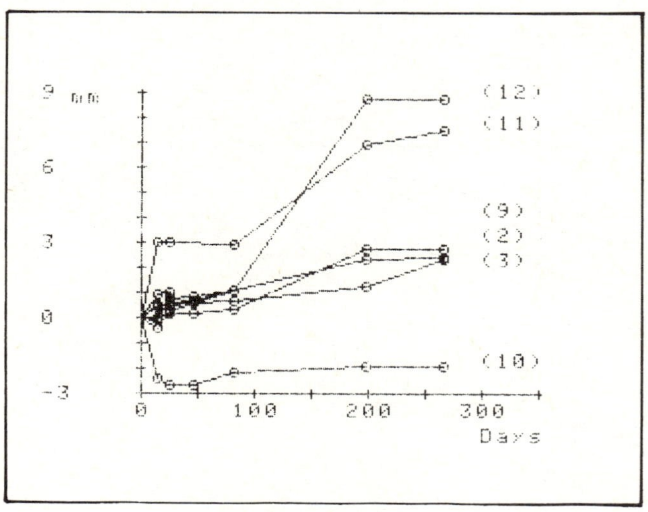

Figure 8 Packing compression, Ring 15, versus time since installation and joint number

Figure 10 Model concrete segmental lining (a) before and (b) after loading

RELATIONS BETWEEN CLOSURE AND SUPPORT IN DEEP MINE EXCAVATIONS

Philip D Shelton

Department of Mining Engineering
University of Newcastle upon Tyne

SUMMARY. The mechanics of rock deformation around underground excavations in Coal Measures rocks are discussed with reference to yield zone solutions and the brittle-ductile behaviour of seatearths and sheared mudstones. The problems of obtaining accurate and representative data on rock properties and limitation to the practical applications of these theories however, require the use of a more flexible, observational approach to the design of supports in deforming ground. The possible use of the technique known as the New Austrian Tunnelling Method has been discussed with reference to the excavation and support of a shaft inset where similar construction procedures have been adopted.

INTRODUCTION

Recent interest in the behaviour of rock around underground coal mine openings including: shafts, insets, major tunnels and roadways and even temporary roadways associated with longwall faces, and the influence of supports on the stability of such excavations has led to the development of design methods based on the formation of a "yield" zone around the excavation (see for instance Wilson, 1980, 1980*). These methods assume brittle failure of the rock immediately after excavation and the rapid development of a fractured zone capable of mobilising significant residual strength to raise the stresses in the "yield" zone to a level where conditions for further fracture of intact rock cease to exist. This can be illustrated in two ways, either as a stress path or through representation of stresses along a radial line through the tunnel axis, Figure 1.

A problem in using this approach, particularly in deep excavations in Coal Measures rocks is that failure is rarely brittle or rapid. It can be shown that under the relatively low strain rates brought into action during excavation, the majority of fireclay seatearths and mudstones - particularly where these are sheared or listricated - will deform in a manner that is at least partly ductile (see Gilbert and Farmer, 1981, Farmer (1980). Figure 2 after Price (1979), shows a typical brittle - ductile transition for an anhydrite marl (similar mechanically to Coal Measures seatearth) during increases in confining pressure from 0 - 42 MN/m^2 the latter is equivalent to the vertical geostatic stress at a depth of 1500 m.

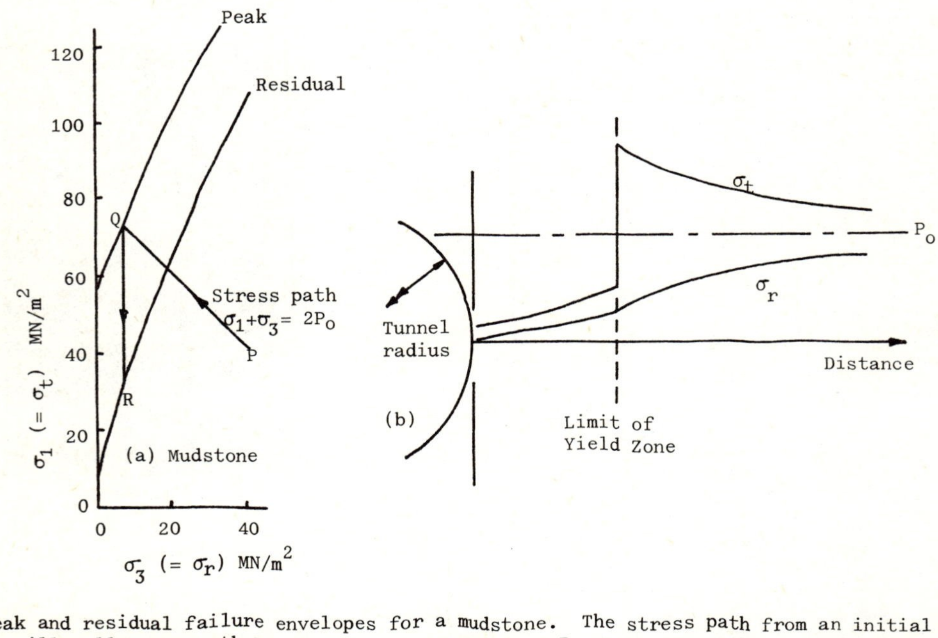

Figure 1(a) Peak and residual failure envelopes for a mudstone. The stress path from an initial hydrostatic stress P_o will follow the path PQ, where $\sigma_1 + \sigma_3$ equals 2 P_o until the peak strength of the mudstone is exceeded, where upon the stress path will drop to the residual level.

(b) This can also be illustrated by considering the radial (σ_r) and tangential (σ_t) stresses, where these are equivalent to σ_3 and σ_1 respectively, in the vicinity of a yield zone surrounding a circular opening.

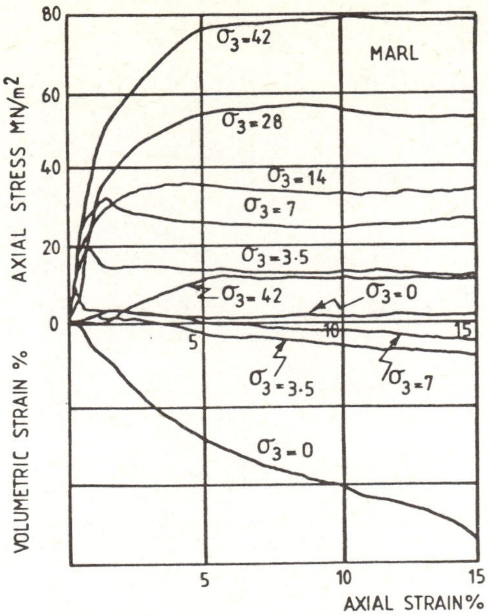

Figure 2 Axial stress and volumetric strain versus axial strain curves for a range of confining pressures, for an anhydritic marl, illustrating a brittle-ductile transition

Figure 3 shows two typical examples of ductile deformation of seatearths in underground roadways. It is quite evident that this deformation, typical of extreme floor heave is neither brittle nor instantaneous.

in very ductile rocks such as evaporites, where rocks will not deform under constant loads provided they have an ability to mobilise shear and retain the ability to redistribute stresses. A more suitable model, discussed by Price and Farmer (1980) and Farmer and Gilbert (1981), illustrates the relationship between the deviatoric and octahedral stresses and the ratio between changes in volumetric and shear strain at varying strain rates, this describes a family of yield curves defining rock reaction. An extension to the concept of yield zone analysis by Ladanyi (1974) gives a method of estimating pressure build up, with time, on various support types by considering the long term strength parameters of rocks.

These models, whilst describing the phenomenon may however be insufficiently precise to do more than illustrate the basic mechanics of deformation in what is a very complex material. They do, however, together with observation provide a basis for an empirical characterisation of rock deformation around an excavation. This can be described in three stages:

(a) rapid deformation due to stress redistribution and any resultant fracture immediately following excavation,

(b) time related deformation following an asymptotic curve with time and continuing at a reducing rate to an insignificant level,

(c) minor time related deformation over a long period.

SUPPORT OF DEFORMING EXCAVATIONS

By defining the deformation process around an excavation - which may vary if different rocks are present in the excavation circumference - it is possible to develop an approach to support design which may bypass some of the more complex analytical approaches. This requires examination of each deformation stage.

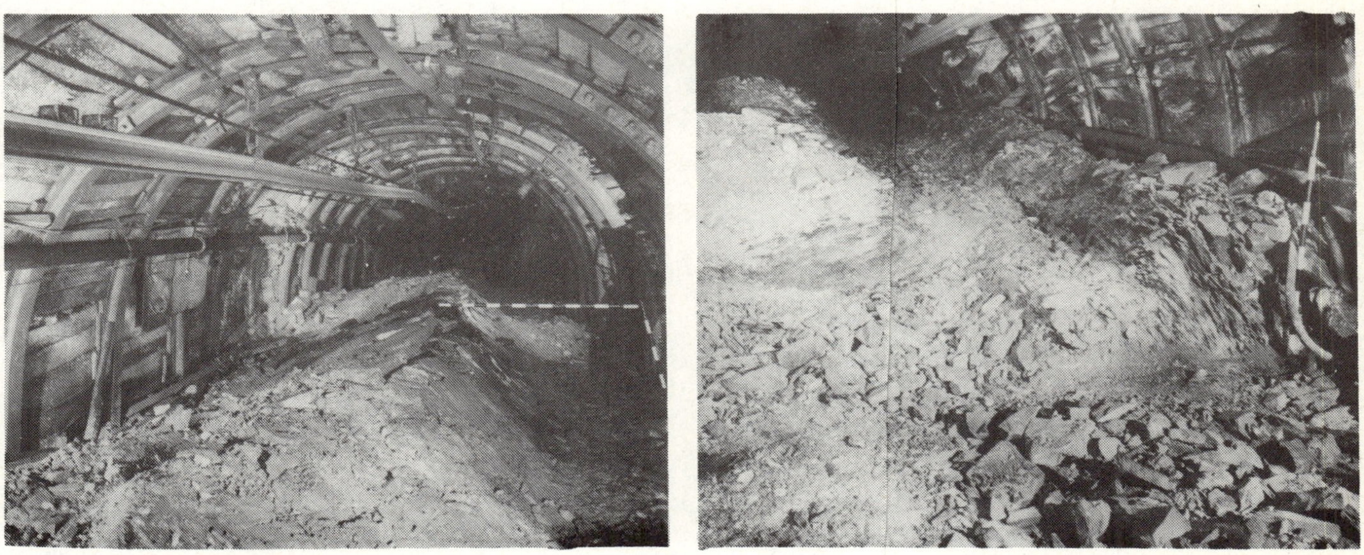

Figure 3 Ductile deformation of a gateroad floor in the Deep Hard Seam at Cotgrave Colliery at a depth of 570 m (after Thomas, 1970)

MODELS OF DUCTILE DEFORMATION

The phenomenon of ductile or partly ductile deformation is characterised by large or small scale deformation of an excavation carcass which takes place over a period of time, usually at a decreasing rate. This is sometimes described, from laboratory tests conducted at constant load, as creep (see Jaeger and Cook, 1968). This is not a satisfactory concept, even

The rapid deformation due to stress redistribution and its consequential effects has been shown by Daemon and Fairhurst (1971), Figure 4, to be approximately 50% complete at the excavation face and almost fully developed in circular tunnels after an advance of one tunnel diameter. It is therefore difficult to see how this can be realistically supported, at this stage they should be designed to control the effects of stress redistribution - such as spalling or detaching blocks - rather than prevent it.

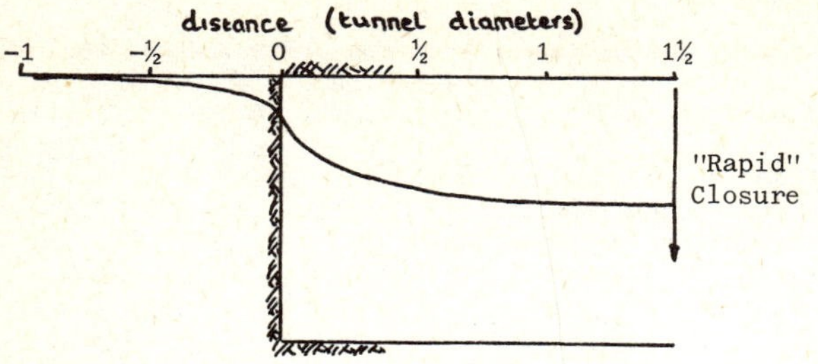

Figure 4 Percentage of rapid deformation resulting from the initial stress redistribution prior to and just following excavation

The main time related deformation will vary with the type of rock and excavation. The design problem is not dissimilar to that facing alpine tunnelling engineers in a similar geotechnical and stress environment where the method known as the New Austrian Tunnelling Method (see Brown, 1981) has been developed. This relies basically on initial, temporary support of the excavation with steel rings or arches, rock bolts or reinforced shotcrete, which can deform with the peripheral rocks, whilst maintaining sufficient structural integrity to provide safe and efficient support. During deformation the rate of closure is monitored daily until a level is reached which is, in the opinion of the engineer sufficiently low to be considered insignificant. This will essentially be a rate which even if it continues over many years, will not lead to major radial stress build up on the lining. At this stage a permanent lining - usually shotcrete - is installed. The time delay prior to the placement of the final support, will depend upon tunnelling conditions, but Brown (1981) and (Braun (1980) suggest that typically this might be after 4 to 5 diameters advance (see Figure 5). Such a situation would cause little interruption to systematic excavation methods. Water pressures can be allowed for in the final lining design or reduced by drainage.

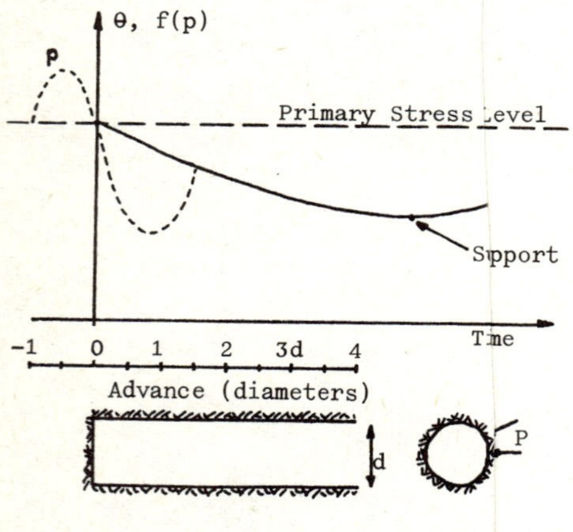

Figure 5 Stress changes and closure behind an advancing tunnel face (after Brown (1980)

At Stillingfleet shaft insets (Borsah, Shelton and Tomlin, 1982) a similar system of support was demonstrated - albeit unintentionally. Figure 6 illustrates roof and floor deformation following excavation of part of the inset. At a point 2 diameters from the end of the inset it can be seen that deformation is rapid at first but levels out - even in the weak floor - after about 8-10 days. The roof was supported during this time by 30 x 24 mm RSJ section arches which, whilst providing adequate support for safety were not designed to prevent roof movement, the floor was unsupported.

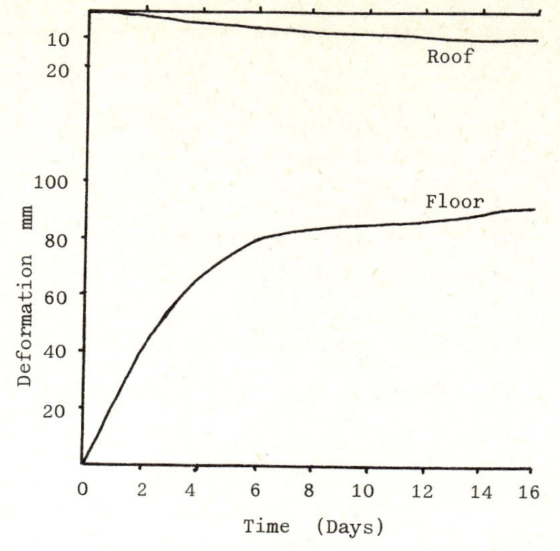

Figure 6 Roof and floor deformations following excavation in 8 x 4 m arched inset. Advance rate was approximately 1.5 m per day

The reinforced concrete lining was installed behind shutters after about 25 days. The build up of strain described by Borsah et al (1982), an example of which is shown in Figure 7 - was equivalent after 100 days to a radial stress of less than 100 kN/m^2 or about 4 m of rock load. Since the concrete requires about 7 days to resist deformation the same effect would have probably been observed if the permanent lining had been installed sooner. There are no indications of any further significant increases in strain in the lining due to further minor long term deformation of the rock.

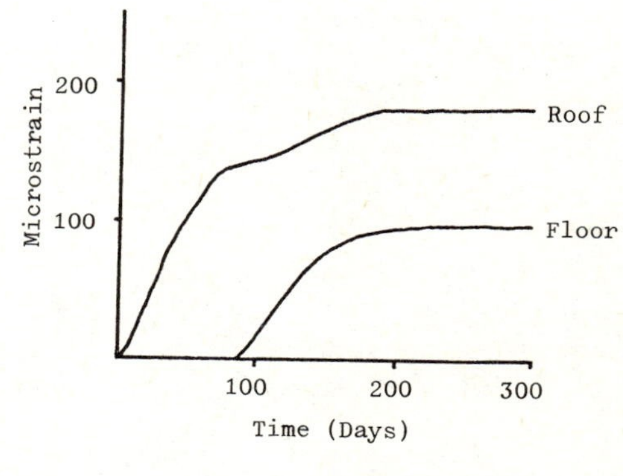

Figure 7 Strains measured in the finished concrete lining in the roof (a) and floor (b)

CONCLUSIONS

The deformation of rocks in the vicinity of many underground excavations in Coal Measures rocks is undoubtedly time dependant. Therefore the loads which must be sustained by any support system will depend upon its stiffness and the state of stress redistribution in the peripheral rocks. It is suggested in the paper that typical time dependant deformation can be described as occurring in three stages. The first, which includes the elastic response to excavation, can be said to be 50% complete at the face and is therefore practically unsupportable. This is followed by a period of time dependant deformation characterised by initially high rates of closure which reduce to less significant proportions as the face is further advanced. Minor time related deformations may then occur over a long period of time.

Current estimates of magnitude of closure using the criterion of yield zone formation and the description of the ductile behaviour of rocks under high confinement do little other than illustrate the mechanics of deformation. A better approach to excavation support design may be the observational procedures adopted by users of the New Austrian Tunnelling Method.

REFERENCES

Borsah, J K, Shelton, P D and Tomlin, N (1982). Deformation of a deep shaft inset in coal measures rocks during and after excavation. Symp. on Strata Mechanics. Newcastle upon Tyne.

Braun, W M (1980). Application of the NATM in deep tunnels and difficult rock formations. Tunnels and Tunnelling, March, 17-20.

Brown, E T (1981). Putting the NATM into perspective. Tunnels and Tunnelling, November, 13-17.

Daemon, J J and Fairhurst, C (1971). Influence of failed rock properties on tunnel stability. Dynamic Rock Mechanics, Amer. Inst. of Min., Metal and Petroleum Engineers, New York, 855-875.

Farmer, I W (1980). Face and roadway stability in underground coal mines; - geotechnical criteria. Report No EUR 7298EN to ECSC Luxembourg.

Farmer, I W and Gilbert M J (1981). Time dependant strength reductions of rocksalt. Proc 1st Symp. on Mechanical Behaviour of Rock salt Penn. State Univ.

Gilbert, M J and Farmer, I W (1981). A time dependant model for lining pressure based on strength concepts. Preprints Int. Symp. weak rock, Tokyo, 1, 130-136.

Jaeger, J C and Cook, N G W (1969). Fundamentals of rock mechanics, Methuen, London.

Ladanyi, B (1974). Use of the long-term strength concept in the determination of ground pressure on tunnel linings. Advances in Rock Mech., Vol II, proc. 3rd Cong. ISRM., Denver, 1150-1156.

Price, A M (1979). The effect of confining pressure on the post-yield deformation characteristics of rocks. PhD Thesis, University of Newcastle upon Tyne.

Price, A M and Farmer, I W (1981). The Horslev Surface in rock deformation. Int. Jour. Rock Mech. Min. Sci. and Geomech. Abs., 18, 229-234.

Wilson, A H (1980). The stability of underground workings in the soft rocks of the Coal Meausres. PhD Thesis, University of Nottingham.

Wilson, A H (1980)* A method of estimating the closure and strength of lining required in drivages surrounded by a yield zone. Int. Jour. Rock mech. Min. Sci. and Geomech. Abs., 17, 349-355.

Thomas, L J (1970). Cotgrave Colliery: the effects of changing pack lengths and the use of a ripping machine in the Deep Hard Seam. Colliery Guardian, August, 399-407.

AN EXPERIMENTAL STUDY OF BEAM BUILDING MECHANISMS
USING FULLY GROUTED BOLTS IN BEDDED MINE ROCK

V. W. Snyder

R. L. Krohn

Department of Mining Engineering
Michigan Technological University
Houghton, Michigan 49931 USA

SUMMARY. An experimental analysis was conducted on beam models containing one row of mechanical or fully grouted bolts. The models were composed of six layers of 1/4 inch thick commercial grade plaster board, 5 1/4 inch wide and lengths of 25 and 36 inches. Strain gages were mounted on each layer of the beam to measure the strain distribution and the deflection at center was recorded. The results indicated the fully grouted bolts are much more effective in beam building than mechanical bolts. The strain gage data showed the beam building to be more complete at the center of the beam than at the ends. When the experimental values of reinforcement were compared to the present theory, the theory failed to accurately predict the reinforcement.

INTRODUCTION

Rapid advances made in underground mining in the United States are the result of better ground control. The improvement on the stability of the roof can be contributed mainly to the introduction of the steel roofbolt about thirty years ago. Initially point anchor bolts, often referred to as mechanical bolts, were inaugurated to underground mining, producing wide unrestrictive entries. About fifteen years ago fully grouted nontension bolts, sometimes called dowels, started to be used underground. Today, the fully grouted bolt constitutes about half of the domestic use of bolts.

It is generally accepted that the success of ground control using roof bolts depends on three types of mechanisms for support, namely; 1) keying of blocks, 2) suspension, and 3) beam building. The work of L. Panek (1956) and (1962) shows how point anchor (mechanical) bolts reinforce the mine back. Fairhurst and Singh (1973) and (1974), provide a theory to explain the support action of fully-grouted bolts. In a study by Gerdeen and others (1979), it was pointed out that a better theory of fully grouted bolts in soft rock was needed.

The primary goal of this study was to assess the beam building mechanism using fully grouted bolts in soft rock. A comparison of the effectiveness of mechanical and fully grouted bolts in reducing maximum deflection was made using small beam models. Strain gages were mounted on the models to give an indication of the strain distribution in each layer of the beam.

EXPERIMENTAL MODEL

Observations made by the author in a series of USA coal mines indicated that the roof was generally composed of soft shales and mudstone. These observations suggest the need for a physical soft modeling material and models composed of thin laminar sheets. A commercial grade plaster board was selected because of its availability and ease of constructing models.

The paper backing on the plaster board was scribed (cut) to negate its strength in tension. The mechanical properties of the plaster board (1/2 inch thick) with the paper scribed were determined to be:

$E = 4.5 \times 10^5$ psi
Tensile Strength = 155 psi
Compressive Strength = 1500 psi.

Care was exercised to assure that all models were cut from the same board. The literature suggested that beam building occur with fully grouted bolts when shear resistance is substantial and slip between layers is eliminated. Thus, the shear stiffness of the bolt-rock layer combination may play a very important role in the support mechanism of beam building. Double shear models of the plaster board and bolts were tested using Duro-Woodhill epoxy resin. Figure 1 illustrates a linear behavior between hole size and shear stiffness with some scatter for smaller holes.

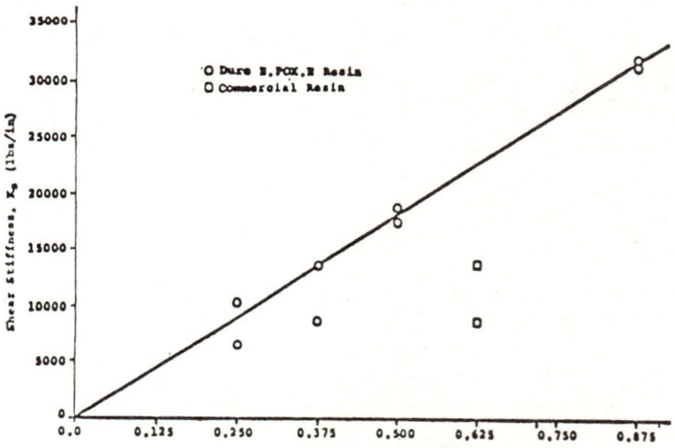

FIGURE 1. Correlation Between Hole Diameter and Shear Stiffness from the Double Shear Tests.

Since 1/4 inch holes were to be used in the beam models a shear stiffness of 10,000 lbs/in can be obtained from Figure 1. This is much lower than those reported in the literature.

Construction of Models

All models consist of six layers of 1/2 inch thick plaster board and were scribed every inch across the span for the full length of the beam. The beam models were supported at each end by clamping it between two 1/2 inch thick steel plates to insure clamped ends.

Each beam was cut from the same plaster board sheet and all layers cut in the same direction to eliminate varying properties between layers. The layers were stacked and marked for a bolting pattern consisting of 3/16 inch bolts in 1/4 inch holes every six inches along the span. Each layer of the model was individually scribed top and bottom.

Strain gages were mounted directly onto the plaster board by removing the paper. The gages were of type BAE-06-500BB-120 and room temperature cure epoxy used to mount them. With the 24 inch span the gages (4 per layer) were mounted top and bottom of each layer at the left clamped end and the mid span of the beam. On one of the 36 inch span beams, gages were mounted at the clamped end, 12 inches from the clamped and at the mid span.

General Testing Procedure

The models were tested as clamped end beams with a uniform load applied across the span. The loading was accomplished by small sandbags. At each increments (layer of sandbags) the mid span deflection was measured and strain gage data recorded. The load was then removed, and the test repeated to verify the deflection and strain gage data.

Since a nondestructive test was being performed a test schedule was set up to first test mechanical bolts, then replacing them with epoxy grouted bolts. For example in the 24 inch model the test schedule was:

with out bolts
with 2 mechanical bolts
with 4 mechanical bolts
with 2 grouted bolts.
with 4 grouted bolts.

The mechanical bolts were 1/8 inch diameter steel bolts, mounted with washers at each end and torqued to 4.8 inch-pounds.

An upper bound on the bending effect of the layer beam was obtained by glueing six layers together. The beams were constructed as above, scribed in the same manner, but each layer was glued to the other by a white glue (Elmer's glue). Deflection data at the mid span was recorded for each increment of loading.

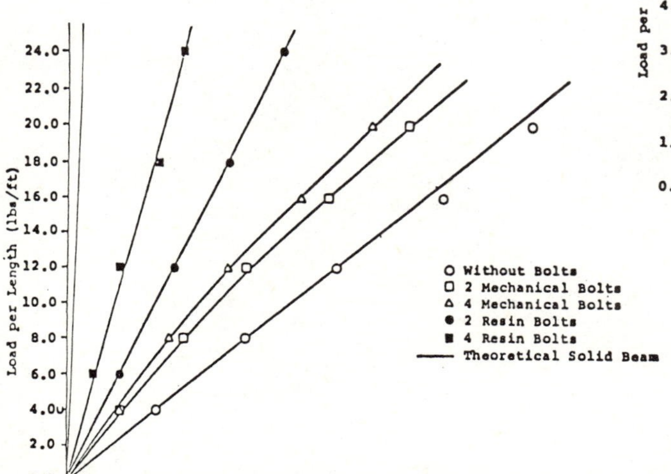

FIGURE 2. Model A Load Versus Deflection For All Bolting Patterns Tested.

TEST RESULTS

The test results for the 24 inch and 36 inch span beams are illustrated in Figure 2 and Figure 3 respectively. These figures show the mid span deflection versus the load per length.

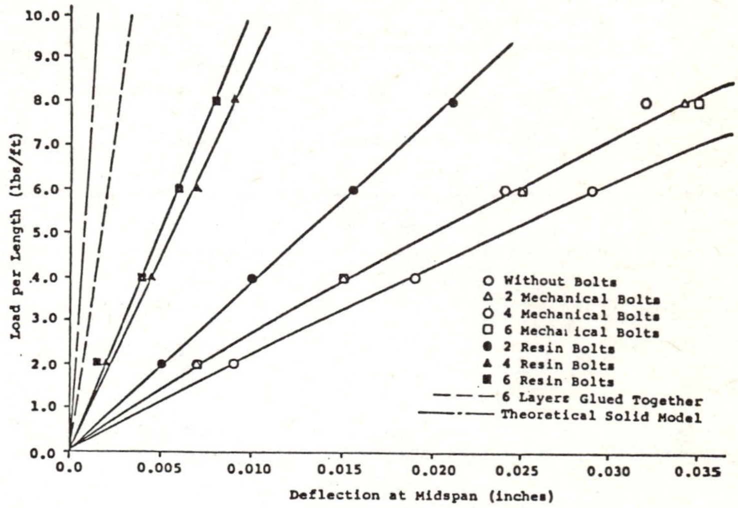

FIGURE 3. Model B Load Versus Deflection For All Bolting Patterns Tested.

In both cases bolt holes were drilled at 6 inch spacing. Bolting started at clamped end and proceeded towards the mid span. For example, 2 bolts in the 24 inch span means only the bolts 3 inches from each clamped end and then 18 inches between them.

Another 36 inch span model was used and tested with 1, 3, 5 and 7 grouted bolts. This test schedule effectively provided testing with 18, 12 and 6 inch bolt spacing. The results are shown below in Figure 4.

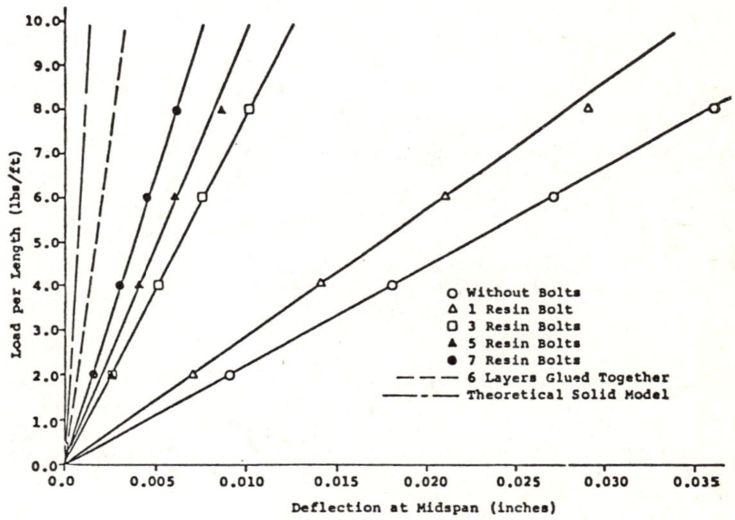

FIGURE 4. Model C Load Versus Deflection For All Bolting Patterns Tested.

Only fully grouted bolts were used in the test illustrations in Figure 4.

Examination of the strain gage data and the corresponding strain distribution reveals some erratic results in the first and sixth layers. In general most of the layers wanted to act independent of the other layers. Figure 4 and 5 show the strain distribution on the 36 inch span for loading of 8 lbs/ft.

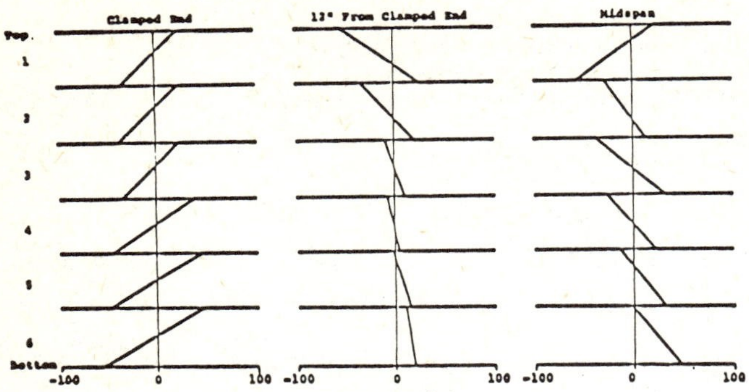

FIGURE 5. Model C Strain Distribution With 3 Resin Bolts at 8 lbs/ft.

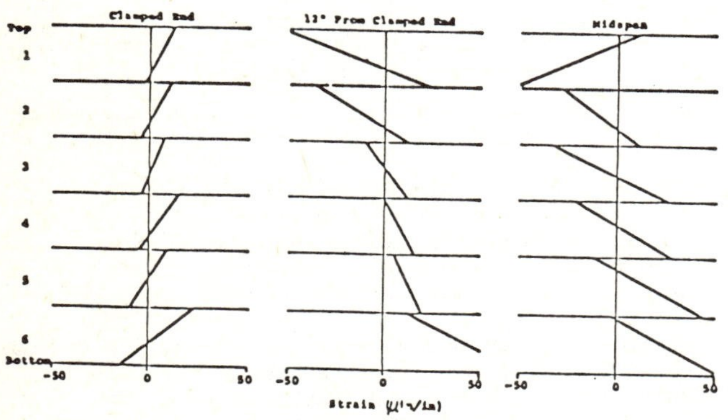

FIGURE 6. Model C Strain Distribution With 5 Resin Bolts at 8 lbs/ft.

In both cases there appears to be little reinforcement (layers acting together as a beam) at the clamped end. In Figure 5, if one disregards the top layer, there is some overall beam action shown at the mid span. In Figure 6, there is evidence of overall beam building at 1/3 point. Also, disregarding the top layer there is some evidence of beam building at midspan.

DISCUSSION OF TEST RESULTS

The fully grouted bolts were more effective in the reduction of the mid span deflection than the mechanical bolts as illustrated in Figure 2 and 3. The reinforcement factor is defined as

$$R = \delta/\delta_{bolt}$$

where δ in the midspan deflection of the unbolted beam and δ_{bolt} in the bolted beamed deflection. For the model with the 24 inch span the fully grouted bolts were about twice as effective as the mechanical bolts (Figure 2). For the 36 inch span model there was not much improvement between two, four or six mechanical bolts (Figure 3).

The questions of where to bolt and is there an optimum spacing of the bolts still remains. These questions can be examined by looking at Figure 2, 3 and 4. In both of these tests, the bolts were just placed near the clamped end. As additional bolts are placed near the center of the beam little additional reinforcement is gained for the mechanical bolts. In Figure 3 there is little improvement between four and six fully grouted bolts. In Figure 4, where the bolting has been shown to increase the density of bolting near the clamped ends, the above saturation of bolts does not appear. The reinforcement obtained with six equally spaced bolts (Figure 3) is the same as the reinforcement obtained with five bolts with the variable spacing.

The theoretical value obtained if the entire layer beam would act together is much higher than any of the bolted beams. In Figure 3 and 4, the beam where all layers were glued together is shown and is less than a the theoretical results by a factor of about 2.

The strain distribution obtained from the strain gage data did indicate some beam building near the mid span and at the 1/3 points of the span (Figure 5 and 6). At the clamped ends, the layers acted independently. The author (1979) reported observing delamination between layers near the ribs in actual mine environment. The possible cause of this delamination may be the high shear encounter at clamped ends.

CONCLUSIONS

The results indicate beam building occurs to a considerably higher degree of reinforcement with fully grouted bolts than with mechanical bolts. The strain gage data showed the beam building effect occurred at the center of the beam and not at the clamped ends. The delamination observed in coal mines did occur in the laboratory models. Increasing the bolt density near the clamped ends (high shear area) was more effective in beam building than placing them in the center of the span.

REFERENCES

Fairhurst, C., Singh, B., 1974, "Roof Bolting in Horizontally Laminated Rock", Engineering and Mining Journal, Vol. 185, February, pp. 80-90.

Gerdeen, J.C., Snyder, V.W., and Viegelahn, G.L. 1979, "Factors Governing the Effectiveness of Roof Bolting Systems Using Fully Resin-Grouted Nontensioned Bolts". Proceedings of 20th U.S. Symposium on Rock Mechanics, June 4-6, pp. 607-614.

Singh, B., Fairhurst, C., Christiano, P.P., 1973, "Computer Simulation of Laminated Roof Reinforced with Grouted Bolts", Symposium on Rock Mechanics and Tunnelling Problems, Indian Geotechnical Society, Regional Engineering College, Kurukshetra, Haryana, India, December 17-18, pp. 41-47.

Gerdeen, J.C., Snyder, V.W., Viegelahn, G.L. and Parker, J., 1979, "Design Criteria for Roof Bolting Plans Using Fully Resin Grouted Nontensioned Bolts to Reinforce Bedded Mine Roof", USBM Contract No. J0366004.

Panek, L.A., 1956, "Principles of Reinforcing Bedded Mine Roofs with Bolts", USBM RI 5156.

Panek, L.A., 1962, "The Effects of Suspension in Bolted Bedded Mine Roof", USBM RI 6138,

AN ELASTICITY SOLUTION FOR THE SHEAR STIFFNESS OF THE FULL-GROUTED RESIN ROOFBOLT

V. W. Snyder
P.M. Schwab
J.C. Gerdeen

Department of Mining Engineering
Michigan Technological University
Houghton, Michigan 49931 USA

SUMMARY. The bolt is modeled as a beam on an elastic foundation where the theory of elasticity in polar coordinates is employed to determine the foundation modulus (spring stiffness) of the resin-rock combination. Parameter studies are presented showing the effects of the rock and resin elasticity on the shear stiffness values. The results obtained are compared to values obtained using a three-dimensional finite element study. The agreement between the elasticity solution and the three-dimensional finite element model is close enough to preclude the need for using the finite element model in this type of an analysis.

INTRODUCTION

The initial attempt at an analytical solution for the shear stiffness of the fully grouted roofbolt was made by Fairhurst and Singh (1974). The model they chose is shown in Figure 1. Here it was assumed that the lateral deflection due to shear decays out in an effective length of 3/4 of the bolt diameter d_1, that the rock was rigid, and no allowances were made for the properties of the resin or the thickness of the resin annulus surrounding the bolt. Thus this model can be considered equivalent to a cantilever beam of length 3/4 d_1 and loaded on the end by a force P. From this configuration the shear stiffness is found to be given by:

$$K_s = \frac{P}{2\delta} = \frac{\pi E_B d_1}{31.5} \tag{1}$$

where δ is the deflection of the free end of the beam due to P, and E_B is the modulus of elasticity of the bolt. The term 2δ appears in the denominator to account for the deflection of both halves of the bolt at the joint.

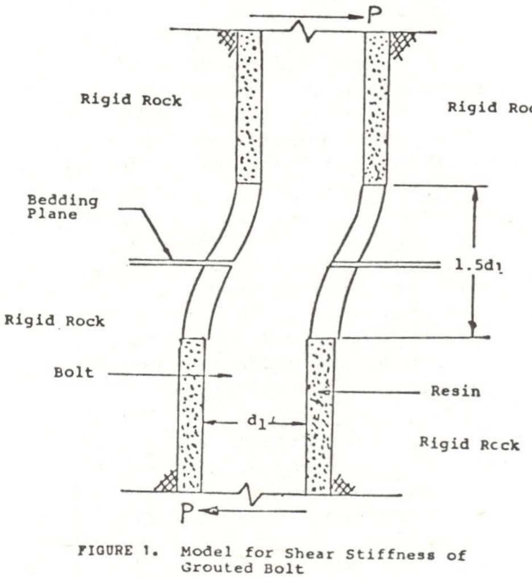

FIGURE 1. Model for Shear Stiffness of Grouted Bolt

In order to account for more of the variables present in the system, Gerdeen (1975) treated the bolt as a beam resting on an elastic foundation, namely the resin. The rock was still assumed rigid but now the resin properties and the thickness of the resin annulus appeared in the results. Using this approach it was found that:

$$K_s = \frac{P}{2\delta} = E_B I_B \beta^3 \tag{2}$$

where β is equal to $(K4E_B I_B)^{.25}$ and I_B is the area moment of inertia of the bolt cross section. The variable K which appears in the expression for the foundation constant β is the elastic foundation modulus of the resin. By giving the bolt a known deflection into its resin foundation and assuming the resin to have little or no strength in tension for a semi-infinite bolt, Gerdeen (1975) found that:

$$K = \frac{2E_R}{\frac{d_2}{d_1} - 1} \tag{3}$$

where E_R is the resin modulus of elasticity and d_2 is the diameter of the borehole. In addition this analysis allows an actual calculation of a decay length L for the shear deflection as:

$$L = \frac{Ln(20)}{\beta} \approx \frac{3}{\beta} \tag{4}$$

As an example, let $E_B = 30 \times 10^6$ PSI, $E_R = 7 \times 10^5$ PSI, the bolt diameter $d_1 = 0.75$ inches and the borehole diameter $d_2 = 1.125$ inches; then Equation (1) gives a value of 2.2×10^6 lbs/in while Equation (2) yields a value of 6.3×10^5 lbs/in. Equation (4) gives a decay length of 2.7 inches while it was assumed that the decay length was 3/4 d_1 or 0.56 inches in the theory used in Equation (1). Experimental values are hard to obtain in the literature; but looking at the work of Hass (1975) an estimated value of 8.6×10^5 lbs/in can be obtained for 1.0 inch diameter bolts and $E_R = 1.5 \times 10^6$ PSI.

ELASTICITY SOLUTION

In the present study the analysis as a beam on an elastic foundation has been retained with the differences being that Equation (2) has been modified to include the effect of the joint opening and that the elasticity of the rock plays a role in the foundation modulus K. Equation (2) now has the form:

$$K_s = \frac{3E_B I_B \beta^3}{3 + 3\beta h + 2h^3\beta^3} \qquad (5)$$

where h is 1/2 of the opening at the joint. For h = 0, Equation (5) reduces to Equation (2). The theory of elasticity in polar coordinates is used to determine the foundation modulus K from the configuration shown in Figure 2. The bolt is treated as a rigid insert in an elastic annulus (the resin) which is, in turn, immersed in an infinite elastic region, namely the rock. The

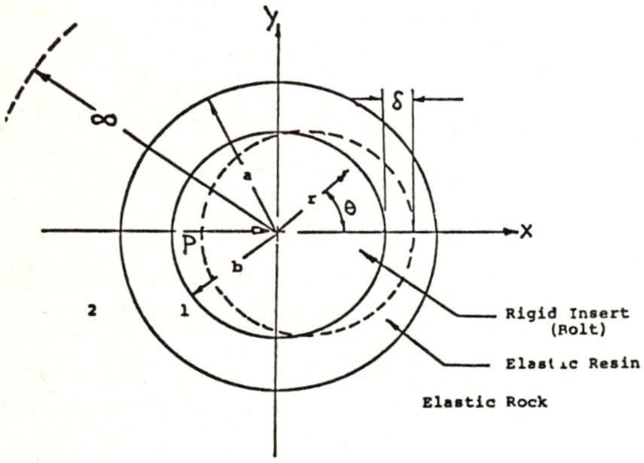

FIGURE 2. Configuration for Determining the Elastic Foundation Modulus by Elasticity Theory

bolt is given a prescribed deflection of δ towards positive x and the resulting stresses are determined in the resin and rock. The load P is then related to the displacement δ through these stresses and the foundation modulus determined. The final solution to the resulting system of equations is performed numerically since the closed form expressions for the needed coefficients are too long and messy to be convenient.

The system depicted in Figure 2 involves two elastic regions where at the interface the stresses and displacements must be equated. The stress and displacement fields within these regions will be described by two stress functions ϕ_1 and ϕ_2 satisfying the homogeneous biharmonic equations. The solution to the biharmonic equations can be obtained by choosing functions which satisfy the governing equations and imposing the boundary conditions. A table of biharmonic functions with their resulting stresses and displacements has been tabulated by Dundurs (1964). For the problem at hand, the biharmonic functions chosen are

$$\phi_1(r,\theta) = Ar\theta\sin\theta + BrLnr\cos\theta + Cr^3\cos\theta + D\frac{\cos\theta}{r}$$

and

$$\phi_2(r,\theta) = Er\theta\sin\theta + FrLnr\cos\theta + H\frac{\cos\theta}{r} \qquad (7)$$

where A,B,C,D,E,F, and H are constants to be determined from the boundary conditions.

Using plane strain analysis, requiring continuity of stresses and displacements at the various interfaces, and vanishing stresses at $r = \infty$ the constant coefficients in Equations 6 and 7 are determined. The solution of the resulting system of equations was performed numerically for specific values of the quantities involved since the longhand solution would result in very burdensome expressions. The foundation modulus is found to be given by:

$$K = \frac{2\pi A}{\delta}$$

Where A is the leading coefficient in Equation (6). The value obtained for K is then used to formulate a new value for β and K_s is then determined from Equation (5).

The actual solution for the seven constants were obtained by using a digital computer program. The program permitted a parameter study to be made. As an example, for a resin with $E = 7.0 \times 10^5$ PSI and $\nu = 0.210$, a rock with $E = 6.8 \times 10^6$ PSI and $\nu = 0.3$, a bolt diameter of 0.75 in., and a resin thickness of 0.1875 in.; the work of Fairhurst and Singh (1974) predicts 2.24×10^6 lb/in., Gerdeen (1975) yields 6.32×10^5 lb/in., and this analysis predicts 1.220×10^6 lb/in. for the shear stiffness.

RESULTS

The effect of the thickness of the grout annulus on the shear stiffness is shown in Figure 3 for 3/4 inch bolt. The figure indicates that a small as possible annulus should be used. Also, the effect of the rock elasticity rapidly disappears as the grout (resin) thickness is increased.

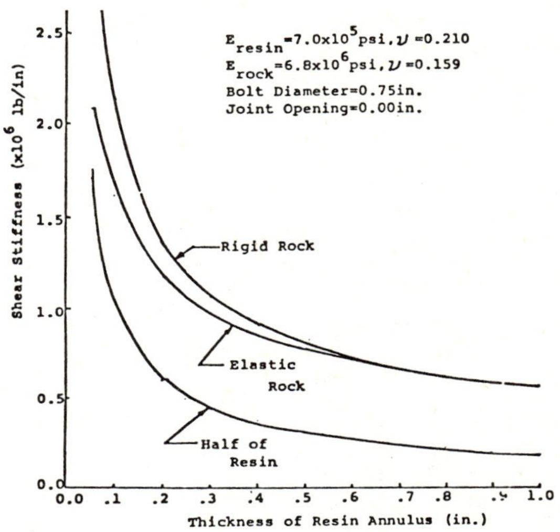

FIGURE 3. Effect of Resin Annulus Thickness on Shear Stiffness

Shown in Figure 4 is the dependence of the shear deflection decay length L on the thickness of the grout annulus. The predictions made by the theory are less than the experimental results of Kwitowski (1975) which show decay to be in approximately 4 inches rather than 2 inches as results from the theory. The shorter decay lengths consequently yield a higher stiffness than is actually observed. This occurs because the analytical solution assumes competent bonding at all interfaces and perfectly elastic behavior. In practice a perfect bond is virtually impossible to achieve. The texture of the surfaces, moisture content, and the degree of competence in the rock surrounding the grout all play a role. For this reason the theory should be treated as an upper bound to the true shear stiffness. One again notices from the plot that the importance of the rock's elasticity disappears as the resin thickness is increased.

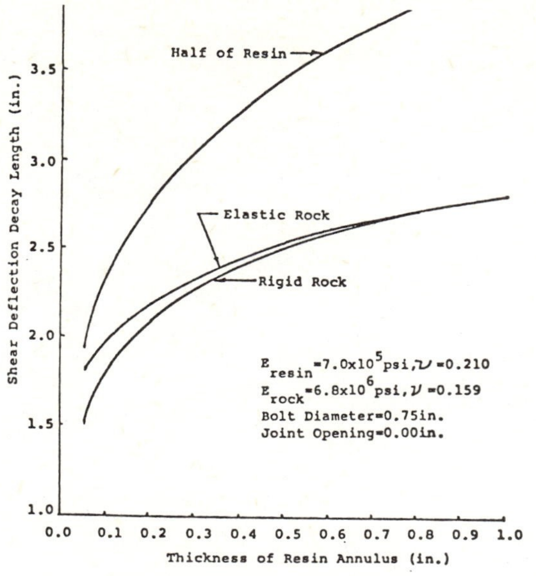

FIGURE 4. Effect of Resin Annulus Thickness on the Shear Deflection Decay Length

Unlike the axial stiffness, which is a function of only the bolt, resin, and rock properties and their respective dimensions, the shear stiffness as it is defined in the literature also depends upon the degree of separation which exists at the joint. Obviously the shear force is constant across a joint but the bending moment is not. It is due to the bending moment that the half gap width h appears in the expression for the shear stiffness, equation (3). Figure 5 illustrates the influence joint opening has on the shear stiffness. As a joint begins to open up the stiffness decreases very rapidly with the rate of decrease slowing as the opening becomes larger.

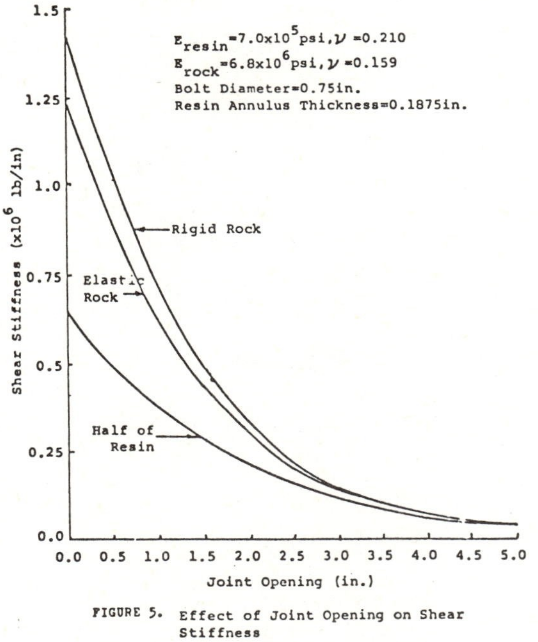

FIGURE 5. Effect of Joint Opening on Shear Stiffness

The bolt diameter also has a marked effect on the shear stiffness. As the bolt diameter is increased one would expect the stiffness to correspondingly increase as is shown in Figure 6. Beyond a diameter of 1.0 inch the solution rises quite steeply and could lead one to believe that the simplest way to achieve increased shear resistance would be to use larger bolts. Cost may make such a course of action impossible, however.

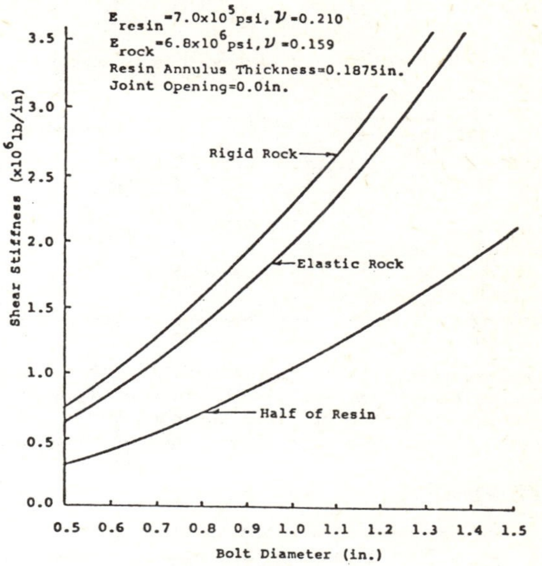

FIGURE 6. Effect of Bolt Diameter on Shear Stiffness

The effect of Poisson's ratio was also examined. The maximum effect of Poisson's ratio of the rock on the shear stiffness was about three percent. There was no sufficient change in shear stiffness when the Poisson's ratio of the resin or grout was considered. For example, with a 3/4 inch diameter bolt in a 1 1/8 hole the change could be as large as 5 percent.

FINITE ELEMENT MODEL

As both a check on the analytical solution and as an entity in itself, a three-dimensional finite element model of the bolt, grout, and rock system was developed. The model was designed to investigate the behavior under transverse loading of a fully grouted roofbolt. A variable node solid isoparametric element was used. The model used had about 3200 degrees of freedom with about 240 elements. Each finite element run required about 45 minutes on UNIVAC 1110 with 64K words of available core.

The comparison between the "closed" form analytical study and the finite element studies is tabulated in Table 1.

Table 1

Shear Stiffness Versus Annulus Thickness

Annulus Thickness (in.)	Shear Stiffness (lb.in.)		Deviation (%)
	Analytical	Finite Element	
0.10	1.612×10^6	1.592×10^6	-1.2%
0.1875	1.220×10^6	1.245×10^6	2.0%
0.30	9.765×10^5	1.044×10^6	6.9%
0.50	7.665×10^5	8.450×10^5	10.5%

Table 2 is a comparison between the elasticity solution and the finite element model for the effect of the bolt diameter on shear stiffness.

Table 2

Shear Stiffness Versus Bolt Diameter

Bolt Diameter (in.)	Shear Stiffness (lb.in.)		Deviation (%)
	Analytical	Finite Element	
0.50	6.335×10^5	6.828×10^5	7.7
0.60	8.472×10^5	9.039×10^5	6.7
0.75	1.220×10^6	1.245×10^6	2.1
1.00	1.987×10^6	1.836×10^6	-7.6

The above table shows that the finite element model and the elasticity solution is in good agreement and is close enough to preclude the need for using the finite element model in these types of analyses.

CONCLUSIONS

It has been shown that an elasticity solution can be used to determine the shear stiffness of a fully grouted roofbolt. The effect of the elastic rock is to lower the shear stiffness but the assumption of a rigid rock will give an upper bound. The decay length will increase slightly with the elastic rock, but in either case the theory is about 1/2 the experimental values.

REFERENCES

Fairhurst, C. and Singh, B., 1974 "Roofbolting in Horizontally Laminated Rock," Eng. Min. Jnl., Vol. 175, February, pp. 80-90.

Gerdeen, J.C., 1975, "Deflection of Resin-Bonded Roof Bolts Under Shear Loading", Advanced Strength of Materials, EM 541--Unpublished Notes, Michigan Technological University, Houghton, Michigan.

Kwitowski, A.J. and Wade, L., 1975, "Laboratory Data of Shear and Bending of Resin Bolts", U.S. Bureau of Mines Princeton, Unpublished Data, September.

Haas, C.J., Keith, H.S., Nitzache, R.N., Dave, J., Ioffmann, R.L., and Patrick, W.C., 1975, "An Investigation of the Interaction of Rock and Types of Rock Bolts for Selection Loading Conditions", 2nd Annual Report Rock Mechanics and Explosives Research Center, University of Missouri-Rolla, U.S. Bureau of Mines Contract H0122110, March.

Dundurs, J., 1964, "Catalog of Stresses and Displacements in Polar Coordinates for Elementary Biharmonic Functions", Unpublished Notes, North Western University.

ROCK CLASSIFICATION AND ASSESSMENT FOR RAPID EXCAVATION

Robert J. Fowell

and

Simon T. Johnson

Department of Geotechnical Engineering

University of Newcastle upon Tyne, NE1 7RU

England

SUMMARY

The use of a boom type tunnelling machine equipped with drag tools is the principal means of mechanically excavating rock tunnels in Britain. This paper outlines the main factors influencing the rate of advance of mechanised drivage operations and describes methods of assessing the factors involved. Comment is also made on the interpretation of borehole information at the site investigation stage.

INTRODUCTION

A number of papers have been presented on specific factors influencing roadheader performance, but few have attempted to identify a comprehensive list. This paper sets out to identify those factors which, in the author's experience, have to be considered, and discuss their significance in predicting machine performance.

Results from a number of sites where machine monitoring of performance has been conducted are used to illustrate the approach adopted for this continuing research programme in the University of Newcastle upon Tyne.

FACTORS INFLUENCING MACHINE PERFORMANCE

These factors are listed in Table 1. Present investigations are concerned with evaluating these parameters. There are two principal reasons for this:

1. It is essential to be able to evaluate machine performance exclusive of any of the factors which dictate utilisation (Table 2), in order to allow any meaningful comparison between different machines operating under a variety of conditions. The choice of machine is of prime importance with the tunnelling system and organisation being at least partly dependent on the machine.
2. Machine-Rockinteraction is possibly more complex for roadheaders than for any other tunnelling machine and, as such, provides parameters requiring quantification and discrete definition for future progress in both the design and use of such machines. The prediction of Cutting Rate and Tool Consumption is fundamental to any assessment of machine performance.

Cutting Rate and Tool Consumption

In predicting cutting rate and tool consumption, the influence of the principal variables; cuttability, cutting wear and the rock mass properties must be assessed. Quantification of the rock material variables is achieved through a number of index tests, the results of which have been correlated with in-situ performance. Classification of the rock mass properties is of particular interest in the present research.

It should be noted that these are 'rock only' factors and that the machine characteristics have not been included at this stage.

Rock Mass Properties

Classifying Rock Mass Properties involves decisions influenced by both experience and analysis. The influence on machine performance of the rock mass can be visualised as a modifer to a prediction using only the rock material properties. Thus, in most cases, the rock mass will advantageously modify the prediction. In general, the machine must be capable of cutting the rock in the massive condition, unless the presence of discontinuities can be guaranteed beyond reasonable doubt. In some cases, the use of light blasting may be introduced to induce fractures that the machine can exploit where difficulties are experienced.

In general discontinuities, their frequency, orientation, position in the face and, to a lesser extent than support, shear strength describe for excavation the geomechanical reaction of the rock mass to applied forces. The problem with using support classification methods for excavation is the inclusion of intact rock strength which oversimplifies the rock material classification for machine performance.

Machine Characteristics

The important requirement in mechanised rock excavation is that the cutting element is capable of taking a reasonable depth of cut to gain the advantages of lower specific energy, dust make and tool consumption, compared with shallow rubbing cuts. The key is to have a stable cutting machine and adequate forces to hold the cutting machine and adequate forces to hold the cutting tools into the rock, avoiding bouncing of the cutting head. Slow head speeds are of advantage in difficult ground reducing impact damage and rate of heat build-up within the pick, which reduces tool life.

Cutting head lacing patterns are of considerable importance. A balanced head design is required to reduce torque fluctuations which are so destructive to gearboxes and other mechanical components. Multiple sequences of tools should be staggered to reduce highly inefficient coring and grooving situations.

For optimum excavation rates the sequence of cutting should be designed to exploit strata weaknesses, employing undercutting modes in preference to overcutting. Profiling devices should be used where possible to avoid overbreak, which may result in poor strata control. Sumping should be confined to the centre of the face, avoiding the confinement at the periphery of the tunnel.

Table 1 Factors influencing machine performance

MAIN FACTOR	VARIABLES		
ROCK PARAMETERS	INTACT PROPERTIES	CUTTABILITY	
		CUTTING WEAR	ABRASIVITY
			IMPACT RESISTANCE
			THERMAL PROPERTIES
		DURABILITY	SLURRY MAKE
	MASS PROPERTIES	DISCONTINUITIES	VOLUMETRIC INTENSITY
			ORIENTATION
			SHEAR STRENGTH
		MIXED FACE CONDITIONS	
		DEGREE OF VARIATION IN STRATA (along line of tunnel)	
	ENVIRON-MENT	WATER	FROM WITHIN ROCK MASS
			FROM DUST DUPPRESSION
		TUNNEL GEOMETRY	SIZE
			SHAPE
			GRADIENT
		IN-SITU STRESSES	
MACHINE PARAMETERS	CUTTING HEAD	NO. OF TOOLS	
		TOOL TYPE	RADIAL/FORWARD ATTACK
			TIP GEOMETRY
			CARBIDE GRADE IN TIP
		LACING PATTERN	
	WEIGHT	SLEWING + LIFTING FORCES	
		HEAD SPEED	
		HEAD POWER	
		RIGIDITY OF MACHINE CONSTRUCTION	
	OPERATIONAL CHARACTER	PROFILING	
		GUIDANCE	

Table 2 Factors influencing machine utilisation

MAIN FACTOR	VARIABLES	
DOWNTIME	PLANNED	MAINTENANCE
		SPARES AVAILABILITY
	UNPLANNED	STAFF AVAILABILITY
		SPARES AVAILABILITY
		CONDITIONS IN THE TUNNEL
SUPPORT	TYPE AND AMOUNT REQUIRED	
	ERECTION SYSTEM	
	DEGREE OF MECHANISATION	
	ANCILLARY OPERATIONS	GROUTING
		LAGGING BOARDS
DEBRIS DISPOSAL	AT THE FACE	CLEANING UP
		GENERAL MUCKING
		SECONDARY BREAKAGE
	BEHIND THE FACE	CONVEYORS
		MINE CARS
		WATER PRESENCE
ANCILLARIES	VENTILATION	
	DUST EXTRACTION	WATER/PUMPING EQUIPMENT
		AIR DUCTS FOR EXTRACTION
	EXTENSIONS	TRACK
		POWER CABLES
		TELEPHONE
		CONVEYOR
LABOUR	AVAILABILITY	
	EXPERIENCE/SKILL	
	TRANSPORT TO FACE	DISTANCE/TIME
		METHOD
ORGANISATION	MANAGEMENT	BONUS SCHEMES
		COMMUNICATION
	SHIFT TIMES	TOTAL PAYABLE TIME
		PRODUCTION TIME
FINAL USE	ENGINEERING TOLERANCES	GRADE
		ALIGNMENT
INTER-GRATION	IS TUNNELLING THE ONLY ON-SITE ACTIVITY, OR IS IT COMPETING FOR RESOURCES WITH OTHER OPERATIONS (i.e. MINING)?	
water	A PROBLEM OF DISPOSAL AND DRAINAGE.	

UTILISATION

To gain a reliable estimate of advance rate the machine must be considered to be integrated into a tunnelling system. The interaction of the tunnelling operation with other activities has also to be considered.

Table 2 has been included to provide in concise form those factors that influence utilisation of the machine.

It should be realised that increased utilisation is the greatest potential area for improvement in overall advance rates. For example, divorcing the support setting from cutting operations by the use of temporary machine mounted supports has considerable appeal.

Advance Rate

For estimation purposes most costs are designated per metre of tunnel. Once the average machine cutting rate is determined for a given face section and the expected utilisation assessed, then a calculation involving the predicted cutting rates and tool consumption can be made to obtain a predicted advance rate, necessary for effective and realistic tendering.

The calculation is of the form:

$$\text{Predicted Advance Rate (m/h)} = \frac{\text{*Predicted Average Cutting Rate}(m^3/h)}{\text{Face Area}(m^2)} \times \frac{\text{Utilisation \%}}{100} \quad \text{..... (1)}$$

* This value of Predicted Average Cutting Rate will take into account both tool consumption and Rock Mass Structure and/or Mixed Strata Faces.

RESULTS FROM FIELD STUDIES

Present investigations are concerned with the appraisal and standardisation of laboratory tests for excavation performance prediction and determination of a method for obtaining a modifier for the influence of the Rock Mass Structure. Details of the laboratory tests for cuttability and abrasivity are given in McFeat Smith and Fowell (1977) and Bougard et al. (1974).

The relationships obtained between in-situ cutting rate, laboratory specific energy and machine weight are given in Figure 1 for massive face conditions. The relationships between abrasivity coefficient and in-situ tool consumption is given in Figure 2 for massive mixed faces. The abrasivity coefficient is based on the Cerchar scratch test. The Cerchar number for each strata type in a mixed face are simply weighted by the proportion of that strata in the face. Thus:

$$\text{Abrasivity (A) Coefficient} = \frac{\text{Cerchar No.} \times \text{\% of strata in face}}{10} \quad \text{..... (2)}$$

This mixed face condition is very common where roadheaders are used and its treatment by weighting the rock material and rock mass properties of the lithologies on the face is a consideration in the present research programme.

An appreciation of the relationship between rock mass parameter and cutting rate is given in Figure 3. This is a plot of the Geomechanics Classification system against cutting rates for the heavyweight class of boom tunnelling machines. The only modification made has been in the rating of orientation since, for excavation in general, an inverse relationship exists between support requirements and ease of excavation.

FIG. 1 — IN SITU CUTTING RATE VS LAB SPECIFIC ENERGY

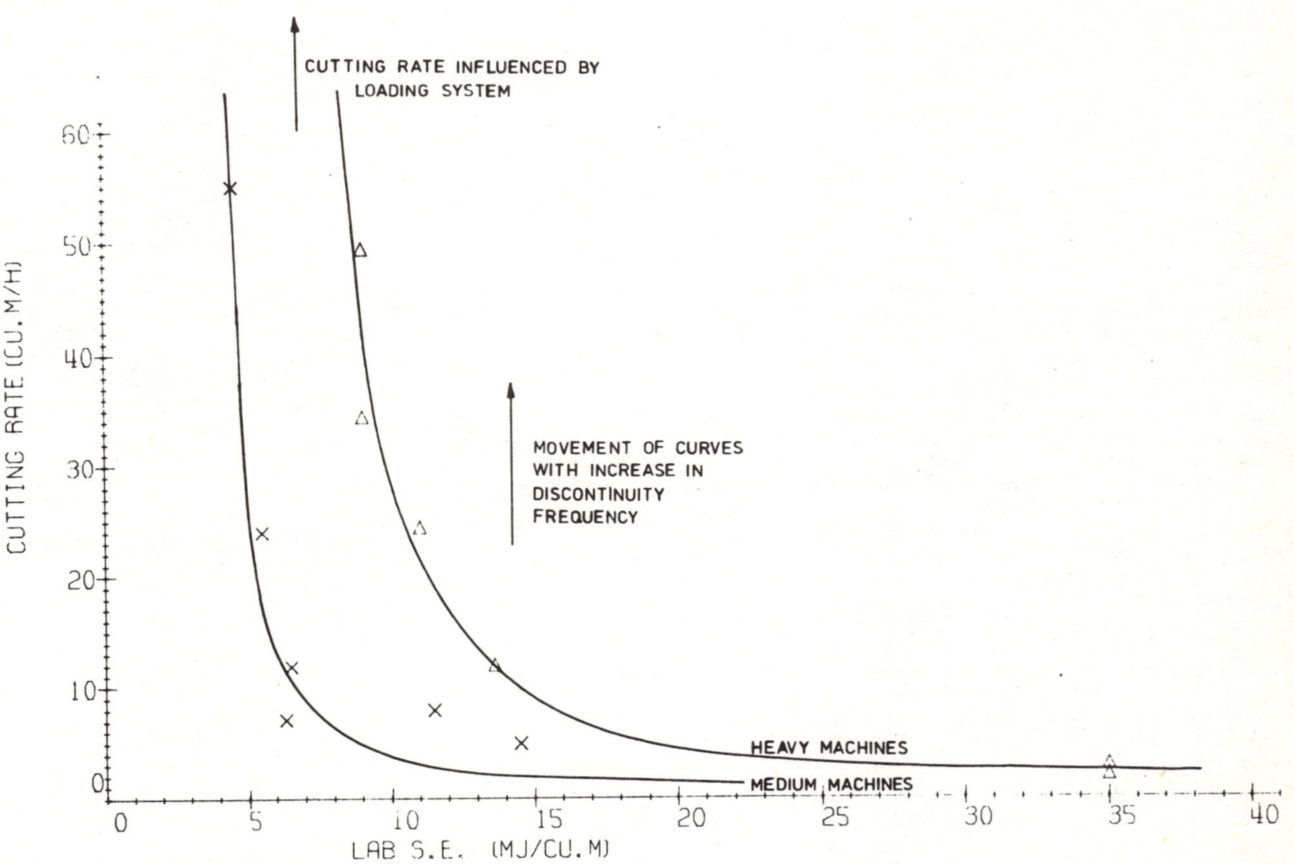

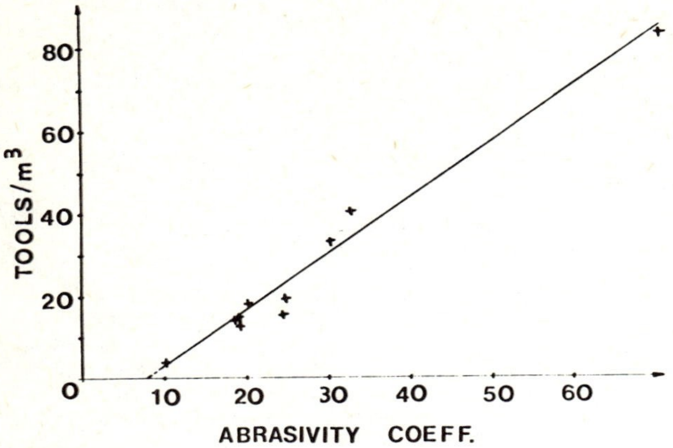

FIGURE 2.

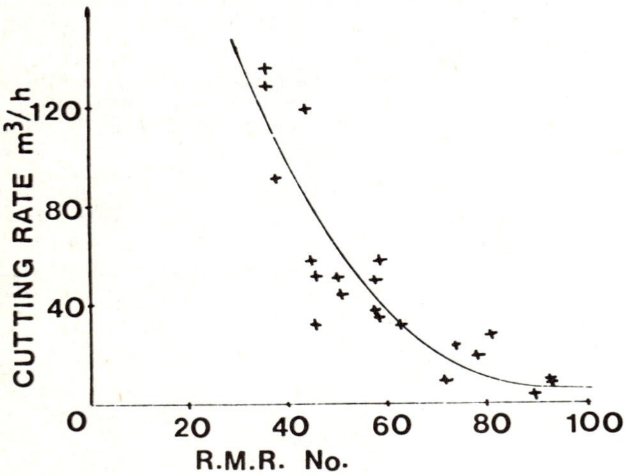

FIGURE 3.

SITE INVESTIGATION

It is considered of relevance that having outlined the factors influencing boom tunnelling machine performance a note should be added on the methods of obtaining the relevant information.

It is common practice to put down boreholes to obtain core samples of the strata to be excavated. These not only provide information on the lithologies, bed thicknesses and mechanical and physical properties, but also an indication of the discontinuities present in the rock mass. Here a note of caution must be introduced. Site investigations have two functions when considering machine drivage; one, to provide information on strata support requirements and two, to provide data for the estimation of machine application and performance. For support the worst conditions in terms of stability need to be identified, for machine tunnelling the most difficult condition is usually found to be the opposite of those required for strata support. Massive face conditions where the machine gains no assistance from the discontinuities in the rock mass may be good for support but provide difficulties for a boom tunnelling machine.

RQD (Rock Quality Designation) is often used to describe the quality of the rock mass for support requirements; care is required in its interpretation for tunnelling machine applications. The question that always must be asked is how was the RQD defined for use in the site investigation? Cores, by their nature, are prone to failure during the drilling process and subsequent handling. Careful inspection of the failure surfaces is required to determine those fractures occurring naturally in the rock mass from those induced by the extraction process. Independent logging of the fractures from a machineability point of view is recommended so that such influences as micaceous laminations that may well influence cutting can be accounted for. Fracture surveys of the borehole walls may be employed in marginal application cases.

CONCLUSIONS

The factors influencing boom tunnelling machine performance have been presented and a method of predicting machine performance in terms of cutting rate and tool consumption from intact rock cutting and abrasive properties moderated by the influence of discontinuities in the rock mass for different classes of machine. For overall project assessment, a Utilisation Factor must also be determined and the parameters that influence this factor have been given.

ACKNOWLEDGEMENTS

The authors wish to thank the management and men of collieries in the North East Area, National Coal Board, and Civil Engineering Contractors and Consultants throughout the country who have given access to machines and generous assistance during the course of these investigations. The financial support of the National Coal Board is gratefully acknowledged.

REFERENCES

Bieniawski Z.T.: "Engineering Classification of Jointed Rock Masses" Trans.S.A.Instn. of Civ. Engrs., Vol.15, No.12, Dec. 1972.

Bourgard et al.: "Propositions Relatives aux Mesures et Essais a Effectuer Dans Le Cadre D'un Chantier De Creusement Mechanique". Tunnels et Ouvrages Souterrains No.5, Sept/Oct., 1974.

Fowell R.J. and Pycroft A.S.: Rock Machineability Studies for the Assessment of Selective Tunnelling Machine Performance. 21st U.S. Symp. Rock Mechs., Rolla, Missouri.

Gehring K.: "Anpassung von Teilschnittmaschinen an durch von Gebirge und Bauwerk Vorgegebene Bedingungen." Proc. 4th Congr. of I.S.R.M. Montreux, 1979.

McFeat-Smith I.: "Effective and Economic Excavation by Roadheaders". Tunnels and Tunnelling, pp.43, Jan. 1978.

McFeat-Smith I. and Fowell R.J.: "Correlation of Rock Properties and the Cutting Performance of Tunnelling Machines". Proc. Conf. Rock Engr., pp.581-602, University of Newcastle upon Tyne, 1977.

Matusek Z and Vasek J.: "Untersuchungen Über den Einsatz von Streckenvortnebsmaschinen beim Nachreiben von Gestein." Glükauf-Forschungshafte, Oktober 1976.

SIZE AND SHAPE EFFECT IN PILLAR DESIGN

Y. Tsur-Lavie and S.A. Denekamp

Geotechnical and Mineral Engineering Department
Technion, Haifa, Israel.

SUMMARY. Size and shape effect in pillars is approached theoretically using energy considerations. Size effect is quantitatively expressed by a relation between load at failure (P) and cross sectional area (A): $P \propto A^n$, where n expresses size effect and has a theoretical range of 0.75 to 1.00. Size effect as expressed by empirical pillar strength formula is found to correspond to the above limits. Energy considerations are also found to provide a partial explanation for shape effect. Empirical pillar strength formula point at a definite relationship between size effect and shape effect. This relationship is explained by theoretical considerations linking both size and shape effect to fundamental material properties of the rock.

INTRODUCTION

Pillar design may be based on laboratory and field testing of large and small rock samples and on the analysis of pillar failures in mines. Laboratory testing of small rock samples is economical and expedient. However, the smaller the samples tested the larger the significance of size effect in the true interpretation of results in terms of the strength of full sized pillars. Hence the considerable importance of size effect in pillar design.

Weilbull's (1939) "Weakest link theory" attributes failure to statistically distributed flaws or inhomogeneities. The number of flaws and hence also the probability of failure increases with the volume of the rock body. Mathematically the theory is expressed by a linear relationship between log σ and log V where σ and V represent uniaxial compressive strength and volume of the tested rock body. Though many have contested Weilbull's theory as a physical explanation of size effect, it frequently fits experimental data, and has been widely used.

More recent theories of size effect link unit strength with the cross sectional area of the rock body (Reichmuth 1968, Brook 1977) giving a relationship of the form of

$$P = C \times A^n \qquad (1)$$

where

P = load at failure

A = characteristic cross section of rock specimen

C,n = constants.

Pointload test data plotted by Brook (1974) for specimens of varied shape (discs, blocks, cores, cubes) show a linear relationship between log A and log P where A represents the cross sectional area of rock specimens between loading points. Brook suggests, on basis of energy considerations, that n-values should approximate 0.75. Variations from this value were attributed by Brook to the influence of shape and rock type. Energy considerations were also used by Wagner and Schümann (1971) for the analysis of stampbearing test results.

Tsur-Lavie and Denekamp (1982) proposed a theoretical basis for size effect which resembles the energy approach as suggested by Brook for pointload tests and by Wagner and Schümann for stampbearing tests. Its formulation is however much broader and is therefore applicable to a range of mechanical testing techniques. In the present paper existing empirical strength formula for pillars, are analysed in view of the above mentioned theoretical approach to size effect. An attempt is also made to use an energy approach to shape effect.

THEORETICAL APPROACH TO SIZE EFFECT

Size effect denotes a decrease in unit strength accompanying an increase in size of rock bodies of equal shape. An explanation of size effect may be based on considerations of elastic energy release as applying to two extreme modes of failure.

a. Failure along a well defined plane, within the rock body expending all stored elastic strain energy through rupture along this plane (perfectly brittle behaviour).

b. Failure involving a volume of material (perfectly elastoplastic behaviour).

Considering failure mode a, the energy stored within a strained elastic body (U) may be expressed as:

$$U = \sigma \cdot \varepsilon \cdot V \tag{2}$$

where:

ε - strain

V - volume of rock body.

Expressing strain by stress and modulus of elasticity (E) we obtain:

$$U = K \cdot \frac{\sigma^2}{E} \cdot a^3 \tag{3}$$

where:

a - linear dimension

K - constant expressing shape.

It may be assumed that expended energy per unit area of failure plane (U_p) is constant and independent of size in equally shaped rock bodies:

$$U_p = \frac{U}{A} = K \cdot \frac{\sigma^2}{E} \cdot a \tag{4}$$

Expressing stress by force (P) and area (A) we obtain

$$P = C \cdot A^{0.75} \tag{5}$$

Considering failure mode b, involving a volume of material, the energy consumption per unit volume may be assumed to be constant and independent of the size of the rock body. Therefore:

$$U_p = \frac{U}{V} = K \cdot \frac{\sigma^2}{E} \tag{6}$$

Expressing stress by force and area we obtain:

$$P = C \cdot A^{1.00} \tag{7}$$

The actual mode of failure may be assumed to lie between the two extreme modes of failure. Size effect may therefore be expressed by an equation similar to that used by Reichmuth and Brook (Eq. 2).

$$P = C \cdot A^n \tag{8}$$

where n ranges between 0.75 and 1.0. Theoretically $n = 0.75$ would represent maximal size effect whereas $n = 1.00$ would represent lack of size effect.

Analysis of published results by Tsur-Lavie and Denekamp (1982) for uniaxial, triaxial, shear, Brazilian, pointload and stamp-bearing tests show that size effect indeed ranges between $n = 0.75$-1.00. Maximal size effect, expressed by n-values closer to 0.75 were found for pointload tests and Brazilian tests which are associated with the creation of single failure planes. Smaller size effects were found for uniaxial compression tests, which are associated with a more complex mode of failure, frequently involving also crushing of material, these observations also support the theoretical approach as presented above.

THEORETICAL APPROACH TO SHAPE EFFECT

Shape effect in pillars denotes a difference in unit strength in pillars of different shape but of equal cross section.

A change in mode of failure is one apparent cause of shape effect. Slender pillars tend to fail by means of a limited number of fractures. For wide pillars the geometrical feasibility of the development of single continuous fractures planes is more limited and failure of the pillar must be by crushing of the pillar material. Hence an increase in pillar strength. Energy considerations may provide an additional explanation to shape effect. Where failure is by means of single fractures it is assumed that a certain amount of stored elastic energy per unit area of potential failure plane must be exceeded for failure to occur. The stored elastic energy of a pillar of height (h) may be obtained from equations (3) and (4). From these equations the following relation between pillar strength and pillar height may be obtained

$$\sigma = B \times h^{-0.5} \tag{9}$$

where B = constant.

When failure takes place through crushing of a volume of material, equations (3) and (6) yield the following relation for pillar strength,

$$\sigma = B \times h^0 \tag{10}$$

For this case therefore no shape effect exists.

It follows from the above considerations that the tendency of a material to develop single fractures is an important contributing factor to shape effect. A strong relation between shape effect and brittleness of the rock material would therefore be expected to exist.

EMPIRICAL EXPRESSIONS OF PILLAR STRENGTH

Formula for the prediction of pillar strength are based on the analysis of pillar failures (mostly in coal mines) and on mechanical testing of large rock samples. Formulas of two forms are in frequent use.

$$\sigma_p = \sigma_c [A + B \left(\frac{W}{h}\right)] \tag{11}$$

$$\sigma_p = K \frac{W^a}{h^b} \tag{12}$$

where:

σ_p = pillar strength

σ_c = uniaxial compressive strength of a cube

W = pillar width

h = pillar height

A,B = constants given in units of pillar strength, their sum (A+B) equals uniaxial compressive strength of a cube.

a,b = dimensionless constants

K = constant. For $a = b$ K equals the uniaxial compressive strength of a cube of unit dimensions.

Table 1 summarizes published expressions of pillar strength given in the form of Eq. (11).

Table 1. Constants A and B in Eq. (11).

Source	Formula	W/H
Bunting (1911)	$0.700 + 0.300\frac{w}{h}$	0.5 - 1.0
Obert et al. (1960)	$0.778 + 0.222\frac{w}{h}$	0.5 - 2.0
Bieniawski (1968)	$0.556 + 0.444\frac{w}{h}$	1.0 - 3.1
Van Heerden (1973)	$0.704 + 0.296\frac{w}{h}$	1.14-3.4
Sorensen & Priseau (1978)	$0.693 + 0.307\frac{w}{h}$	0.5 - 2.0

Equation (12) may be written in the form:

$$P = K \times (\frac{w}{h})^b \times A^{\frac{2+a-b}{2}} \qquad (13)$$

In this equation $\frac{2+a-b}{2}$ represents size effect since it applies to cross section A, and b represents shape effect since it applies to the shape of the specimens $(\frac{w}{h})$. Comparison between equations (8) and (13) gives:

$$n = \frac{2+a-b}{2} \qquad (14)$$

Table 2 represents a and b values used by various authors in expressions of pillar strength given in the form of Eq. (12) and n values computed by means of Eq. (14).

Table 2. Constants a and b used by different authors and corresponding n-values.

Source	Formula	a	b	n
Streat (1954)	$\sigma = kh^{-1.00}w^{0.5}$	0.5	1.00	0.75
Holland-Gaddy (1962)	$kh^{-1.00}w^{0.5}$	0.5	1.00	0.75
Greenwald et al. (1939)	$kh^{-0.83}w^{0.5}$	0.5	0.833	0.833
Hedley & Grant (1972)	$kh^{-0.75}w^{0.5}$	0.5	0.75	0.875
Salamon & Munro (1967)	$kh^{-0.66}w^{0.46}$	0.46	0.66	0.90
Bieniawski (1968)	$kh^{-0.55}w^{0.16}$	0.16	0.55	0.805
Morrison et al.	$kh^{-0.5}w^{0.5}$	0.5	0.5	1.00
Zern (1926)	$kh^{-0.5}w^{0.5}$	0.5	0.5	1.00
Hazen & Artler (1976)	$kh^{-0.5}w^{0.5}$	0.5	0.5	1.00
Holland (1956)	$kh^{-0.5}w^{0.5}$	0.5	0.5	1.00

Table 3 shows the relative weight of shape effect in the different pillar strength formula. A comparison is made between strength values for pillars with a w/h ratio of 0.5 and for pillars with a w/h ratio of 1.0. Table 3 demonstrates that differences in predicted strength between pillars of different shape, as given by formulas in the form of Eq. (11) are larger than those given by formulas in the form of Eq. (12). They are therefore less conservative.

Table 3. Comparison of shape effect in different pillar strength formula.

Eq. (11)	$\frac{C_p}{C_c}$	Eq. (12)	$\frac{C_p}{C_c}$
Bunting	0.85	Greenwald et al.	0.56
Obert et al.	0.89	Hedley & Grant	0.59
Bieniawski	0.78	Salamon & Munro	0.63
Van Heerden	0.85	Bieniawski	0.68
Sorensen et al.	0.85	Hazen & Artler	0.71

DISCUSSION

Pillar strength formula in the form of Eq. (12) express both size effect (n) and shape effect (b). Table 2 shows that n values vary between 0.75 and 1.00 which agrees with the theoretically predicted range of n-values. Empirical pillar strength values obtained from analysis of pillar failure in mines show relatively small size effects, ranging between 0.875 - 1.00 (Hedley and Grant 1972, Salamon and Munro 1967, Holland and Morison 1956). Empirical values obtained from testing large model pillars show higher values of shape effect ranging between 0.805 - 0.833.

Table 2 shows values of shape effect (b) ranging between 0.5 - 1.00 versus a much smaller range of b values as based on theoretical energy considerations alone (0.0 - 0.5). The larger empirical values underline the fact that apart from energy considerations, the varying mode of failure in differently shaped pillars also contributes significantly to shape effect.

Values for b and n presented in Table 1 are plotted in Fig. 1. The results point at a distinct relationship; the larger the shape effect (expressed by b) the larger also is the size effect (expressed by n). The relationship seems to indicate that both shape and size effect depend on a common fundamental material property. This agrees with theoretical considerations linking both size and shape effect to brittleness.

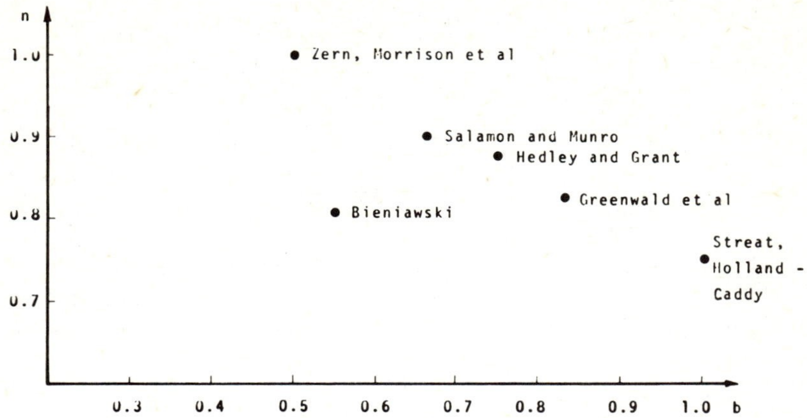

Fig. 1: Size effect versus shape effect

REFERENCES

Bieniawski, Z.T., 1968, "The effect of specimen size on compressive strength of coal", Int. J. of Rock Mechanics and Mining Sciences, 5, 325-335.

Brook, N., 1974, "The use of irregular specimens for rock strength tests", Int. J. of Rock Mechanics and Mining Sciences, 14, 193-202.

Bunting, D., 1911, "Chamber-pillars in deep anthracite mines", Trans. Am. Inst. of Mining Engineering, 42, 235-245.

Greenwald, H.P., Howarth, H.C., Hartman, I., 1939, "Experiments on the strength of small pillars of coal in the Pittsburg bed", U.S. Bureau of Mines Technical Paper 605, 22.

Hazen, G., Artler, L., 1976, "Practical coal pillar design problem", Mining Congress Journal, June. 86-92.

Hedley, D.G.F. and Grant, F., 1972, "Stope and pillar design for the Elliot Lake uranium mines", CIM Bulletin, July, 37-44.

Holland, C.T., 1956,"Some aspects of permanent support of overburden on coal beds", Proc. West Virginia Coal Mining Inst., 45-56.

Holland, C.T., 1962, "Design of pillars for overburden support", Mining Congress Journal, March, 24-28.

Hudson, J.A., Brown, E.T. and Fairhurst, C., 1971, "Shape of the complete stress-strain curve for rock," 13th Symp. on Rock Mechanics, 773-795.

Morrison, R.G.K., Corlett, A.V. and Rice, H.R., "Report of the special committee of mining practices at Elliot Lake", Ontario Dept. of Mines, Bulletin 155, 85.

Obert, L., Duvall, W.I. and Merril, R.H., 1960, "Design of underground openings in competent rock", U.S. Bureau of Mines, Bulletin 587.

Reichmuth, D.R., 1968, "Point load testing of brittle materials to determine tensile strength and relative brittleness", 9th Symp. on Rock Mechanics, 134-159.

Salamon, M.G.D. and Munro, A.H., 1967, "A study of the strength of coal pillars", J. of the South African Inst. of Mining and Metallurgy, 68, 55-67.

Singh, B., 1980, "Rock mechanics investigation in India, practical application in coal mines and tunnels", 21st Symp. on Rock Mechanics, 806-819.

Sorenson, W.K. and Pariseau W.G., 1978, "Statistical analysis of laboratory compressive strength and Young's modulus data for the design of production pillars in coal mines", 19th Symp. on Rock Mechanics, 30-37.

Streat, F.A., 1954, "Strength and stability of pillars in coal mines", J. Chemical Metallurgical and Mining Soc. of South Africa, 54, 309-325.

Tsur-Lavie, Y., and Denekamp, S.A., 1982, "Comparison of size effect for different types of strength tests", Rock Mechanics (accepted for publication).

Van Heerden, W.L., 1974, "In situ determination of complete stress-strain characteristics of 1.4 m. square specimens with width to height ratios up to 3.4", CSIR Research Report, ME 1265, 30.

Wagner, H., Schümann, E.H.R., 1971,"The stamp-load bearing strength of rock, an experimental and theoretical investigation", Rock Mechanics, 3, 185-207.

Weibull, W., 1939, "The phenomenon of rupture in solids", Ingvetensk Akad. handli, No. 153.

Zern, E.N., 1926, "Coal Miners Pocketbook", 11th edition, New York, McGraw-Hill.

DESIGN OF HIGH EXTRACTION PANELS IN A DEEP POTASH MINE

R.F. Cook

SUMMARY. The paper describes a rock mechanics investigation into the 'design' of narrow high extraction pillars in a deep potash mine and the problems associated with such pillars.

INTRODUCTION

High extraction mining of potash using 'narrow' panels consisting of a number of roadways separated by small pillars which are allowed to yield has been successfully applied in Canadian mines at depths of upto 1100m. Recently, the technique has been adopted in U.K. for the extraction of a potash seam situated below 1200m under geological conditions, different from those existing in the Canadian Potash mines.

The basic concept behind the yield pillar technique concerns the redistribution of stresses away from the worked-out panels and onto adjacent barriers. The redistribution is brought about by the mining of intermediate pillars within the panels sufficiently narrow that they are unconfined and yield under the high uniaxial loads resulting from the overburden. Yielding of the pillars within a panel creates large deviatoric stress conditions above and below the abutments where shear failure will occur. The stresses in the immediate roof and floor strata are thereby relieved and a destressed zone is created which will protect roof and floor conditions. The success of the technique at an underground location will depend upon a number of factors including the mining layout, sequence and rate of mining operations which are all factors within the control of mine planning and geological and rock mechanics factors which are constraints outside of any control.

This paper describes a rock mechanics investigation concerned with the design of 'narrow' high extraction panels for the mining of potash under the geological conditions existing in the U.K. Firstly, the potential rock mechanics problems that could affect mining are identified and then the geometry for high extraction panels which minimises the effects from the rock mechanics problems is established using linear elastic theory for far seam strata and the concepts of soft rock mechanics as proposed by Wilson (1977) for the near seam strata. The design approach for the high extraction mining panels is necessarily simplistic, treating the strata as possessing idealised mechanical properties based upon laboratory measurements and assumed boundary conditions. However, the restraints imposed upon mining require that attempts should be made to predict strata behaviour and estimate optimum opening dimensions.

POTENTIAL ROCK MECHANICS PROBLEMS IN HIGH EXTRACTION WORKINGS

The geology of the potash mine (see Fig. 1) has been described in detail by Woods (1973) and only geological conditions considered relevant to the rock mechanics problems associated with mining will be treated in this paper.

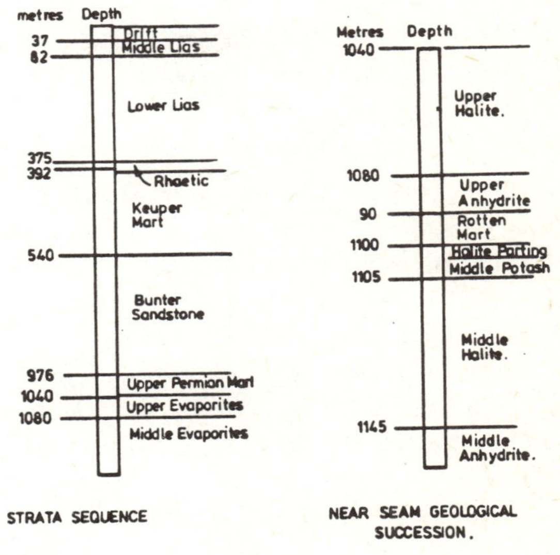

Figure 1. Strata Sequence

These for high extraction workings include:

1. Rapid deterioration of the rotten marl brought about through contact with air and moisture. The level of the marl/potash interface is variable and since it is uneconomical to reduce the height of extraction of the ore beyond certain limits the exposure of the rotten marl remains a potential source for poor local roof conditions.
2. Poor local roof conditions occur in areas of shaley potash ore due to its inherent low strength.
3. The water in the sandstone, some 125m above the workings represents a potential problem to mining of the potash. With the previously worked room and pillar system of mining in which the pillars were capable of accepting the full overburden, the subsidence above the mined area was uniform and gradual, being dependent upon the time-dependent movement of the pillars; also, fracturing, although extensive about roadways was localised. These factors ensured that the strata barrier between the water and the mine workings remained unbroken and impenetrable. The introduction of high extraction narrow panels increased the effective size of an opening and stress changes within the strata therefore affect a much greater area. Should the stresses exceed limiting material failure criteria then fracturing of strata will develop. Fracturing could provide pathways from the sandstone to the workings and enable water to enter the mine. Also the alternation of areas of high extraction with barrier pillars will induce strains within the strata which could lead to fracturing. The two thick strata between the sandstone and the workings are the Permian Marl and the Upper Halite. The marl is known to be fractured whereas the creep properties of the halite ensure that this, provided it is undisturbed, remains unfractured. The halite therefore, represents the most reliable barrier preventing water from entering the workings.
4. Due to environmental reasons, surface subsidence is strictly controlled in the mining area. This does not pose a direct problem since subsidence will be small owing to the small extraction thickness and the depth to the mine workings. However, shallow room and pillar mined-out ironstone workings are present in the area and large surface subsidence would result from pillar collapse in these workings. It is therefore, important that the shallow workings are not disturbed by tilting or differential settlements caused by the deeper workings.

DESIGN OF HIGH EXTRACTION WORKINGS

The design approach proposed in this paper for the high extraction workings is as follows:

1. estimate the induced stress distribution about an opening having overall panel dimensions using a linear elastic model.
2. determing the zones about the opening where deviatoric stresses exceed a specified failure criteria.
3. analyse the rock mass bounded by the 'failed' zones assuming that shear forces are mobilised along these boundaries in order to determine the required support pressure across the panel width to maintain static equilibrium.
4. use the post-failure characteristics of potash and the required panel support pressure to determine the internal panel dimensions including yield pillar width and roadway span.

The approach adopted for the design of the excavations in the deep potash deposit has ignored the traditional concepts concerning the creep behaviour of evaporites. The time-dependent behaviour of the rock has been considered as a combination of the time-dependent behaviour of the rock in a post-failure state together with the creep effect of the evaporite material caused by stress relief and stress gradient. Actual underground behaviour will depend upon prevailing conditions. Under low deviator stress states, the creep effects of the evaporite material will dominate overall time dependent behaviour, whereas under high deviatoric stress conditions as will exist around openings in deep evaporite mines, fracture of the rock will occur and behaviour will be controlled by the post-failure characteristics of the material.

Panel geometry

A two dimensional plane strain linear elastic finite element analysis has been used to establish the stress distributions and deformations caused by excavations within the potash seam. Material failure criteria have been applied to these results in order to estimate the extent of material failure about the openings. The studies have established the overall panel and barrier dimensions that ensure that surface settlements remain small and that the extent of the 'failed' zone above the openings does not provide pathways for the water in the sandstone into the workings.

The material parameters for the strata sequence were established by Patchet (1970) in an extensive laboratory test programme carried out on core samples. The material failure criteria under triaxial stress states are illustrated for the near seam strata in Fig. 2.

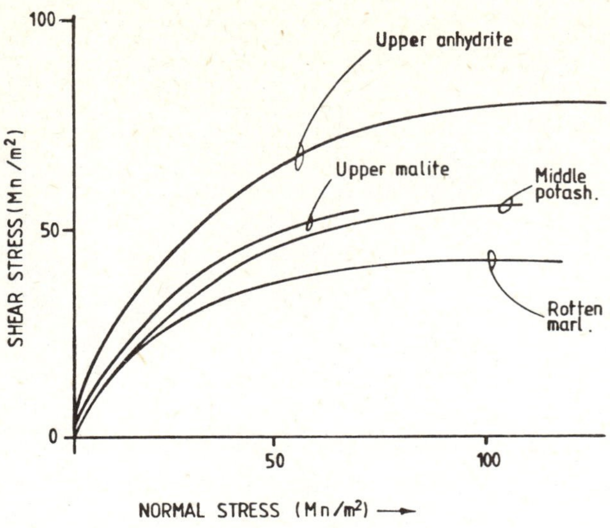

Figure 2. Mohr Envelopes for Near-Seam Strata (after Patchet)

The analyses indicated that in order to limit surface subsidence, maintain a barrier between the sandstone and the workings and minimise interaction between adjacent panels, panel widths should be a maximum of 50m wide and barrier pillars a minimum of 100m wide. With this geometry, the analysis indicates that surface subsidence extends laterally about 1500m from the edge of underground workings, giving an angle of draw of 53 degrees; maximum subsidence is about 0.1 times seam closure and near-seam tilt strains are negligible and their impact upon the near-surface ironstone workings is minimal. Material failure is expected upto about 50m above the openings which would allow a 10-15m barrier of intact halite.

Panel support pressures

Once the overall panel geometry has been established the panel support loads which will govern the internal geometry of the panel can be investigated. The zone identified in which material failure occurs can simply be considered as providing a dead weight to the yield pillars but this takes no account of the shear resistance that can be mobilised by a fractured or highly deformed material.

An alternative approach, is to consider the vertical stress distribution above a yielding base in a material for which frictional resistance is mobilised along the 'failed' zones identified from the linear elastic analysis. The basis for the method was proposed by Terzaghi (1943) and is shown in Fig. 3. The zone above the yielding base in which frictional resistance is mobilised is approximated to a rectangular wedge, bounded by the excavation roof and the 'failed'/intact material interface and by vertical shear zones starting from the edges of the excavation.

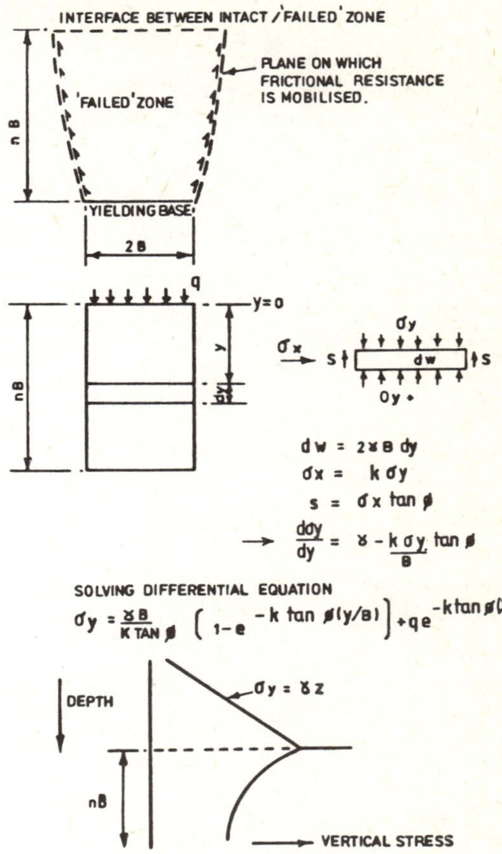

Figure 3. Frictional Model for Immediate Roof Strata

The static equilibrium within the wedge is then considered under the action of the shear and gravity stresses and the upper boundary condition. A simple Mohr Coulomb criteria has been adopted to determine the frictional resistance mobilised. The vertical stress distribution within the failed zone is given in Fig. 3.

The friction angle ϕ has been estimated as 45 degrees from laboratory triaxial testing by Farmer and Gilbert (1981) of the post-failure characteristics of halite which represents the thickest layer within the zone under consideration. K, which represents the ratio of the horizontal to vertical stresses within the zone has been assumed to equal sin ϕ. The vertical stress at the failed/intact interface which represents the upper boundary conditions for the problem is the overburden stress.

Using these parameters, the vertical stress distribution above a 50m wide opening in which shear stresses are mobilised along 'failed' zones upto 50m above the opening can be assessed. The results indicate a support pressure across the panel width of 7000 kN/m^2 and the yield pillars within a panel should be designed to provide such a load bearing capacity.

Yield pillars

The concept behind the design of a yielding pillar is based upon the ability of a material under load to sustain some level of load even after the peak strength of the material has been exceeded.

Laboratory tests were carried out by Hebblewhite (1977) to determine the post-failure characteristics of potash at various width/height ratio and test strain rates. The results which are shown in Fig. 4, have been used to determine the optimum dimensions for yield pillars and to estimate their load-bearing capacity.

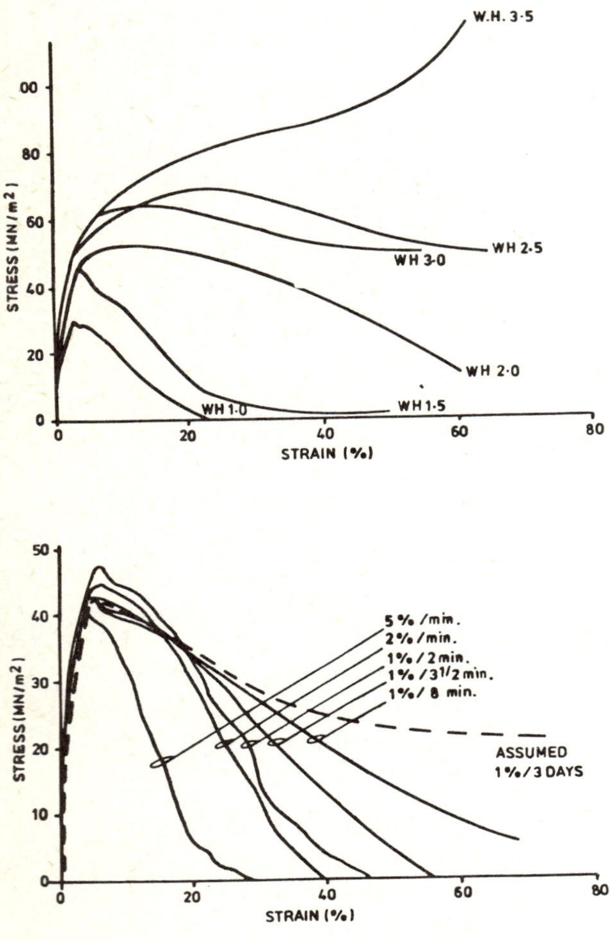

Figure 4. Complete Stress/Strain Curves for Potash (after Hebblewhite)

They indicate that samples with width/height ratios greater than 2.5 do not fail whereas samples having width/height ratios in the range 1.0-2.0 yield and possess sufficient residual strength capacity to carry loads for a period of time after failure. Tests on samples having a constant width/height ratio of 1.5 were carried out at different strain rates in order to evaluate the residual strength characteristics relevant to yielding pillars. The results indicate that at low strain rates, the post-failure load/strain curve is shallow implying that a high proportion of peak strength load can be carried and the deformation of the material can be large without total disintegration of the sample. At high strain rates, the sample rapidly loses its ability to sustain load and disintegrates at a relatively low strain.

Applying these results to the underground workings indicates that for a seam extraction height of 3.5m, the width of yielding pillars could be expected to lie within the range 3.5m-7.0m. In order to maximise production and maintain small width panels the smaller pillar dimension is more likely to be adopted. Long term panel stability is achieved by slow pillar strain rates which imply high residual strengths. Fig. 4. shows that for pillars with a width/height ratio of 1.5, the residual strengths are greater than 75% of the peak strengths for low strain rates and strains upto 25%. However, since an overestimate for the load-bearing capacity of a pillar will lead to fast strain rates and produce excavation instability it is important to limit the support loads applied to the pillars. On this basis the maximum load-bearing capacity of a pillar has been taken as 50% of peak pillar strength.

Roadway width

The maximum roadway width is established from the required panel support pressure (7000 kN/m^2) and the allowable design load capacity of the yield pillars. A 5.0m wide yield pillar can provide sufficient support for a maximum span of 14m which is equivalent to a roadway width of 9.5m. Obviously, the seam geology will dictate local roof conditions and since these are generally poor, roadway width may need to be reduced in order to maintain an integral potash beam above the roadway.

SUMMARY AND CONCLUSIONS

The geology at the U.K. potash mine has been shown to present difficult mining conditions with the main strata control problems being:

1. immediate roadway roof conditions which can deteriorate due to variable seam thickness and weak near- seam strata.
2. presence of water-bearing sandstone about 125m above the workings.
3. restrictions imposed upon surface subsidence for environmental reasons.

The initial room and pillar system of mining ensured that an impenetrable barrier to the migration of water was maintained. High local stresses however, were created about roadways which deteriorated due to the weakness of the near-seam rocks. This led to the reassessment of the mining method and the introduction of high extraction narrow panels with yielding internal pillars.

A design method for these high extraction panels has been outlined in this paper and has adopted a linear elastic analysis to describe far seam strata behaviour and a frictional model to determine near-seam strata behaviour. Material failure criteria have been used to designate zones of shear and the post-failure characteristics of the potash have been used to determine yield pillar dimensions. Interaction between adjacent panels and between workings and water bearing strata as well as estimates of surface subsidence have been evaluated.

The rock mechanics design has proposed the following limiting mine geometry; the maximum panel width to be 50m, with minimum barrier dimensions of 100m. The internal geometry of the panel has been estimated using an extraction ratio of 3.5m to give a yield pillar dimension of 5.0m with a roadway width of 9.5m. However, the variable potash seam thickness will also be significant in determining roadway spans and it is considered that optimum roadway width should be assessed further by underground observation.

The design has considered that the time-dependent movements of strata are primarily caused by fracture under high deviatoric stress conditions and has ignored the effects of 'creep' of evaporites. Creep would tend to reduce the induced deviatoric stresses about an opening and so gradually improve conditions and secondly, where shear had occurred, to close and heal fractures. Creep could therefore be beneficial although the effects would possibly be long term.

REFERENCES

Farmer, I.W. and Gilbert, M.J., 1981, "Time dependent strength reduction of rock salt". 1st Conf. Mech. Behaviour Rock Salt, Penn. State.

Hebblewhite, B., 1977, "Underground potash mine design based upon rock mechanics principles and measurements", Ph.D. Thesis, University of Newcastle upon Tyne.

Patchet, S.J., 1970, "Rock mechanics studies associated with the development of a deep potash mine", Ph.D. Thesis, University of Newcastle upon Tyne.

Terzaghi, K., 1943, "Theoretical Soil Mechanics", J. Wiley.

Wilson, A., 1977, "The stability of tunnels in soft rock at depth", Proc. of Conf. Rock Engineering, University of Newcastle upon Tyne.

Woods, P.J.E., 1973, "Potash Exploration in Yorkshire Boulby Mine Pilot Borehole", Trans. Inst. Mining and Metallurgy, Vol. 82.

A NUMERICAL PROCEDURE FOR ROCK PRESSURE PROBLEMS IN LEVEL SEAMS

P. R. Sheorey, M. N. Das and B. Singh

Central Mining Research Station
Dhanbad-826001, India.

SUMMARY. A numerical method is presented for estimation of rock pressure over workings in level seams. The method employs the finite difference equations describing the deflection of a plate supported elastically. For a realistic assessment of rock pressure distribution in a given situation, methods are given for calculating the input parameters defining the behaviour of the roof-seam contact plane, from underground observations. The procedure is proposed as a simpler alternative to other existing numerical techniques of a similar nature. A case study of shaft pillar stability and another of pillars in an irregular array are given as examples.

INTRODUCTION

Several strata mechanical problems involve the estimation of rock pressure over the seam. Barring a few stray cases where analytical closed-form solutions may be feasible, numerical methods are the only way out in practical mining situations. Some obvious examples of such geometries are: pillars in irregular array or in a panel whose width is comparable with the depth from surface, longwall panel abutments, shaft pillars and so on. The shear size and complexity of the geometry many times precludes the use of the finite element method and attempts have, therefore, been made in the past to formulate such problems of "in-seam" stress analysis so as to restrict considerations to the seam level. These methods were based on the theory of thin slots in infinite media, Berry and Sales (1962), harmonic functions in the theory of elasticity, Salamon (1964) and Thompson (1975) and Volterra's theory of dislocations in elastic media, Sinha (1979).

An alternative numerical method is given here for problems which involve the estimation of rock pressure over the seam only. This method is in no way meant to detract from the value of the other past methods, but is proposed as a simpler alternative which can give satisfactory solutions.

THE THEORY OF PLATES FORMULATION

Although "discretisation" of the plate deflection equation into the finite difference form is not a new method, it needs to be considered in the light of in-seam problems. The mathematics of the finite difference method, and hence programmability, are simpler than other methods.

In a horizontal seam (dip $< 20^o$), the middle plane of the seam can be taken to remain plane after excavation. In other words, the displacements about the middle plane are symmetrical, making the shear stresses τ_{xz} and τ_{yz} zero on the middle plane, z-axis being vertical in the coordinate system (x,y,z). If the strata are modelled as a plate supported elastically by Winkler foundations (seam or goaf materials), the underlying assumption is that the shear stresses at the roof-seam contact plane are zero rather than at the middle plane. This is tantamount to assuming that the seam thickness can be neglected when compared with the depth from surface.

The shear stresses, which vary through the thickness of a plate, twist the originally plane cross-sections of the plate, causing the vertical displacements to vary through the thickness. The vertical deflection at any point at a height z is thus a sum of the deflections due to bending and twisting (or shear strain). At the roof-seam contact, the shear stresses being zero, the shear strain is zero. So the vertical deflection of the plate is expected to be due to bending alone at the roof-seam contact, Sheorey (1975). Such deflections are described by the thin plate equation, whose finite difference form (for elastically supported plates) is

$$\left(\frac{k_i \lambda^4}{D} + 20\right) w_i - 8(w_{i+1} + w_{i-1} + w_{i+m} + w_{i-m}) + 2(w_{i+m+1} + w_{i+m-1} + w_{i-m+1} + w_{i-m-1}) + w_{i+2} + w_{i-2} + w_{i+2m} + w_{i-2m} = \frac{\lambda^4 q}{D} \qquad (1)$$

in which w_i is the deflection of a node i in a grid on the roof-seam contact plane, Figure 1. The other notations are:

q = loading on the plate i.e. cover pressure which is not necessarily unfform,

k_i = nodal stiffness,

D = flexural rigidity of the strata,

m = number of nodes in a row along x ,

λ = size of an "element" or distance between two adjacent nodes.

Boundary conditions

The condition away from working areas is

$$w_i = w_v = \frac{q}{k_v}$$

where the suffix v refers to virgin seam conditions. The mesh is generally laid so that this condition is satisfied at nodes sufficiently far away from the area of interest. Even if unworked portions of the seam are not available at the boundary nodes, virgin conditions can still be assumed so long as these nodes are

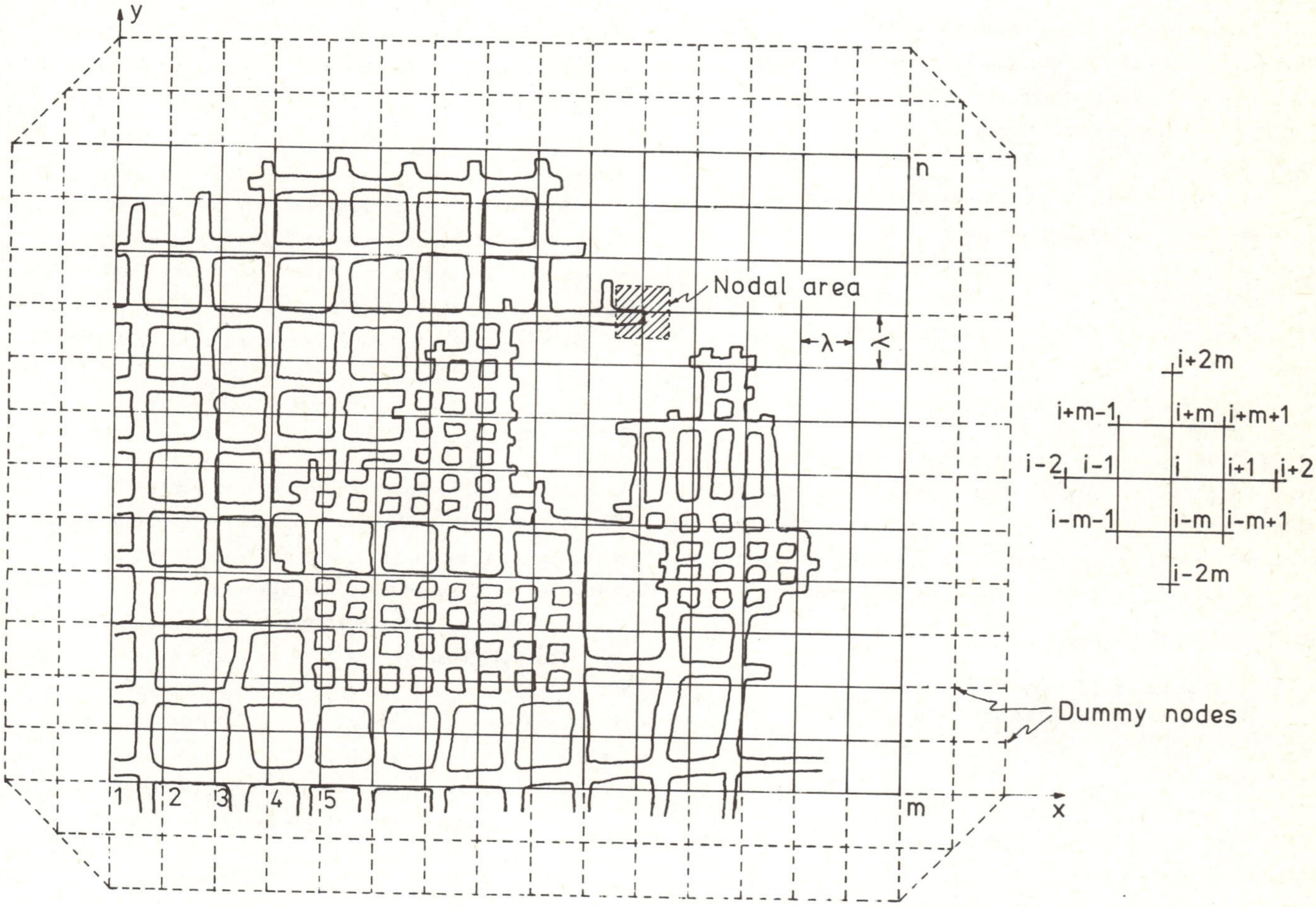

SCHEME OF A FINITE DIFFERENCE GRID

Figure 1

The grid is generally rectangular in shape and consists of equal square elements. Alternatively, rectangular elements may also be used changing Eq.(1), but this will be at the expense of accuracy in one direction.

For all the nodes n within the grid we get n simultaneous linear algebraic equations of the form of Eq.(1) giving nodal displacements w_i upon solution. The pressures are then given by $p_i = k_v w_i$ or $k_s w_i$ as the case may be, where k_v is virgin seam stiffness and k_s is goaf stiffness as defined later.

sufficiently far away.

The other condition frequently applicable is that of symmetry. In this case the deflections w on either side of a symmetric axis are equal at distances equal from the axis.

For the application of these boundary conditions, two rows of dummy nodes are also laid out all around the grid, Figure 1. The deflection equation (1) are simply modified for the boundary nodes while forming the matrix of equations.

Computer procedure

The nodes in a grid are numbered consecutively row by row. Since the matrix band width is (4m + 1), where m is the number of nodes in a row along x, choosing the shorter side of the grid as the x-axis will reduce the band width. But this is immaterial since the band width is always made up of only 13 non-zero matrix elements for any size of problem and it is necessary to store only these non-zero elements. Another property of the matrix is that it is amenable to solution by an interation method, being diagonally dominant.

The present computer program consists of generating the matrix, storing it peripherally with the known right hand vector and solving by the Gauss-Seidel iteration method. Since only one row is dealt with at a time in iteration, this procedure requires very low central memory on the computer. Large problems can therefore be solved on relatively small machines.

The solution generally converges in about 15 iterations or less without over-relaxation if

$$\lambda^4 \geqslant \frac{6D}{k_v}$$

where k_v is the virgin seam stiffness. This puts a lower limit on the element size λ and mesh formation should consider this fact. If λ is smaller than this over-relaxation is necessary, otherwise the solution may not converge even after hundreds of iterations.

DETERMINATION OF NODAL STIFFNESS

AND FLEXURAL RIGIDITY

The following formulae are used for calculating nodal stiffness for various nodal locations.

Working area

$$k_i = \frac{2E_c}{h(1-\mu_c^2)} \quad (1 - e_i) \tag{2}$$

where E_c, μ_c = elastic constants of the seam,
h = working height in the seam,
e_i = extraction ratio within the nodal area i.

Virgin seam

$$k_i = \frac{2E_c}{h(1-\mu_c^2)} = k_v \tag{3}$$

since $e_i = 0$.

Goaf

$$k_i = \frac{2E_s}{h} = k_s \tag{4}$$

with E_s = compressibility of the goaf material.

Multilift extraction in a thick seam

$$k_i = \frac{E_c/(1-\mu_c^2)}{\frac{h_1}{1-e_i^1} + \frac{h_2}{1-e_i^2} + \ldots + \frac{h_n}{1-e_i^n}} \tag{5}$$

where h_n and e_i^n denote the n th lift.

Nodal areas partly over the goaf (e.g. stowed pack) and partly over coal are assigned stiffness values assuming no goaf material since the seam will offer much greater resistance than the goaf material at the goaf edge.

The elastic constants of the seam E_c, μ_c can be determined experimentally in situ.

Determination of the flexural rigidity D of the roof strata is much more difficult since a whole range of factors like order of stiffness of the roof beds, cohesion between beds, depth of cover, joints are likely to influence the flexural behaviour. In fact the term flexural rigidity in the sense of a plate is a misnomer because here it is really meant to define the deflection behaviour of the roof-seam contact plane.

To eliminate uncertainties and to simulate realistic roof behaviour, the value of D can be directly estimated from underground observations. The distance L at which the displacements and stresses due to the influence of a face become zero is related to D by

$$L = \pi \left(\frac{k_v}{4D} \right)^{-\frac{1}{4}}$$

according to the theory of beams (and plates) on elastic support. This gives us the value of D since k_v is known from Eq.(3). The distance L can be conveniently measured, if not already available, using stress-metres or extensometers.

The stiffness of the goaf material k_s can be determined from Eq.(4) measuring the compressibility E_s of the stowing material in the laboratory. In caving this is not feasible. Whatever the goaf material, the better way to obtain k_s is by measuring L_s, the distance to which face influence travels on the goafside. Then the relation

$$L_s = \pi \left(\frac{k_s}{4D} \right)^{-\frac{1}{4}}$$

will yield the value of k_s.

Using these methods to obtain the various mechanical parameters involved, this numerical technique should give reliable results.

EXAMPLES

Two examples which made use of the numerical method outlined are given here with results.

Jitpur Colliery

The shaft pillar area, which is at a depth of 450 m at this mine, is 240 x 240 m approximately with two panels taken by three-lift longwall with hydraulic stowing and one panel by single lift surrounding the shaft pillar area. The seam is the 8.8-m thick XIV seam of Jharia coalfield. Development within the shaft pillar has been done for essential roadways,

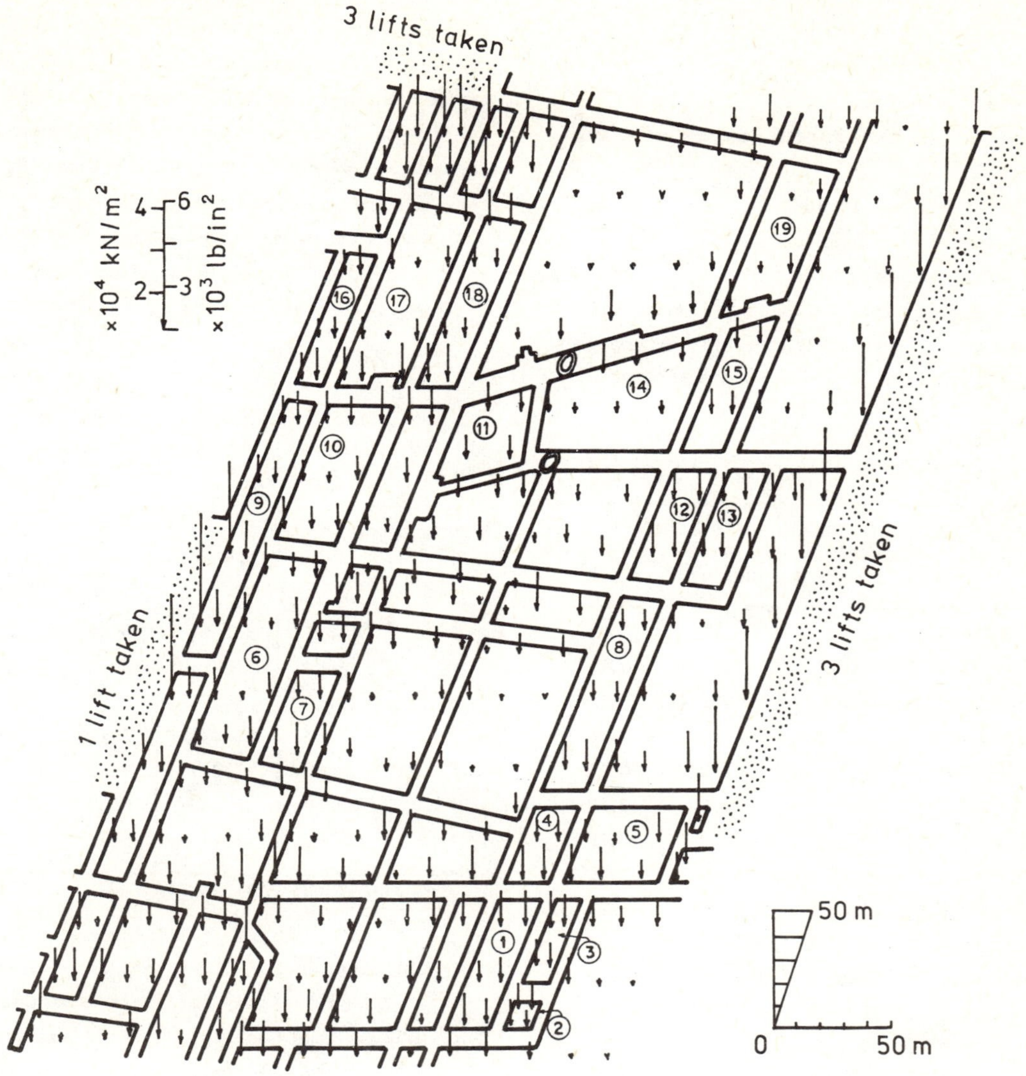

ROCK PRESSURE DISTRIBUTION OVER SHAFT PILLAR AREA AT JITPUR

Figure 2

coal bunker, power substation etc. The shaft pillar area in most parts has been exhibiting high rates of convergence, some roof falls, floor heaving and side spalling.

A grid of 546 nodes (21 x 26 nodes, λ = 18 m) was laid for obtaining the rock pressure distribution over the shaft pillar area. The results are shown in the isometric plot in Figure 2. The safety factors of individual pillars, numbered 1 to 19, were estimated from the rock pressure values, and using the following formulae for pillar strength, Wilson (1972):

(a) when $W_1 > 0.003\ hH$

$$S = \frac{4\gamma H}{W_1 W_2}\left[W_1 W_2 - 1.5\,(W_1 + W_2)\,hH \times 10^{-3} + 3\,h^2 H^2 \times 10^{-6}\right] \text{ tons/ft}^2 \tag{6}$$

(b) when $W_1 \leqslant 0.003\ hH$

$$S = \frac{1333\gamma W_1^2}{W_1 W_2 h}\left(\frac{3W_2 - W_1}{6}\right) \text{ tons/ft}^2$$

and, Sheorey et al. (1980):

$$S = RK\left(\frac{4A}{C}\right)^{0.46} / h^{0.66} \text{ lb/in}^2 \tag{7}$$

where W_1, W_2 = mean pillar dimensions in ft, $W_1 < W_2$,
H = depth from surface, ft,
h = working height, ft, = 12 ft (3.6 m),
γ = rock density, 0.0707 tons/ft^3,
R = long-term strength factor,
K = mean in situ strength of 1-ft cube of seam,
= 445 lb/in^2 (3075 kN/m^2),
A = pillar plan area, ft and
C = perimeter of pillar, ft.

The safety factors calculated using Eq.(6) and (7) and the pressure values from the computer run are given in Table 1. The wide disparity in the last two columns is apparent. Since Eq.(7) is known to underestimate

Table 1. Safety factors of pillars in the shaft pillar area at Jitpur (Figure 2)

Pillar No.	Mean size m x m	Average pressure MN/m^2	Safety factor	
			Eq.(6)	Eq.(7)*
1	18.0 x 34.5	17.75	1.17	0.25
2	10.5 x 12.0	16.23	0.64	0.18
3	12.0 x 21.0	17.82	0.77	0.20
4	18.0 x 21.0	16.16	1.11	0.25
5	21.0 x 39.0	15.33	1.52	0.31
6	28.5 x 48.0	13.74	2.02	0.39
7	18.0 x 25.5	17.13	1.12	0.25
8	18.0 x 48.0	15.81	1.39	0.29
9	15.0 x 63.0	20.51	0.95	0.22
10	28.5 x 34.5	15.40	1.50	0.33
11	19.5 x 30.0	16.64	1.27	0.27
12	18.0 x 30.0	18.23	1.10	0.24
13	18.0 x 30.0	17.13	1.18	0.25
14	24.0 x 60.0	17.68	1.52	0.29
15	18.0 x 31.5	14.92	1.36	0.29
16	15.0 x 33.0	15.81	1.13	0.26
17	30.0 x 43.5	14.23	1.96	0.38
18	18.0 x 42.0	15.68	1.37	0.29
19	27.0 x 33.0	13.12	1.92	0.38

* Long-term strength factor R was neglected.

the strength of flat pillars ($W_1/h > 5$), such low safety factors can be expected, especially since the in situ coal strength is poor. On the otherhand it would also seem that Eq.(6) (which indicates stability when the safety factor is 1.0 or greater) somewhat overestimates the strength of these pillars, because there is physical evidence like severe spalling, as well as some experimental evidence from borehole extensometers that the pillars were practically completely crushed. It is felt that there is a need to develop a reliable formula for strength of flat pillars.

Pindra Colliery

Pindra Khas seam at this mine is 4.2 m thick, dips at 1 in 3.5 (less than 20°) and the working height is 3 m. The minimum depth is 18 m in the area under consideration, the surface being roughly level. The long-term stability of pillars 1 to 9 shown in Figure 3 was in question, as some of these pillars adjoin the main haulage roadway. A 17 x 21 (357)-node grid with λ = 11.4 m was solved to give the rock pressure over these pillars. The cover pressure q was varied linearly along the dip.

The rock pressure values and Eq.(7) (R =0.625 and K = 11600 kN/m^2 for this seam) were used to obtain safety factors as in Table 2.

Table 2. Safety factor values for pillars 1 to 9 at Pindra (Figure 3).

Pillar No.	Plan area m^2	Perimeter m	Strength kN/m^2	Pressure kN/m^2	Safety Factor
1	9.00	13.21	4315	3140	1.35*
2	34.56	26.43	5785	2550	2.27
3	45.33	32.44	5980	2550	2.33
4	2.88	7.21	3335	2745	1.23*
5	36.69	34.84	5200	2450	2.11
6	1.44	4.80	2940	2355	1.26*
7	2.88	7.21	3335	1960	1.74*
8	33.12	22.83	6080	2060	3.00
9	36.69	23.43	6275	2060	3.06

* Unstable pillars, stabilised by bolting.

While using Eq.(7), pillars donot posses long-term stability if the safety factor is less than 2.0. Four such pillars were declared unstable and were stabilised by fully grouted pretensioned through-bolting.

DISCUSSION

For estimation of rock pressure the numerical procedure outlined is simpler than other existing methods which deal with the behaviour of the roof-seam contact. Because of mathematical simplicity it is easily programmable and also gives rise to a matrix of simultaneous equations which has a very narrow effective band width of 13 elements regardless of the grid size. The matrix can be solved iteratively row by row, making it possible to deal with large problems on relatively small computers.

For some situations the solution may not converge, if the element size is too small, even after hundreds of iterations. Even with a properly chosen element size, this may happen if large areas of the grid are taken up by goaves. The obvious solution is to redesign the grid. In general, for ensuring convergence direct matrix solution like Gaussian elimination is clearly a better choice, but this will mean large storage capacity on the computer.

Reliable estimates of rock pressure should be obtained if the parameters defining roof, seam and goaf

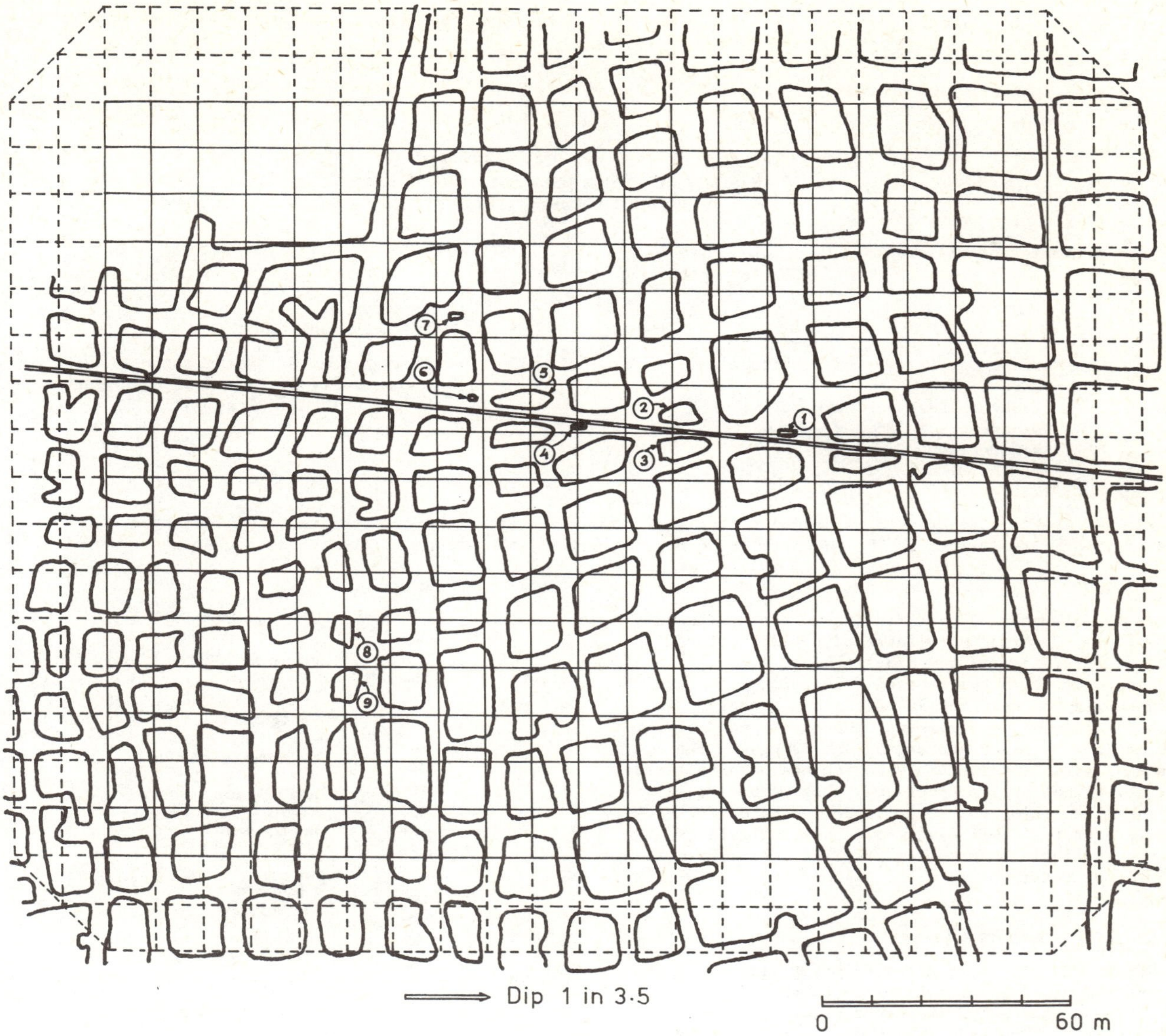

IRREGULAR PILLAR LAYOUT AT PINDRA COLLIERY
UNDER A FINITE DIFFERENCE MESH

Figure 3

behaviour are determined from in situ observations as mentioned. A possible modification of the present computer procedure will be to incorporate post-failure seam behaviour, changing the stiffness k iteratively.

Acknowledgement - Thanks are due to the Director, CMRS for permission to present this paper. The opinions expressed are those of the authors and not necessarily of CMRS.

REFERENCES

Berry, D.S. and Sales, T.W., 1962, "An elastic treatment of ground movement due to mining", Journal of Mechanics and Physics of Solids, 10, 73-83.

Salamon, M.D.G., 1964, "Elastic analysis of displacements and stresses induced by mining seam or reef deposits", Journal of South African Institute of Mining and Metallurgy, 65, 319-338.

Sheorey, P.R., 1976, Author's reply to discussion by A.K. Chugh on "An approach to thick beam analysis for roof strata", International Journal of Rock Mechanics and Mining Sciences and Geomechanics Abstracts, 13, 340-341.

Sheorey, P.R., et al., 1980, "Research in pillar design problems with special reference to partial extraction and stability of pillar arrays", Seminar on Protection of Surface Features, Dhanbad (in print).

Sinha, K.P., 1979, "Displacement discontinuity technique for analysing stresses and displacements due to mining in seam deposits, Ph.D. Thesis, University of Minnesota.

Thompson, M., 1975, "The face element methods : numerical solution of elastostatic problems by the use of harmonic functions", Ph.D., Thesis, University of Newcastle upon Tyne.

Wilson, A.H., 1972, "Research into the determination of pillar size", Mining Engineer, 131, 409-417.

THE COLLAPSE OF ROOM AND PILLAR WORKINGS
IN A SHALEY GYPSUM MINE DUE TO
DYNAMIC LOADING

P. Mottahed*

A. Szeki**

*Potash Corporation of Saskatchewan,
Sedco Centre,
Saskatoon Research Park,
Saskatoon S7K 3X5,
Saskatchewan, Canada.

**Department of Mining Engineering,
University of Newcastle upon Tyne,
Newcastle upon Tyne, NE1 7RU,
England.

SUMMARY: The time sequence of a sudden collapse of room and pillar workings confirms the hypothesis that the cause of the pillar failure is the uncontrolled energy release by the strong strata in the higher roof.

INTRODUCTION

At the Stamphill Gypsum Mine, near the village of Kirkby Thore, Cumbria, a collapse of pillars occured at an abandoned section of the mine in June, 1973. A second collapse occurred in the same district six weeks later. The two collapses resulted in mechanical damage of no serious consequence underground; the collasped district was used only as water standage. The damage on the surface affected farmland only, which has been restored since to agricultural production. Rock mechanics investigations prior to the collapse indicated instability of pillars at the "well district", back analysis of the data and available information resulted in an improvement of design procedure of room and pillar workings in this deposit.

GEOLOGY OF THE DEPOSIT

Gypsum and anhydrite are widely distributed in the U.K. and are mined in various parts of the country. In Cumbria gypsum and anhydrite occur within the Upper Permian strata. The beds, which dip in an Easterly direction, are contained in the lower part of the Eden strata and outcrop from Cotehill to Kirkby Thore, (Fig. 1).

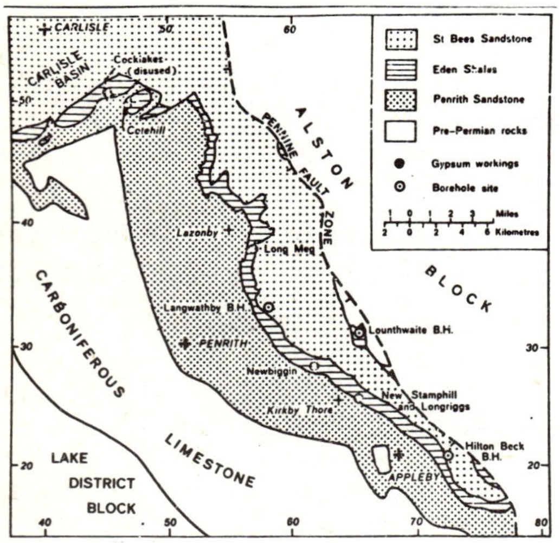

Fig. 1 Location of gypsum workings in the Vale of Eden.

There are four main gypsum/anhydrite beds, A, B, C, and D, in the Kirkby Thore area, (Fig. 2). Only the two lower beds, A and B, are being mined. Although B-bed has been worked extensively in the past, present mining operations

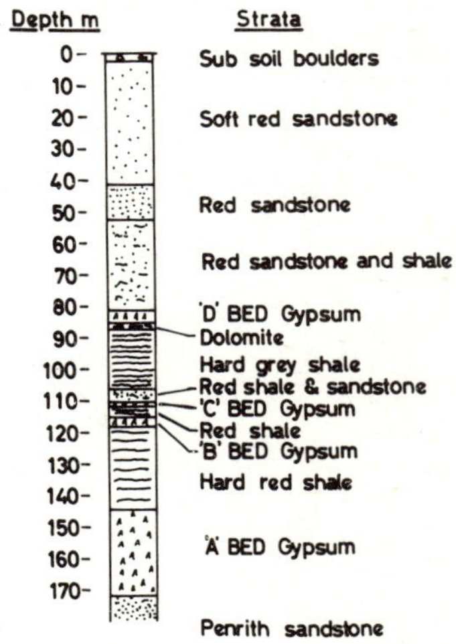

Fig. 2 Borehole section, Well-District.

are confined chiefly to A-bed, which consists of 30-40m of thinly bedded gypsiferous grey mudstone with satinspar and dolomitic siltstone. The gypsum/anhydrite deposits in this area are contained between the Penrith-Sandstone below and the St. Bees-Sandstone above. Both sandstone beds are known to be water bearing and supply water to the Vale of Eden.

METHOD OF MINING

The working of the gypsum bed is by the panelled bord and pillar mining method, having a variable extraction rate between 50 and 75 per cent on an area base. A feature of the A-bed deposit is that the economic horizon changes along the vertical plane within short distances of advance. This changing quality has resulted, at least in the older parts of the mine, in the formation of pillars having variable height to width ratios.

The thinly bedded nature of the gypsum deposit excluded the formation of a flat roof in the underground roadways and a semi-arched roof configuration was adopted which resulted in stable roof conditions (Fig. 3).

The depth of the mine workings varied from 30m at the shallow parts of the mine down to 270m in other parts. Consequently the limiting factor of the underground room and pillar dimensions was different in the shallow and deeper parts of the mine. In the deeper parts the ratio of the ultimate compressive strength of the pillars to the imposed stress could be regarded as the dominant factor affecting the overall stability

Fig. 3 Semi-arched roadways.

of the mine. The thick and competent strata in the higher roof has a dominant contributory effect on the magnitude of the load, and consequently on the stress imposed upon the supporting pillars.

INVESTIGATIONS INTO PILLAR STRENGTH

Laboratory Investigations

In order to assess the in-situ strength of the pillars only laboratory rock testing was carried out, for practical and economic reasons. Becuase the pillar strength in compression is a function of its height to width ratio, a series of conventional compressive strength tests were conducted on 76.2mm dia. cores of gypsiferous strata, having different lengths. Visual examination of the cores revealed that the specimens were composed of numerous layers of different rock materials, varying from 100% gypsiferous marl to 100% amorphous gypsum. This variation in petrological characteristics of the specimens resulted in a wide scatter of compressive strength values superimposed on the variation due to height-width ratio effect. The graphical presentation of the results is shown in (Fig. 4).

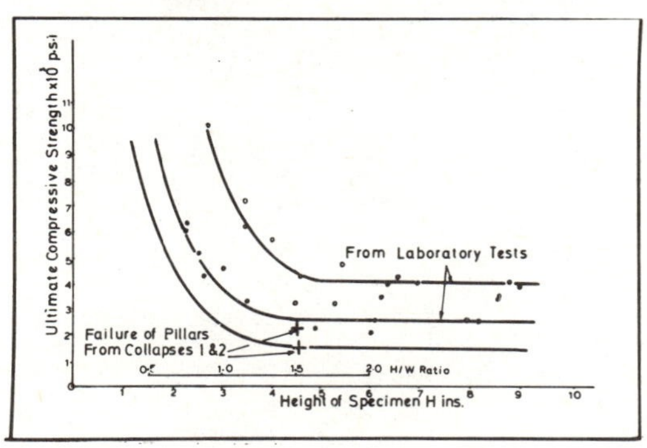

Fig. 4 Compressive strength of gypsum from the Well-District.

It can be seen quite clearly from the plotted results that the value of compressive strength gradually decreases as the height to width ratio increases approaching assimp-totically a constant value at a H/W ratio of 1.5. The ultimate minimum mean value can be taken as 15.4 MN/m^2 (2200 psi).

It was realised from the outset, that the strength curve for the gypsiferous strata could not provide the true compressive strength of a pillar of a given H/W ratio. Nevertheless, it was recognised that a correlation between laboratory strength results and in-situ pillar strength does exist. This correlation, however, could not be estabilished by direct tests for economic reasons.

In-situ Investigations

The depth of the workings in the A-bed gypsum varied from 60m to 270m in the Well District where the collapse occurred, hence the load on the pillars varied also. The district was worked out by 4.8m wide roadways and 7.2m square pillars; the working height varied between 3.0m and 9.0m depending on the quality of the seam. At one stage of the investigation it was decided to install displacement measuring stations in and around pillars.

The site for instrumentation was chosen at the deepest section of the Well District, in an area where the full effect of the over-load would be transmitted to the instrumented pillars. It was hoped then that it would be possible to infer from the measurements obtained the state of stability of the instrumented pillars and with the aid of the already available laboratory data the state of stability of the pillars within the boundaries of the district.

The instrumentation scheme was chosen to determine the effect of the re-distribution of the existing load due to the partial extraction of large pillars supporting the main haulage road, and hece the stability of the resulting smaller pillars. The instrumentation was designed to record the lateral pillar deformation and the stress changes as mining proceeded. The extenso-meter instrumentation and four stages of extraction are shown in Fig. 5. In pillars D and G rigid inclusion

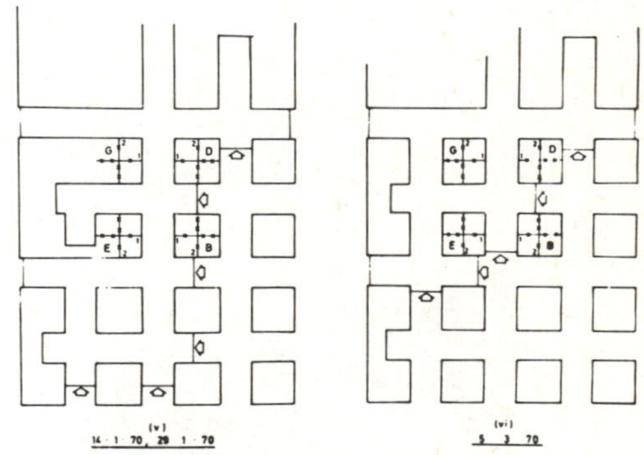

Fig 5. Instrumentation scheme in pillars along the main haulage road.

Potts-stressmeters were also installed which were calibrated in the laboratory prior to installation. Stressmeter outputs and a typical lateral pillar deformation are shown in Figs. 6 and 7. From the

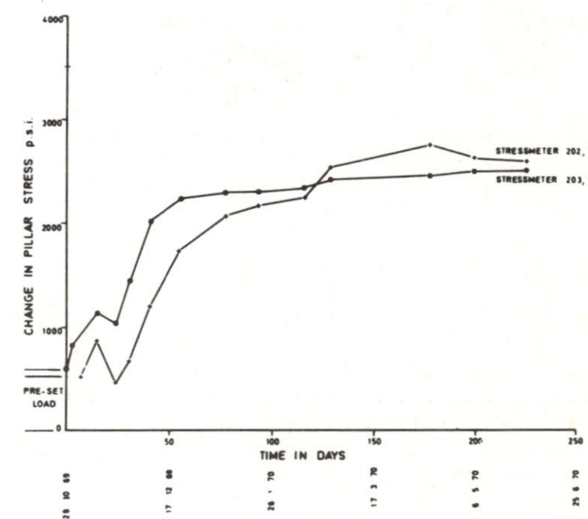

Fig 6 Graph showing stress-meter output.

measurements a number of significant factors became evident:

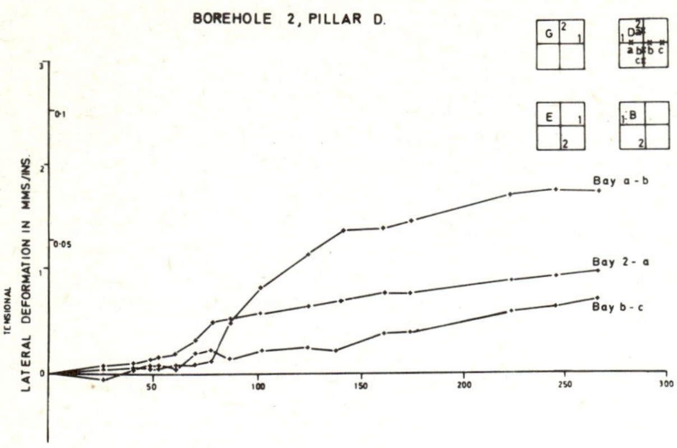

Fig 7. Graph showing lateral pillar deformation

a) the pillars are expanding laterally along their full width.

b) the magnitude of the overall lateral deformation was significant for all instrumented pillars

c) the induced lateral strains in the outer section of the pillars were only marginally higher than in the central zone, hence a stable core at the centre of the pillars was non existent.

From these factors, it was concluded that the state of stability of this district had to be viewed with concern.

At a later stage the visual signs of instability of the pillars, even in the shallower part of the Well District became conspicuous (Fig. 8).

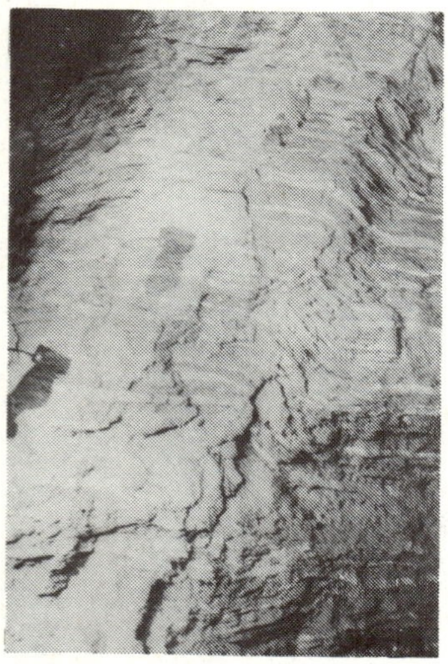

Fig 8. Pillar showing sign of shear deformation.

THE COLLAPSE

The failure of the pillars within the boundaries of the district occurred in three stages. At about 8.00 am on Sunday, June 3rd, miners who were working in the main drift had noticed a sudden gust of air through the adit. Because at the time of this event the main ventilation fan was not running due to maintenance, they attributed this sudden flow of air to a test run of the fan by the electrician. At about 10.30 am it became more noticeable and a sudden gust of airborne dust was blown through the adit. At the same time the reversal of the normal flow of ventilation through the intake drift was reported on the surface. A search for the cause of this unusual occurrence lead to the discovery of partial collapse of the Well District (Fig. 9). These two events interrupted the pumping operations and water at a rate of 2.5m^3/min, started to fill up the lower sections of the district. The accumulation of water adversely affected the pillars which according to pillar measurements were slowly deforming already. The aggregation of circumstances

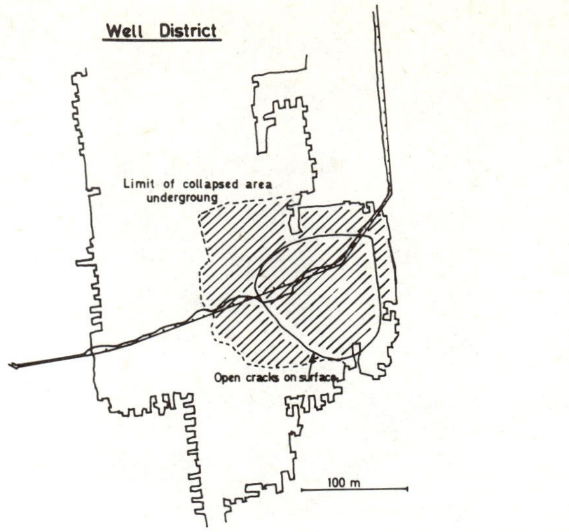

Fig 9. Plan of the collapsed zone after the 2nd event.

cumulated in the third event six weeks after the first two, on Sunday 15th July.

The surface affected by the 2nd event was an area of 3.43 hectares (8.5 acres). There was a wide crack extending for 60-75m directly above the solid rib-side. The dimension of this crack was about 2.5-3.0m deep and 1.5-2.5m wide. The fences were distored in compression, but further away from the centre of the subsidence trough there were tension cracks clearly visible (Fig.

Fig 10. Fences distorted in the compressional zone on the surface.

The third event covered a much larger area and destroyed well over 500 pillars. The air-blast as the result of the sudden collapse was for more violent than those after the first and second events. The extent of the shock wave was so great that a 0.9m diameter, 1.5m long ventilation duct was thrown about 200m along the roadway. This collapse was registered at Edinburgh Observatory as an earth tremor, with a magnitude of 3.5 on the Richter-Scale. A total of 16.2 hectares of land was affected by these collapses.

Cause of the Sudden and Violent Collapse of Pillars

A further series of laboratory tests of the strength properties of the gypsiferous strata by servo-control testing systems demonstrated conclusively that the rock is fairly ductile, i.e. losing its strength gradually after reaching it speak strength when the axial straining rate is 17$\mu\varepsilon$ per second (Fig. 11). Because a period of 2. hours and 6 weeks had elapsed between the 1st and 2nd, and the 2nd and 3rd events, this fact must be taken into consideration in speculation on mechanism of collapse.

It can be assumed that the pillars had sustained up to the time of collapse, the full or greater part of the

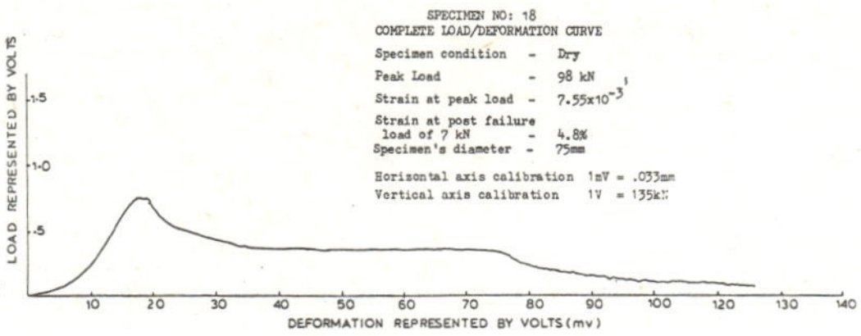

Fig. 11 Complete load-deformation curve of gypsiferous strata.

redistributed cover load which is a function of the depth of the workings and the rate of extraction. Due to adverse factors such as the presence of water, dip of strata, large height to width ratio of pillars and a nearby fault plane some of the weakened and deformed pillars failed under the weight of the strata either below the "hard grey shale" or below the "red sandstone and shale" (see Fig. 2). Until this moment the strong beds in the roof were capable of bridging the area partially be shedding the load of the overburden onto the rib-sides, and a precarious equilibrium existed between the slowly sagging roof strata and the gradually deforming pillars.

When the limited support offered by the pillars was transposed from failed pillars onto nearby pillars after the first event a progressive and fairly rapid load re-distribution took place in the higher roof layers. Because none of the beds in the higher roof were capable of supporting their own weight over a span between rib-sides in the East-West direction, these beds failed in tension. The tensile fracture zone in a thick plate or beam would occur near to the abutment zones inside the span and would slant sub-vertically towards the centre of the span. Because in tensile failure the frictional forces have no retarding effect, the failure would produce a shock-wave when the failing rock mass encountered resistance. There exists very little doubt that the pillars were destroyed in the second and third event by the gravitational energy released by the failure of the higher roof layers spanning the workings between the abutment zones.

Back Analysis of Pillar Strengths

It has already been stated that laboratory testing of rock materials cannot provide the true in-situ strength of the rock mass, because the size of laboratory specimens are lower, by a few orders of magnitude, than the rock mass surrounding underground structures, especially in the case of supporting pillars.

Such a case where a large number of pillars are destroyed simultaneously may be regarded as a large scale in-situ test. In the case of pillar failure at the Well District, the load acting on the pillars could be assessed with reasonable certainty because the span of the panel was large compared with the depth of workings, hence in the centre of the panel the load carried by the individual pillars could be calculated by using the "tributory area principle".

The calculated stresses acting on the pillars are marked on Fig. 4 for the corresponding height to width ratios. If the trend of the curve representing the relationship between the compressive plate strength and H/W ratio for laboratory specimens is accepted as valid for volumes greater by an order of magnitude of five, then a curve through these points and parallel to the laboratory trend curves will represent the in-situ compressive strength of pillars in the gypsum formation in the Kirkby Thore Area.

CONCLUSIONS

The sudden collapse of the Well District on two separate dates both occurred fortuitously on Sundays when nobody was employed in the District on salvage work. Prior to the collapse there were clear indications of unstable conditions of pillars especially in the vicinity of the pumping chamber. The instrumentation in the District however, was not extensive enough in order to predict the actual time of a pending collapse. One gratifying aspect of the laboratory investigation which may be summarised is that the strength of gypsiferous strata (or of any strata or rock formation) determined in the laboratory on small samples is closely related to the in-situ strength of strata, and can be extrapolated to in-situ conditions with circumspection.

The in-situ measurements of pillar deformation are also indicative; the lack of a stable core at pillar centres will become appreciable well before visual sign of pillar deterioration become conspicuous.

The testing of gypsum specimens by servo-control systems revealed that the rock or the pillars do not lose their strength in a brittle manner when the peak load in uniaxial compression is reached and that degradation is a gradual process. Consequently, a sudden failure of specimens (such as in a "soft" testing machine) or a sudden collapse of pillars must be the result of a rapid or continuously maintained load due to release of stored energy from the loading system itself.

In the case of room and pillar mining the stored potential energy is available in the gravitational energy in the strata overlaying the room and pillar panels; hence when the competence of the roof strata breaks down due to gradual deterioration of pillars it would result inevitably in a dynamic loading condition with disasterous consequences.

REFERENCES

ARTHURTON, R.S., Report No. 71/17, Inst. Geol. Soc., 1971.

JONES, B. "Laboratory and underground investigations into the stability of mine workings in gypsum", Ph.D. Thesis, Dec. 1970. University of Newcastle upon Tyne.

MOTTAHED, P., "Correlation between critical design criteria and data from field and laboratory investigations in gypsum mining". Ph.D. Thesis, Feb. 1974. University of Newcastle upon Tyne.

SALAMON, M.D.G. and MUNRO, H. (1967). "The study of strength of coal pillars". Jl. S. African Inst. Min. Metall., Vol 68, pp 55-67.

USE OF ASH TO IMPROVE STRATA CONTROL IN BORD AND PILLAR WORKINGS

James M. Galvin

and

Horst Wagner

Mining Operations Laboratory
Research Organisation
Chamber of Mines of South Africa
Johannesburg.

SUMMARY. Ash produced from South African power stations possesses self-cementing properties when mixed with water. This material is utilized for strata control purposes in bord and pillar workings. In order to quantify the effects of ashfill on strata behaviour, the mode of failure of a coal pillar is examined and model pillar tests are described. These tests indicate that ashfill not only increases pillar strength in excess of 40 per cent, but also has a significant influence on the post-failure behaviour of coal pillars. Practical examples of the application and integration of ashfill into bord and pillar mining operations are given.

INTRODUCTION

Bord and pillar mining is the primary means of extracting coal in South Africa, accounting for over 75 per cent of the 115 million tonnes of coal produced in 1980. About 40 million tonnes of this coal was consumed by power stations located in the coalfields. Since the coal has an average ash content of about 25 per cent, vast tonnages of ash have to be disposed of. In 1973 an hydraulic ash disposal system was commissioned to dispose of 3 000 tonnes of ash per day produced by the Grootvlei Power Station into the bord and pillar workings of the adjacent Springfield Colliery. This system proved successful and disposal of ash into bord and pillar workings is now an accepted practice.

The application of conventional stowing materials, such as sand, is uneconomic in South Africa if these materials are not available on site. Economic constraints also prevent the use of additives, such as Portland cement, in the materials. Power station ash is not affected by these constraints. The additional cost of disposing of ash underground, and thus the cost of the stowing material to a colliery, is as little as R0,01/tonne. Furthermore, about 80 per cent of the ash produced by a power station has self-cementing properties when mixed with water and so there is no need for additives. Thus, the disposal of ash in bord and pillar workings offers the potential to utilize the material for strata control purposes. Previous experiences confirm this potential.

PREVIOUS EXPERIENCES WITH ASHFILL

The placement of ash underground was first performed in South Africa at Koornfontein Colliery, in an attempt to stabilize a low safety factor area in bord and pillar workings. The area had been mined in 1956 to a safety factor of less than 0,72, which corresponds to over a 96 per cent chance of collapse (all safety factors are based on the Salamon Pillar Design Formula, Salamon (1967)). The area began to collapse in late 1961 and continued to collapse intermittently throughout 1962. Ashfilling commenced in 1963, with ashfill being placed hydraulically to an average height of two-thirds of the mining height. Until the colliery closed down in the mid-1970's, no further major collapses had taken place within the ashfilled areas. Underground observations of the effect of ashfill on pillar stability during this period are summarised in Table 1 and illustrate the effectiveness of the ashfill.

Table 1 Ashfill effectiveness at Koornfontein Colliery.

Safety Factor	Geologic Conditions	Pillar Height Ashfilled	Ashfill Effectiveness
0,65	good	40	None
0,40	good	up to 80	None
0,54	poor	up to 80	None
0,46	good	70 - 80	Good
0,60	good	50 - 60	Good

In 1976, the ash disposal system at Springfield Colliery was utilized to stabilize workings in the combined No. 2 and No. 3 Seams. The underlying No. 1 Seam, at a depth of about 190 m, is separated from the No. 2

and No. 3 Seams over a considerable area of the mine by a parting of less than 22m in thickness. Consequently, bord and pillar workings in both the No. 1 Seam and the combined No. 2 and No. 3 Seams need to be designed to a high safety factor. However, mining of the No. 2 and No. 3 Seams had been in progress for over 10 years whilst no consideration had been given to mining the No. 1 Seam. As a result some areas within the No. 2 and No. 3 seams had been mined to a safety factor considered marginal for multi-seam extraction. Therefore, in order to permit mining of the No. 1 Seam, it was decided that all marginal areas in the No. 2 and No. 3 Seams must be ashfilled prior to being under-mined.

Mining of the No. 1 Seam below ashfilled workings has now been in progress for over 4 years, during which time no adverse interaction between the lower and upper horizon workings has been noted. Ashfilling appears to have stabilized the upper No. 2 and No. 3 Seam workings and to have limited strata displacement resulting from interaction of pillar stress profiles between upper and lower mining horizons. Consequently, an additional 5 million tonnes of coal reserves have been generated.

During the last 15 years there have been numerous other occasions when, on a small scale, use has been made of ashfill to stabilize workings. Mine offices, houses, railway lines and the like have been built above ashfilled workings. Workings in an advanced state of collapse have been stabilized. In almost all instances, ashfill has been effective in ensuring stability of bord and pillar workings.

Thus, local experiences have highlighted the fact that ashfill can have an extremely favourable influence on pillar stability. During the last three years, extensive research has been undertaken into quantifying this influence. This research has had two aspects, namely:

i) to establish those parameters which control ashfill quality;

ii) to quantify the effects of good quality ashfill on pillar behaviour, and thus strata control.

PARAMETERS CONTROLLING ASHFILL QUALITY

About 20 per cent of the ash produced at coal-fired power stations is collected from the bottom of the furnaces in the form of clinker, whilst the remaining ash is carried out of the furnaces with the flue gases. Before passing into the atmosphere the flue gases pass through three or four banks of precipitators, referred to as fields, which extract most of the suspended ash. This ash is known as PFA (pulverised fuel ash) or flyash. Over 75 per cent of the PFA is collected by the first field, Field 1, whilst only 1 to 3 per cent is collected by the last field in the system.

Ashfills derived from clinker ash possess no self-cementing properties and are rapid draining. However, PFA based ashfills are self-cementing since PFA's are pozzolans, 'siliceous or siliceous and aluminous materials, which in themselves possess little or no cementitious value but will, in finely divided form and in the presence of moisture, chemically react with calcium hydroxide at ordinary temperatures to form compounds possessing cementitious properties' (ASTM Standard Definitions of Terms Relating to Hydraulic Cement).

Coal burnt at South African thermal power stations produces a PFA which contains a high percentage of calcium hydroxide. Thus, only the addition of water is required to cause this material to cement and to yield a stowing material of significantly higher strength than that which can be achieved from other unmodified stowing materials such as sandfill. Field and laboratory studies have shown that the pozzolanic activity of PFA, slurry concentration and time are the most important parameters controlling ashfill quality (See Fig. 1). Ashfill quality improves markedly with increased slurry concentration and with time. Of particular significance is that ashfill strength may only approach its ultimate value some 6 months after placement. Consequently, ashfilling must be incorporated into mining operations in such a manner that the strength properties of the ashfill are not required to control strata displacement within this critical time period.

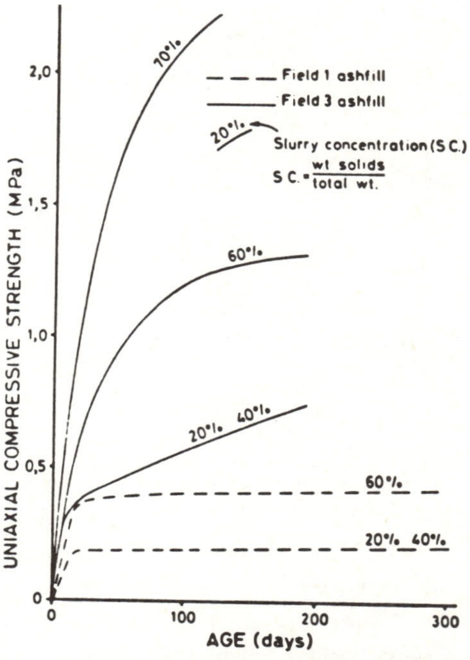

Fig. 1 Influence of pozzolanic activity, slurry concentration and time on ashfill strength.

THE EFFECT OF ASHFILL ON PILLAR BEHAVIOUR

Conceptual model of pillar failure

In order to quantify the effects of ashfill on pillar behaviour it is necessary to examine briefly the mode of failure of a coal pillar. Large scale in-situ coal pillar tests by Wagner (1974), in which a displacement controlled loading system was used, provided a detailed insight into the complex process of pillar failure. Fig. 2 shows the distribution of vertical stresses in the centre plane of a coal pillar at different stages of pillar compression. Also shown in this figure is the stress-compression diagram for the whole pillar. Two points emerge clearly. Firstly, the load bearing capacity of the pillar corners and, to a lesser extent, pillar sides is small compared to that of the centre of the pillar. This highlights the importance of confinement on the strength of a pillar. Obviously, the corners and sides are the least confined portions of the pillar whereas the central portion is subjected to the greatest confinement. Secondly, the central portion of a coal pillar is capable of withstanding extremely high stresses even if the outer portions of the pillar have already failed.

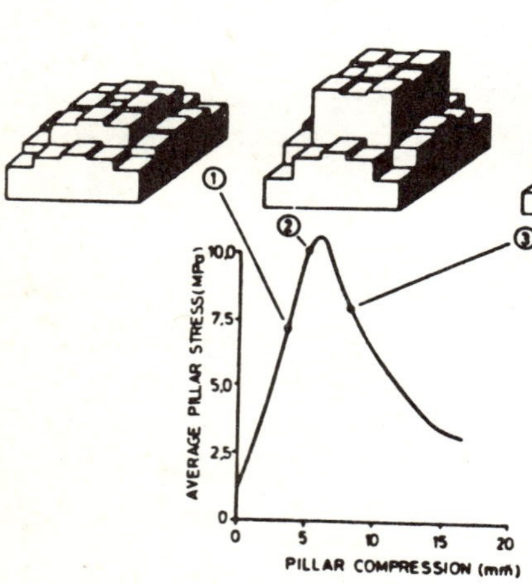

Fig. 2 Distribution of vertical stress in a pillar undergoing compression (Wagner, 1974)

It follows from these observations that even a small amount of confinement could result in a marked increase in overall pillar strength. Confinement to the sides of a coal pillar can be provided in a number of ways: for example, bolting of pillar sides and long wire ropes applied around the perimeter of the pillar. However, the most effective means of confinement is to fill the voids around the pillar. Fill material provides confinement to the pillar in two ways. Firstly, because of its weight, it exerts a certain pressure on the sides of the pillar. This pressure increases linearly with depth of fill but is generally very small because of the limited height of fill and the relatively low density of fill. For ashfill the effective fill pressure is typically of the order of 10 to 50 kPa. Secondly, confinement is provided by the fill material in reaction to lateral expansion of the pillar. This aspect is far more important than the passive fill pressure.

It is well known that if the load acting on a pillar is increased, the pillar will shorten in the direction of the applied load and at the same time the pillar will expand laterally. This process is known as the Poisson effect. As long as the applied stresses are small compared to the strength of the pillar, the lateral deformations of the pillar are small. If the stresses acting on the pillar are increased and fracturing of the unconfined corners and sides of the pillar commences, the lateral deformation behaviour of the pillar changes markedly. This process is known as dilation and manifests itself in the form of slabbing and spalling. A representative example of the lateral deformation behaviour of a coal pillar is given in Figure 3. It can be seen that pillar dilatation commences at a stress level of about one half of the ultimate strength of the pillar and that the rate of dilatation reaches a maximum just prior to pillar failure. Also shown in Fig. 3 is the emission of micro-seisms which are indicative of internal fracturing in the pillar. The close relationship between lateral deformation and frequency of internal fracturing is remarkable.

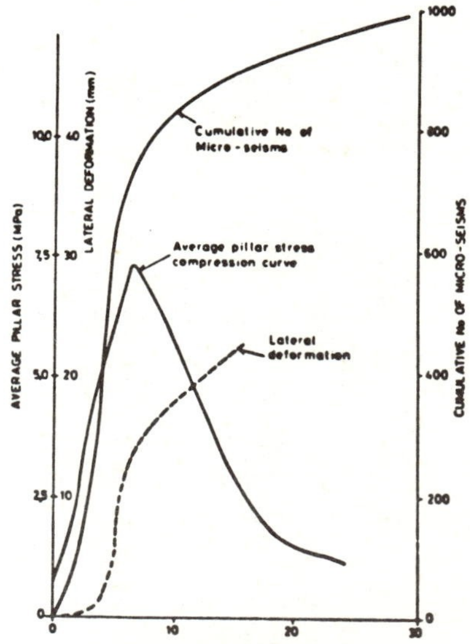

Fig. 3 Pillar behaviour during compression (Wagner, 1974).

From the discussion of the mode of pillar failure it follows that the main objective of fill is to provide confinement to the pillars and not to support the weight of the overburden. The effect of filling bords with ashfill is that the fill material will resist the lateral expansion of the pillar. Figure 4 shows a typical section through an ashfilled bord and pillar panel. In this figure w and B denote the pillar and bord widths, respectively. The lateral expansion of the pillar that occurs after the introduction of fill is denoted by Δw. For reasons of symmetry no horizontal movements can occur in the vertical planes I and II through the centre of the pillar and the centre of the bord, respectively. If $E_{\bar{F}}$ is the effective modulus of deformation of the fill material, then the fill reaction, σ_F, is given by the relationship -

$$\sigma_F = \varepsilon_F E_{\bar{F}} = \frac{\Delta w}{B} E_{\bar{F}} = k_F \, \Delta w, \quad \text{(MPa)} \qquad (1)$$

where ε_F is the lateral fill strain, and

$k_F = \frac{E_{\bar{F}}}{B}$, the fill stiffness ($MN/m^3$).

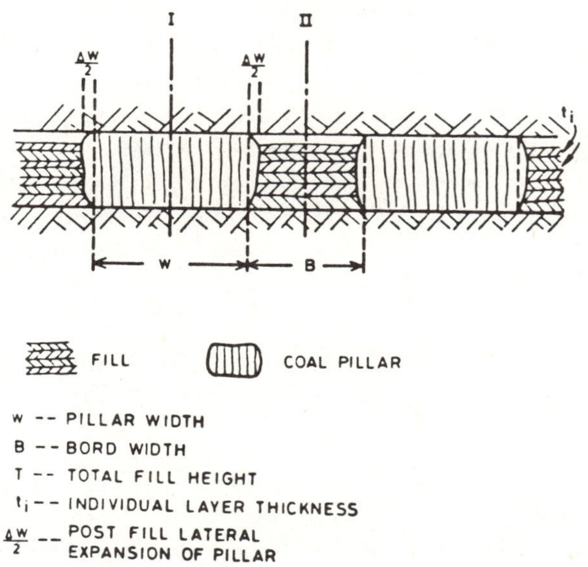

Fig. 4 Diagrammatic section through an ashfilled bord and pillar panel.

Two important conclusions can be drawn from Eqn. (1). Firstly, the effectiveness of filling is directly proportional to the fill stiffness, k_F. In turn, the fill stiffness is directly proportional to the quality of the fill material, as expressed by its effective modulus of deformation, and inversely proportional to the bord width. Secondly, the confinement provided to the pillar by the fill increases with lateral expansion of the pillar. Since most of the lateral expansion of a coal pillar is caused by slabbing and spalling of pillar sides, fill is most effective where and when it is needed most.

Further important conclusions can be deduced from Eqn. (1) if the lateral pillar displacement, Δw, is expressed in terms of lateral pillar strain, $\varepsilon_{\ell p}$, as -

$$\sigma_F = \frac{E_{\bar{F}} \cdot w \cdot \varepsilon_{\ell p}}{B} = k_F \, w \cdot \varepsilon_{\ell p} \quad \text{(MPa)} \qquad (2)$$

This equation highlights that the effectiveness of filling is also directly proportional to pillar width. Thus, the greatest benefits will be gained from ashfill at depth, where pillar sizes are large.

Because of the confinement provided to the pillar by the roof and floor strata, the lateral expansion of a pillar, in particular that of failing pillars, is usually greatest at mid-height, unless a very weak band is present in the seam or at some other horizon. Therefore, for filling to be most effective, the height of fill must be more than half the pillar height. Filling should preferably extend up at least two-thirds of the total height of the pillar. Early placement of fill is also important since the fill reaction does not depend on the total lateral expansion of the pillar but only on that portion, Δw, which occurs after filling of the bords.

Model pillar tests

To assess quantitatively the effects of ashfill on pillar strength a number of model pillar tests have been conducted. In these tests coal samples of 150 mm diameter and width : diameter ratios of 1 and 2, were centred in a steel pipe of 235 mm internal diameter and 6 mm wall thickness. The internal circumference of the pipe represents the vertical plane of symmetry through the centre of an ashfilled bord. The wall thickness of the pipe was designed so that this plane of symmetry would displace by less than 0,03 per cent of the pipe diameter, or 0,07 mm. The cavity between the model and the steel pipe was filled with ashfill.

After allowing the ashfill to cure for 100, 200 and 400 days, model coal pillars were loaded under uniaxial compression at a displacement rate of 6mm/hr. Vertical and lateral pillar deformation, and lateral ashfill pressure have been monitored continuously during testing.

These tests confirm that ashfill can increase pillar strength significantly at 200 days after ashfill placement. The strength of model coal pillars with a width : height ratio of 1 was increased by over 50 per cent, and those with a width : height ratio of 2 by about 40 per cent. At 400 days these values were of the order of 60 per cent and 50 per cent respectively. Furthermore, the pressence of ashfill has a significant influence on the post-failure behaviour of the coal pillars. (Figure 5). Not only is the failure controlled but also, after sufficient post-failure compression, the load carrying capacity of a pillar stops decreasing and increases with further compression.

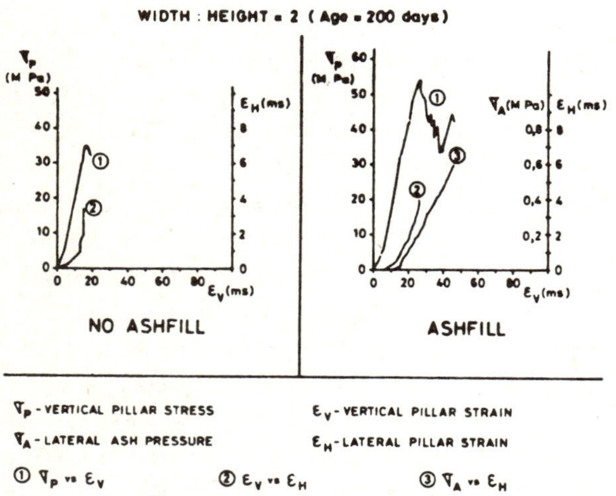

Fig. 5 Load compressor characteristics of confined and unconfined model coal pillars.

In-situ experiences

During 1979 a full scale mining experiment was completed in the combined No. 2 and No. 3 Seams at Springfield Colliery where the seam thickness exceeded 12 m.

The experimental mining was conducted at a depth of 175 m in a bord and pillar section which measured 150 x 200 m and which had been mined to a safety factory of 1,3. The average width of pillars and bords was 16 m and 6 m respectively. Pillars had a width : height ratio of 4,27. The aim of the experiment was to successively top-coal and ashfill the section until no coal remained in the roof, thereby reducing the width : height ratio and thus the nominal safety factor of the coal pillars.

The safety factor of the experimental area was ultimately reduced to less than 0,6, corresponding to a 99,94 per cent chance of collapse in the absence of ashfill. Although the upper 3 m of the workings was not ashfilled, minimal deterioration occurred within the area. The maximum lateral pillar displacement measured was only 2 mm, whilst the pressure build-up in the ashfill was, typically, o,o5 MPa. This corresponds to an effective ashfill modulus of deformation of 150 MPa. Laboratory testing of samples collected from underground produced a similar effective modulus (130 to 185 MPa).

With the exception of a major intersection fall, attributed to an excessive local bord width of 8,5 m, no support problems were encountered during mining. Slabbing and spalling of the exposed upper 3 m of pillars since the completion of mining has also been minimal. In addition to providing lateral confinement to the coal pillars, the ashfill acted as a good working platform. No load-bearing or dust problems were encountered and the smooth, level surface resulted in easily and well-maintained travelling ways.

It may be assumed that the lack of ashfill confining the upper 3 m of the 12 m high pillars had no significant effect on pillar stability since ashfill extended to at least two-thirds of the pillar height. Had the entire mining height been ashfilled and had ashfill increased pillar strength by 50 per cent the ultimate effective safety factor of the workings would have been 0,9. The lack of strata displacement which occurred indicates that the final effective safety factor was probably much greater.

POTENTIAL AREAS OF APPLICATION OF PFA IN BORD AND AND PILLAR WORKINGS

From the discussion of the effects of flyash fill on the strength of coal pillars it is evident that this readily available fill material offers tremendous potential for improving percentage extraction, stabilization of low safety factor areas and protection of surface structures.

USE OF ASHFILL TO STABILISE OLD WORKINGS AND TO PROTECT SURFACE STRUCTURES.

As was indicated earlier the most obvious application of PFA in bord and pillar workings is the stabilization of low safety factor areas. In this instance the pillar layout is pre-determined. To obtain maximum benefit from the ashfill, careful consideration must be given to the method of placement and in particular drainage of excess water from the workings. Levelling of the working prior to filling is essential to determine the correct position of ashing points and the position of ash dams and drainage channels.

In many cases collapse of low safety factor areas in itself is not considered dangerous provided it can be confined to the old areas and does not effect vital mining excavations. Effective barrier pillars can be formed by filling the bords between adjacent rows of pillars. To be effective the minimum number of rows of pillars should be three. This ensures that the pillars in the central row are fully surrounded by ashfill. This method of isolating low safety factor areas has been employed successfully and regional pillar collapses have been arrested by these artificial barriers.

In South Africa up to 30 percent of the mining areas are effected by mining restrictions due to protection of surface structures of one kind or another. Presently surface objects are protected by leaving larger blocks of coal unmined. The size of these protective pillars is defined by the size of the surface structure on the one hand and the depth of mining on the other. Generally pillar dimensions increase by $\frac{2D}{2,7}$ where D is the depth of mining and the factor 2,7 determined by the angle of draw. Valuable coal reserves could be utilised by undermining these surface structures with bord and pillar workings designed to a safety factor of about 2 and by filling the bords with high quality ashfill. According to Salamon (1967) the probability of failure of pillars designed to a safety factor of 2 is less than 0,5 percent and for all practical purposes can be ruled out. If filling of the bords is carried out close to the roof then the possibility of surface subsidence due to bord collapse is eliminated as well. This type of subsidence is quite common in very shallow bord and pillar workings and should not be ignored.

Measurements of surface subsidence above bord and pillar workings have shown that subsidence caused by pillar compression is typically of the order of a few millimetres and for all practical purposes insignificant. Indeed the differential subsidence when changing from normal bord and pillar workings to monolytic protective pillars is greater than is the case with protective pillars using ashfill.

As far as South African collieries are concerned it is estimated that about 35 percent of coal which would be left in monolytic protective pillars could become available by employing a combination of normal pillars and ashfill.

Use of ashfill to improve extraction in bord and pillar workings.

In South Africa many coal seams have been extracted by means of conventional bord and pillar methods. In the past, large areas have been mined without leaving regularly spaced barrier pillars between extraction panels. It is only since the Coalbrook disaster in 1960 that barrier pillars have been employed on a regular basis. Secondary pillar extraction without the barrier pillars is unduly risky and could lead to uncontrolled regional pillar collapses. Systematic use of ashfill to form substantial artificial barrier pillars between extraction panels could lead to a wider application of pillar extraction methods in existing workings in old collieries. Recent successes with the use of continuous miners in pillar extraction operations make this approach an attractive proposition.

Of even greater significance is the use of ashfill in new collieries. In order to obtain the maximum benefit from ashfill material and to avoid hazardous situations it is important that some basic principles are adhered to. First, studies of the strength properties of ashfill have shown that because of the slowness of the pozzolanic reaction the fill develops its full strength potential only several months after placement. Second, the strength properties of ashfill are strongly dependent on the slurry concentration and drainage.

Because of the slowness of pozzolanic reaction, mining systems using ashfill have to be designed in such a manner that the ashfill is given opportunity to develop its strength potential before the pillars are being subjected to high stresses by either mining activities in their vicinity or by reducing their size. To illustrate the principles involved, consider the use of ashfill in a thick seam. In this instance the seam could be extracted either by mining to the full seam height before introducing ashfill or by extracting the seam in a series of alternate stages of mining and filling until the full seam height is reached. In the latter case the ashfill is given the opportunity to cure before the height of the pillar is increased or the width to height ratio decreased. Since the strength of a coal pillar, σ_p, of given width, W, decreases with height, h, according to the following relationship

$$\sigma_p = K. \frac{w^{0,46}}{h^{0,66}} = K.\ k_w . \frac{1}{h^{0,66}}$$

where $K_w = w^{0,46}$

it follows that the ashfill is given opportunity to cure while the pillar is strongest, Table 2.

Table 2. Change of Safety Factor as a result of increasing pillar height in a number of mining steps.

Pillar Height (m)	Safety Factor
2	2,5
4	1,55
6	1,2

Depth of mining : 100 m
Bord width : 6 m
Pillar width : 11,7 m

Taking into account the effect of ashfill on the strength of a pillar a final nominal safety factor of about 1,2 is considered adequate.

A further advantage of this mining method is that mining operations can be conducted on top of the fill, that is close to the roof.

Galvin (1981) has calculated the additional percentage extraction that could be achieved from South African coalfields by incorporating repeated cycles of ashfilling and top-coaling into bord and pillar mining in thick coal seams. The results of this calculation are given in Table 3.

Table 3. Estimate of additional percentage extraction resulting from the use of ashfill in thick S.A. coal seams.

Seam thickness (m) \ Depth (m)	50	100	150	200
6	7,9	10,8	11,8	11,9
8	20,6	17,6	15,5	13,2
11	30,0	21,9	16,5	12,8

To obtain maximum benefit from the use of ashfill in narrow coal seams requires a departure from conventional bord and pillar layouts. In this instance pillars have to be developed to their final dimension before the bords can be filled. Since the nominal safety factor of these pillars is about 1,2 it is important that the dimensions of the extraction panel are chosen in such a manner that the pillars are not exposed to the full weight of the overburden and the stiffness of the mining layout is such that the possibility of an unstable pillar collapse can be excluded, (Salamon (1970). Both criteria can be met by mining panels of narrow span and filling them immediately after mining. An example of such a sequence of mining and filling is shown in Figure 6. Naturally there are many variations that are possible but the basic principles remain the same.

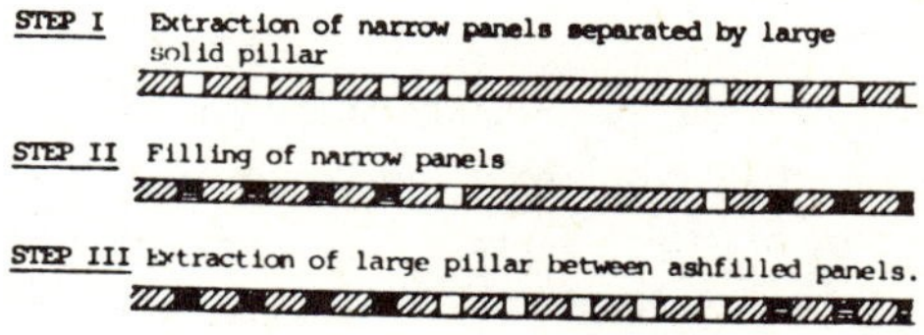

Fig. 6 Example of a mining system employing ashfill.

CONCLUSIONS

The availability and unique properties of pulverised fuel ash make this material well suited to improve the strength and stability of bord and pillar workings. Ashfill offers great potential to stabilise low safety factor mining areas, to form artificial barrier and protective pillars and to improve percentage extraction in bord and pillar workings.

REFERENCES

Galvin, J.M., 1981, "The mining of South African thick coal seams - Rockmechanics and mining considerations." PhD. Thesis, University of the Witwatersrand.

Salamon, M.D.G. and Munro, A.H.; 1967, "A study of the strength of coal pillars" Journal S.A. Inst. Min. Metall., 67, pp. 55-67.

Salamon, M.D.G., 1970, "Stability, Instability and the design of pillar workings". Inst. J. Rock Mech. Min. Sc., Vol. 7, pp 613-631.

Wagner, H., 1974, "Determination of the Complete bord deformation characteristics of coal pillars". Proc. 3rd Inst. Congre. Int. Soc. Rock Mech., Denver, National Accademy of Sciences, pp 1076-1081.

POST FAILURE EXAMINATION OF MODEL EVAPORITE PILLARS

R.W. McCrae

Dept. of Mining Engineering, University of Newcastle upon Tyne.

H.D.S. Miller

Dept. of Mining Engineering, University of British Columbia.

R.F. Cook

Dept. of Civil Engineering, Michigan Techincal University

SUMMARY. The paper describes the preliminary stages of an investigation into the deformation patterns found in model exaporite pillars subjected to high strains. Typical structures found in both laboratory specimens and actual mine pillars are described.

INTRODUCTION

In evaporite mines, particularly those at great depth, it may be necessary to include yielding pillars in the extraction panels. The principle underlying the design of such pillars being that they can still sustain some level of load after the peak strength of the material has been exceeded (Cook 1982).

The stress/strain characteristics of such "post-failure" behaviour has been rigorously investigated in the laboratory and a considerable number of results are available (Miller, 1980). A number of problems are found, however, in taking these results and applying them to the underground situation.

Firstly, the strain rates used in the laboratory are considerably higher than those measured in actual mines. Secondly, different samples, though tested at the same strain rate, often show very different deformation and load bearing properties. This paper describes the preliminary stages of an investigation into how the actual material, and particularly the individual crystals, behave during the "post-failure" deformation and whether this can help to explain the stress/strain relationship. To accomplish this a number of thin rock sections were prepared, from both tested and untested specimens, to determine the changes that occured. As well as examining sections taken from laboratory specimens, samples were also collected and examined from failed pillars.

PREPARATION AND TESTING OF ORIENTATED SPECIMENS

A 100mm diameter core of halite was obtained and three orientations of cubic model pillar samples were cut from it. The halite displayed an apparent planar anisotropic fabric which was attributed to post-depositional flow and was used as a reference plane to orientate the samples (Figure 1). These were all tested at a constant strain rate of 16 micro-strain per second, until a total deformation of 10mm, giving a total strain of 16%. The complete set of curves is given in Figure 2.

Thin sections were made in the same set of orthogonal planes as referred to above, i.e. parallel and at right angles to the direction of flow, from all three specimens after testing as well as from an untested sample from the same piece of core.

An inspection of these thin sections was made in order to investigate the effect of the loading and straining on the rock fabric. The technique used was similar to geotechnical line logging. With the slide correctly mounted on a petrographic microscope stage, it was slowly traversed across and all discontinuities in it were recorded along with their distance from the left hand side of the mounted section. This was taken as the zero datum point.

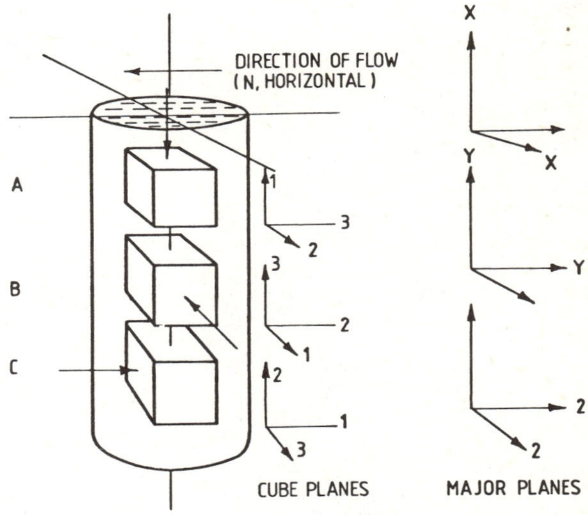

Figure 1. Sample orientation.

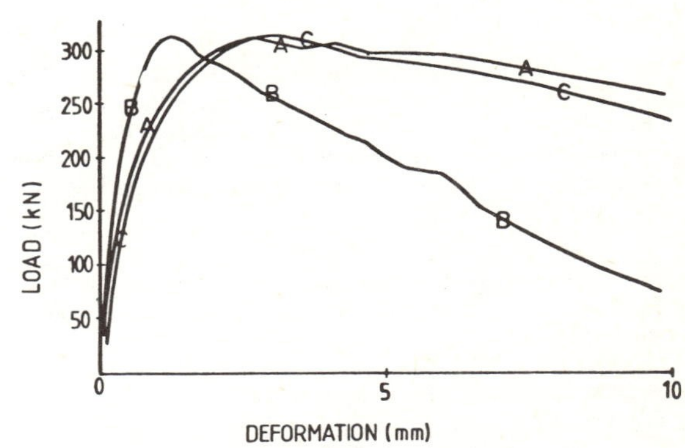

Figure 2. Load deformation curves for Halite specimens.

The discontinuities were divided into two categories; grain (or crystal) boundaries (G) and fractures (F). In addition to distinct boundaries, which were often annealed, major cleavage partings, often with clay inclusions, were included in the category of grain boundaries.

The objective was to place all the features due to sedimentary and diagenetic processes in the first category. All other fractures were included in the second category. These were predominantly cleavage partings which were assumed to be due to tectonic stresses and the test loading. In addition to cleavage partings, which were dominantly planar, a small number of conjugate Reidel type shears were found to be present in the tested specimens. All crystals less than 0.5mm and greater than 5mm in length were rejected from the sampling so that all crystals studied were within a similar range of crystal sizes. In total this meant that 0.55% of the total rock volume was not considered. The results of the sampling are given in Table 1.

CRYSTAL DEFORMATION MECHANISMS

Though the study of orientated specimens had shown that there would appear to be a relationship between fracturing and post failure load bearing capacity of evaporite model pillars, it was not necessarily the only process occuring. A number of further specimens were therefore sectioned after testing to try to identify the mechanical deformation processes occuring.

In all specimens examined the mean fracture frequency was considerably higher, as previously mentioned, after loading with new fractures developing parallel to pre-existing sets (in both halite and potash). When examined under crossed polars, with a mica ¼ plate placed in the optical path it was noticeable that considerably greater areas of the crystals, particularly at sharp corners, showed interference colours after testing. This is evidence of disordering of the crystal structure and deformation occuring, though this phenomena will produce insignificant strains compared to the fracturing processes.

Table 1. Grain and fracture spacings in orientated specimens

	SPEC. NO.	MEAN NO OF FRACTURES PER GRAIN	MEAN GRAIN SEPARATION,G	MEAN FRACTURE SPACING, F	MEAN NO OF FRACTURES PER GRAIN
UNLOADED	Un-tested	x - y	14.32	12.15	1.18
		y - y	15.368	14.600	1.05
		x - x	14.303	13.111	1.09
LOADED SAMPLES	A	A_{23}	15.00 (5)	5.690 (46)	2.636 (223)
		A_{13}	14.50 (6)	6.868 (47)	2.111 (208)
		A_{12}	15.111 (6)	6.476 (49)	2.333 (214)
	B	B_{13}	13.00 (7)	8.210 (67)	1.58 (134)
		B_{23}	14.286 (7)	7.500 (51)	1.905 (190)
		B_{12}	17.000 (18)	5.333 (40)	3.188 (292)
		C_{13}	13.632 (5)	6.023 (49)	2.263 (192)
		C_{12}	13.478 (6)	5.849 (45)	2.304 (211)
		C_{23}	14.941 (3)	6.350 (48)	2.353 (224)

The figures in brackets are the % change from the untested sample.

Despite the microscopic flow fabric the untested specimen displayed a marked isotropic mean crystal spacing distribution which, as expected, was similar to that in the tested specimens. The mean fracture spacing in specimens A and C was also isotropic and in both cases approximately 45% less than in the unloaded specimen. However sample B showed a marked anisotropic spacing.

If this is compared with the stress/strain history (Figure 2) it can be seen that specimens A and C are very similar but specimen B obtained 10% strain at a much lower load bearing capacity, due to the considerably higher fracturing in the plane normal to loading direction.

The major orientations of the fracture planes were measured and are shown in Table 2. These show that there is a definite preferred fracture orientation, in all the specimens, striking approximately E-W and dipping between 50° and 70° to both N and S. There is not an apparent new fracture set induced by the testing, suggesting that fracturing has taken place on pre-existing weakness planes, probably induced in geological time by loading and flow of the material.

Table 2. Orientations of principal fracture planes

SPEC. NO.	PLANE (DIP/AZIMUTH)			
UN-LOADED	50/004,	73/250,	60/170	
A	50/038,	66/158,	70/128	80/024
B	50/030,	70/170,	72/224,	80/240
C	70/360,	70/156,	70/176,	84/296

The fracturing processes involved in the crystal deformation are not however the only major factors leading to deformation. Though fracturing is the first process to occur as the strain magnitude increases shearing tends to produce a more ovaloid form of grain. This is caused by shearing off at the corners and rotation of the long axes, particularly in the smaller grains, so that they lie normal to the loading direction (Figure 3.) Larger crystals tend to break up prior to rotation but, the required loads to produce similar deformation, were found to be higher than for the finer grained material. Eventually in extremely highly strained samples (> 25%) the rock is reduced to a cataclasic texture similar to those found due to dynamic metamorphism.

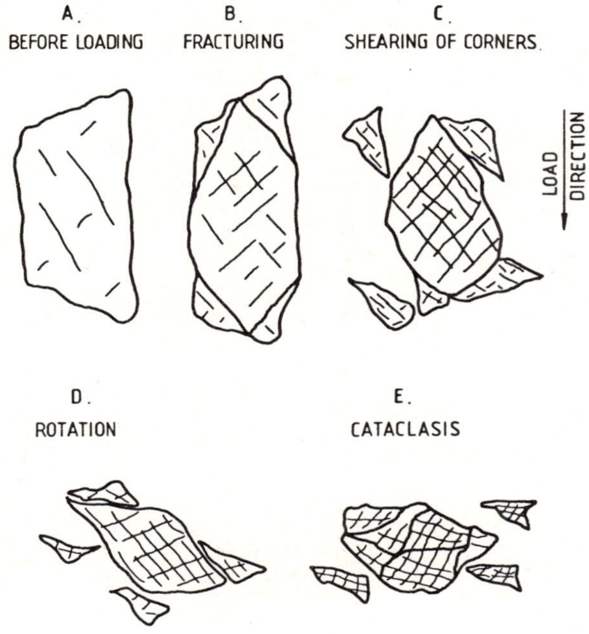

Figure 3. Idealised crystal deformation mechanisms with increasing strain.

Where marked bands or variations occurred in the composition and crystal size of the material, the mechanical behaviour was found to differ through the sample. Where bands of fine grained materials occurred, as layers considerably thinner (< 5 times) than the thickness of the surrounding material, necking and boundinage structures developed. Within these zones initially, Reidel shears followed by displacement shearing parallel to the layering, with increasing load. In the specimens in which the highest strains were produced, the lenses began to develop a sigmoidal appearance. As deformation was measured over the whole specimen, and the strain distribution is inhomogenous, the absolute strains in the layers are not known.

These observations are very similar to structures found in much larger tectonic shear zones (Skempton, 1966). Similar observations were also made where bands of marl or clay occurred in the specimens, and are distinctly related to higher deformation magnitudes.

Marl inclusions, even when not in bands, were found to be regions of higher strains with marked crystal disruption around them. This is almost certainly related to the observations of (Gilbert, 1981) that higher strain rates are found in halite specimens, used for creep testing, when the marl content is increased.

To check whether these observations were comparable to behaviour of in-situ evaporite mine pillars samples were taken from pillars with a highly strained core and distinct shear fractures. The material away from the core did not vary considerably from the surrounding country rock though similar phenomena to those found in the laboratory were observed in samples taken from the core. Complete cataclasis had occurred and a planar fabric developed parallel to the mine roof. Vugs of marl were completely flattened into very thin planar lenses. Fracturing in the crystals was restricted to two perpendicular planes corresponding to crystal cleavage planes. The whole appearance was very reminiscent of the fabric developed in thrust zones in high grade metamorphic terrains.

CONCLUSIONS

1) Studies of thin sections taken from laboratory tested specimens have been used to observe the deformation processes that occur. Comparison with samples taken from actual pillars show that, particularly in highly strained cores, similar phenomena occur.

2) The deformation initiates through fracturing (orientated parallel to pre-existing structures) followed by shearing rotation normal to the direction of the loading, and finally cataclasis. The degree of deformation is effected by the grain size and bands of find grained material show considerably higher strains, indicative of zones of varying deformation within the sample. Inclusions of weaker material will also undergo greater deformation.

3) The structures developed during post-failure testing of evaporite samples are often very similar in appearance to those found in metamorphic rocks.

4) This preliminary study has identified the mechanisms of deformation occurring and how they may be altered by variation of a number of material parameters. Future work is now required to apply this to actual in-situ conditions and asses how material variation will effect the stress/strain behaviour of evaporite mine pillars.

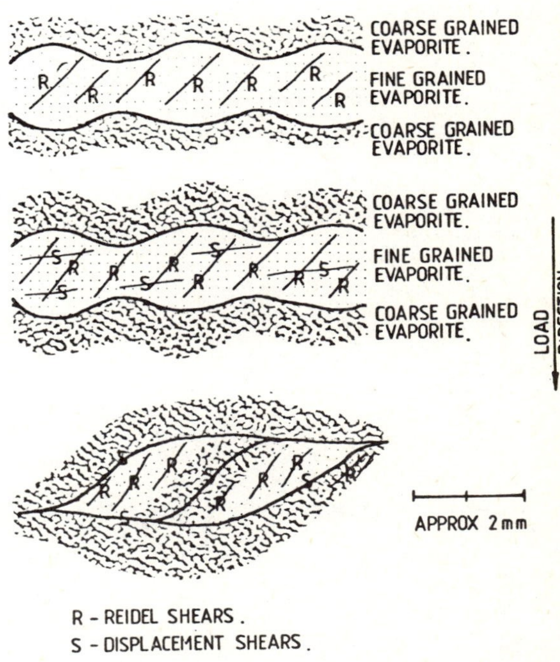

Figure 4. Development of Boudins and displacement shears at high strains.

REFERENCES

Cook, R.F., (1982), "Design of High Extraction Panels in a Deep Potash Mine", Conf. on Strata Mechanics, Newcastle upon Tyne.

Gilbert, L., (1981), "Correlation Between the Petrological Features of Cheshire Rock Salt and its Creep Characteristics", M.Sc. Thesis, Univ. Newcastle upon Tyne.

Miller, H.D.S., (1980), "Stability of Excavations and Design of Mining Layouts at Boulby Mine", Ph.D. Thesis, Univ. Newcastle upon Tyne.

Skempton, A.W., (1966), "Some observations on tectonic shear zones", Proc. 1st Int. Cong. Rock Mech., Lisbon.

INTERACTION BETWEEN PILLAR WORKINGS AT ELLINGTON COLLIERY

David L. Hodkin

G.H.Hill & Sons,
45, Kingsway,
Manchester, M19 2LL,
England.

SUMMARY. The paper, which is to be presented to the Symposium on Strata Mechanics at the University of Newcastle Upon Tyne, April 5th.-7th. 1982, describes an investigation into strata movements in coal measures rocks as an old room and pillar working was undermined by a similar new development with a small inter-seam separation.

INTRODUCTION

The area in which this investigation was carried out at Ellington Colliery lies some 8km. off the coast of North-east England. The Main (F) Seam, with a cover to the seabed between 105m and 120m, was extracted in 1972 by room and pillar working. The 23m square pillars and 5.5m roadways gave a 35% extraction ratio. The overlying High Main (E) Seam was taken on completion of the Main Seam working, giving a total excavated height of 4.2m.

The interval between the Main and Yard (G) Seams varies between 15m and 3m. Under the rules at the time of this investigation working was permitted to within 6m and an exemption had been granted to work within 4.5m. The Yard Seam was also worked with 23m square pillars and 5.5m roadways, the extracted height being 2.4m. The Main and Yard Seam workings were superimposed, i.e. pillar under pillar. This investigation was designed to study the feasibility of working the Yard Seam to within 3m of the Main Seam, thereby increasing the reserves of the colliery by some two million tonnes.

UNDERGROUND INSTRUMENTATION

The location of the instrumented sites in Panel GM18 are shown in Figure 1. Shell-type extensometer anchors were installed in the vertical roof boreholes at the centres of the junctions between 6 road, the main supply road, and G,J, and L stentons, and in a horizontal borehole in the pillar to the right of the supply road half-way between the planned positions of J and K stentons. The objective of this instrumentation was to observe roof strata movements relative to the exposed surface at various distances into the roof and movements and strains occurring in the pillar as it was isolated.

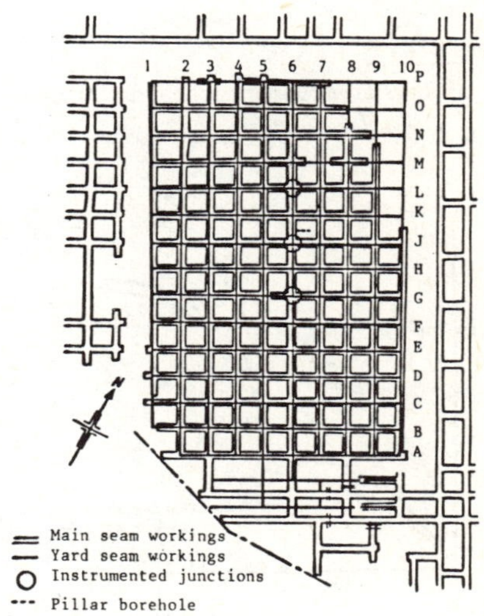

Figure 1 Instrumentation locations in Panel GM18

Vertical boreholes

A core taken at the top of the belt drift from the Main Seam to GM18 showed the strata to be approximately 1m of mudstone above the top of the Yard Seam overlain by seatearth.

The proposed method of installing the anchors in the roof was to drill a single borehole at the centre of each junction to within 1m of the floor of the Main Seam and to install the first anchor at this point. At this level, however, a band of loose water-bearing strata was encountered in which it proved impossible to fix the anchors. Rather than try to drill through this into a more solid layer, with the attendant risk of holing through into the upper workings, in each case the rear anchor was located just below this band.

At junction G/6 the anchors were located at 3.5m, 2.4m and 1.2m above the roof, at junction J/6 at 3.0m and 1.5m, and at junction L/6 at 3.2m and 1.5m. The hole at junction J/6 was the only one from which water was dripping, albeit very slowly.

In addition to the roof anchors, a single anchor was fixed in the floor at each location in order to monitor convergence between the roof and floor by means of a steel tape and an extensometer.

Horizontal borehole

The horizontal borehole was drilled and instrumented in order to monitor lateral movements in a pillar as it was isolated. The hole was located half-way between the planned positions of J and K stentons. The rear anchor was located at a depth of 11.4m, the centre of the projected pillar, and the other anchors were at depths of 6.1m, 2.9m and 1.5m from the mouth of hole station. An anchor was to have been installed at 3m depth but it failed to grip, suggesting a fault or joint there, and so it was withdrawn until it gripped firmly at 2.9m depth.

RESULTS

Vertical boreholes

Graphs of anchor movements against elapsed time for junctions G/6, J/6 and L/6 are given in Figure 2.

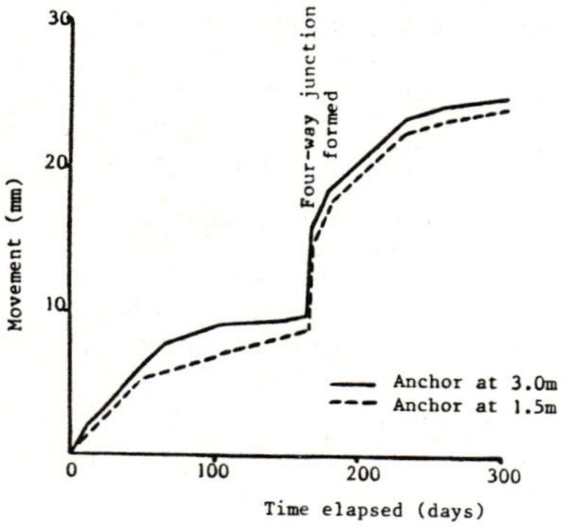

a) Junction J/6 anchor movements

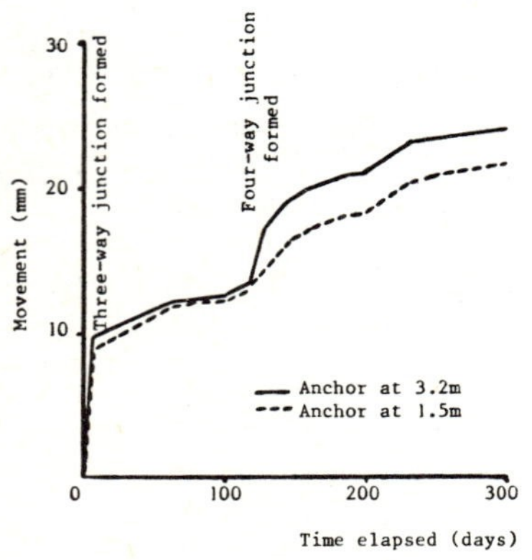

b) Junction L/6 anchor movements

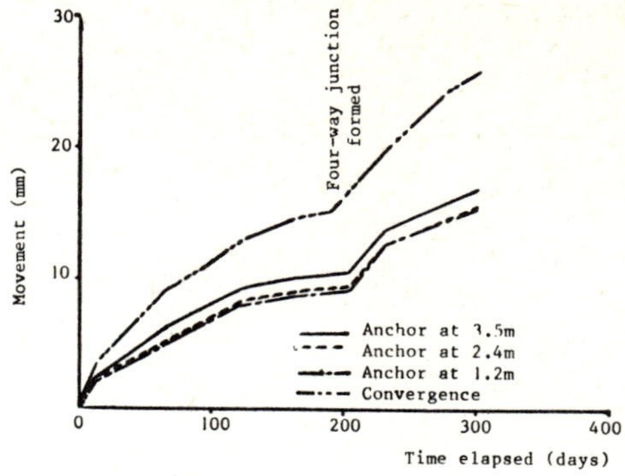

c) Junction G/6 anchor movements and convergence

Figure 2 Roof anchor movements vs. elapsed time

All three sets of results show similar behaviour, the closeness of the lines indicating that the anchors were moving with very little bed separation occurring between them. The majority of the movement took place between the mouth of hole station and the first anchor in each hole, i.e. in the first 1.2m of roof at G/6 and the first 1.5m at J/6 and L/6. At each location there was a gradual lowering of this strata, with more rapid movements as three- and four-way junctions were formed.

At J/6 and L/6 the graphs tend to level off again after the four-way junctions had been formed showing that a stable situation had again been established. At G/6, however, the graph showed no sign of levelling off after the formation of the four-way junction. At about 100days there was a lessening of the slope until the four-way junction was formed. Following the rapid movement after holing through, the curves regained the slope prior to the change at about 100days. This indicates that the lowering of the immediate roof was still occurring but there was no sign of excessive weighting on the supports when the final set of readings was taken.

Convergence

Unfortunately the floor stations at J/6 and L/6 were both lost. Despite being set into the floor and covered with steel caps they were both damaged by machinery passing over them.

The floor station at G/6 was not damaged, however, and the results therefrom are plotted on Figure 2(c). As can be seen, the curve follows closely that of the anchors in the roof but with movements of greater magnitude. If this extra movement was due to floor heave, it is unlikely that the curves would follow each other so closely. It is more probable, therefore, that the anchors in the roof strata were also moving downwards, thereby reducing their movement relative to the mouth of hole station. Thus, although the majority of the movement was occurring in the first 1.2m of the roof, the results indicate that there was lowering of the strata above this level also.

Horizontal borehole

The graphs showing the anchor movements relative to the rear (pillar centre) anchor against elapsed time are given in Figure 3.

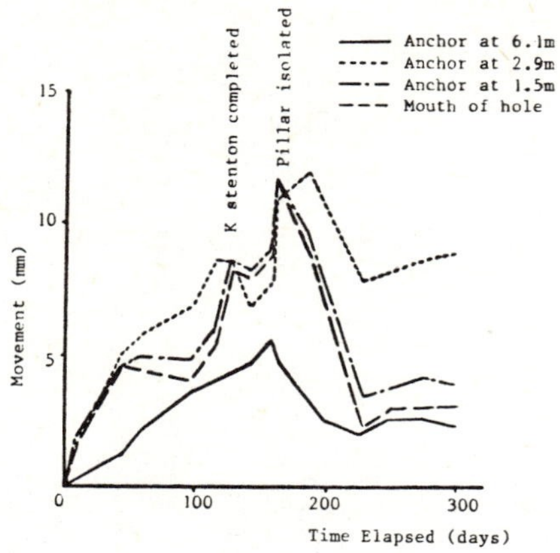

Figure 3 Horizontal borehole anchor movements relative to rear anchor

In general the results show a gradual dilation of the pillar with increasing load as the pillar was isolated. As the neighbouring pillars were isolated, the pillar returned to nearly its original width, indicating decreasing load.

The anchor at 2.9m depth followed the general trend of the other anchors but its residual movement was approximately double that of the others. The known discontinuity at 3m depth could have caused the delay before recording contraction following isolation of the pillar, possibly due to stick-slip movement on this plane of weakness. If there was another discontinuity between this anchor and that at 1.5m depth, then the isolated block could have moved independently of the rest of the pillar, resulting in the additional dilation recorded. A visual inspection of the pillar revealed no sign of any discontinuities, however, as the continuous miner leaves a shaped and abraded profile. The presence and actual size and location of any discontinuities can only be conjectured, therefore.

DISCUSSION

From the vertical borehole results it can be deduced that support in this panel was only required for the immediate roof and that, therefore, in this instance the distance between the seams had little influence on the roadway stability.

It was unfortunate that no core testing could be carried out, only the descriptive log of the strata between the seams being available when the instrumentation programme was being planned. The bending of the strata was typical of where a thin member underlies a thicker one and it is possible, therefore, that a different support system could be beneficially utilised. Rock bolts and the roof truss are two possible alternatives, an investigation into the use of the latter system having already been carried out, also at Ellington Colliery: Hodkin (1975). As salvage operations are carried out on the existing props and girders system, however, the economic viability of using non-recoverable support systems would have to be considered as well as the safety aspect.

As has already been noted, one of the horizontal borehole anchors gave anomalous readings, possibly due to inhomogeneity of the coal. The results do show, however, that there was no excessive stress increase in the pillar on isolation. It is not thought that the results obtained should be considered typical of pillar behaviour without further investigatory work in similar districts, however.

CONCLUSIONS

The results of this investigstion indicate that movement was occurring in the first metre or so of roof only and so the actual inter-seam separation had little effect on the stability of the Yard Seam roof.

The results from the pillar borehole showed that only a small increase in the loading of the pillar occurred as it was isolated.

REFERENCE

Hodkin D.L., 1975, Report to the National Coal Board "The Roof Truss Installation in FCP 71B at Ellington Colliery", Dept of Min. Eng., University of Newcastle Upon Tyne.

ROCK ENGINEERING STUDIES WITH REFERENCE

TO THE IN-SITU STRENGTH OF BAUXITE PILLARS

Evan K. S. Passaris

Department of Geotechnical Engineering,
University of Newcastle upon Tyne.

SUMMARY: An empirical approach originating from coal mining industries is employed for the determination of the in-situ strength of bauxite pillars. The work is based on data obtained from a survey of actual mining conditions carried out in panel 790 of the "51KM" bauxite mine in Greece. In order to arrive at an estimate of the average pillar stress, use was made of an improved formula which takes into account several rock engineering factors otherwise ignored by the classical tributary area theory. A comparison of the actual data with the estimated values indicated that the method used in this paper is a satisfactory approach in predicting the strength of bauxite pillars.

INTRODUCTION

In the development and planning of underground mining systems the room and pillar mining method retains an important role. The essence of the technique is that mining excavations must be supported by pillars of unmined mineral on either side. The concept of using pillars to support overburden rock dates back further towards man's ancient history to times when underground excavations were first considered.

The room and pillar technique expanded with the evolution of mining technology and nowadays is considered as a method of working that is applied, with many variations as to detail, in a wide range of sedimentary and massive mineral and coal deposits.

The results of the investigations which are described in this paper, have been restricted to the study of panel 790 worked by the room and pillar method in the "51KM" bauxite mine, near Delphi, Greece, operated by the Bauxites Parnasse Mining Company.

The design of conventional room and pillar systems is based on the classical engineering concept that the stresses acting on any member of the structure should be always less than their failure stress. The failure load of a bauxite pillar is that total load on the pillar which will cause the pillar to collapse, disintegrate or yield excessively. In an analogous manner the failure stress of a bauxite pillar is given by the failure load of the pillar divided by the nominal cross-sectional area of the pillar.

In view of the importance of the failure stress of the bauxite pillars when utilizing a realistic procedure for the determination of mining dimensions in room and pillar methods, a study was initiated to develop a relationship capable of defining the in-situ strength of the pillar. The present paper considers the methods and main results of the investigations conducted during this study.

GEOLOGY:

The location of the "51KM" bauxite mine is part of the west limb of the geosyncline of the Parnassus-Giona mountain chain, whose axis has a north-south orientation.

The area is characterised by outcrops of white non-crystalline autochthonous limestones stratigraphically belonging to Maastrichtian corresponding to the top of the Cretaceous in Europe. These limestone formations are generally thin-bedded with the exception of the upper layers where they become thick-bedded with typical occurrences of nodules of hornfelses.

In the same area and at the level of panel 790 the limestone roof becomes dark-coloured and is formed by rudistids-bearing bituminous limestones stratigraphically belonging to Turonian corresponding to the European mid-upper Cretaceous.

Three bauxite horizons can be distinguished in this area, and the oreborly under investigation geologically belongs to the upper or third horizon. The mineral composition of the bauxite deposits in the Parnassus-Giona mountain chain vary substantially with respect to their content of boehmite and diaspore. However, the specific bauxite horizon encountered in pannel 790 consisted mainly of the diasporic insoluble type, with pisolithic texture and redish-brown colour.

The floor of the bauxite deposit consists of lower Cretaceous white limestone which dips at a different angle to the overlying Turonian limestone, thus establishing a chracteristic angular unconformity.

PILLAR STRENGTH ANALYSIS

The need for a reliable method to calculate the strength of rock pillar arose when it became apparent that the strength of a pillar is not equal to the strength of the pillar materials.

In the past, numerous attempts have been made to derive suitable pillar strength formulae (Salamon 1970, Wilson 1972, Holland 1973, Hardy and Agapito 1975) with an emphasis on the empirical approach. As a general comment one can state that all these formulae indicated that the strength of a pillar increases with width, w, to height, h, ratio but decreases with size or volume.

In comparing the pillar strength formulae published by various authors for coal, Salamon and Munro (1967) have shown that the pillar strength σ_{ps} could be expressed in terms of a general power law:

$$\sigma_{ps} = K h^{a} w^{b} \qquad (1)$$

where K is the strength of unit cube of coal (expressed in MPa for 1 metre cube) and a and b are constants.

A review of the available data from coal pillars (Greenwald et al 1939, Steart 1954, Holland and Gaddy 1957, Salamon and Munro 1967, Beniawski 1968) indicated that the average values for the constants a and b are -0.765 and 0.405 respectively.

In an attempt to rationalize the empirical pillar design (Hoek and Brown 1980: 206-210) plotted in the same diagramme (reporduced here as fig. 1) for a range of pillar volumes, the relationship between w/h and σ_{ps}/K as derived from eqn (1) using the aforementioned average values of a and b (solid line curves) and as derived from stress analysis techniques (dashed curves).

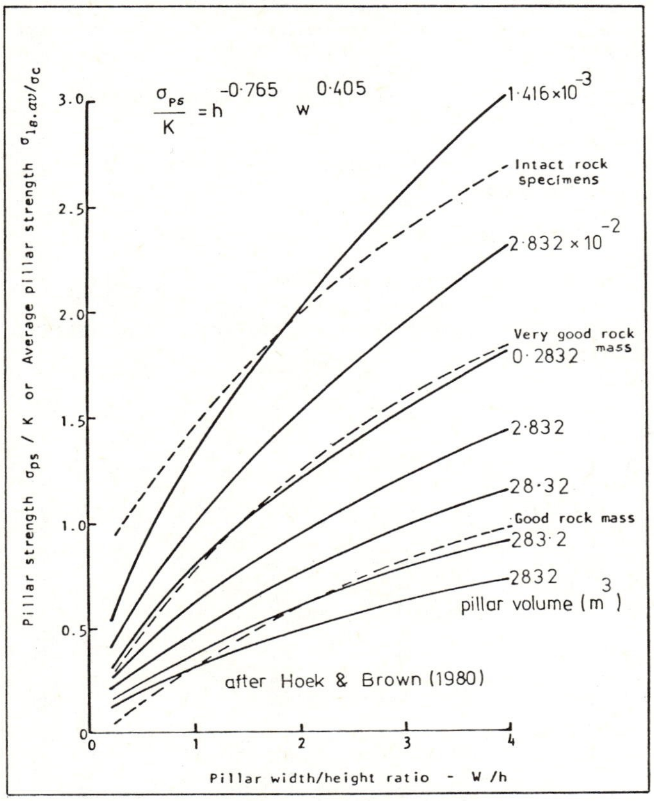

Fig. 1 Influence of pillar size and shape upon pillar strength.

It is interesting to note in Fig. 1 that the trends predicted by the dashed "theoretical" curves are very similar to those suggested by the empirical approach. In fact Hoek and Brown (1980) state that in view of the evidence shown in Fig. 1 one is entitled to use eqn (1) for the design of pillars other than coal, provided that appropriate values for the constants K, a and b can be determined. Consequently, on the basis of these arguments it was decided to employ eqn (1) for the estimation of the in-situ strength of the bauxite pillars of panel 790 of the "51KM" bauxite mine.

PILLAR LOADING

In calculating the average pillar stress σ_p use is often made of the tributary area theory (Obert and Duval 1967: 540-541) by assuming that the pillars uniformly support the entire load of the superincumbent strata overlying both the pillars and excavated area. In this case the average pillar stress is expressed as a function of the area extraction ratio R, defined as the ratio of the excavated-to-total area and of the vertical geostatic stress σ_v:

$$\sigma_p = \sigma_v \left(\frac{1}{1 - R}\right) \qquad (2)$$

However, with reference to the room and pillar system in panel 790, eqn (2) was considered unsuitable for the calculation of σ_p as it ignores geometrical properties such as the span or breadth of the mining zone with respect to its depth, the height and breadth of the pillars and their location within the mining zone. Furthermore, such factors as the nature of the in-situ geostatic stress field and the interaction between pillars and the roof rocks are not included

A more satisfactory solution given by Coates (1970: 4.7 - 4.12) is based on solving the statically indeterminant net deflection of the roof resulting from the mining excavations. This net deflection at the pillars is considered as a measure of the increase of σ_p resulting from the mining development. The components of the roof deflection are thought to be made up of: the convergence caused by the mineral extraction, the reverse deflection resulting from the average roof stress related to the increase of σ_p, the stress relief that results from eliminating the lateral constraint on the pillars and the local penetration of the pillars related to the average roof stress concentration. The equation for determing the increase in the average pillar stress σ_p is given by the following expression:

$$\left(\frac{\sigma_p}{\sigma_v} - 1\right) = \frac{(2Rs - h(1-C_r)k - C_b hqk}{hq + 2s(1-R)(1+1/n) + 2wR(1-C_r)/\pi} \qquad (3)$$

where s = the span or breadth of the mining opening
$C = \nu/(1 - \nu)$
ν = Poisson's ratio
$k = \sigma_h/\sigma_v$
σ_h = horizontal geostatic stress
$q = E_r(1 - \nu_b^2)/E_b(1 - \nu_r^2)$
E = Modulus of deformation
and subscripts r and b denote roof rock and bauxite correspondingly.

Eqn (3) is valid where the pillar breadths are all approximately equal and equally spaced apart. However, since in panel 790 the breadths of the pillar and their spacing varies considerably, it was decided to employ an improved version of eqn (3). This was achieved by replacing the expression (1-R)(1 + 1/n) in the denominator by $1/(1 + W_o/w)$ where W_o is the average breadth of adjacent openings surrounding a pillar. Consequently eqn (3) becomes:

$$\sigma_p = \frac{(2Rs - h(1 - C_r)k) - C_b hqk\,\sigma_v}{hq + 2s/(1 + W_o/w) + 2wR(1-C_r)/\pi} + \sigma_v \qquad (4)$$

The average unit weight γ of the overlying limestone in the area of panel 790 is 26 kN/m^3 and for a depth z = 140m corresponding to the limestone-bauxite contact, the vertical stress component becomes:

$$\sigma_v = z\gamma = 140 \times 26 = 3640 \text{ kPa} \qquad (5)$$

The modulus of deformation and Poisson's ratio of the limestone forming the roof and of the bauxite constituting the pillars were determined from a series of laboratory tests as being:

$$E_r = 60 \text{ GPa}, \ \nu_r = 0.34, \ E_b = 70 \text{ GPa, and } \nu_b = 0.30 \qquad (6)$$

Furthermore, from the 790 plan the extraction ratio was estimated as being

$$R = 0.85 \qquad (7)$$

and in the absence of any in-situ stress measurements the horizontal stress component was estimated using the following relationship suggested by Terzaghi and Richart (1952)

$$\sigma_h = \frac{\nu_r}{1 - \nu_r}\sigma_v$$

By substituting the values of σ_v and ν_r from eqns (5) and (6) respectively the above expression gives:

$$\sigma_h = 1875.15 \text{ kPa}$$

resulting in a ratio of the horizontal to vertical stress equal to

$$k = \sigma_h/\sigma_v = 0.515 \qquad (8)$$

THE STRENGTH OF BAUXITE PILLARS

The factor of safety of any pillar in panel 790 is defined as the ratio of the in-situ strength of the bauxite pillar, as opposed to the strength of bauxite, to the average stress acting on the pillar:

$$F_s = \frac{\sigma_{ps}}{\sigma_p} \qquad (9)$$

where the expression σ_{ps} and σ_p are determined by eqns (1) and (4) respectively.

A survey of 81 bauxite pillars in panel 790 produced for each pillar all the relevant mining dimensions used in eqn (4), i.e: height and width of pillar, the average breadth of the openings surrounding the pillar and the span of the mining opening where the pillar is located. By utilizing the collected data while employing eqn (4), the average pillar stress was calculated for the bauxite pillar and the results are shown in Fig. 2 where the histogram of σ_p for the pillars of panel 790 is shown.

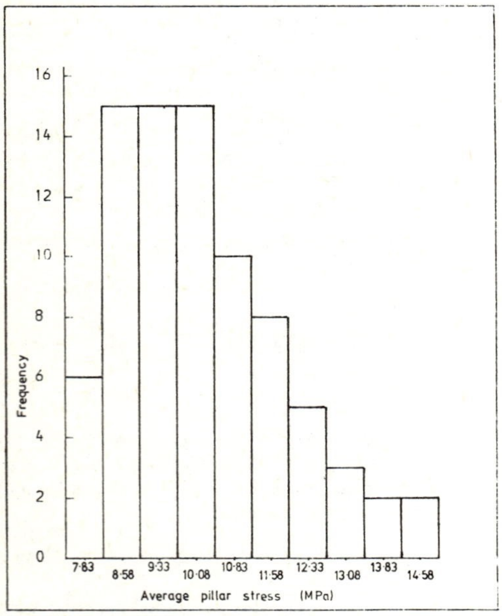

Fig. 2 Distribution of the average pillar stress in panel 790 at the "51KM" bauxite mine.

The mean value of σ_p for the 81 pillars of panel 790 was 10.15 MPa (7.83 MPa minimum, 14.81 MPa maximum) with a standard deviation of 1.66 MPa.

During this survey, in addition to the geometric characteristics of each individual bauxite pillar, the state of its stability was assessed and classified, accordingly as being:

- Competent pillar with $F_s > 1$

or - Pillar in a transition phase of stability with $F_s \simeq 1$

or - Unstable or failed pillar with $F_s < 1$

Consequently, for the unstable and transition phase pillars the combination of eqns (1), (4) and (9) yields:

$$Kh^a w^b = \frac{\sigma_v\left[(2Rs - h(1 - C_r)k) - C_b hqk\right]}{hq + 2s/(1 + W_o/w) + 2wR(1 - C_r)/\pi} + \sigma_v \qquad (10)$$

In order to determine the values of K, a and b a numerical analysis was conducted employing a computer program based on the Harwell FORTRAN subroutine VA 05A and eqn (1) was numerically fitted to the collected data. The subroutine was originally developed by the Atomic Energy Research Establishment and is based on an algorithm produced by Marquardt (1963), but with modifications which improve the performance of the method in certain circumstances yet which required negligible extra computer time and storage.

The result of the computer analysis gave the following values

$$K = 15 \text{ MPa}, \quad a = -0.658 \text{ and } b = 0.461 \qquad (11)$$

Substituting the above values in eqn (1) the in-situ strength of the bauxite pillars of panel 790 were calculated and plotted in the form of a histogram in fig. 3.

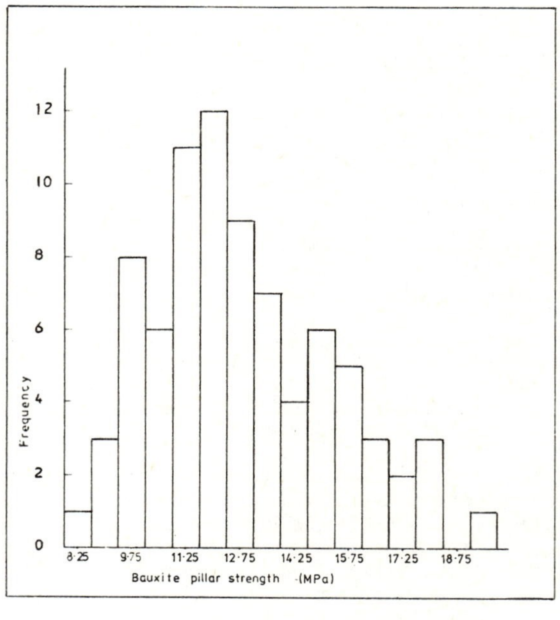

Fig. 3 Histogram of the in-situ strength of bauxite pillar in panel 790 at the "51KM" bauxite mine.

The value of σ_{sp} for panel 790 varied from a minimum of 8.25 MPa to a maximum of 19.5 MPa resulting in a mean value of 12.79 MPa with a standard deviation of 2.54 MPa. It is interesting to notice that if σ_p is plottd against the expression $(h^a w^b)$ for the bauxite pillars of panel 790 (as shown in Fig. 4), the transition phase pillars are clustered very close to the line corresponding to K = 15 MPa.

In addition one can see that the unstable pillars are above the line K = 15 MPa while all the competent pillars (with the exception of only one pillar) are located below this line. The conclusion drawn from fig. 4 is that the calculated value of 15 MPa expresses with adequate statistical confidence the value that K should have when used in eqn (1) to describe the strength of bauxite pillars.

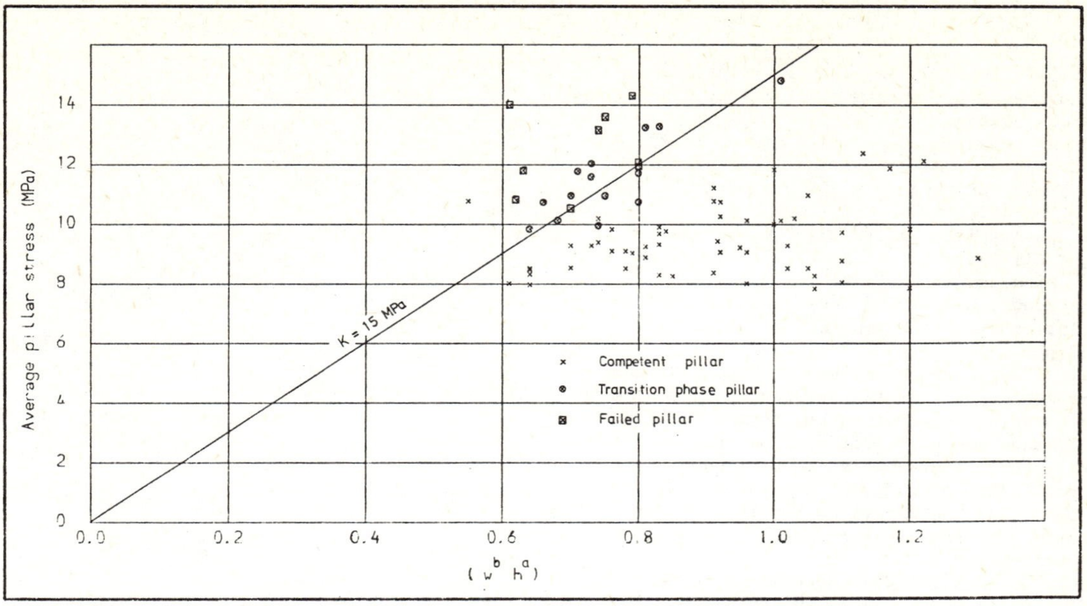

Fig. 4 Scatter diagramme of average pillar stress vs $(w^b h^a)$ for the bauxite pillar of panel 790.

CONCLUSIONS

The method followed in this paper for the determination of the in-situ strength of bauxite pillars indicates that the critical safety factors corresponding to the transition phase pillars, should be scattered about a median value of unity. This implies that for a given value of strength the points corresponding to the transition phase pillars should be clustered on either side of a line having a slope with a tangent equal to 1 which is shown as $F_s = 1$ in fig. 5.

The fact that this seems to be the case in fig. 5 may be interpreted as a clear indication that the approach used in this paper for the prediction of the strength of bauxite pillars is satisfactory. In the preceeding analysis the variation in bauxite quality and local geological parameters were regarded as random variations around an average condition. Such an assumption implies that the approach used in this paper is only capable of predicting the strength of pillar composed of an "average" type of bauxite and surrounded by "average" strata.

It is recognised therefore that this method tends to be a rather phenomenological approach as opposed to a method based on the actual physical mechanism of the strength behaviour of pillars. Consequently, the results of such an analysis may not describe exactly the real situation with reference to the strength of bauxite pillars, however they appear to offer a most reasonable approximation. In fact, by examining the results shown in fig. 5 it is obvious that for all mining design and operational control purposes the difference between predicted and actual behaviour is statistically so small that it is of no practical significance.

REFERENCES

Bieniawski, Z.T. 1968, The effect of specimen size on the compressive strength of coal, Intnl. J. Rock Mechanics & Min. Sci. 5: 325-335.

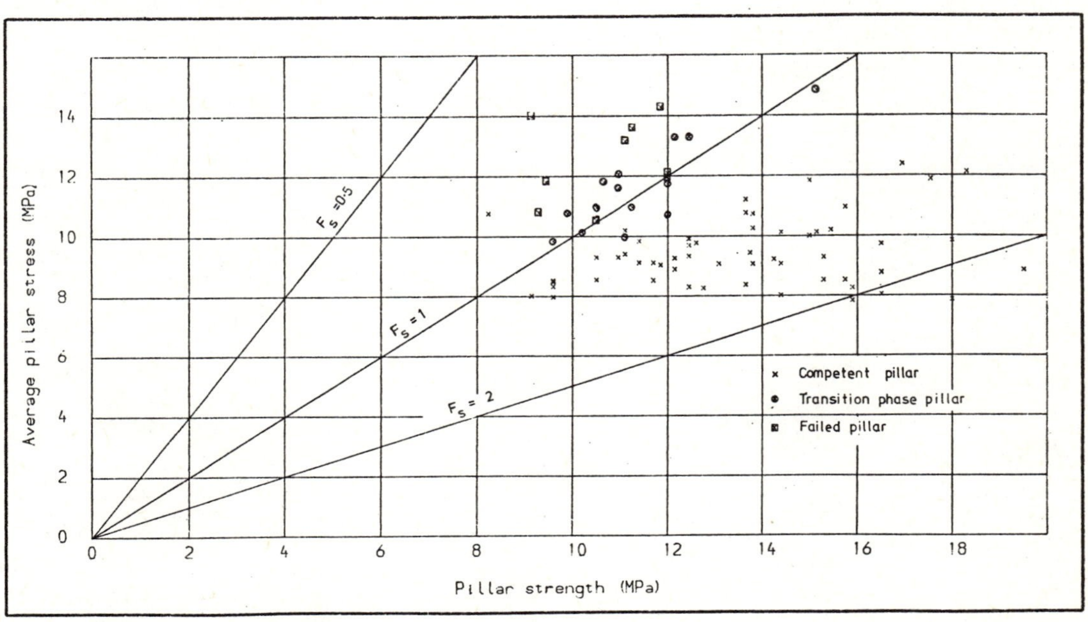

Fig. 5 Bauxite pillar strength vs average pillar stress, panel 790 at the "51KM" bauxite mine.

Coates, D.F. 1970, Rock mechanics principles, Mines Branch Monograph 874, Department of Energy Mines and Resources, Ottowa.
Greenwald, H.P., Howarth, H.W. and Hartman, I. 1939, Experiments on the strength of small pillars of coal in the Attsburg bed, U.S. Bureau of Mines Tech. Rep. 605: 22pp.
Hardy, M.P. and Agapito, J.F.T. 1975, Pillar design in underground oil shale mines, Proc. 16th Symp. on Rock Mechanics, Minneapolis: 257-266.
Hoek, E. and Brown, E.T. 1980, Underground Excavations in Rock, Institution of Mining and Metallurgy, London: 527pp.
Holland, C.T. 1973, Pillars design for permanent and semi-permanent support of the overburden in coal mines, Proc. 9th Canadian Rock Mechanics, Montreal: 113-139.
Holland, C.T. and Gaddy, F.L. 1957, Some aspects of permanent support of overburden on coal beds, Proc. W. Virginnia Coal Mining Institute: 43-66.
Marquardt, D.W. 1963, An algorithm for least squares estimation of non-linear parameters, J. Soc. Industr. & Appl. Maths. 11: 431-441.
Obert, L. and DuVall, W.I. 1967, Rock mechanics and the design of structures in rock, J. Wiley & Sons, Inc., N. York: 650 pp.
Salamon, M.D.G. 1970, Stability, instability and design of pillar working, Intnl. J. Rock Mechanics & Min. Sc. 7: 613-631.
Salamon, M.D.G. and Munro, A.H. 1967, A study of the strength of coal pillars, J. South African Inst. of Mining and Metallurgy 68 (2): 55-67.
Steart, F.A. 1954, Strength and stability of pillars in coal mines, J. Chem. Metall. Min. Soc. South Africa 54: 309-325.
Terzaghi, K. and Richart, F.E. 1952, Stresses in rock about cavities, Geotechnique 3: 57-90.
Wilson, A.H. 1972, Research into the determination of pillar size - Part 1, The Mining Engineer 131: 409-417.

SUBSIDENCE BEHAVIOUR OF COAL MEASURES ABOVE BORD AND PILLAR WORKINGS

N.C. Saxena and B. Singh

Central Mining Research Station
Dhanbad-B26001, India.

SUMMARY. The angle of draw in Indian coalfields is positive and varies between 4° and 31°. The non-effective width varies between 0.3 and 1.0 times the depth. It has not been possible to establish general relationships between different subsidence parameters and there is need for further and more detailed investigations. Variation in the proportion of sandstone from 64 to 95% in the Coal Measures Succession, and its nature, is a contributing factor in the variation of subsidence behaviour. Safe limits of surface slope and strains for various categories of surface features and structures are proposed.

INTRODUCTION

Ground movements due to extraction of coal seams underground tend to damage surface features and structures and endanger underground workings underneath water bodies. Due to the lack of the knowledge of the subsidence behaviour of strata overlying coal seams in India more than 1500 million tonne of coal is sterilised. Although, the importance of subsidence research was realised long ago subsequent observations were limited in precision and scope. Further information on the subsidence behaviour of Indian Coal Measures is required and the application of foreign know-how has not been successful.

Realising the need, the Central Mining Research Station (CMRS), Dhanbad started systematic subsidence investigations in the Indian coalfields in 1964 with the following objectives:

1. To develop an understanding of surface movements due to underground extraction of coal seams.
2. To develop relations between different subsidence parameters.
3. The development of methods of estimating subsidence movements.
4. To define safe limits of subsidence for different categories of surface feature and structure.
5. To develop systems of underground extraction for minimising surface movements.

Investigations have been completed over 22 panels/ areas and they are in progress over 19 more as part of a 15 year programme of extensive investigations to cover 14 major coalfields. 40 problems related to subsidence have been studied to help the coal mining industry to conserve more than 4 million tonnes of coal by way of extraction underneath and in the vicinity of surface features and structures.

NATURE OF SUBSIDENCE

In general, multiple seams in close proximity are being extracted by a bord and pillar system with and without stowing at very shallow to medium depths in Indian coalfields. Application of longwall systems has been on a very limited scale. Under these conditions the development of a subsidence trough on the surface has a stepped as well as smooth profile, as shown in Figure 1. The stepped troughs generally develop suddenly over caved bord and pillar workings at shallow depths in thicker seams. In such cases wide cracks develop on the surface almost along the perimeter of workings underground.

In areas where smooth subsidence troughs have been observed, the subsidence movements on the surface start after a certain minimum width has been extracted. This has been termed the non-effective width of extraction. The angle of draw in this type of subsidence has been observed to be positive. All the observations by the CMRS are concerned with this type of subsidence, and the main findings are discussed later. The width-depth ratio of the workings, over which investigations were conducted, varies between 0.36 to 3.4. The critical width of extraction has been found to be about 1.7 times the depth at two collieries.

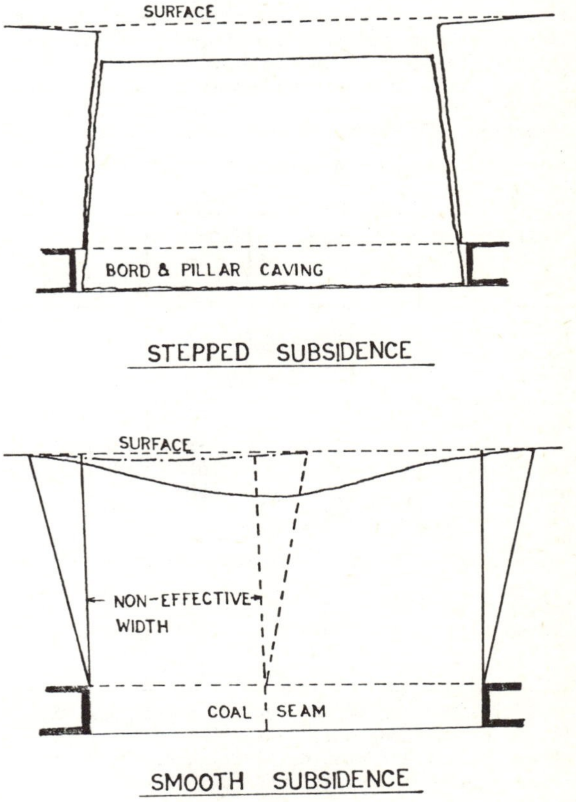

Figure 1. Nature of subsidence in Indian coal measures.

ANGLE OF DRAW

Until a few years ago it was believed that the angle of draw in Indian Coal Measures was negative, signifying that the area affected on the surface was less than that extracted underground. The investigations have shown that the area affected on the surface is more than that extracted underground, and therefore the angle of draw is positive. The observed values of the angle of draw from the vertical, in different coalfields are given below:

Jharia coalfield		
Sudamdih Project		
(a seam dipping at 30°)	-	38°-40° dip side
		13°-15° rise side
		18°-20° strike
Moonidih Project	-	14°
Other collieries	-	5°-27°
Ranigunj coalfield		
Rana	-	12°-13°
Jamuria	-	15°-17°
Dhemomain	-	5° -21°
Shivdanga Incline	-	10°-22°
Amritnagar	-	15°-18°
Ningah	-	9°
Moira	-	25°-30°
Korba coalfield		
Surakachhar	-	7° -10°
Pench & Kanhan coalfield		
North Chandametta	-	23°-29°
Sukri	-	20°
Karanpura coalfield		
Gidi A Pilot Mine	-	14°-31°

These values of angle of draw are for a limiting vertical movement of 5mm, and in conditions of general settlement of ground the values were determined from subsidence profiles. The variations in the values of draw in different coalfields and limited observations warrents intensification of investigations. However, the observations have been useful in designing protection pillars for shafts and other surface features and structures, and assessing the feasibility of extraction of coal seams underground.

NON-EFFECTIVE WIDTH OF EXTRACTION

During subsidence investigations in Indian coalfields it has been observed that the movements on the surface start after a certain width has been extracted under-ground in the case of caving or with hydraulic sand stowing. This is a unique feature of the Indian Coal Measures, and has been termed as non-effective width of extraction. The observations given in Table 1 show:

Table 1. Observed non-effective width of extraction

Colliery	Seam	Depth m	N.E.W.	Method of working
Sudamdih	XI/XII	90	0.3	LW-S
Moonidih	XVII-Top	310	0.4	LW-C
North Chandametta	No.1	72	0.7	BP-C
Surakachhar (12 panels)	G III	40-150	0.7	LW-C&S
Rana	Rana	101	0.7	BP-C
Bijuri	No.1	80	0.7	LW-C
Banki	G III	110	1.0	LW-C
Sawang (2 panels)	Kargali-top	170-180	0.36 & 0.43	LW-S
Ningah (2 panels)	Disergarh	180*	0.6 & 0.7**	LW-C

N.E.W. - Non-effective width of extraction with respect to depth, LW-Longwall, BP-Bord & Pillar, C-Caving, S-Hydraulic sand stowing, *-Thickness of parting, **-Observations in overlying Sripur seam.

that the non-effective width varies between 0.3 and 1.0 times the depth. The variation is mainly due to the nature of superincumbent strata in which the percentage of sandstone varies from 64 to 95 and its thickness in beds from a few centimetres to more than 50m. The observations have been helpful in designing underground workings for the protection of surface features and structures at Sudamdih and Moonidih.

SUBSIDENCE AND WIDTH-DEPTH RATIO

Figure 2 shows the maximum percentage subsidence observed over longwall and bord and pillar workings with or without stowing plotted against width-depth ratio. The observations do not lead to any relationship between the two parameters as the scatter of the points in Figure 2 is very large. In order to arrive at any definite relation further observations are necessary.

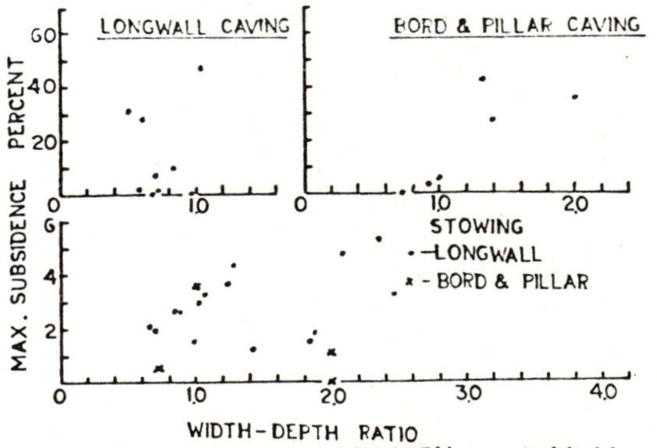

RELATIONSHIP BETWEEN MAXIMUM SUBSIDENCE AND WIDTH-DEPTH RATIO

Figure 2

The scatter of the points in Figure 2 can be attributed again to the variation in superincumbent strata. In case of bord and pillar workings the observations may also be affected by the percentage of coal left underground during final extraction.

RELATION BETWEEN SUBSIDENCE PARAMETERS

Analysis of the observations established the following general relationships:

$$T = K_1 \frac{S}{h} \quad (1)$$

$$E_{(-)} = K_2 T \quad (2)$$

$$E_{(+)} = K_3 T \quad (3)$$

where S is maximum subsidence in mm, h is average depth in m, T is maximum slope in mm/m, E is maximum strain in mm/m, (-) being compressive and (+) being tensile, and K_1, K_2, and K_3 are constants. The estimated values of the constants from the observations in different coalfields are given in Table 2, which shows wide variation and indicates that generalisation is not possible at this stage.

Table 2. Values of K_1, K_2 and K_3 on the basis of observations in different coalfields

Coalfield/ colliery	Width-depth ratio	K_1	K_2	K_3	Method of working
(Jharia)) Sudamdih)	1.05-1.63	1.96-2.65	0.25-1.32	0.39-0.59	LW-S
Moonidih	0.60	5.7	0.28	0.14	LW-C
(Ranigunj)					
Rana	0.73	1.63	2.56	0.56	LW-C
Jamuria	0.53	1.03	1.89	1.65	LW-C
Amritnagar	0.93	3.37	0.27	0.25	BP-C
Shivdanga	2.00	1.89-2.55	0.31-0.49	0.31-0.65	BP-C
Dhemomain	0.84	5.40	0.21	0.22	BP-C
Moira	1.00	2.92	0.23	0.17	BP-S
(Korba)					
Surakachhar	1.76	2.38	0.74	0.45	LW-S

Table 2. (continued)

Coalfield/ colliery	Width-depth ratio	K_1	K_2	K_3	Method of working
(Karanpura)					
Gidi	1.10	2.06	0.95	0.40	LW-C
(Pench & Kanhan)					
North Chandametta	1.38	2.26	0.70	0.65	BP-C
Sukri	1.60	1.91	0.76	0.40	BP-C

LW- Longwall, BP - Bord & Pillar, C- Caving, and S - Stowing.

SAFE LIMITS OF SURFACE MOVEMENTS

For ascertaining the feasibility of extraction of coal seams and the protection of underground workings and surface features and structures it is essential to know in advance the safe limits of subsidence movements. Such limits of slope and strains have been defined by many countries for their conditions, but are mostly lacking for Indian conditions. On the basis of investigations and experience the following safe limits of subsidence movements have been set for different categories of surface features and structures:

Railway Line -

At Sudamdih and Surakachhar collieries extraction underneath and in the vicinity of railway lines has been carried out after recommendations from CMRS. Studies for such extraction were also made for Kustore, Burragarh, and Salanpur collieries. In the initial studies for Sudamdih the safe limit of surface strain for the railway line was taken as 3mm/m as per Polish standards. Until now the railway line has been able to sustain a surface strain of 2.9mm/m without any disturbances in its normal operation. The observations are continuing. In other studies the same safe limit of surface strain has also been used. The safe limit of surface slope has been taken as 1 in 100 (10mm/m), which is the limiting operating gradient for railways. The railway line at Sudamdih had suffered a maximum surface slope of 6.75mm/m due to subsidence without any problems. At Surakachhar the railway line was subjected to a maximum subsidence of 55mm and strain of 1.44mm/m.

Buildings

The extent and type of damage to buildings due to subsidence movements depends upon the size and nature of construction, and the nature and magnitude of the movements to which the building is subjected. Polish standards classify buildings in four categories and allow 15mm/m and 9mm/m of maximum slope and strain respectively for kutchha/timber buildings, 10mm/m and 6mm/m respectively for small buildings, 5mm/m and 3mm/m respectively for water tanks, and 2.5mm/m and 1.5mm/m respectively for monuments, churches, and very high and important buildings. In the U.K. the damage to buildings has been classified according to the change in the length due to movements. This depends on the length of building and the magnitude of strain to which it is subjected. For Indian conditions, Polish standards were used initially, but at present U.K. standards of the N.C.B. are being used and a maximum movement of 60mm is being allowed in buildings. Extraction underneath built-up areas has been carried out using these standards, at Moonidih, and Loyabad collieries. The buildings were damaged slightly due to ground movements which were subsequently repaired at a cost of not more than Re 1/- per tonne of coal extracted from underneath.

Water bodies

Extraction underneath water bodies poses a danger to underground workings aganst inrush of water. The water bodies may be on the surface, in overlying worked out or developed seams, and in the intervening strata. For Indian conditions maximum allowable tensile strain has been taken as 3mm/m due to the face that the percentage of shales in Indian Coal Measures varies between almost nil to 35. With this condition, extraction has been carried out at Surakachhar underneath highest flood level and surface water bodies. If in an area the percentage of shales is found to be greater than 35% a higher limit of strain may be considered.

ACKNOWLEDGMENT

The authors are thankful to the Director, CMRS, for according permission for presenting this paper in this Symposium.

The views expressed in the paper are those of the authors and not necessarily of the CMRS.

AUTHOR INDEX

SUBJECT INDEX